T0181637

Lecture Notes in Artificial Intelligence   13073

Subseries of Lecture Notes in Computer Science

Series Editors

Randy Goebel
  University of Alberta, Edmonton, Canada

Yuzuru Tanaka
  Hokkaido University, Sapporo, Japan

Wolfgang Wahlster
  DFKI and Saarland University, Saarbrücken, Germany

Founding Editor

Jörg Siekmann
  DFKI and Saarland University, Saarbrücken, Germany

More information about this subseries at https://link.springer.com/bookseries/1244

André Britto · Karina Valdivia Delgado (Eds.)

Intelligent Systems

10th Brazilian Conference, BRACIS 2021
Virtual Event, November 29 – December 3, 2021
Proceedings, Part I

Springer

Editors
André Britto ⓘ
Universidade Federal de Sergipe
São Cristóvão, Brazil

Karina Valdivia Delgado ⓘ
Universidade de São Paulo
São Paulo, Brazil

ISSN 0302-9743   ISSN 1611-3349 (electronic)
Lecture Notes in Artificial Intelligence
ISBN 978-3-030-91701-2   ISBN 978-3-030-91702-9 (eBook)
https://doi.org/10.1007/978-3-030-91702-9

LNCS Sublibrary: SL7 – Artificial Intelligence

© Springer Nature Switzerland AG 2021
This work is subject to copyright. All rights are reserved by the Publisher, whether the whole or part of the material is concerned, specifically the rights of translation, reprinting, reuse of illustrations, recitation, broadcasting, reproduction on microfilms or in any other physical way, and transmission or information storage and retrieval, electronic adaptation, computer software, or by similar or dissimilar methodology now known or hereafter developed.
The use of general descriptive names, registered names, trademarks, service marks, etc. in this publication does not imply, even in the absence of a specific statement, that such names are exempt from the relevant protective laws and regulations and therefore free for general use.
The publisher, the authors and the editors are safe to assume that the advice and information in this book are believed to be true and accurate at the date of publication. Neither the publisher nor the authors or the editors give a warranty, expressed or implied, with respect to the material contained herein or for any errors or omissions that may have been made. The publisher remains neutral with regard to jurisdictional claims in published maps and institutional affiliations.

This Springer imprint is published by the registered company Springer Nature Switzerland AG
The registered company address is: Gewerbestrasse 11, 6330 Cham, Switzerland

Preface

The Brazilian Conference on Intelligent Systems (BRACIS) is one of Brazil's most meaningful events for students and researchers in artificial and computational intelligence. Currently in its 10th edition, BRACIS originated from the combination of the two most important scientific events in Brazil in artificial intelligence (AI) and computational intelligence (CI): the Brazilian Symposium on Artificial Intelligence (SBIA), with 21 editions, and the Brazilian Symposium on Neural Networks (SBRN), with 12 editions. The conference aims to promote theory and applications of artificial and computational intelligence. BRACIS also aims to promote international-level research by exchanging scientific ideas among researchers, practitioners, scientists, and engineers.

BRACIS 2021 received 192 submissions. All papers were rigorously double-blind peer reviewed by an international Program Committee (with an average of three reviews per submission), which was followed by a discussion phase for conflicting reports. At the end of the reviewing process, 77 papers were selected for publication in two volumes of the Lecture Notes in Artificial Intelligence series, an acceptance rate of 40%.

We are very grateful to Program Committee members and reviewers for their volunteered contribution in the reviewing process. We would also like to express our gratitude to all the authors who submitted their papers, the general chairs, and the Local Organization Committee for supporting the conference during the COVID-19 pandemic. We want to thank the Artificial Intelligence and Computational Intelligence commissions from the Brazilian Computer Society for the confidence they placed in us serving as program chairs for BRACIS 2021.

We are confident that these proceedings reflect the excellent work in the artificial and computation intelligence communities.

November 2021

André Britto
Karina Valdivia Delgado

Organization

General Chairs

Reinaldo A. C. Bianchi   Centro Universitário FEI and C4AI, Brazil
Zhao Liang   Universidade de São Paulo and C4AI, Brazil

Program Committee Chairs

André Britto   Universidade Federal de Sergipe, Brazil
Karina V. Delgado   Universidade de São Paulo, Brazil

Steering Committee

Leliane Barros   Universidade de São Paulo, Brazil
Denis Deratani Maua   Universidade de São Paulo, Brazil
Felipe Rech Meneguzzi   Pontifícia Universidade Católica do Rio Grande do Sul, Brazil
Jaime Sichman   Universidade de São Paulo, Brazil
Maria Viviane de Menezes   Universidade Federal do Ceará, Brazil
Tatiane Nogueira Rios   Universidade Federal da Bahia, Brazil
Solange Rezende   Universidade de São Paulo, Brazil
Gina Oliveira   Universidade Federal de Uberlândia, Brazil
Anisio Lacerda   Universidade Federal de Minas Gerais, Brazil
Helida Salles   Universidade Federal do Rio Grande, Brazil
João Xavier-Jr.   Universidade Federal do Rio Grande do Norte, Brazil
Ricardo Prudêncio   Universidade Federal de Pernambuco, Brazil
Renato Tinós   Universidade de São Paulo, Brazil

Program Committee

Adenilton da Silva   Universidade Federal de Pernambuco, Brazil
Adriane Serapião   Universidade Estadual Paulista, Brazil
Adriano Oliveira   Universidade Federal de Pernambuco, Brazil
Adrião Duarte Dória Neto   Universidade Federal do Rio Grande do Norte, Brazil
Alexandre Delbem   Universidade de São Paulo, Brazil
Alexandre Ferreira   Universidade Estadual de Campinas, Brazil
Aline Paes   Universidade Federal Fluminense, Brazil

Aluizio Arańjo	Universidade Federal de Pernambuco, Brazil
Alvaro Moreira	Universidade Federal do Rio Grande do Sul, Brazil
Amedeo Napoli	LORIA, France
Ana Carolina Lorena	Instituto Tecnológico de Aeronáutica, Brazil
Ana Cristina B. Kochem Vendramin	Universidade Tecnológica Federal do Paraná, Brazil
Anísio Lacerda	Centro Federal de Educação Tecnológica de Minas Gerais, Brazil
Anderson Soares	Universidade Federal de Goiás, Brazil
André Britto	Universidade Federal do Sergipe, Brazil
André Coelho	Alana AI, Brazil
André Ponce de Leon F. de Carvalho	Universidade de São Paulo, Brazil
André Rossi	Universidade Estadual Paulista, Brazil
André Ruela	Universidade de São Paulo, Brazil
André Grahl Pereira	Universidade Federal do Rio Grande do Sul, Brazil
Andrés Eduardo Coca Salazar	Universidade Tecnológica Federal do Paraná, Brazil
Anna Costa	Universidade de São Paulo, Brazil
Anne Canuto	Universidade Federal do Rio Grande do Norte, Brazil
Araken Santos	Universidade Federal Rural do Semi-árido, Brazil
Ariane Machado-Lima	Universidade de São Paulo, Brazil
Aurora Pozo	Universidade Federal do Paraná, Brazil
Bianca Zadrozny	IBM Research, Brazil
Bruno Nogueira	Universidade Federal de Mato Grosso do Sul, Brazil
Bruno Pimentel	Universidade Federal de Alagoas, Brazil
Carlos Ferrero	Instituto Federal de Santa Catarina, Brazil
Carlos Ribeiro	Instituto Tecnológico de Aeronáutica, Brazil
Carlos Silla	Pontifícia Universidade Católica do Paraná, Brazil
Carlos Thomaz	Centro Universitário FEI, Brazil
Carlos Alberto Estombelo Montesco	Universidade Federal de Sergipe, Brazil
Carolina Paula de Almeida	Universidade Estadual do Centro-Oeste, Brazil
Celia Ralha	Universidade de Brasília, Brazil
Cesar Tacla	Universidade Tecnológica Federal do Paraná, Brazil
Cleber Zanchettin	Universidade Federal de Pernambuco, Brazil
Clodoaldo Lima	Universidade de São Paulo, Brazil

Daniel Araújo	Universidade Federal do Rio Grande do Norte, Brazil
Daniel Dantas	Universidade Federal de Sergipe, Brazil
Danilo Sanches	Universidade Tecnológica Federal do Paraná, Brazil
Denis Fantinato	Universidade Federal do ABC, Brazil
Denis Mauá	Universidade de São Paulo, Brazil
Dennis Barrios-Aranibar	Universidad Católica San Pablo, Peru
Diana Adamatti	Universidade Federal do Rio Grande, Brazil
Diego Furtado Silva	Universidade Federal de São Carlos, Brazil
Edson Gomi	Universidade de São Paulo, Brazil
Edson Matsubara	Fundação Universidade Federal de Mato Grosso do Sul, Brazil
Eduardo Borges	Universidade Federal do Rio Grande, Brazil
Eduardo Costa	Corteva Agriscience, Brazil
Eduardo Gonçalves	Escola Nacional de Ciências Estatísticas, Brazil
Eduardo Palmeira	Universidade Estadual de Santa Cruz, Brazil
Eduardo Spinosa	Universidade Federal do Paraná, Brazil
Edward Hermann Haeusler	Pontifícia Universidade Católica do Rio de Janeiro, Brazil
Elizabeth Goldbarg	Universidade Federal do Rio Grande do Norte, Brazil
Elizabeth Wanner	Centro Federal de Educação Tecnológica de Minas Gerais, Brazil
Emerson Paraiso	Pontificia Universidade Catolica do Parana, Brazil
Eraldo Fernandes	Universidade Federal de Mato Grosso do Sul, Brazil
Erick Fonseca	Real Digital, Brazil
Evandro Costa	Universidade Federal de Alagoas, Brazil
Everton Cherman	Universidade de São Paulo, Brazil
Fabiano Silva	Universidade Federal do Paraná, Brazil
Fabrício Enembreck	Pontifícia Universidade Católica do Paraná, Brazil
Fabricio Olivetti de França	Universidade Federal do ABC, Brazil
Fábio Cozman	Universidade de São Paulo, Brazil
Felipe França	Universidade Federal do Rio de Janeiro, Brazil
Fernando Osório	Universidade de São Paulo, Brazil
Fernando Von Zuben	Universidade Estadual de Campinas, Brazil
Flavio Tonidandel	Centro Universitário FEI, Brazil
Flávio Soares Corrêa da Silva	Universidade de São Paulo, Brazil
Francisco Chicano	University of Málaga, Spain
Francisco De Carvalho	Universidade Federal de Pernambuco, Brazil
Gabriel Ramos	Universidade do Vale do Rio dos Sinos, Brazil
George Cavalcanti	Universidade Federal de Pernambuco, Brazil

Gerson Zaverucha	Universidade Federal do Rio de Janeiro, Brazil
Giancarlo Lucca	Universidade Federal do Rio Grande, Brazil
Gina Oliveira	Universidade Federal de Uberlãndia, Brazil
Gisele Pappa	Universidade Federal de Minas Gerais, Brazil
Gracaliz Dimuro	Universidade Federal do Rio Grande, Brazil
Guilherme Barreto	Universidade Federal do Ceará, Brazil
Guilherme Derenievicz	Universidade Federal do Paraná, Brazil
Guillermo Simari	Universidad Nacional del Sur in Bahia Blanca, Argentina
Gustavo Giménez-Lugo	Universidade Tecnológica Federal do Paraná, Brazil
Gustavo Paetzold	University of Sheffield, UK
Heitor Lopes	Universidade Tecnológica Federal do Paraná, Brazil
Helena Caseli	Universidade Federal de São Carlos, Brazil
Helida Santos	Universidade Federal do Rio Grande, Brazil
Heloisa Camargo	Universidade Federal de São Carlos, Brazil
Huei Lee	Universidade Estadual do Oeste do Paraná, Brazil
Humberto Bustince	Universidad Pública de Navarra, Spain
Humberto César Brandão de Oliveira	Universidade Federal de Alfenas, Brazil
Ivandré Paraboni	Universidade de São Paulo, Brazil
Jaime Sichman	Universidade de São Paulo, Brazil
Joéo Balsa	Universidade de Lisboa, Portugal
Joéo Bertini	Universidade Estadual de Campinas, Brazil
Joéo Papa	Universidade Estadual Paulista, Brazil
Joéo C. Xavier-Júnior	Universidade Federal do Rio Grande do Norte, Brazil
Joéo Luís Rosa	Universidade de São Paulo, Brazil
Jomi Hübner	Universidade Federal de Santa Catarina, Brazil
Jonathan Andrade Silva	Universidade Federal de Mato Grosso do Sul, Brazil
José Antonio Sanz	Universidad Pública de Navarra, Spain
José Augusto Baranauskas	Universidade de São Paulo, Brazil
Jose Eduardo Ochoa Luna	Universidad Católica San Pablo, Peru
Juan Pavón	Universidad Complutense Madrid, Spain
Julio Nievola	Pontifícia Universidade Católica do Paraná, Brazil
Karla Lima	Universidade de São Paulo, Brazil
Kate Revoredo	Vienna University of Economics and Business, Austria
Krysia Broda	Imperial College, UK
Leandro dos Santos Coelho	Pontifícia Universidade Catálica do Paraná, Brazil
Leliane Nunes de Barros	Universidade de São Paulo, Brazil

Leonardo Emmendorfer	Universidade Federal do Rio Grande, Brazil
Leonardo Matos	Universidade Federal de Sergipe, Brazil
Leonardo Filipe Ribeiro	Technische Universitat Darmstadt, Germany
Li Weigang	Universidade de Brasília, Brazil
Livy Real	B2W Digital and Universidade de São Paulo, Brazil
Lucelene Lopes	Universidade de São Paulo, Brazil
Luciano Digiampietri	Universidade de São Paulo, Brazil
Luis Antunes	Universidade de Lisboa, Portugal
Luis Garcia	Universidade de Brasília, Brazil
Luiz Carvalho	Universidade Tecnológica Federal do Paraná, Brazil
Luiz Coletta	Universidade Estadual Paulista, Brazil
Luiz Henrique Merschmann	Universidade Federal de Lavras, Brazil
Luiza de Macedo Mourelle	Universidade Estadual de Rio de Janeiro, Brazil
Marcela Ribeiro	Universidade Federal de São Carlos, Brazil
Marcella Scoczynski	Universidade Tecnológica Federal do Paraná, Brazil
Marcelo Finger	Universidade de São Paulo, Brazil
Marcilio de Souto	Université d'Orléans, France
Marco Cristo	Universidade Federal do Amazonas, Brazil
Marcos Domingues	Universidade Estadual de Maringá, Brazil
Marcos Quiles	Universidade Federal de São Paulo, Brazil
Maria do Carmo Nicoletti	Universidade Federal de São Carlos, Brazil
Marilton Aguiar	Universidade Federal de Pelotas, Brazil
Marley M. B. R. Vellasco	Pontifícia Universidade Católica do Rio de Janeiro, Brazil
Marlo Souza	Universidade Federal da Bahia, Brazil
Mauri Ferrandin	Universidade Federal de Santa Catarina, Brazil
Márcio Basgalupp	Universidade Federal de São Paulo, Brazil
Moacir Ponti	Universidade de São Paulo, Brazil
Murillo Carneiro	Universidade Federal de Uberlândia, Brazil
Murilo Naldi	Universidade Federal de São Carlos, Brazil
Myriam Delgado	Universidade Tecnológica Federal do Paraná, Brazil
Nádia Felix	Universidade Federal de Goiás, Brazil
Norton Roman	Universidade de São Paulo, Brazil
Nuno David	Instituto Universitário de Lisboa, Portugal
Patrícia Tedesco	Universidade Federal de Pernambuco, Brazil
Patricia Oliveira	Universidade de São Paulo, Brazil
Paulo Cavalin	IBM Research, Brazil
Paulo Ferreira Jr.	Universidade Federal de Pelotas, Brazil

Paulo Quaresma	Universidade de Évora, Portugal
Paulo Henrique Pisani	Universidade Federal do ABC, Brazil
Paulo T. Guerra	Universidade Federal do Ceará, Brazil
Petrucio Viana	Universidade Federal Fluminense, Brazil
Priscila Lima	Universidade Federal do Rio de Janeiro, Brazil
Rafael Bordini	Pontifícia Universidade Católica do Rio Grande do Sul, Brazil
Rafael Giusti	Universidade Federal do Amazonas, Brazil
Rafael Gomes Mantovani	Universidade Tecnológica Federal do Paraná, Brazil
Rafael Parpinelli	Universidade do Estado de Santa Catarina, Brazil
Rafael Rossi	Universidade Federal de Mato Grosso do Sul, Brazil
Reinaldo Bianchi	Centro Universitário FEI, Brazil
Renata Wassermann	Universidade de São Paulo, Brazil
Renato Assuncao	Universidade Federal de Minas Gerais, Brazil
Renato Krohling	Universidade Federal do Espírito Santo, Brazil
Renato Tinos	Universidade de São Paulo, Brazil
Renê Gusmão	Universidade Federal de Sergipe, Brazil
Ricardo Cerri	Universidade Federal de São Carlos, Brazil
Ricardo Marcacini	Universidade de São Paulo, Brazil
Ricardo Prudêncio	Universidade Federal de Pernambuco, Brazil
Ricardo Rios	Universidade Federal da Bahia, Brazil
Ricardo Suyama	Universidade Federal do ABC, Brazil
Ricardo Tanscheit	Pontifícia Universidade Católica do Rio de Janeiro, Brazil
Ricardo Fernandes	Universidade Federal de São Carlos, Brazil
Roberta Sinoara	Instituto Federal de Ciência, Educação e Tecnologia de São Paulo, Brazil
Roberto Santana	University of the Basque Country, Spain
Robson Cordeiro	Universidade de São Paulo, Brazil
Rodrigo Barros	Pontifícia Universidade Católica do Rio Grande do Sul, Brazil
Rodrigo Wilkens	University of Milano-Bicocca, Italy
Ronaldo Prati	Universidade Federal do ABC, Brazil
Ronnie Alves	Instituto Tecnologia Vale, Brazil
Roseli Romero	Universidade de São Paulo, Brazil
Rui Camacho	University of Porto, Portugal
Sandra Sandri	Instituto Nacional de Pesquisas Espaciais, Brazil
Sandra Venske	Universidade Estadual do Centro-Oeste, Brazil
Sandro Rigo	Universidade do Vale do Rio dos Sinos, Brazil
Sarajane Peres	Universidade de São Paulo, Brazil

Sílvia Maia	Universidade Federal do Rio Grande do Norte, Brazil
Sílvio Cazella	Universidade Federal de Ciências da Saúde de Porto Alegre, Brazil
Solange Rezende	Universidade de São Paulo, Brazil
Sylvio Barbon Junior	Universidade Estadual de Londrina, Brazil
Tatiane Nogueira	Universidade Federal de Bahia, Brazil
Teresa Ludermir	Universidade Federal de Pernambuco, Brazil
Thiago Covoes	Universidade Federal do ABC, Brazil
Thiago Pardo	Universidade de São Paulo, Brazil
Tiago Almeida	Universidade Federal de São Carlos, Brazil
Valdinei Freire	Universidade de São Paulo, Brazil
Valerie Camps	Paul Sabatier University, France
Valmir Macario	Universidade Federal Rural de Pernambuco, Brazil
Vasco Furtado	Universidade de Fortaleza, Brazil
Viviane Torres da Silva	IBM Research, Brazil
Vladimir Rocha	Universidade Federal do ABC, Brazil
Wagner Botelho	Universidade Federal do ABC, Brazil
Wagner Meira Jr.	Universidade Federal de Minas Gerais, Brazil
Yván Túpac	Universidad Católica San Pablo, Peru

Additional Reviewers

Alexis Iván Aspauza Lescano
André Carvalho
Antonio Parmezan
Bernardo Scapini Consoli
Caetano Ranieri
Cristina Morimoto
Daniel Pinheiro da Silva Junior
Daniela Vianna
Daniela Fernanda Milon Flores
Dimmy Magalhães
Eliton Perin
Eulanda Santos
Felipe Serras
Felipe Zeiser
Fernando dos Santos
Guillermo Simari
Hugo Valadares Siqueira
Italo Oliveira
Jefferson Souza
Jhonatan Alves

Joel Costa Júnior
Juliana Wolf
Kristofer Kappel
Lucas Evangelista
Lucas Navarezi
Lucas Rodrigues
Luciano Cota
Luiz Fernando Oliveira
Maicon Zatelli
Marcia Fernandes
Mariane Regina Sponchiado Cassenote
Matheus Pavan
Maurício Pamplona Segundo
Mohamed El Yafrani
Murilo Falleiros Lemos Schmitt
Nauber Gois
Nelson Sandes
Newton Spolaôr
Rafael João
Rafael Katopodis

Rafhael Cunha
Rodolfo Garcia
Sérgio Discola-Jr.
Tauã Cabreira
Thiago Homem

Thiago Miranda
Vítor Lourenço
Wesley Santos
Wesley Seidel

Contents – Part I

Contents – Part II

Text Mining and Natural Language Processing

Agent and Multi-agent Systems, Planning and Reinforcement Learning

A Conversational Agent to Support Hospital Bed Allocation

Débora C. Engelmann[1,2](✉) [ID], Lucca Dornelles Cezar[1] [ID],
Alison R. Panisson[3] [ID], and Rafael H. Bordini[1] [ID]

[1] School of Technology, PUCRS, Porto Alegre, RS, Brazil
{debora.engelmann,lucca.cezar}@edu.pucrs.br, rafael.bordini@pucrs.br
[2] DIBRIS, University of Genoa, Genoa, Italy
[3] Department of Computing, UFSC, Araranguá, SC, Brazil
alison.panisson@ufsc.br

Abstract. Bed allocation in hospitals is a critical and important problem, and it has become even more important since last year because of the COVID-19 pandemic. In this paper, we present an approach based on intelligent-agent technologies to assist hospital staff in charge of bed allocation. As part of this work, we developed a web-based simulation of hospital bed allocation system integrated with a chatbot for interaction with the user. As a core component in our approach, an intelligent agent uses the feedback of a plan validator to check if there are any flaws in a user-made allocation, communicating any detected problems to the user using natural language through the chatbot. Thus, our resulting application not only validates bed allocation plans but also interacts with hospital professionals using natural language communication, including giving explainable suggestions of better alternative allocations. We evaluated our approach with professionals responsible for bed allocation in two local hospitals and a doctor who provides consultancy to another local hospital. The version of the system reported in this paper addresses all the suggestions made by the specialists who evaluated its previous version.

Keywords: Hospital bed allocation · Intelligent agent · PDDL plan validation · Linear programming · Chatbot

1 Introduction

Bed allocation is a challenge faced by hospitals because hospital beds are scarce, and when poorly managed, it can lead to long queues or chaos in emergency rooms [9]. This is even more critical considering that developing countries face growing financial constraints making planning and efficient allocation of hospital beds increasingly difficult [11,20]. Also, the area responsible for bed allocation needs to be concerned with several restrictions during bed allocation, such as the type of medical speciality, whether the patient is surgical or clinical, gender,

© Springer Nature Switzerland AG 2021
A. Britto and K. Valdivia Delgado (Eds.): BRACIS 2021, LNAI 13073, pp. 3–17, 2021.
https://doi.org/10.1007/978-3-030-91702-9_1

and age of the patient [17]. This makes bed management an essential part of the planning and controlling of the operational capacity, and an activity that demands the efficient use of available resources [18].

Effective management of hospital beds has been the focus of much research such as the IMBEDS model that uses artificial neural networks and multi-attribute value theory for decision making [9]; statistical and data mining approaches [19]; an optimisation model based on evolutionary algorithms is used for bed allocation in [13]; and also literature reviews have been carried out [2,11]. Although all that work seeks to improve bed management, they do not provide easy interaction nor decision support so that the professional in charge retains full control over allocations.

Keeping professionals in charge is important, given that this is a domain of sensitive decision-making, hence there is much resistance to replacing human operators with automated systems. In domains like healthcare, a mixed-initiative system, which supports human-computer interaction, becomes not just useful [10] but essential. In this domain, wrong decisions can lead to the loss of lives. That is why it is important that intelligent systems can support the decisions being made, but a human must make the final decisions.

We have developed an intelligent agent to validate bed allocation plans and to communicate with hospital professionals using natural language. Although humans still make the final decisions, an intelligent agent checks if all allocation rules are being complied with and warn the user if they are not. To make this possible, we use plan validation techniques and a chatbot that provides the agent with the ability to communicate with the human operator through natural language. Our approach performs the validation of the bed allocation plan using a planning domain built based on the allocation rules used in each of the local hospitals that cooperated with our work. To use different rules for other hospitals, it is only necessary to adapt the planning domain and the way the planning problem is generated. We developed our own PDDL plan validator in Java, so that it interacts better with the chatbot developed using the Jason platform [4]. We also created an automated generator of linear programming models for generating optimal bed allocations, which we can be used to provide suggestions of better alternative allocations when requested by the human operators, for example when the chatbot detected errors in the allocations they created themselves. We can generate optimal allocations much quicker with a solver such as GLPK than with a PDDL planner.

We have evaluated our approach with professionals responsible for bed allocation in two local hospitals. Also, the approach was evaluated by a doctor who provides consultancy to a local hospital. With their feedback, we were able to assess the efficacy and acceptance of our approach by those professionals; they considered our approach extremely useful and usable in the daily routine of the hospital. The version of the chatbot described in this paper in fact addresses all the suggestions made by those professionals to increase the usefulness of the chatbot for their work.

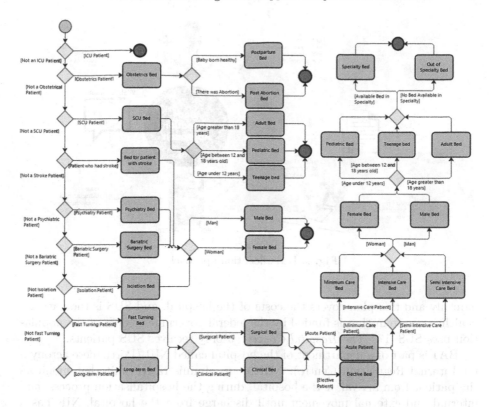

Fig. 1. Bed allocation rules.

The remainder of this paper is structured as follows. Section 2 shows the bed allocation scenario and our approach. Section 3 presents the results obtained in the evaluation of our approach. Section 4 explains the improvements made. Section 6 provides some conclusions.

2 Materials and Methods

2.1 Eliciting Bed Allocation Rules

We interviewed a person responsible for bed allocation at *Hospital Conceição* in Porto Alegre - RS, Brazil. This professional (referred to as BA1) has more than seven years experience in this work. Our objective was to understand the real scenario of hospital bed allocation.

Hospital Conceição is the largest unit of the Conceição Hospital Group (GHC). It offers all the specialities of a general hospital in its outpatient clinic, as well as an emergency room and inpatients. It maintains a medical emergency service with doors open 24 h a day and has 784 beds [8].

For contextualisation, in Brazil there are three types of health systems: private, where the patient pays; private health insurance, where the patient pays

Fig. 2. Bed allocation approach.

monthly and this plan covers the costs of the hospital; and SUS is the national healthcare system that is funded by the federal government. 80% of the population uses SUS [17]. The *Hospital Conceição* only receives SUS patients.

BA1 is part of a department of the hospital called NIR (Portuguese acronym for Internal Regulation Centre). NIR is an administrative unit that monitors the patient from arrival at the hospital, during the hospitalisation process, and internal and external movement until discharge from the hospital. NIR has a manual created in cooperation with the Health Ministry to guide SUS managers and better conduct the process of creation and running of NIR units [12].

The NIR has full control over hospital beds. The function of the operational nurse (BA1) is the real-time management of free beds. BA1 is responsible for authorising new admissions from the requested reservations, the exchanges, and the blocking required according to the demand and availability.

BA1 currently uses a locally developed system that has a feature for bed management. This system only shows which patients need hospitalisation and which beds are currently available. The rules for bed allocation are in a document outside the system, so the person responsible for allocation needs to look at the document, or in case of BA1, memorised all those rules. Bed allocation errors often occur because some rule is overlooked. Some errors can cause delays in a patient's accommodation since it is necessary to wait for a new allocation, besides all the unnecessary patient movement.

Based on the information obtained from BA1 and the NIR manual [12], we created a diagram with the main bed allocation rules (see Fig. 1). The diagram shows a large number of rules that need to be considered for an adequate allocation, prioritising good care, privacy, and the psychological state of patients. When there is overcrowding in the hospital, some rules may be ignored. In contrast, others can never be relaxed – for example, placing a patient who needs isolation in a non-isolation area.

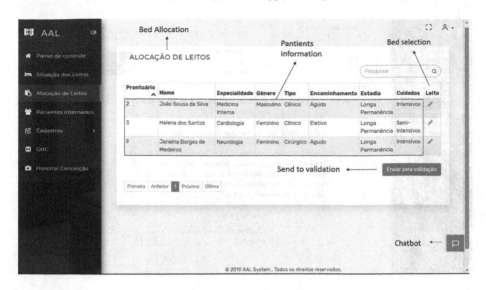

Fig. 3. Web simulator – bed allocation screen.

2.2 Developing an Approach to Bed Allocation

We propose an approach for helping the professionals ensure that all bed allocation rules have been followed (see Fig. 2). In our approach, healthcare professionals inform which patient and which bed they want to allocate the patient for the
system to validate the allocation in accordance to a particular set of rules. This
is done as follows: our system, based on the allocation information (bed and
patient), automatically generates a PDDL (Planning Domain Definition Language) [1] problem file and a PDDL plan file; the particular hospital rules in
question are represented as a PDDL domain specification. The part of the system with patient and hospital information we call "web simulator",[1] because it
is implemented as a web system to simulate a hospital information system. Our
web simulator was developed to facilitate interaction with the user when testing
our approach. Figure 3 shows the simulator bed-allocation screen.

Our approach also has an intelligent BDI agent constructed with Jason [4],
which has access to the PDDL problem file and a PDDL plan file and uses them
to validate the allocation, checking if any rules are being broken.[2]

To do this validation, the Jason agent originally used the VAL validator [7,
10], which outputs a LaTeX report saved to a folder that will be analysed by the
agent; we later mention how this has been reworked to improve performance.
After the report is saved, the agent reads and processes that file. VAL is a plan
validator that in our case checks if any bed allocation rule has been broken in the

---

[1] The web simulator code is available at https://github.com/smart-pucrs/bed-
allocation-simulator.
[2] The agent code is available at https://github.com/smart-pucrs/jason_assistant_
to_bed_allocation.

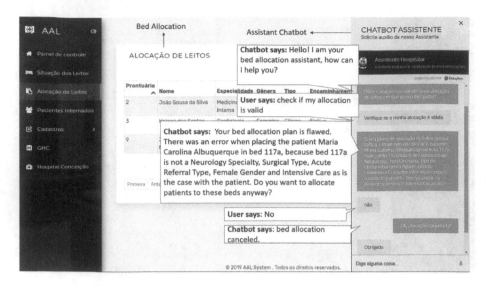

Fig. 4. Web simulator – bed allocation screen with chatbot.

user allocation. VAL receives three files as input: 1) the plan file containing the information of the allocation made; 2) the problem file contains the current world state and the goals that should be achieved; 3) the domain file[3] that describes the rules that must be followed to perform the allocation.

This domain file has predicates that describe the characteristics of beds and patients. The actions contained in the domain file are responsible for determining the behaviour according to the real hospital rules.

After analysing the validation report, our agent creates the response to the user, saying whether or not the plan is valid, and if it is not valid, it also tells the user which rule was being broken. One of the limitation of VAL, which led to the development we explain later, is that it stops the validation as soon as a broken rule is found, so if there are more broken rules, our agent would still report only the first one found. This response is sent to the user through a chatbot using the DialogFlow platform,[4] which in addition to responding, asks whether the user wants to confirm the allocation (see Fig. 4). If the answer is positive, the chatbot itself triggers the routine on the system responsible for completing the allocation. If the answer is no, the chatbot just cancels the allocation, leaving the system's validation history saved.

When the user requests the chatbot to "check if my allocation is valid", an intention is triggered on DialogFlow; we call it "Get Validation Result". This intention calls through the webhook a function in Cloud Functions that searches in the database the last validation carried out by our Jason intelligent agent and

---

[3] The domain file with plans and problems examples are available at https://github.com/smart-pucrs/hospital-domain-PDDL.

[4] https://dialogflow.com/.

returns to Dialogflow the answer elaborated by it. In addition to answering, the chatbot also asks if the user wants to "confirm the allocation". For the chatbot to wait for the user's response, we use Dialogflow contexts.

For the integration between the web simulator, Dialogflow, and Firestore, we use Cloud Functions[5] that runs on a NodeJs platform. Furthermore, to integrate the intelligent agent with Cloud Functions, we use an API developed in Asp.Net Core.[6] The information returned by the API, as well as the representation of the plan and problem, are saved in the database so that the chatbot can access it when prompted. Once this data is saved, the simulator issues a warning informing the user that it can already request the validation results to the chatbot.

3   Evaluation

We evaluated our approach in three phases. The first phase was carried out with two professionals responsible for bed allocation at *Hospital Conceição*; one is the same professional who previously informed us of the bed allocation rules. The second phase was with two professionals also responsible for bed allocation but at *Hospital São Lucas da PUCRS (HSL)*. Finally, we evaluated our chatbot with the help of one of the doctors who assisted in the construction of the NIR manual [12].

For all evaluations, we fed the system with simulated data about beds, doctors, and patients. Based on the data in the system, we asked that professionals use the simulator to check out the simulated hospital situation and ask the chatbot to validate the bed allocation they created and then evaluate the feedback that the chatbot gave. Furthermore, we performed a semi-structured interview in order to collect the feedback of these professionals about the use of the system. All professionals signed a consent form for participation.

3.1   First Phase

We evaluated the approach with each professional individually. The first professional (BD1), who informed us about the allocation rules, has 7 years of experience in bed allocation. The second professional has more than 4 years experience in bed allocation. We highlight some points reported by those professionals:

- Chatbot and allocation rules:
  - Interaction with chatbot is extremely easy.
  - The information about bed allocation problems that the chatbot points out are very useful and really what they use in practice.
  - They agree that it is not viable a system that allocates alone without a final decision made by a human operator.
  - They would be willing to use our system in their daily activities.

---

[5] https://cloud.google.com/functions/.
[6] https://www.asp.net/core/overview/aspnet-vnext.

– Suggestions:
  • Consider patients' priorities.
  • The validation performed by the system is useful, but it would be even more valuable if it suggested how to correct the allocation if any rules were broken, providing allocation suggestions that do not break the rules.

3.2 Second Phase

We then evaluated our approach with two professionals at the same time. Both professionals have more than one year of experience in the bed allocation function. We also asked questions related to the allocation process to understand the differences between the reality of this hospital and *Hospital Conceição*. The main difference is that *HSL* serves the three types of health systems. We highlight some important considerations reported in this interview:

– Chatbot and allocation rules:
  • They would not like a system that allocates beds autonomously, and prefer one that only make recommendations when requested.
  • They believe that the agent can help them in the daily routine.
  • They are willing to use our system in their daily activities if it remains not necessary to talk much to the robot or chatbot.
– Suggestions:
  • Their current system does not distinguish between SUS beds, private health insurance, and private; it would be important that the chatbot had this knowledge.
  • Considering the rules of patient prioritisation is extremely important to them.
  • They suggested that the system should give priority to the relocation of a patient released from the ICU rather than other patients who need a bed of that type.
  • Interpret natural language written texts in the patient's evolution to retrieve relevant information.
  • Generate warnings when a patient has been in the ICU more than 8 days or more than 30 days in the same bed.
  • Based on the patients' discharge plan for the following day and the scheduling of procedures, the agent could advise if beds will be missing.
  • Tell when a patient was discharged for more than 30 min but the bed has not yet been vacated.
  • Tell when a bed is interdicted for more than 24 h.
  • The agent should also knows the business rules (regarding private health insurance) so that it can validate them too.

3.3 Third Phase

We performed an evaluation with the doctor who graduated over 10 years ago and has more than a year of experience as Consultant Medical Doctor. We highlight some important considerations reported in this interview:

- Chatbot and allocation rules:
  - The rules are well in line with what are current practices.
  - It is feasible, and it is even necessary to reduce the variability in the conduction of bed allocation processes.
  - They informed us that they would use our system, because this type of software would make the process very fast, in addition to the accuracy and availability to work 24 h a day, seven days a week.
- Suggestions:
  - Define the bed typology and not ask the doctor to define it, but validate the typology.
  - Suggest an allocation to the user. Not that the agent should make an allocation alone, but it would be good to give suggestions that someone would confirm.
  - Offer options of frequent-asked questions, instead of user typing or talking to the chatbot all the time.
  - When there is a list of patients in the emergency room waiting for a bed and a bed that is suitable for any of these emergency patients becomes vacant, the system should warn the operator.
  - Generate a bed availability forecast, for example, for the next six hours according to the discharge forecasts contained in the system.

4   Improvements Based on the Evaluation

Some suggestions made during the evaluation were considered of major importance and, in order to achieve them, we have carried out some improvements in our system. To facilitate communication between the Jason agent and Dialogflow, we started using Dial4JaCa[7] [5,6], a general integration between JaCaMo and Dialogflow that has recently become available. To solve the limitation imposed by the VAL validator, which did not return a complete list with all the allocation problems that the plan has, we implemented a new PDDL validator that will be presented next. Moreover, to enable our agent to make optimised suggestions for bed allocations, we implemented an optimal bed allocation technique using linear programming, which will also be presented in more detail below.

4.1   Bed Allocation Optimisation

In order to allow our agent to make optimal suggestions, we developed an approach to bed allocation based on linear programming. After a linear programming problem is generated for a particular bed allocation instance, we use the GLPK solver to find an optimal solution.[8] The optimisation tries to allocate as many patients as possible while attempting to decrease as much as possible the distance of critical patients to the office where nurses are based.

---

[7] https://github.com/smart-pucrs/Dial4JaCa.
[8] https://en.wikibooks.org/wiki/GLPK.

The bed allocation optimisation program takes the database's restrictions and converts them into three types of linear restrictions: equality, relative equality, and negation. Consider for example the following constraint.

```
(2*Q112[p])+(abs(speciality[p]-2)/2) <= 2;
```

The equality constraint above requires that every patient allocated to the bedroom has a specific characteristic. In this case, we want the patient to have the speciality 2. Q112[p] is a Boolean value equal to 1 if patient p is in the bedroom; gender [p] is an integer value referring to the patient's speciality, hospitals will usually have specific rooms for specific types of specialities. The left side of the sum results in 2 if the patient has been allocated to that bedroom; otherwise, it results in 0. The right-hand side results in 0 if the patient's speciality is 2; otherwise, it results in some number greater than zero. The result of the sum must be less than or equal to 2, being the only case in which this does not happen when the patient is allocated to the bedroom, but the speciality is not 2.

```
(Q112[p1]+Q112[p2])+(abs(gender[p1]-gender[p2])/2) <= 2;
```

Relative equality is used to require, for example, that if two patients are allocated to the same bedroom, they must have the same characteristic (in the case above, the same gender). The sum's left-hand side has a maximum value of 2 when the two patients are in room 112. The right-hand side results in a value greater than 0 when the two patients' genders are different. The restriction would only be violated when the two patients are in the same bedroom, but the genders are different, which must not be allowed.

```
(Q112[p]) - abs(isolation[p]-1) <= 0;
```

Negation is used for example when we require that every patient allocated to the room does not have a specific characteristic. If the patient is in the room, but the isolation characteristic is 1, this will result in $1 - 0 = 1$, which does not comply with the restriction. If the patient is not allocated to the room or the isolation value is not 1, the result is less than or equal to zero, according to the bed rules. If the room is not designed for isolation, it will not allow patients that must be isolated under the normal circumstances under which the optimiser normally operates.

In this work, we used the GLPSol solver of GLPK[9] (GNU Linear Programming Kit), which is a free open source software for solving linear programming problems. One of the implemented algorithms is the simplex method where, after assembling a geometrical figure, the program chooses a point and recursively chooses the relative maximum (or minimum) point, approaching the global optimal result at each iteration. GLPSol allows the user to set certain limits, such as a time limit. When the limit is reached, the process returns the best result

---

[9] http://winglpk.sourceforge.net/.

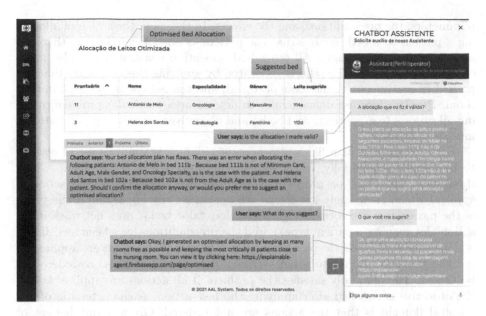

Fig. 5. Optimiser output in web simulator

found so far. This allows the program to generate some possibly useful allocation suggestions without having to arrive at the optimal result, which can be an
extremely time-consuming process in some instances.

We have already integrated the optimiser with Jason Platform, allowing us
to execute tests with our complete structure. Figure 5 presents the result of an
allocation made using the optimiser.[10]

4.2 Plan Validator

When real-world problems can be modelled in a planning language, it is possible
to use a plan validator for example to tell the human operator whether a given
plan is feasible or not [10]. Behnke *et al.* [3] define plan validation as "the task
of determining whether a plan is a solution to a given planning problem". A
plan validator can be used in a wide range of applications. The application that
interests us is the validation of bed allocation plans prepared by the user.

We developed a new plan validator[11] using Java to facilitate the integration
with Jason. Like VAL, our plan validator also receives three PDDL files as input:
a file containing the domain, a file containing the problem, and a file containing
the plan to be validated.

The domain file establishes some basic rules, such as the types of objects
and possible actions. The actions are generally divided into three parts, the

---

[10] All patient data in our tests are fictitious.

[11] The validator code is available at https://github.com/smart-pucrs/PDDL-plan-
validator.

parameters, the preconditions, and the effects. In this case, the action of allocating a patient to a bed requires that the patient is not allocated to another bed and that the bed is empty. All other bed allocation constraints are also modelled as preconditions. The effects generated by applying this action are that the patient is now allocated, the room occupied, and the patient is allocated in that room. Problem files use domain rules to determine a particular problem, making all objects (e.g., patients and beds) and objectives explicit (e.g., all patients must be allocated).

For validators, a plan file is also necessary, which is simply a set of actions to be applied sequentially, which lead to the objective of the problem. Given these files, the validation process is straightforward, checking if each action is applicable and then apply its effects. An action is only applicable if the types of the parameters are correct (e.g., the action "aloc bed3" does not work if the action aloc requires a patient type) and if the preconditions have been met. If any action is not applicable, the plan is considered flawed. If all actions are applicable, but the problem's objective is not satisfied, then the plan is also considered to be flawed. The plan only satisfies the problem if all actions are applicable and the objective is satisfied after applying the last action. A characteristic of the hospital domain is that the actions are not ordered. Given a suitable set of allocations, any order of execution will solve the problem, which, in general, would not be the case in other domains. This means that for us it is particularly useful to detect all errors in the plan and not just the first.

In our plan validator, the user has the possibility of printing the plan and the validation result in the terminal through the *planTest* function. The user still has the possibility, through the *valOut("filename")* function, to obtain a LaTeX file that generates a PDF with the validation report,[12] the result can also be analysed by our intelligent agent to give an answer to the user in natural language.

As the validator was designed so far thinking about a specific type of problem, we have made, for now, certain restrictions to facilitate the implementation of an initial version of the software. In our initial version, we have some limitations that do not negatively affect the results of the tasks for which we are using the validator, but that need to be resolved to make our validator available to the research community to use it. Although the PDDL language is modular, some specific options are practically universally accepted, however not all of them have been implemented yet. In total, our validator currently supports three options:

- "STRIPS", which allows for actions to add or remove effects, and is required in our validator;
- "typing", which allows the use of types and sub-types; and
- "equality", which allows the use of equality comparisons.

Another actual limitation is that the plan validator was made to be used in automatically generated problems. It considers that the PDDL files are

---

[12] https://github.com/smart-pucrs/PDDL-plan-validator/blob/main/latexOutput Validator.pdf.

semantically and syntactically correct. We have not yet implemented any pre-processing to look for errors automatically, but we developed a callable method to perform this check. The current version of our plan validator is available on GitHub.[13]

5 Related Work

Oliveira et al. [13] developed an iterative Simulation-Optimisation approach using an evolutionary algorithm to optimise hospital-bed allocation. They applied their approach using datasets from DATASUS of Minas Gerais, Brazil, where the public health system assists nearly 80% of the patients.

Grübler et al. [9] proposed a model called IMBEDS to allocate patients to beds. IMBEDS is a hybrid model that aids in the bed selection process, using some techniques that work together to manage a waiting list of emergency patients and scheduled patients. The model uses an artificial neural network (ANN) and multiattribute value theory (MAVT), a technique used for decision making and conflict resolution in projects with multiple criteria, using contextual information about patients and beds. They evaluated the model in a Hospital in Porto Alegre, Brazil, that receives only patients from private insurance plans.

In Teow et al. [19], the authors extracted data from a hospital in Singapore and applied statistical and data mining approaches to identify the patterns behind bed overflow. Their main objective was to help the hospital devise strategies to reduce bed overflow and improve patient care.

Differently from the work described above, not only does our approach suggest and validate bed allocations through optimisation techniques but it also supports natural language communication using chatbots to interact with the users. We believe our application makes progress towards existing needs in the application of AI systems in domains like healthcare, in which agents provide natural language explanations, e.g., allocation failure in our scenario, supporting users' decision making rather than being the decision makers.

6 Conclusions

This paper presented an approach to support hospital staff during the process of bed allocation. Although there is some work in the literature that intends to make completely automated allocations, this area of application has resistance to replacing human operators with automated systems. Therefore, in our approach, we only seek to support the decision-making of human professionals.

Our approach was built based on the expertise of professionals in the field, mainly based on an interview we conducted with a professional responsible for bed allocation in a local hospital. We evaluated our approach at *Hospital Conceição* and *HSL*, the latter having practices with significant differences since it serves not only SUS but also private patients. Also, we interviewed a doctor that provides consultancy to a local hospital in the area of bed allocation.

---

[13] https://github.com/smart-pucrs/PDDL-plan-validator.

Among the points highlighted by the professionals about our approach, it has been mentioned that it is easy to interact with the chatbot and the information contained in the validation of allocation are very useful for their routine. Some good ideas have also been raised to improve our system so that it performs useful tasks in the bed-allocation routine. As our approach was developed based only on the existing practices at *Hospital Conceição*, the analysis made by the professionals of *HSL* yielded many ideas for our agent to be more useful in its planned use in those hospitals.

Through this research, we created a domain knowledge and developed a planning domain for PDDL and HTN planners that can also be useful for other projects that involve automated planning and plan validation.

As future work, we intend to investigate argumentation techniques [14–16] to implement more interactive agents and explain the allocation suggestions as requested by the interviewees. We believe that with the use of argumentation techniques, intelligent agents will be able to reason about beds and patients' relations, thus providing useful explanations that will help the users by saying why a particular suggestion is being made. We intend to integrate our approach with the Hospitals' information systems, which will allow us to make several queries as suggested by the experts during the evaluation phase. Moreover, we aim to use natural language models to allow the chatbot to answer questions about a patient's evolution.

Acknowledgements. This research was partially funded by CNPq and CAPES – Finance Code 001.

References

1. Aeronautiques, C., et al.: PDDL— the planning domain definition language. Technical Report (1998)
2. Ahmadi-Javid, A., Seyedi, P., Syam, S.S.: A survey of healthcare facility location. Comput. Oper. Res. **79**, 223–263 (2017)
3. Behnke, G., Höller, D., Biundo, S.: This is a solution! (... But is it though?) - verifying solutions of hierarchical planning problems. In: 27th International Conference on Automated Planning and Scheduling, pp. 20–28 (2017)
4. Bordini, R.H., Hübner, J.F., Wooldridge, M.: Programming Multi-agent Systems in AgentSpeak using Jason. Wiley, Hoboken (2007)
5. Engelmann, D., et al.: Dial4JaCa – a demonstration. In: Dignum, F., Corchado, J.M., De La Prieta, F. (eds.) PAAMS 2021. LNCS (LNAI), vol. 12946, pp. 346–350. Springer, Cham (2021). https://doi.org/10.1007/978-3-030-85739-4_29
6. Engelmann, D., et al.: Dial4JaCa – a communication interface between multi-agent systems and chatbots. In: Dignum, F., Corchado, J.M., De La Prieta, F. (eds.) PAAMS 2021. LNCS (LNAI), vol. 12946, pp. 77–88. Springer, Cham (2021). https://doi.org/10.1007/978-3-030-85739-4_7
7. Fox, M., Howey, R., Long, D.: Validating plans in the context of processes and exogenous events. In: 20th National Conference on Artificial Intelligence and the Seventeenth Innovative Applications of Artificial Intelligence Conference, pp. 1151–1156 (2005)

8. GHC: Hospital Conceição (2019). https://www.ghc.com.br/default.asp?idMenu=unidades&idSubMenu=1
9. Grübler, M.D.S., da Costa, C.A., Righi, R., Rigo, S., Chiwiacowsky, L.: A hospital bed allocation hybrid model based on situation awareness. Comput. Inform. Nurs. **36**, 249–255 (2018)
10. Howey, R., Long, D., Fox, M.: VAL: automatic plan validation, continuous effects and mixed initiative planning using PDDL. In: 16th International Conference on Tools with Artificial Intelligence, pp. 294–301 (2004)
11. Matos, J., Rodrigues, P.P.: Modeling decisions for hospital bed management - a review. In: 4th International Conference on Health Informatics, pp. 504–507 (2011)
12. Ministério da Saúde: Manual de implantação e implementação : núcleo interno de regulação para Hospitais Gerais e Especializados. Technical report, Ministério da Saúde, Secretaria de Atenção à Saúde, Departamento de Atenção Hospitalar e de Urgência (2017)
13. e Oliveira, B., de Vasconcelos, J., Almeida, J., Pinto, L.: A simulation-optimisation approach for hospital beds allocation. Int. J. Med. Inform. **141**, 104174 (2020)
14. Panisson, A.R., Bordini, R.H.: Knowledge representation for argumentation in agent-oriented programming languages. In: 2016 Brazilian Conference on Intelligent Systems, BRACIS (2016)
15. Panisson, A.R., Bordini, R.H.: Towards a computational model of argumentation schemes in agent-oriented programming languages. In: IEEE/WIC/ACM International Joint Conference on Web Intelligence and Intelligent Agent Technology (WI-IAT) (2020)
16. Panisson, A.R., Engelmann, D., Bordini, R.H.: Engineering explainable agents: an argumentation-based approach. In: International Workshop on Engineering Multi-Agent Systems (EMAS) (2021)
17. Pinto, L.R., de Campos, F.C.C., Perpétuo, I.H.O., Ribeiro, Y.C.N.M.B.: Analysis of hospital bed capacity via queuing theory and simulation. In: Proceedings of the Winter Simulation Conference 2014, pp. 1281–1292. IEEE (2014)
18. Proudlove, N.C., Gordon, K., Boaden, R.: Can good bed management solve the overcrowding in accident and emergency departments? Emerg. Med. J. **20**, 149–155 (2003)
19. Teow, K.L., El-Darzi, E., Foo, C., Jin, X., Sim, J.: Intelligent analysis of acute bed overflow in a tertiary hospital in Singapore. J. Med. Syst. **36**, 1873–1882 (2012)
20. Tsai, J.C.H., et al.: Adjusting daily inpatient bed allocation to smooth emergency department occupancy variation. In: Healthcare, vol. 8, p. 78. Multidisciplinary Digital Publishing Institute (2020)

A Protocol for Argumentation-Based Persuasive Negotiation Dialogues

Mariela Morveli-Espinoza[1]([⊠])[iD], Ayslan Possebom[2][iD],
and Cesar Augusto Tacla[1][iD]

[1] Program in Electrical and Computer Engineering (CPGEI),
Federal University of Technology of Parana (UTFPR), Curitiba, Brazil
morveli.espinoza@gmail.com, tacla@utfpr.edu.br
[2] Paraná Federal Institute of Education, Science and Technology (IFPR),
Paranavai, Brazil

Abstract. Argumentation-based persuasive negotiation is a form of negotiation dialogue in which agents, with different interests and goals, exchange proposals that are supported by rhetorical arguments such as threats, rewards, or appeals. Besides rhetorical arguments, additional kinds of illocutions may also be exchanged during the dialogue, for instance, agents may ask for explanations, give explanations, or attack (or contradict) previous arguments. This paper presents a formal protocol for argumentation-based persuasive negotiation dialogues in which a proponent agent tries to persuade his opponent to perform a given action and the opponent tries to maintain his position. The protocol is modelled as a dialogue game (i.e. the interactions between the proponent and the opponent are governed by a set of rules) and the outcome of the dialogue is determined by applying an argumentation semantics. We prove the soundness and completeness of our proposal and illustrate the proposed protocol by using an example.

Keywords: Persuasive negotiation · Intelligent agents · Dialogue protocol · Argumentation-based dialogues · Rhetorical arguments

1 Introduction

Negotiation is a key form of interaction among agents that can be used for resolving conflicts and reaching agreements. Formal argumentation is a process based on the construction and comparison of arguments considering the conflicts that may emerge among them. Such conflict are called attacks. The idea is to determine set(s) of non-conflicting arguments (called extensions), which are considered acceptable or justified. The function in charge of calculating the extensions is called semantics [13].

Some works on negotiation argue that argumentation – using explanatory arguments – allows that an agent acquire additional information about his opponents, which can be used for attacking his opponent's proposals or justifying his

Supported by CAPES.

© Springer Nature Switzerland AG 2021
A. Britto and K. Valdivia Delgado (Eds.): BRACIS 2021, LNAI 13073, pp. 18–32, 2021.
https://doi.org/10.1007/978-3-030-91702-9_2

own proposals (e.g., [3,12,26,29]). Besides explanatory arguments, there exist other kinds of arguments that can be used in negotiation dialogues and act as persuasive elements aiming to force or convince an opponent to accept readily a given proposal, these are called *rhetorical arguments* (e.g., in [20–22]). According to Ramchurn et al. [27], a negotiation involving these kinds of arguments is called persuasive negotiation.

We can describe the rhetorical arguments as follows: (i) *threats*, which try to persuade an opponent agent by using the argument that something negative will happen to him if he does not accept to do the requirement sent by the proponent; (ii) *rewards*, which try to persuade an opponent by using the argument that something positive will happen to him if he accepts to do the requirement sent by the proponent; and (iii) *appeals*, which try to persuade an opponent in the same form than rewards, but this positive event will depend on the opponent; hence, appeals can be seen as self-rewards [4].

Rhetorical arguments have been studied in terms of speech acts (e.g., [27]) and in other articles, a logical formalization has been given (e.g., [5,6]). It was also studied how to evaluate their strength values (e.g., [6,21,22]). However, to the best of our knowledge, no study about a protocol involving these arguments has been proposed.

In order to better understand the problem, imagine a scenario where two agents, Maria (M) and Carlos (C), are discussing about household chores. Maria is trying to persuade Carlos to do the cleaning of their apartment. The following dialogue shows how agreement is reached:

(1) M: Carlos, could you please do the cleaning?

(2) C: No, I can not, I have to work.

(3) M: If you do the cleaning, I could help you with your reports and you can finish your work early.

(4) C: You can not help me.

(5) M: Why?

(6) C: Because these reports are about a topic you do not know.

(7) M: Well, if you do not do the cleaning, I will not go to your mother's house on Saturday.

(8) C: If you will not go to my mother's house, I will not talk to her about the work for your brother.

(9) M: That is not longer necessary, my brother got a job yesterday.

(10) C: OK. You win!

In this example, Maria succeeds in persuading her husband Carlos to do the cleaning of their apartment. On the first attempt to persuade Carlos, Maria uses a reward (line 3), which is not accepted, resulting in an attack to her reward (line 4). In this settings, an attack is a contradictory statement. Since her reward was not successful, she uses a more powerful argument, i.e. a threat(line 7), and then he also answers with another threat (line 8), which we can call a counter-threat. She answers attacking Carlos' counter-threat (line 9). Notice that this attack is

not a counter-threat, which indicates that there is more than one way to attack
a rhetorical argument. Finally, Carlos accepts to do the cleaning.

Besides rhetorical arguments and their corresponding attacks, we can notice
that an explanation is required during the conversation (line 5), this means that
agents can make questions to each other and can use explanatory arguments
to justify their opinions. Figure 1 shows the outline of the dialogue in terms of
rhetorical arguments, attacks, and other illocutions.

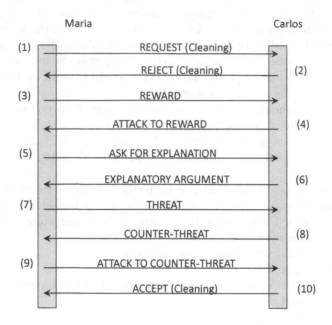

Fig. 1. Outline of the dialogue between Maria and Carlos, which ends successfully for
Maria.

From this scenario, we can observe that during a persuasive negotiation dialogue agents can exchange rhetorical arguments, attacks to rhetorical arguments,
questions, explanatory arguments, and attacks to explanatory arguments.

Most of the research about protocols in literature is focused on negotiation
(e,g., [1,11,16,28]), some others on persuasion (e.g., [10,17]), and some others
on argumentation-based dialogues (e.g., [8,9,14,23,25,30,31]). Although some
of these protocols take into account explanatory arguments and attacks among
them, these are not embedded in a persuasive negotiation dialogue (according
to the definition of [27]) and do not interact with other kinds of arguments.

Thus, the research questions that are addressed in this paper are:

1. How can rhetorical arguments, explanatory arguments, attacks, and other
   illocutions be combined in a coherent dialogue?
2. How can argumentation techniques be used in this kind of dialogue?

In order to address the first question, we propose a protocol that has a dialogue game form where utterances are viewed as moves in a game, which is guided according to a defined set of rules. According McBurney and Parsons [18], formal dialogue games allow sufficient flexibility of expression while avoid state-space explosion. Regarding the second question, an important part of a dialogue is the outcome. In this case, the outcome has to determine the final status of the dialogue (that is, it ends with an agreement or not), the winner (if there is an agreement), and the set of commitments the winner agent has to fulfil after the dialogue, for example, Maria will go to the house of the mother-in-law since Carlos will do the cleaning. We will use argumentation semantics to determine the outcome of the dialogue.

Next section presents the main concepts about argumentation and argumentation semantics. Section 3 concerns with the type of arguments and attacks. Section 4 presents the proposed protocol, that is, the rules that govern the interactions among the agents, the argumentation framework that determines the outcome of the dialogue and the main properties of the approach. In Sect. 5, we illustrate our new protocol by applying it to the example given in Introduction. Finally, Sect. 6 is devoted to conclusions and future work.

2 Background

In this section, we present the concepts of argumentation framework and the acceptability semantics for linear dialogues.

An argumentation framework consists of a set of arguments and a attack relation between them. The following definitions were extracted from [2] and [13].

Definition 1. *(Argumentation framework) An argumentation framework is an ordered pair* $\mathcal{AF} = \langle \mathrm{ARG}, \mathcal{R} \rangle$*, where* ARG *is a finite set of arguments and* $\mathcal{R}$ *a binary relation on* ARG *(i.e.,* $\mathcal{R} \subseteq \mathrm{ARG} \times \mathrm{ARG}$*). We call* $\mathcal{R}$ *an attack relation and* (A, B) *means that argument A attacks argument B.*

Before presenting the acceptability notion, it is important to study linear discussions. A linear discussion is a sequence of arguments such that each argument attacks the argument preceding it in the sequence. This sequence will determine which arguments can be considered acceptable and which cannot.

Definition 2. *(Linear discussions) Let* $\mathcal{AF} = \langle \mathrm{ARG}, \mathcal{R} \rangle$ *be an argumentation framework and* $A \in \mathrm{ARG}$*. A linear discussion for A in* ARG *is a sequence* $s = \langle A_1, ..., A_n \rangle$ *of elements of* ARG *(where n is a positive integer) such that* $A_1 = A$ *and* $\forall i \in \{2, 3, ..., n\}$ $(A_i, A_{i-1}) \in \mathcal{R}$*.*

Next we present the semantics, which determines what arguments are considered acceptable or justified. First, it is necessary the notion of supporters. Given three arguments A_1, A_2, and A_3. If $(A_2, A_1) \in \mathcal{R}$ and $(A_3, A_2) \in \mathcal{R}$, then A_3 supports A_1. The acceptability notion is directly related to argument A because it represents the central point of the discussion. Thus, given a sequence s, we can say that $\forall A_i \in s$, A_i supports A if i is odd and it attacks A if i is even. Let $\text{sup}(A)$ return all the supporters of A_1 and $\text{att}(A)$ return all its attackers.

Definition 3. *(Semantics) Let $\mathcal{AF} = \langle \text{ARG}, \mathcal{R} \rangle$ be an argumentation framework, $A \in \text{ARG}$, s a sequence for A, and n the length of s. It holds that (i) if n is odd, then $A \cup \text{sup}(A)$ are acceptable, or (ii) if n is even, then $\text{att}(A)$ are acceptable.*

Let $\text{SEM}(\mathcal{AF})$ be a semantics function that returns the set of acceptable arguments.

3   Building Blocks

In this section, we present the topic language used to represent the content of illocutions exchanged by agents. We also present the definitions of the kinds of arguments that can be exchanged and study the possible attacks to each of them.

According to Van Veenen and Prakken [30], formal dialogue games have a topic language $\mathcal{L}_t$, expressed in a certain logic, and a communication language $\mathcal{L}_c$ with a protocol $\mathcal{P}$, which specifies the allowed moves at each point in a dialogue. We can say that a persuasive negotiation dialogue happens between a proponent agent P and an opponent agent O about a topic $t \in \mathcal{L}_t$. In this work, the **topic language** $\mathcal{L}_t$ is specified using the classical logical language. Symbols $\wedge, \vee$ and $\neg$ denote the logical connectives conjunction, disjunction, and negation, respectively. Besides, $\vdash$ stands for the classical inference and $\equiv$ logical equivalence. From $\mathcal{L}_t$ we can distinguish the three following sets of formulas:

- $\mathcal{G}$ contains the goals the agent pursues;
- $\mathcal{GO}$ contains what the agent believes the goals of the other agent are (that is, his opponent's goals); and
- $\mathcal{K}$ is the knowledge base of the agent, which gathers the information the agent has about the environment.

Goals and opponent's goals are represented with positive literals[1] from $\mathcal{L}_t$. Besides, $\mathcal{G}$ and $\mathcal{GO}$ are finite sets such that $\mathcal{G} \cap \mathcal{GO} = \emptyset$.

Now, let us present the definitions of arguments, explanatory arguments, rhetorical arguments, and attacks involved during a persuasive negotiation dialogue. In what follows, for a given argument, the function SUPP returns all the beliefs of $\mathcal{K}$ (called support) used to build the argument and CONC returns its conclusion. The first one is a basic definition where any element of $\mathcal{K}$ can be an argument whereas explanatory and rhetorical arguments have a deductive form. Indeed, a fact or a goal is entailed from the support.

---

[1] A literal is either an atomic formula or the negation of an atomic formula. When a literal is an atomic formula, we say that it is a positive literal, and when a literal is the negation of an atomic formula, we say it is a negative literal.

Definition 4. *(Argument* [19]*)* Let $\mathcal{K}$ be a knowledge base. An argument A is φ if $\varphi \in \mathcal{K}$ with $\text{SUPP}(A) = \{\varphi\}$, $\text{CONC}(A) = \varphi$.

Explanations are the most common category of arguments. They represent the reasons to believe in a fact.

Definition 5. *(Explanatory argument* [7]*)* Let $\mathcal{K}$ be the knowledge base of an agent. An explanatory argument is a tuple $A = \langle S, h \rangle$ such that (i) $S \subseteq \mathcal{K}$, (ii) $S \vdash h$, and (iii) S is consistent and minimal.[2] With: $\text{SUPP}(A) = S$ and $\text{CONC}(A) = h$.

Unlike explanatory arguments, rhetorical arguments are also made up by goals and opponent's goals.

Definition 6. *(Rhetorical arguments* [7]*)* Let $\mathcal{K}$ be the knowledge base of an agent, $\mathcal{G}$ be his goals base, and $\mathcal{GO}$ be his opponent's goals base. A threat, reward, or appeal is a triple $A = \langle S, g, go \rangle$ such that (i) $S \subseteq \mathcal{K}$, (ii) $g \in \mathcal{G}$, (iii) $go \in \mathcal{GO}$, and (iv) S is consistent and minimal. Besides:

- In the case of **threats**, it holds that $S \cup \{\neg g\} \vdash \neg go$
- In the case of **rewards and appeals**, it holds that $S \cup \{g\} \vdash go$.

With: $\text{SUPP}(A) = S$, $\text{CONC}(A) = g$, and $\text{OPGOAL}(A) = go$ *returns the goal that is being threatened, rewarded, or appealed.*

It is also necessary to define the attacks each kind of argument may receive. An attack relation between two arguments A and B denotes the fact that these arguments cannot be accepted simultaneously since they contradict each other. In the case of explanatory arguments, two kinds of attacks can be determined, undercuts and rebuttals. An undercutting argument is an argument whose conclusion contradicts some of the elements of the support of another argument, and a rebutting argument is an argument whose conclusion is the negation of the conclusion of another argument. Formally:

Definition 7. *(Attacks to explanatory arguments* [7]*)* Let $\langle S, h \rangle$ and $\langle S', h' \rangle$ be two explanatory arguments:

- $\langle S', h' \rangle$ **undercuts** $\langle S, h \rangle$ iff $\exists h'' \in S$ such that $h' \equiv \neg h''$.
- $\langle S', h' \rangle$ **rebuts** $\langle S, h \rangle$ iff $h \equiv \neg h'$.

Regarding attacks to rhetorical arguments, we have distinguished three types of attacks. The first one occurs when a threat is attacked by another threat (we can call it a counter-threat). In this case, there is no a logical contradiction but we can notice that the goal threatened by an agent is used by the opponent to construct another threat as can be observed in lines (7) and (8) of the example given in Introduction.

---

[2] Minimal means that there is no $S' \subset S$ such that $S \vdash h$ and consistent means that it is not the case that $S \vdash h$ and $S \vdash \neg h$, for any h [15].

Definition 8. *(Counter-threat) Let $\mathcal{G}$ be the set of goals of an proponent agent P, $th_P = \langle \mathcal{S}, g, go \rangle$ be a threat of P, and $th_O = \langle \mathcal{S}', g', go' \rangle$ be a threat of opponent agent O. We say that th_O counter-threatens th_P when $g' = go$ and $go' \in \mathcal{G}$.*

The second type of attack occurs when the opponent answers disesteeming his threatened/rewarded/appealed goal and denies his interest in achieving it. For example, line (9) of the example given in Introduction. In line (8), Carlos threatens Maria with not talking with his mother about a job for Maria's brother and in line (9), she says that her brother already got a job. This attack has the form of an explanatory argument.

Definition 9. *(Disesteemation) Let $A = \langle \mathcal{S}, g, go \rangle$ be a rhetorical argument. An argument $\langle \mathcal{S}', g' \rangle$ disesteems A when $g' = \neg go$.*

To the best of our knowledge, the two previous types of attacks were not studied before. On the other hand, a rhetorical argument can also be undercut.

4   The Proposal

In this section, we present the language and the rules for the dialogue game. Besides, we present the argumentation framework that represents the dialogue and determines its outcome.

4.1   The Proposed Protocol

The game is mainly based on the following ideas. Each move in the dialogue, except the initial one, replies to the previous move of the other agent (we refer to the previous move as its target). In [24], the author proposes the idea of *attack* and *surrender* as a categorization of the possible replies to previous moves during the dialogue. A reply is a surrender when it is not against the previous move; otherwise, it is an attack. In summary, a reply can either be an attack or a surrender.

Table 1 presents the persuasion **communication language** $\mathcal{L}_c$, which takes into account the attacks and the surrenders. In this table, A, B, C, D, E and C' are arguments. Let us recall that an attack relation between two arguments denotes the fact that these arguments cannot be accepted simultaneously since they contradict each other. Following this idea, we can say that rejecting a request is an attack because it is a contradiction. We can also say that a rhetorical argument attacks a rejection because it defends a different position. Thus, we will consider an attack those defined in previous section – which are more related to logical inconsistency – and also those that support a different position during the dialogue. It is reasonable to think that all the illocutions used by the proponent P aim to support his position, which is contrary to the position of the opponent O. In the attack column, besides the possible attacks, the conditions that relate the attacked with its attacker are stated. In order to standardize the content of speech acts, all of them are arguments; whether they are explanatory, rhetorical, or the basic ones according to Definition 4.

The idea is that the proponent agent uses rhetorical arguments to try to convince his opponent. Thus, in this first version of the protocol only the proponent can use rhetorical arguments, that is threats, rewards, and appeals. Counter-threats act as attacks, in this sense, these can be used by the opponent. After a rhetorical argument, the opponent can accept the proposal, send an attack, or withdraw from the dialogue. The last case, may happen when he has no attack to send but does not want to accept the proposal or when he has an explanation for a questioning. Note that attacks can also be attacked and can be questioned in an element of their support. Only counter-threats cannot be questioned because it is based on goals and not in beliefs.

Table 1. Speech acts and possible replies in $\mathcal{L}_c$

Speech act	Attack	Surrender
$request(A)$	$reject(A)$	$accept(\varphi)$
		$(\varphi = \text{CONC}(A))$
$reject(A)$	$threat(B), reward(B)$, or $appeal(B)$	withdraw
	$(\text{CONC}(B) = \text{CONC}(A))$	
$threat(B)$	$counter\text{-}threat(C) \mid \text{CONC}(C) = \text{OPGOAL}(B)$	$accept(\varphi)$
	$undercut(C) \mid \neg\text{CONC}(C) \in \text{SUPP}(B)$	$(\varphi = \text{CONC}(B))$
	$disesteemate(C) \mid \text{CONC}(C) = \neg\text{OPGOAL}(B)$	withdraw
$reward(B)$	$undercut(C) \mid \neg\text{CONC}(C) \in \text{SUPP}(B)$	$accept(\varphi)$
$appeal(B)$	$disesteemate(C) \mid \text{CONC}(C) = \neg\text{OPGOAL}(B)$	$(\varphi = \text{CONC}(B))$
		withdraw
$counter\text{-}threat(C)$	$counter\text{-}threat(C') \mid \text{CONC}(C') = \text{OPGOAL}(C)$	$accept(\varphi)$
	$undercut(C') \mid \neg\text{CONC}(C') \in \text{SUPP}(C)$	$(\varphi = \text{CONC}(C))$
	$disesteemate(C') \mid \text{CONC}(C') = \neg\text{OPGOAL}(C)$	
	$threat(B), reward(B), appeal(B)$	withdraw
$undercut(C)$	$undercut(C') \mid \neg\text{CONC}(C') \in \text{SUPP}(C))$	withdraw or
$rebuttal(C)$	$rebuttal(C') \mid \neg\text{CONC}(C') \equiv \neg\text{CONC}(C))$	$concede(\varphi)$
$disesteemate(C)$	$why(D) \mid \text{CONC}(D) \in \text{SUPP}(C))$	$(\varphi = \text{CONC}(C))$
	$threat(B), reward(B), appeal(B)$	
$why(D)$	$explanation(E) \mid \text{CONC}(E) = \text{CONC}(D)$	withdraw
$explanation(E)$	$undercut(C) \mid \neg\text{CONC}(C) \in \text{SUPP}(E))$	
	$rebuttal(C) \mid \neg\text{CONC}(C) \equiv \neg\text{CONC}(E))$	
	$why(D) \mid \text{CONC}(D) \in \text{SUPP}(E))$	$concede(\varphi)$
	$threat(B), reward(B), appeal(B) \mid$	$(\varphi = \text{CONC}(E))$
	$(\text{CONC}(B) = \text{CONC}(A))$	withdraw
$accept(\varphi)$		
$concede(\varphi)$	end of dialogue	
withdraw		

The third component of a dialogue game is the protocol, which specifies the allowed moves at each point in a dialogue. Thus, let us define first of all what a move is.

Definition 10. *(Move) A move is a tuple $m = \langle id, sd, tg, sp \rangle$ where:*

- *$id \in \mathbb{N}$ is the identifier of the move;*
- *$sd \in \{P, O\}$ is the sender of the message, i.e. the agent that makes the move;*
- *$tg \in \mathbb{N}$ is target of the move, i.e. a previous move to which it is directed. The target of a move is the identifier of some earlier move in the dialogue;*
- *$sp \in \mathcal{L}_c$ is an speech act.*

Let $\mathcal{M}$ be a set of moves. As for notation, we use $id(m)$, $sd(m)$, $tg(m)$, and $sp(m)$ to refer to each of the components of a given move m. For the sake of simplicity, when we want to refer to the i-th move in a sequence, we use m_i. Besides, we use ARGUM($sp(m)$) to refer to the argument associated to a given speech act.

A dialogue can be seen as a set of moves, which fulfil some conditions. Let us now present the formal definition of dialogue.

Definition 11. *(Dialogue) A dialogue D between two agents P and O is a finite sequence $\langle m_1, ..., m_n \rangle$, such that:*

- *$m_1 = \langle 1, P, -, request(A) \rangle$. It means that the first utterance is sent by the proponent agent and has to be a request;*
- *The content of m_k is $request(A)$ iff $k = 1$. It means that a request can only be sent in the first move;*
- *$tg(m_1) = 0$. It means that the first utterance has no target;*
- *$\forall k > 1$, it holds that $tg(m_k) = j$, for $j = k - 1$. It means that the target of a move is always the previous move.*

Let $\mathcal{D}$ stand for the set of all dialogues.

In the illustrative example, we can see that the proponent and the opponent agents take the turn to speak one after another. These moves are controlled by a function that determines which of the agents will make the next move. Take into account that such move must agree with the possible replies defined in Table 1. Thus, a **turn-taking function** is a mapping $T : \mathcal{D} \rightarrow \{P, O\}$, such that given a dialogue $D = \langle m_1, ..., m_i \rangle$, it holds that (i) $T(\emptyset) = P$, (ii) $T(D) = P$ if i is even, and (iii) $T(D) = O$ if i is odd. We can notice that our definition of turn-taking forces a strict interleaving between agents P and O.

Next, we define our protocol in terms of legal moves the agents can perform. In Table 1, we can notice that the answer for a speech act $why(A)$ is an explanation for it; however, there is a need for a stop condition COND in order to avoid infinite questioning. This condition can be a maximum number of rounds.

Definition 12. *(Legal-move function) A legal-move function is a mapping $\mathcal{P} : \mathcal{D} \rightarrow 2^{\mathcal{M}}$ such that, given $D = \langle m_1, ..., m_n \rangle \in \mathcal{D}$, for all $m \in \mathcal{P}(D)$, the following rules must be satisfied:*

- $R_1 : sd(m) = T(D)$;
- $R_2 : sp(m)$ is a legal speech act after D (considering Table 1);
- $R_3 :$ If $\exists m_i | tg(m_i) = m_k$ - for $1 < i \leq n$, $k < i$ - then $\nexists m_j | tg(m_j) = m_k$, for $i \neq j$;
- $R_4 :$ If $sp(m) = threat(A)$, $sp(m) = reward(A)$, or $sp(m) = appeal(A)$, then $sd(m) = P$;
- $R_5 :$ If $sp(m) = why(A)$ AND COND $==$ true, then the sender agent has to change his move and use another speech act.

Rule 1 says that the sender of a move has to obey the turn-taking function. Rule 2 has to do with the valid answers to speech acts. Rule 3 means that there is no move with the same target in a dialogue. Rule 4 ensures that only the proponent can send a rhetorical argument. Finally, rule 5 concerns with avoiding infinite questions.

Besides the rules related to legal moves, it is important to define some rules about the beginning and the end of the dialogue.

- $R_6 :$ If $id(m) = 1$, then $sd(m) = P$;
- $R_7 :$ If $id(m) = 1$, then $sp(m) = request(A)$;
- $R_8 :$ If $sp(m) = accept(\varphi)$, then D ends with an agreement;
- $R_9 :$ If $sp(m) = concede(\varphi)$, then D ends with an agreement
- R_{10} If $P(D) = withdraw$, then D ends without an agreement;

Rule 6 says that the proponent agent always begins the dialogue and rule 7 asserts that the first movement is a request. Rules 8, 9, and 10 have to do with the termination of a dialogue.

A dialogue system also has effect or **commitment rules**, which specify the effects of moves on the participants' commitments. A commitment store gathers the statements each agent have made and the challenges they have issued. Commitment rules define how these commitment stores have to be updated and whether particular illocutions can be uttered at a particular time.

Let $CS(P)$ and $CS(O)$ be the commitment stores of the proponent and the opponent agent, respectively. The set CR of commitment rules is the followings:

1. $CR_1 :$ If $sp(m_i) = why(A)$ and $sd(m_i) = P$ then $CS_i(P) = CS_{i-1}(P)$
2. $CR_2 :$ If $sp(m_i) = reward(A)$ and $sd(m_i) = P$ then $CS_i(P) = CS_{i-1}(P) \cup$ SUPP$(A) \cup \{$CONC$(A)\} \cup \{$OPGOAL$(A)\}$
3. $CR_3 :$ If $sp(m_i) = threat(A)$ and $sd(m_i) = P$ then $CS_i(P) = CS_{i-1}(P) \cup$ SUPP$(A) \cup \{$CONC$(A)\} \cup \{\neg$OPGOAL$(A)\}$
4. $CR_4 :$ If $sp(m_i) = explanation(A)$ and $sd(m_i) = P$ then $CS_i(P) = CS_{i-1}(P) \cup$ SUPPORT$(A) \cup \{$CONC$(A)\}$
5. $CR_5 :$ If $sp(m_i) = accept(\varphi)$ and $sd(m_i) = P$ then $CS_i(P) = CS_{i-1}(P) \cup \{\varphi\}$

It also holds that: (i) CR_2 also holds $appeal(A)$ (ii) CR_2 also holds counter-threat(B); (iii) CR_4 also holds for undercut(B), rebuttal(B), and disesteemate(B); (iv) CR_5 also holds for concede(φ); and (v) these rules hold for $CS(O)$. Finally, it holds that both $CS(O)$ and $CS(P)$ are consistent.

4.2 Argumentation Framework and Dialogue Outcome

In this subsection, we present the argumentation framework for a persuasive negotiation dialogue and how to determine the outcome of the dialogue based on the semantics defined in Sect. 2.

Definition 13. *(Dialogue Argumentation Framework) An AF for a negotiation persuasive dialogue is a tuple* $\mathcal{DAF} = \langle \mathtt{ARG}, \mathcal{R}, D, \mathcal{CS}(P), \mathcal{CS}(O) \rangle$ *such that:*

- $D \in \mathcal{D}$ *is a dialogue constructed under the rules of protocol* $\mathcal{P}$ *and the commitment rules* $\mathcal{CR}$*;*
- $\mathcal{CS}(P)$ *and* $\mathcal{CS}(O)$ *are the commitments sets of the proponent and the opponent, respectively;*
- $\mathtt{ARG} = \{\mathtt{ARGUM}(sp(m_i)) \mid m_i \in D, \text{ for } 1 \leq i < n\}$*;*
- $\mathcal{R} = \{(B, A) \mid A = \mathtt{ARGUM}(sp(m_{j-1})) \text{ and } B = \mathtt{ARGUM}(sp(m_j)), \text{ for } 2 \leq j \leq n - 1\}$*;*

Recall that the speech acts that end a dialogue are not associated with an argument, which is reflected in the set of arguments. We can notice that $\mathtt{ARG}$ and $\mathcal{R}$ form a linear discussion with a sequence $s = \langle A_1, ..., A_{n-1} \rangle$ where n is the number of movements of D, and $A_1 = \mathtt{ARGUM}(sp(m_1))$. This means that we can apply the semantics given in Definition 14 in order to define the acceptable arguments and based on these arguments, we can define the outcome of the dialogue, that is, the winner of the dialogue. If the proponent wins the dialogue (that is, he persuades his opponent), then the opponent has to perform the required action and the proposed threat, reward, or appeal has to be fulfilled. Note that the proponent can send more than one rhetorical argument, in this case, the last rhetorical argument sent during the dialogue is the one that has to be fulfilled. On the other hand, when the proponent loses the dialogue, the opponent does not have to perform the required action and no offer has to be fulfilled. Before define the outcome of the dialogue, we have to make a modification on the semantics due to the condition of the last movement.

Definition 14. *(DAF Semantics) Let* $\mathcal{DAF} = \langle \mathtt{ARG}, \mathcal{R}, D, \mathcal{CS}(P), \mathcal{CS}(O) \rangle$ *be a dialogue argumentation framework,* $A \in \mathtt{ARG}$*,* s *a sequence for* A*, and* $n = |\mathtt{ARG}| + 1$*:*

- *If* n *is odd and* $sp(m_n) = accept(\varphi)$ *(or* $sp(m_n) = concede(\varphi)$*), then* $\mathtt{att}(A)$ *are acceptable.*
- *If* n *is even and* $sp(m_n) = accept(\varphi)$ *(or* $sp(m_n) = concede(\varphi)$*), then* $A \cup \mathtt{sup}(A)$ *are acceptable.*
- *If* $sp(m_n) = withdraw$*, then there are no acceptable arguments.*

Let $\mathtt{SEM_{DAF}}(\mathcal{DAF})$ *be a semantics function that returns the set of acceptable arguments.*

We can now define the outcome of the dialogue.

Definition 15. *(Dialogue Outcome) Let* $\mathcal{DAF} = \langle \mathrm{ARG}, \mathcal{R}, D, \mathcal{CS}(P), \mathcal{CS}(O) \rangle$ *be a dialogue argumentation framework,* $A \in \mathrm{ARG}$ *is an argument that represents a required action, and* P *and* O *the proponent and the opponent agent, respectively:*

- *If* $A \in \mathrm{SEM}_{\mathrm{DAF}}(\mathcal{DAF})$ *then* P *wins the dialogue, and* O *has to perform* $\mathrm{CONC}(A)$.
- *If* $A \notin \mathrm{SEM}_{\mathrm{DAF}}(\mathcal{DAF})$ *then* P *loses the dialogue.*

4.3   Properties of the Proposal

In this section, we will study some properties of our proposal. The aim is to evaluate its legality in the sense of fulfillment of the rules and the soundness and completeness of the argumentation process.

The first proposition concerns with the **legality** of the moves exchanged during the dialogue.

Proposition 1. *Given* $D = \langle m_1, ..., m_n \rangle$, *and* $D' = \langle m_1, ..., n_m \rangle$, *where* $1 \le m \le n$ *and considering that* $\mathcal{P}(\langle m_1, ..., m_m \rangle) = m_{m+1}$ *is compatible with* $\mathcal{T}$ *and fulfils all of the previously established rules. We can say that if* $\forall m$, D' *is a legal dialogue, then* D *is also a legal dialogue.*

Next propositions concerns with the **soundness and completeness** of the argumentation process.

Proposition 2. $\mathcal{DAF} = \langle \mathrm{ARG}, \mathcal{R}, D, \mathcal{CS}(P), \mathcal{CS}(O) \rangle$ *be a dialogue argumentation framework:*

- *If* $\forall A \in \mathrm{ARG}$, *if* $A \in \mathrm{SEM}(\mathcal{DAF})$, *then* $\forall A'$ *such that* $A' = \mathrm{ARGUS}(sp(m'))$, $A = \mathrm{ARGUS}(sp(m))$, *and* $sd(m) = sd(m')$, $A' \in \mathrm{SEM}(\mathcal{DAF})$.
- $\forall A, A' \in \mathrm{SEM}(\mathcal{DAF})$ *such that* $A' = \mathrm{ARGUS}(sp(m'))$ *and* $A = \mathrm{ARGUS}(sp(m))$, $sd(m) = sd(m')$.

The first item say that if an argument sent by one the agents is acceptable, then all the arguments sent by the same agent have to be acceptable as well. The second item says that all acceptable arguments were sent by the same agent.

5   Applying the Proposal to the Illustrative Example

In this section, we evaluate if the example given in Introduction fulfills the rules of the protocol. Besides, we determine the outcome of the dialogue.

We use P to refer to Maria because her role in the dialogue is to be the proponent and O to refer to Carlos because his role in the dialogue is to be the opponent. Next, we have the set of moves:

m_1: $\langle 1, P, 0, request(\langle \{cleaning\}, cleaning \rangle) \rangle$
m_2: $\langle 2, O, 1, reject(\langle \{cleaning\}, cleaning \rangle) \rangle$
m_3: $\langle 3, P, 2, reward(\langle \{cleaning \rightarrow can\_help, can\_help \rightarrow finish\_work\},$
       $cleaning, finish\_work \rangle) \rangle$

m_4: $\langle 4, O, 3, undercut(\langle\{\neg can\_help\}, \neg can\_help\rangle)\rangle$

m_5: $\langle 5, P, 4, why(\langle\{\neg can\_help\}, \neg can\_help\rangle)\rangle$

m_6: $\langle 6, O, 5, explanation(\langle\{work, work \rightarrow \neg cleaning\}, \neg cleaning\rangle)\rangle$

m_7: $\langle 7, P, 6, threat(\langle\{\neg cleaning \rightarrow \neg going\_mother\_house\}, cleaning, going\_mother\_house\rangle)\rangle$

m_8: $\langle 8, O, 7, counte-threat(\langle\{\neg going\_mother\_house \rightarrow \neg talking\_about\_brother\_work\}, going\_mother\_house, talking\_about\_brother\_work\rangle)\rangle$

m_9: $\langle 9, P, 8, disesteemate(\langle\{brother\_has\_work \rightarrow \neg talking\_about\_brother\_work\}, \neg talking\_about\_brother\_work\rangle)\rangle$

m_{10}: $\langle 10, O, 9, concede(\neg talking\_about\_brother\_work)\rangle$

We have a dialogue $D = \{m_1, m_2, m_3, m_4, m_5, m_6, m_7, m_8, m_9, m_{10}\}$ where the request is the first move and the target of every move is the previous move. Regarding the rules of the legal-move function, we can say that all the moves of D follow these rules.

Let us now present the commitments sets:

- $\mathcal{CS}(P) = \{cleaning \rightarrow can\_help, can\_help \rightarrow finish\_work, cleaning, finish\_work, \neg cleaning \rightarrow \neg going\_mother\_house, \neg going\_mother\_house, brother\_has\_work \rightarrow \neg talking\_about\_brother\_work, \neg talking\_about\_brother\_work\}$.

- $\mathcal{CS}(O) = \{\neg can\_help, work, work \rightarrow \neg cleaning, \neg cleaning, \neg going\_mother\_house \rightarrow \neg talking\_about\_brother\_work, \neg talking\_about\_brother\_work\}$

We can notice that both commitment sets are consistent. Note also that $\mathcal{CS}(P)$ includes the requested action as a positive literal ($cleaning$) whereas $\mathcal{CS}(O)$ includes the requested action as a negative literal ($\neg cleaning$).

Now, let us define the dialogue argumentation framework: $\mathcal{DAF} = \langle \text{ARG}, \mathcal{R}, D, \mathcal{CS}(P), \mathcal{CS}(O)\rangle$ where $\text{ARG} = \{A_1, A_2, A_3, A_4, A_5, A_6, A_7, A_8, A_9\}$ such that each argument is associated to the number of the move in the dialogue D; $\mathcal{R} = \{(A_2, A_1), (A_3, A_2), (A_4, A_3), (A_5, A_4), (A_6, A_5), (A_7, A_6), (A_8, A_7), (A_9, A_8)\}$; and D and the commitment sets were presented above.

The result of applying the semantics is: $\text{SEM}_{\text{DAF}}(\mathcal{DAF}) = \{A_1, A_3, A_5, A_7, A_9\}$. We can now determine the outcome of the dialogue. We can notice that $A_1 \in \text{SEM}_{\text{DAF}}(\mathcal{DAF})$, this means that P (Maria) wins the dialogue and O (Carlos) has to do the cleaning of the apartment ($cleaning$).

6   Conclusions and Future Work

In this paper, we have presented a protocol for a persuasive negotiation dialogue. In the resulting dialogue game, agents can exchange rhetorical and explanatory arguments, can utter attacks for such arguments, can question an element of explanatory arguments, and also can use negotiation speech acts to request, reject, and finish the dialogue. The use of additional kinds of illocutions enriches the dialogue and allows the agents to not only try to persuade the other party but to defend their positions. The proposed protocol is also flexible since it allows for different alternative replies, which were categorized as attacks and surrenders.

For future research, we propose two possible directions: (i) the first one is to improve the protocol itself, for example, allowing that the target of an utterance to be any of the earlier moves and (ii) the second one is extending the protocol for more than two agents, which also has consequences on the possible attacks for arguments.

References

1. Aknine, S., Pinson, S., Shakun, M.F.: An extended multi-agent negotiation protocol. Auton. Agents Multi-Agent Syst. **8**(1), 5–45 (2004)
2. Amgoud, L., Ben-Naim, J.: Ranking-based semantics for argumentation frameworks. In: Liu, W., Subrahmanian, V.S., Wijsen, J. (eds.) SUM 2013. LNCS (LNAI), vol. 8078, pp. 134–147. Springer, Heidelberg (2013). https://doi.org/10.1007/978-3-642-40381-1_11
3. Amgoud, L., Parsons, S., Maudet, N.: Arguments, dialogue, and negotiation. In: Proceedings of the 14th European Conference on Artificial Intelligence, vol. 10, pp. 338–342. IOS Press (2000)
4. Amgoud, L., Prade, H.: Threat, reward and explanatory arguments: generation and evaluation. In: Proceedings of the ECAI Workshop on Computational Models of Natural Argument, pp. 73–76 (2004)
5. Amgoud, L., Prade, H.: Formal handling of threats and rewards in a negotiation dialogue. In: Proceedings of the Fourth International Joint Conference on Autonomous Agents and Multiagent Systems, pp. 529–536. ACM (2005)
6. Amgoud, L., Prade, H.: Handling threats, rewards, and explanatory arguments in a unified setting. Int. J. Intell. Syst. **20**(12), 1195–1218 (2005)
7. Amgoud, L., Prade, H.: Formal handling of threats and rewards in a negotiation dialogue. In: Parsons, S., Maudet, N., Moraitis, P., Rahwan, I. (eds.) ArgMAS 2005. LNCS (LNAI), vol. 4049, pp. 88–103. Springer, Heidelberg (2006). https://doi.org/10.1007/11794578_6
8. Belardinelli, F., Grossi, D., Maudet, N.: A formal analysis of dialogues on infinite argumentation frameworks. In: 24th International Joint Conference on Artificial Intelligence (IJCAI-15), pp. 861–867 (2015)
9. Bentahar, J., Moulin, B., Chaib-draa, B.: Specifying and implementing a persuasion dialogue game using commitments and arguments. In: Rahwan, I., Moraïtis, P., Reed, C. (eds.) ArgMAS 2004. LNCS (LNAI), vol. 3366, pp. 130–148. Springer, Heidelberg (2005). https://doi.org/10.1007/978-3-540-32261-0_9
10. Boella, G., Hulstijn, J., Van der Torre, L.: Persuasion strategies in dialogue. In: Proceedings of the ECAI Workshop on Computational Models of Natural Argument (CMNA 2004), Valencia (2004)
11. Calvaresi, D., et al.: Multi-agent systems' negotiation protocols for cyber-physical systems: results from a systematic literature review. In: ICAART (1), pp. 224–235 (2018)
12. Dimopoulos, Y., Moraitis, P.: Advances in argumentation based negotiation. In: Negotiation and Argumentation in Multi-agent Systems: Fundamentals, Theories, Systems and Applications, pp. 82–125 (2011)
13. Dung, P.M.: On the acceptability of arguments and its fundamental role in nonmonotonic reasoning, logic programming and n-person games. Artif. Intell. **77**(2), 321–357 (1995)

14. Heras Barberá, S.M., Botti Navarro, V.J., Julian Inglada, V.J.: Case-based argumentation framework. Dialogue protocol (2011)
15. Hunter, A.: Base logics in argumentation. In: COMMA, pp. 275–286 (2010)
16. Ito, T., Klein, M., Hattori, H.: A multi-issue negotiation protocol among agents with nonlinear utility functions. Multiagent Grid Syst. **4**(1), 67–83 (2008)
17. Letia, I.A., Vartic, R.: Defeasible protocols in persuasion dialogues. In: Proceedings of the 2006 IEEE/WIC/ACM International Conference on Web Intelligence and Intelligent Agent Technology, pp. 359–362. IEEE Computer Society (2006)
18. McBurney, P., Parsons, S.: Dialogue games for agent argumentation. In: Simari, G., Rahwan, I. (eds.) Argumentation in Artificial Intelligence, pp. 261–280. Springer, Boston (2009). https://doi.org/10.1007/978-0-387-98197-0_13
19. Modgil, S., Prakken, H.: The ASPIC+ framework for structured argumentation: a tutorial. Argument Comput. **5**(1), 31–62 (2014)
20. Morveli-Espinoza, M.: Persuasive negotiation dialogues using rhetorical arguments. In: Proceedings of the 16th Conference on Autonomous Agents and MultiAgent Systems, AAMAS 2017, pp. 1845–1846. International Foundation for Autonomous Agents and Multiagent Systems (2017)
21. Morveli-Espinoza, M., Nieves, J.C., Tacla, C.A.: Measuring the strength of threats, rewards, and appeals in persuasive negotiation dialogues. Knowl. Eng. Rev. **35**, 1–27 (2020)
22. Morveli Espinoza, M., Possebom, A.T., Tacla, C.A.: On the calculation of the strength of threats. Knowl. Inf. Syst. **62**(4), 1511–1538 (2019). https://doi.org/10.1007/s10115-019-01399-2
23. Perrussel, L., Doutre, S., Thévenin, J.-M., McBurney, P.: A persuasion dialog for gaining access to information. In: Rahwan, I., Parsons, S., Reed, C. (eds.) ArgMAS 2007. LNCS (LNAI), vol. 4946, pp. 63–79. Springer, Heidelberg (2008). https://doi.org/10.1007/978-3-540-78915-4_5
24. Prakken, H.: Coherence and flexibility in dialogue games for argumentation. J. Logic Comput. **15**(6), 1009–1040 (2005)
25. Prakken, H.: Models of persuasion dialogue. In: Argumentation in Artificial Intelligence, pp. 281–300. Springer, Boston (2009). https://doi.org/10.1007/978-0-387-98197-0_14
26. Rahwan, I., Ramchurn, S.D., Jennings, N.R., Mcburney, P., Parsons, S., Sonenberg, L.: Argumentation-based negotiation. Knowl. Eng. Rev. **18**(04), 343–375 (2003)
27. Ramchurn, S.D., Jennings, N.R., Sierra, C.: Persuasive negotiation for autonomous agents: a rhetorical approach (2003)
28. Saha, S., Sen, S.: An efficient protocol for negotiation over multiple indivisible resources. In: IJCAI, vol. 7, pp. 1494–1499 (2007)
29. Sierra, C., Jennings, N.R., Noriega, P., Parsons, S.: A framework for argumentation-based negotiation. In: Singh, M.P., Rao, A., Wooldridge, M.J. (eds.) ATAL 1997. LNCS, vol. 1365, pp. 177–192. Springer, Heidelberg (1998). https://doi.org/10.1007/BFb0026758
30. van Veenen, J., Prakken, H.: A protocol for arguing about rejections in negotiation. In: Parsons, S., Maudet, N., Moraitis, P., Rahwan, I. (eds.) ArgMAS 2005. LNCS (LNAI), vol. 4049, pp. 138–153. Springer, Heidelberg (2006). https://doi.org/10.1007/11794578_9
31. Wang, G., Wong, T., Wang, X.: A negotiation protocol to support agent argumentation and ontology interoperability in MAS-based virtual enterprises. In: 2010 Seventh International Conference on Information Technology: New Generations (ITNG), pp. 448–453. IEEE (2010)

Gradient Estimation in Model-Based Reinforcement Learning: A Study on Linear Quadratic Environments

Ângelo Gregório Lovatto$^{(\boxtimes)}$ [iD], Thiago Pereira Bueno [iD],
and Leliane Nunes de Barros [iD]

Instituto de Matemática e Estatística, Universidade de São Paulo, São Paulo, Brazil
angelo.lovatto@usp.br

Abstract. Stochastic Value Gradient (SVG) methods underlie many recent achievements of model-based Reinforcement Learning agents in continuous state-action spaces. Despite their practical significance, many algorithm design choices still lack rigorous theoretical or empirical justification. In this work, we analyze one such design choice: the gradient estimator formula. We conduct our analysis on randomized Linear Quadratic Gaussian environments, allowing us to empirically assess gradient estimation quality relative to the actual SVG. Our results justify a widely used gradient estimator by showing it induces a favorable bias-variance tradeoff, which could explain the lower sample complexity of recent SVG methods.

Keywords: Reinforcement learning · Model-based · Machine learning

1 Introduction

Model-Based Reinforcement Learning (MBRL) [16,20] is a promising framework for developing intelligent systems for sequential decision-making from limited data. Unlike in model-free Reinforcement Learning (RL) methods, MBRL agents use collected experiences to fit a predictive model of the environment. The agent can use the model to evaluate potential action sequences, saving costly trial-and-error experimentation in the real world, or to estimate quantities useful for improving its learned behavior. Stochastic Value Gradient (SVG) methods belong to the latter category, using the model to estimate the value gradient. While model-free methods use the score-function estimator of the value gradient, SVG methods can leverage the model to produce gradients via the pathwise derivative estimator, usually found to be more stable in practice [23]. Recently proposed RL agents using the SVG approach have demonstrated its effectiveness in learning robotic locomotion from data with unprecedented sample-efficiency [1,4,8], i.e. with few collected experiences.

Although SVG methods have been validated empirically, some of the algorithm design choices still lack rigorous theoretical or empirical justification. As recent work on model-free methods has shown, there can be a gap between the

© Springer Nature Switzerland AG 2021
A. Britto and K. Valdivia Delgado (Eds.): BRACIS 2021, LNAI 13073, pp. 33–47, 2021.
https://doi.org/10.1007/978-3-030-91702-9_3

theoretical underpinnings of RL methods and their behavior in practice, often due to code-level optimizations [6,11,13]. Moreover, it is common for theoretically promising MBRL algorithms to fail in practice [14], although negative results are not often publicized.

In this work, we take a step towards a better understanding of the core tenets behind SVG methods by analyzing an important algorithm design choice: the gradient estimator computation [17,23]. We aim to identify the key practical differences between the theoretically sound, unbiased formulation from the Deterministic Policy Gradient (DPG) framework [24] and an estimator more often used in SVG methods [1,4,8] that deviates from traditional policy gradient theory. The former represents an approach that's theoretically sound, but lacks empirical validation, while the latter generalizes the methods that deviate from the theory, but achieve state-of-the-art results in continuous control benchmarks.

We evaluate gradient estimation quality and policy optimization performance on Linear Quadratic Gaussian (LQG) regulator environments. The LQG framework is extensively studied in the Optimal Control literature and is a special class of RL with continuous actions (a.k.a. continuous control), where SVG methods seem to excel. LQG has been proposed as a simple, yet nontrivial class of continuous control problems to help distinguish the various approaches to RL [21]. More importantly, LQG has a simple environment formulation, allowing us to compute the ground-truth policy performance and gradient, an ability not present in more complicated, nonlinear benchmarks for continuous control [28]. Thus, we can perform a more rigorous empirical assessment of gradient estimation quality and policy optimization performance within the LQG framework.

Therefore, our main contribution in this paper is a careful empirical analysis, using the LQG framework, of the practical differences between gradient estimators for SVG methods. The rest of this paper is organized as follows. Section 2 outlines related work on SVG methods and the gap between theory and practice of RL. Section 3 introduces the reader to the minimal technical background on RL, LQG and SVG, required to follow our analysis. Section 4 describes the scope of the empirical analysis and the experimental setup developed accordingly. Section 5 contains the main experiments, results, and our analysis thereof. Finally, we summarize our observations in Sect. 6 and point to possible future work to explore remaining gaps in our knowledge of the core tenets of SVG methods.

2   Related Work

Model-based policy gradients have been recently used in a variety of RL algorithms for continuous control. The PILCO algorithm is one of the first to leverage a learned model's derivatives to compute the policy gradient with few samples, but its use of Gaussian processes hinders scalability to larger problems [5]. The original SVG paper introduced gradient estimation with stochastic neural network models using the reparameterization trick, an approach scalable to higher-dimensional problems [9]. Dreamer and Imagined Value Gradients explore SVGs with latent-space models [2,8]. Model-Augmented Actor-Critic (MAAC) and SAC-SVG extend the SVG framework to that of maximum-entropy RL to incentivize exploration and stabilize optimization [1,4]. However, the computation of the gradient is not studied in isolation or contrasted with other gradient estimation methods. Our work analyses, in detail, the gradient computation used in recent SVG-style algorithms that show a high level of performance with efficient use of data [1,4,8].

A few recent works have analyzed the gap between theory and practice in different areas of RL, highlighting our poor understanding of current methods. Reliability and reproducibility concerns regarding modern RL methods have been raised by several works [3,10,12], indicating a disconnect between the theory motivating these algorithms and their behavior in practice. Code-level optimizations have been found to contribute more to successful policy gradient methods than the choice of general training algorithm [6]. Closest to our work is that of Ilyas et al. [11], which shows that model-free policy gradient algorithms succeed in optimizing the policy despite having poor gradient estimation quality metrics in the relevant sample regime. Our work is, to the best of our knowledge, the first to propose a fine-grained analysis of gradient estimation in SVG methods using LQGs to provide solid references of their expected behavior.

3   Background

3.1   A Brief Introduction to RL

We consider the agent-environment interaction modeled as a continuous Markov Decision Process (MDP) [25], defined as the tuple $(\mathcal{S}, \mathcal{A}, R, p^*, \rho, H)$, with each component described in what follows. Interaction with the environment occurs in a sequence of discrete timesteps $t \in \mathcal{T} = \{0, \ldots, H-1\}$, where $H \in \mathbb{N}$ denotes the time horizon, after which an *episode* of interaction is over. At every timestep t, the agent observes the current state $\mathbf{s}_t \in \mathcal{S} \subseteq \mathbb{R}^n$ from the set of possible states of the environment. It must then select an action $\mathbf{a}_t \in \mathcal{A} \subseteq \mathbb{R}^d$ from the set of possible actions to be executed. The environment then transitions to the next state by sampling from the transition probability kernel, $\mathbf{s}_{t+1} \sim p^*(\cdot \mid \mathbf{s}_t, \mathbf{a}_t)$, and emits a reward signal using its reward function, $r_{t+1} = R(\mathbf{s}_t, \mathbf{a}_t)$. The initial state is sampled from the initial state distribution, $\mathbf{s}_0 \sim \rho(\cdot)$.

A *policy* defines a mapping from environment states to actions, $\mathbf{a}_t = \mu(\mathbf{s}_t)$. The objective of an RL agent is to find a policy that produces the highest cumulative reward, or *return*, from the initial state: $J(\mu) = \mathbb{E}[\sum_{t=0}^{H-1} R(\mathbf{s}_t, \mathbf{a}_t)]$. Here, the expectation is implicitly w.r.t. the initial state distribution and the sequential application of $\mathbf{s}_{t+1} \sim p^*(\cdot \mid \mathbf{s}_t, \mu(\mathbf{s}_t))$. The key difference between Optimal Control and RL, both frameworks for optimal sequential decision making, is that in the former the agent has access to the full MDP, while in the latter the agent only knows the state and action space and has to learn its policy by trial-and-error in the environment.

3.2   Linear Quadratic Gaussian Regulator

The LQG is a special class of continuous MDP in which the transition kernel is linear Gaussian and the reward function is quadratic concave [27]:

$$p^*(\cdot \mid s_t, a_t) = \mathcal{N}(\cdot \mid F_s s_t + F_a a_t, \Sigma) \tag{1}$$

$$R(s_t, a_t) = -\tfrac{1}{2}(s_t^{\mathsf{T}} C_{ss} s_t + a_t^{\mathsf{T}} C_{aa} a_t). \tag{2}$$

LQGs are often used as a discretization of continuous-time dynamics described as linear differential equations, such as those of physical systems.

An important class of policies in this context is that of time-varying (a.k.a. *nonstationary*) linear policies:

$$\mu_\theta(s_t) = K_t s_t + k_t, \tag{3}$$

where $K_t \in \mathbb{R}^{d \times n}$ and $k_t \in \mathbb{R}^d$, a.k.a. the *dynamic* and *static gains* respectively in the optimal control literature. Here, θ is the flattened parameter vector corresponding to the collection $\{K_t, k_t\}_{t \in \mathcal{T}}$ of function coefficients. Given any linear policy μ_θ, its *state-value function*, the expected return from each state, can be computed recursively via the Bellman equations [25]:

$$V^{\mu_\theta}(s_t) = R(s_t, \mu_\theta(s_t)) + \mathbb{E}_{s_{t+1}}[V^{\mu_\theta}(s_{t+1})], \qquad t \in \mathcal{T}, \tag{4}$$

where $V^{\mu_\theta}(s_H) = 0$ and the expectation is w.r.t. $p^*(\cdot \mid s_t, \mu_\theta(s_t))$. Since the policy is linear, rewards are quadratic, and the transitions, Gaussian, the expectations in Eq. (4) can be computed analytically by iteratively solving Eq. (4) from timestep $H - 1$ to 0 (a *dynamic programming* method). The solution V^{μ_θ} is itself quadratic and thus the policy return can be computed analytically:

$$\mathbb{E}\left[\sum_{t=0}^{H-1} R(s_t, a_t)\right] = \mathbb{E}_{s_0}[V^{\mu_\theta}(s_0)]. \tag{5}$$

Moreover, LQGs can be solved for their optimal policy μ_θ^*, which is time-varying linear, by modifying the dynamic programming method above to solve for the policy that maximizes the expected value from each timestep based on Eq. (4) (also computable analytically). Algorithm 1 shows the pseudocode of the procedure we use to derive the optimal policy, which can be slightly modified to return the value function for a given policy. LQG is thus one of the few classes of nontrivial continuous control problems which allows us to evaluate RL methods against the theoretical best solutions.

3.3   Stochastic Value Gradient Methods

In the broader RL context, methods that learn parameterized policies, often called policy optimization methods, have gained traction in the recent decade. As function approximation research, specially on deep learning, has advanced, parameterized policies were able to unify perception (processing sensorial input from the environment) and decision-making (choosing actions to maximize

---

Algorithm 1: LQG control

Input: LQG parameters $(\mathbf{F}, \mathbf{f}, \mathbf{\Sigma}, \mathbf{C}, \mathbf{c})$
Output: Optimal dynamic and static gains
1 Set $V^*(\mathbf{s}_H) = 0$
2 **for** $t = H - 1, \ldots, 0$ **do**
    // We solve for all states/actions implicitly
3   Compute $Q^*(\mathbf{s}_t, \mathbf{a}_t) = R(\mathbf{s}_t, \mathbf{a}_t) + \mathbb{E}_{\mathbf{s}_{t+1}}[V^*(\mathbf{s}_{t+1})]$ using $(\mathbf{F}, \mathbf{f}, \mathbf{\Sigma}, \mathbf{C}, \mathbf{c})$
4   Solve $\mu_\theta^*(\mathbf{s}_t) = \arg\max_{\mathbf{a}} Q^*(\mathbf{s}_t, \mathbf{a})$, yielding gains $\mathbf{K}_t, \mathbf{k}_t$
5   Compute $V^*(\mathbf{s}_t) = Q^*(\mathbf{s}_t, \mu_\theta^*(\mathbf{s}_t))$
6 **return** $\{\mathbf{K}_t, \mathbf{k}_t\}_{t=0}^{H-1}$

---

return) tasks [15]. To improve such parameterized approximators from data, the workhorse behind many policy optimization methods is Stochastic Gradient Descent (SGD) [22]. Thus, it is imperative to estimate the gradient of the expected return w.r.t. policy parameters, a.k.a. the *value gradient*, from data (states, actions and rewards) collected via interaction with the environment.

SVG methods build gradient estimates by first using the available data to learn a *model* of the environment, i.e., a function approximator $p_\psi(\cdot \,|\, \mathbf{s}, \mathbf{a}) \approx p^*(\cdot \,|\, \mathbf{s}, \mathbf{a})$. A common approach to leveraging the learned model is as follows. First, the agent collects B states via interaction with the environment, potentially with an exploratory policy β (we use $\mathbf{s}_t \sim d^\beta$ to denote sampling from its induced state distribution). Then, it generates short model-based trajectories with the current policy μ_θ, branching off the states previous collected. Finally, it computes the average model-based returns and forms an estimate of the value gradient using backpropagation [7]:

$$\nabla J(\theta) \approx \nabla_\theta \, \mathbb{E}_{\mathbf{s}_t \sim d^\beta} \left[\sum_{l=0}^{K-1} R(\mathbf{s}_{t+l}, \mu_\theta(\mathbf{s}_{t+l})) + \hat{Q}^{\mu_\theta}(\mathbf{s}_{t+K}, \mu_\theta(\mathbf{s}_{t+K})) \right]. \quad (6)$$

Here, $\hat{Q}^{\mu_\theta}$ is an approximation (e.g., a learned neural network) of the policy's *action-value* function $Q^{\mu_\theta}(\mathbf{s}_t, \mathbf{a}_t) = \mathbb{E}_{\mu_\theta}[\sum_{l=t}^{H-1} R(\mathbf{s}_l, \mathbf{a}_l)]$. We refer to Eq. (6) as the MAAC(K) estimator, as it uses K steps of simulated interaction and was featured prominently in the MAAC paper [4].[1]

We question, however, if Eq. (6) actually provides good empirical estimates of the true value gradient. To elucidate this matter, we compare MAAC(K) to the value gradient estimator provided by the DPG theorem [24]:

$$\nabla J(\theta) = \mathbb{E}_{\mathbf{s}_t \sim d^{\mu_\theta}} \left[\nabla_\theta \mu_\theta(\mathbf{s}_t) \nabla_{\mathbf{a}} \, Q^{\mu_\theta}(\mathbf{s}_t, \mathbf{a})|_{\mathbf{a} = \mu_\theta(\mathbf{s}_t)} \right]. \quad (7)$$

Besides the fact that Eq. (7) requires us to use the on-policy distribution of states d^{μ_θ}, more subtle differences with Eq. (6) can be seen by expanding the

---

[1] Our formula differs slightly from the original in that it considers a deterministic policy instead of a stochastic one.

definition of the action-value function to form a K-step version of Eq. (7), which we call the DPG(K) estimator. Note that DPG(0) is equivalent to Eq. (7).

Figure 1 shows the Stochastic Computation Graphs (SCGs) of the MAAC(K) and DPG(K) estimators [23]. Here, borderless nodes denote input variables; circles denote stochastic nodes, distributed conditionally on their parents (if any); and squares denote deterministic nodes, which are functions of their parents. Because of the $\nabla_{\mathbf{a}} Q^{\mu_\theta}(\mathbf{s}_t, \mathbf{a})|_{\mathbf{a}=\mu_\theta(\mathbf{s}_t)}$ term in Eq. (7), we're not allowed to compute the gradients of future actions w.r.t. policy parameters in DPG(K), hence why only the first action has a link with θ. On the other hand, MAAC(K) backpropagates the gradients of the rewards and value-function through all intermediate actions. Our work aims at identifying the practical implications of these differences and perhaps help explain why MAAC(K) has been used in SVG methods and not DPG(K).

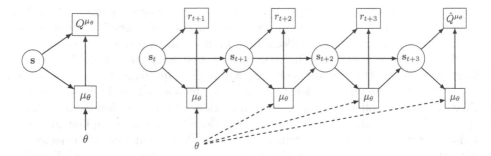

Fig. 1. Stochastic computation graphs for policy gradients. (Left) Model-free DPG. (Right) Model-based DPG: the dashed edges represent the K-step deterministic dependencies of the policy parameters in MAAC(K) for $K = 3$; DPG(K) ignores these dependencies when backpropagating the action gradients.

4 Methodology

We now turn to our research goals and the methods employed to perform the proposed empirical analysis.

4.1 Scope of Evaluation

In this work, we propose a fine-grained analysis of the properties of DPG(K) and MAAC(K) in practice. We simplify our evaluation by using *on-policy* versions of the gradient estimators, i.e., by substituting d^{μ_θ} for d^β where present. We also opted for using perfect models of the environment dynamics and rewards, instead of learning them from data, to focus on the differences between gradient estimators. Thus, we approximate the expectations in Eqs. (6) and (7) via Monte Carlo sampling, using the actual transition kernel p^* and reward function R, to generate (virtual) transitions and compute the bootstrapped returns. One can view this setting as the best possible case in an SVG algorithm: when the model-learning subroutine has perfectly approximated the true MDP, allowing us to focus on the gradient estimation analysis.

We also compute the true action-value function, required for the K-step returns in Eqs. (6) and (7), recursively via dynamic programming (analogously to Eq. (4)). Computing the ground-truth action-value function allows us to further isolate any observed differences between the estimators as a consequence of their properties alone.

4.2   Randomized Environments and Policies

To test gradient estimation across a wide variety of scenarios, we define how to sample LQG instances, $\mathcal{M} = (\mathcal{S}, \mathcal{A}, R, p^*, \rho)$, to run our experiments on. The main configurations for our procedure are: state dimension (n), action dimension (d) and time horizon (H). From these parameters we define the state space $\mathcal{S} = \mathbb{R}^n$, action space $\mathcal{A} = \mathbb{R}^d$ and timesteps $t \in \mathcal{T} = \{0, \ldots, H - 1\}$.

The transition dynamics are stationary, sharing $\mathbf{F}, \mathbf{f}, \boldsymbol{\Sigma}$ across all timesteps. The coefficients $\mathbf{F_s}$ and $\mathbf{F_a}$ are initialized so that the system may be *unstable*, i.e., with some eigenvalues of $\mathbf{F_s}$ having magnitude greater or equal to 1, but always *controllable*, meaning there is a *dynamic gain* $\mathbf{K}$ such that the eigenvalues of $(\mathbf{F_s} + \mathbf{F_a}\mathbf{K})$ have magnitude less than 1. This ensures we are able to emulate real-world scenarios where uncontrolled state variables and costs may diverge to infinity, while ensuring there exists a policy which can stabilize the system [21]. Finally, we fix the transition bias to $\mathbf{f} = \mathbf{0}$ and the Gaussian covariance to the identity matrix, $\boldsymbol{\Sigma} = I$.

The initial state distribution is always initialized as a standard Gaussian distribution: $\rho(\mathbf{s}) = \mathcal{N}(\mathbf{s} \,|\, \mathbf{0}, I)$. As for the reward parameters, we initialize both $\mathbf{C_{ss}}$ and $\mathbf{C_{aa}}$ (see Eq. (2)) as random symmetric positive definite matrices, sampled via the `scikit-learn` library for machine learning in Python [19].[2]

Since we consider the problem of estimating value gradients for linear policies in LQGs, we also define a procedure to generate randomized policies. We start by initializing all dynamic gains $\mathbf{K}_t = \mathbf{K}$ so that $\mathbf{K}$ stabilizes the system. This is done by first sampling target eigenvalues uniformly in the interval $(0, 1)$ and then using the `scipy` library to compute $\mathbf{K}$ that places the eigenvalues of $(\mathbf{F_s} + \mathbf{F_a}\mathbf{K})$ in the desired targets [26, 29].[3] This process ensures the resulting policy is safe to collect data in the environment without having state variables and costs diverge to infinity. The generating procedure also serves to mimic practical situations where engineers devise a policy which can keep a system stable, but is not able to optimize running costs, which is where RL can serve to fine-tune it. Finally, we initialize all static gains as $\mathbf{k}_t = \mathbf{0}$.

---

[2] We use the `make_spd_matrix` function.
[3] We use the `scipy.signal.place_poles` function.

5   Empirical Analysis

We analyze the behavior of each estimator on two main settings: (I) gradient estimation for fixed policies and (II) impact of gradient quality on policy optimization.

5.1   Gradient Estimation for Fixed Policies

Following previous work on model-free policy gradients [11], we evaluate the quality of the gradient estimates, for a given policy, produced by each estimator using two metrics: (i) the average cosine similarity with the true policy gradient and (ii) the average pairwise cosine similarity.

The first metric is a measure of gradient *accuracy* and we denote it as such in the following plots. For a given minibatch size B and step size K, we compute 10 estimates of the gradient, each using B initial states sampled on-policy ($\mathbf{s}_t \sim d^{\mu_\theta}$) and K-step model-based rollouts from each state. Then, we compute the accuracy as the average cosine similarity of each of the 10 estimates with the true policy gradient, obtained as follows. We first compute the true expected return of a policy μ_θ via dynamic programming, following Eqs. (4) and (5). Our implementation in PyTorch [18] then allows us to use automatic differentiation to compute the gradient of the expected return w.r.t. policy parameters.

The second metric is a measure of gradient *precision* and we denote it as such in the following plots. Again, we compute 10 estimates of the gradient in the same manner used in computing the accuracy. Then, we compute the precision as the average pairwise cosine similarity of the 10 estimates (the higher this quantity, the lower the variance).

We first analyze the accuracy of each estimator when given enough states from the policy's distribution d^{μ_θ} to approximate their true expected values. Figure 2 shows the accuracy obtained by DPG and MAAC for different values of K using 50000 states from the policy's distribution. The LQGs considered have state and action spaces of dimension 2 and horizon of length 20. For each value of K, we initialize 10 different environment-policy pairs and compute the accuracy for each, denoted as different markers in each vertical line.[4] Note how all but one of the instances using DPG(K) converged to the true value gradient, indicating that it is indeed an unbiased estimator. On the other hand, MAAC(K) incurs a larger bias with increasing values of K, indicating that the added action action-gradient terms (see Fig. 1) influence the final direction.

Although the results above indicate that MAAC(K) is biased at convergence, most SVG algorithms operate on a much smaller sample regime. Figure 3 shows the accuracy across 10 different environment-policy pairs; this time, however, using smaller sample sizes from the policy distribution. Lines denote the average results and shaded areas, the 95% confidence interval.[5] For $K = 0$, the estimators are equivalent, which is verified in practice. In this more practical sample regime,

---

[4] We use the same 10 random seeds for experiments across values of K.

[5] We use `seaborn.lineplot` to produce the aggregated curves.

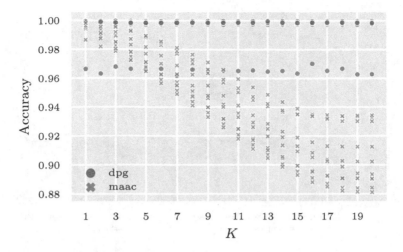

Fig. 2. Gradient accuracy for each estimator near convergence for different virtual rollout lengths (K). We used 50000 states sampled from the policy to approximate the expected value.

we see that MAAC(K) produces more accurate results, specially for larger values of K.

Similar to Fig. 3, Fig. 4 shows the gradient precision in the same setting. We see that the variance of MAAC(K) is lower than that of DPG(K) across all tested values of $K > 0$. Overall, Figs. 2, 3 and 4 illustrate a classic instance of the bias-variance tradeoff in machine learning: MAAC(K) introduces bias, although a small one, in return for a much more stable (less variable) estimate of the gradient, whereas the unbiased DPG(K) demands much more samples to justify its use.

Note that the accuracy and precision metrics only account for differences in gradient direction and orientation. The magnitude may also be important, as it influences the learning rate when used to update policy parameters. Figure 5 shows that MAAC(K) produces gradients with higher norms compared to DPG(K). One should keep this in mind when choosing the learning rate for SGD, as the following experiments show that the gradient norm have a significant impact on policy optimization.

5.2 Impact of Gradient Quality on Policy Optimization

Previous work on model-free policy gradients has shown that policy optimization algorithms can improve a policy despite using poor gradient estimation [11]. It is not clear, however, if better policy gradient estimation translates to more stability or faster convergence in SVG algorithms. We therefore conduct our next experiments comparing the MAAC and DPG estimators for policy optimization.

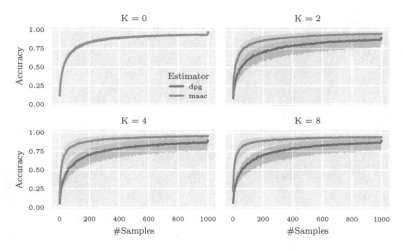

Fig. 3. Gradient accuracy for each estimator for different minibatch sizes ($B = $ #Samples) and virtual rollout lengths (K).

Figures 6 and 7 show learning curves as total cost (negative return) against the number of SGD iterations across several instances of LQGs ($\dim(\mathcal{S}) = \dim(\mathcal{A}) = 2$ and $H = 20$). We use the same hyperparameters for both estimators.[6] The results in Fig. 6 suggest that the better quality metrics observed for MAAC(K) in Figs. 3 and 4 do translate to faster and more stable policy optimization. However, if we normalize the gradient estimates before passing them to SGD, as in Fig. 7, we see that both estimators are evenly matched. These results suggest that the main advantage of MAAC(K) over DPG(K) is in its stronger gradient norm (see Fig. 5), which has been alluded to in previous work as a "strong learning signal" [4], inducing a faster learning rate.

Table 1. Median suboptimality gap, the percentage difference in expected return against the optimal policy, across 10 seeds. LQG dimension refers to the dimension of state and action spaces. We use $K = 8$ and $B = 20$ for both estimators.

Estimator	Time (min)	LQG dimension								
		2	3	4	5	6	7	8	9	10
DPG	1	29.10	218.75	242.94	1730.07	1567.81	4129.88	1100.74	6111.44	7290.04
	3	6.32	53.21	138.89	439.54	465.29	3468.03	552.87	277.38	6445.64
	5	2.66	27.63	91.20	400.31	241.32	2877.18	263.54	2297.37	4830.16
MAAC	1	2.33	20.31	45.05	302.72	255.53	2065.97	340.63	3477.36	5008.28
	3	0.55	3.57	11.28	80.26	38.87	317.76	45.87	1468.44	3568.37
	5	0.38	1.92	6.34	40.13	21.23	290.91	23.23	330.21	2004.51

---

[6] Learning rate of 10^{-2}, $B = 200$, and $K = 8$.

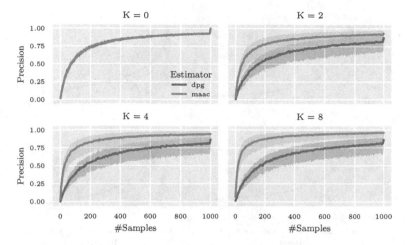

Fig. 4. Gradient precision for each estimator for different minibatch sizes ($B = \#$Samples) and virtual rollout lengths (K).

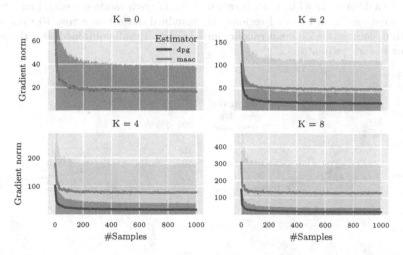

Fig. 5. Gradient norm for each estimator for different minibatch sizes ($B = \#$Samples) and virtual rollout lengths (K).

We also evaluate if our previous findings generalize to higher state-action space dimensions, where sample-based estimation gets progressively harder. Our performance metric is the suboptimality gap, i.e., the percentage difference in expected return between the current policy and the optimal one: $100 \times (J(\mu_\theta^\star) - J(\mu_\theta))/J(\mu_\theta^\star)$.[7] Table 1 summarizes our results with policy optimization with varying LQG sizes and time budgets.[8] We don't normalize gradients in this

---

[7] Recall from Sect. 3 that LQG allows us to compute the optimal policy analytically.

[8] We found that the computation times for both estimators were equivalent.

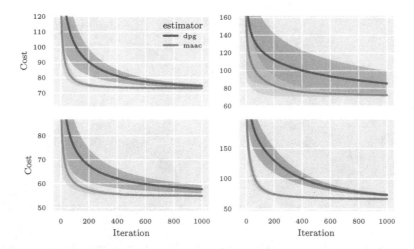

Fig. 6. Policy optimization with unnormalized SVG estimation. Each panel corresponds to a different LQG instance (generated via different random seeds). Lines denote the average results and shaded regions, one standard deviation, across 10 runs of the algorithm, each with a different random initial policy. Results obtained with the 8-step versions of each estimator.

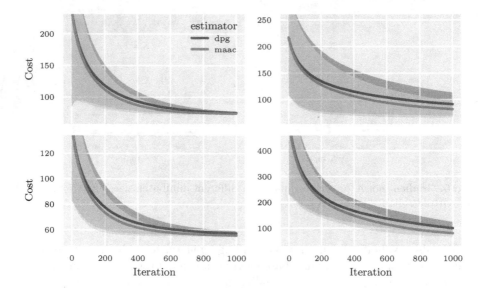

Fig. 7. Policy optimization with normalized SVG estimation. Each panel corresponds to a different LQG instance (generated via different random seeds). Lines denote the average results and shaded regions, one standard deviation, across 10 runs of the algorithm, each with a different random initial policy. Results obtained with the 8-step versions of each estimator. Gradients were normalized before being passed to SGD.

case, as that is not a common practice in SVG algorithms.[9] Our findings show that the performance gap between DPG(K) and MAAC(K) tends to widen with higher dimensionalities, with policies trained via the latter outperforming those using the former. These results further emphasize the practicality of MAAC(K) over DPG(K), justifying the former's use in recent SVG methods [1,4,8].

6   Conclusions and Future Work

In this work, we take an important step towards a better understanding of current SVG methods. Using the LQG framework, we show that the gradient estimation used by MAAC and similar methods induces a slight bias compared to the true value gradient. On the other hand, using a corresponding unbiased estimator such as the K-step DPG one increases sample-complexity due to high variance. Moreover, the MAAC gradient estimates have higher magnitudes, which could help explain the fast learning performance of current methods. Indeed, we found that policies trained with MAAC converge faster to the optimal policies than those using the K-step DPG across several LQG instances.

Future work may further leverage the LQG framework to perform fine-grained analyses of other important components of SVG algorithms. For example, little is known about the interplay between model, value function, and policy learning from data in practice. A study on model and value function optimization metrics and their relation to the gradient estimation accuracy and precision can help in the design of stable and efficient SVG algorithms in the future. Another direction for investigation is analyzing the impact of off-policy data collection for model training. Since models have limited representation capacity, learning the MDP dynamics from the distribution of another policy may not translate to good gradient estimation of the target policy.

Acknowledgments. This work was partly supported by the CAPES grant 88887.339578/2019-00 (first author), FAPESP grant 2016/22900-1 (second author), and CNPq scholarship 307979/2018-0 (third author).

References

1. Amos, B., Stanton, S., Yarats, D., Wilson, A.G.: On the model-based stochastic value gradient for continuous reinforcement learning. CoRR arXiv:2008.1 (2020)
2. Byravan, A., et al.: Imagined value gradients: model-based policy optimization with transferable latent dynamics models. In: CoRL. Proceedings of Machine Learning Research, vol. 100, pp. 566–589. PMLR (2019)
3. Chan, S.C.Y., Fishman, S., Korattikara, A., Canny, J., Guadarrama, S.: Measuring the reliability of reinforcement learning algorithms. In: ICLR. OpenReview.net (2020)
4. Clavera, I., Fu, Y., Abbeel, P.: Model-augmented actor-critic: backpropagating through paths. In: ICLR. OpenReview.net (2020). https://openreview.net/forum?id=Skln2A4YDB

---

[9] We only clip the gradient norm at a maximum of 100 to avoid numerical errors.

5. Deisenroth, M.P., Rasmussen, C.E.: PILCO: a model-based and data-efficient approach to policy search. In: Getoor, L., Scheffer, T. (eds.) Proceedings of the 28th International Conference on Machine Learning, ICML 2011, Bellevue, Washington, USA, 28 June–2 July 2011, pp. 465–472. Omnipress (2011). https://icml.cc/2011/papers/323_icmlpaper.pdf

6. Engstrom, L., et al.: Implementation matters in deep RL: a case study on PPO and TRPO. In: ICLR. OpenReview.net (2020). https://github.com/implementation-matters/code-for-paper

7. Goodfellow, I.J., Bengio, Y., Courville, A.C.: Deep Learning. Adaptive Computation and Machine Learning. MIT Press, Cambridge (2016)

8. Hafner, D., Lillicrap, T.P., Ba, J., Norouzi, M.: Dream to control: learning behaviors by latent imagination. In: ICLR. OpenReview.net (2020)

9. Heess, N., Wayne, G., Silver, D., Lillicrap, T.P., Erez, T., Tassa, Y.: Learning continuous control policies by stochastic value gradients. In: NIPS, pp. 2944–2952 (2015). http://papers.nips.cc/paper/5796-learning-continuous-control-policies-by-stochastic-value-gradients

10. Henderson, P., Islam, R., Bachman, P., Pineau, J., Precup, D., Meger, D.: Deep reinforcement learning that matters. In: AAAI, pp. 3207–3214. AAAI Press (2018)

11. Ilyas, A., et al.: A closer look at deep policy gradients. In: ICLR. OpenReview.net (2020)

12. Islam, R., Henderson, P., Gomrokchi, M., Precup, D.: Reproducibility of benchmarked deep reinforcement learning tasks for continuous control. CoRR arXiv:1708.04133 (2017)

13. Liu, Z., Li, X., Kang, B., Darrell, T.: Regularization matters for policy optimization - an empirical study on continuous control. In: International Conference on Learning Representations (2021). https://github.com/xuanlinli17/iclr2021_rlreg

14. Lovatto, A.G., Bueno, T.P., Mauá, D.D., de Barros, L.N.: Decision-aware model learning for actor-critic methods: when theory does not meet practice. In: Proceedings on "I Can't Believe It's Not Better!" at NeurIPS Workshops. Proceedings of Machine Learning Research, vol. 137, pp. 76–86. PMLR, December 2020. http://proceedings.mlr.press/v137/lovatto20a.html

15. Mnih, V., et al.: Human-level control through deep reinforcement learning. Nature 518(7540), 529–533 (2015)

16. Moerland, T.M., Broekens, J., Jonker, C.M.: Model-based reinforcement learning: a survey. In: Proceedings of the International Conference on Electronic Business (ICEB) 2018-December, pp. 421–429 (2020). http://arxiv.org/abs/2006.16712

17. Mohamed, S., Rosca, M., Figurnov, M., Mnih, A.: Monte Carlo gradient estimation in machine learning. J. Mach. Learn. Res. 21, 132:1–132:62 (2020)

18. Paszke, A., et al.: PyTorch: An imperative style, high-performance deep learning library. In: Advances in Neural Information Processing Systems, vol. 32, pp. 8024–8035. Curran Associates, Inc. (2019). http://papers.nips.cc/paper/9015-pytorch-an-imperative-style-high-performance-deep-learning-library.pdf

19. Pedregosa, F., et al.: Scikit-learn: machine learning in Python. J. Mach. Learn. Res. 12, 2825–2830 (2011)

20. Polydoros, A.S., Nalpantidis, L.: Survey of model-based reinforcement learning: applications on robotics. J. Intell. Robot. Syst. 86(2), 153–173 (2017). https://doi.org/10.1007/s10846-017-0468-y

21. Recht, B.: A tour of reinforcement learning: the view from continuous control. Ann. Rev. Control Robot. Auton. Syst. 2(1), 253–279 (2019). https://doi.org/10.1146/annurev-control-053018-023825, http://arxiv.org/abs/1806.09460

22. Ruder, S.: An overview of gradient descent optimization algorithms. CoRR arXiv:1609.04747 (2016)
23. Schulman, J., Heess, N., Weber, T., Abbeel, P.: Gradient estimation using stochastic computation graphs. In: NIPS, pp. 3528–3536 (2015)
24. Silver, D., Lever, G., Technologies, D., Lever, G.U.Y., Ac, U.C.L.: Deterministic Policy Gradient (DPG). In: Proceedings of the 31st International Conference on Machine Learning, vol. 32, no. 1, pp. 387–395 (2014). http://proceedings.mlr.press/v32/silver14.html
25. Szepesvári, C.: Algorithms for Reinforcement Learning. Synthesis Lectures on Artificial Intelligence and Machine Learning. Morgan & Claypool Publishers (2010). https://doi.org/10.2200/S00268ED1V01Y201005AIM009
26. Tits, A.L., Yang, Y.: Globally convergent algorithms for robust pole assignment by state feedback. IEEE Trans. Autom. Control 41(10), 1432–1452 (1996). https://doi.org/10.1109/9.539425
27. Todorov, E.: Optimal Control Theory. Bayesian Brain: Probabilistic Approaches to Neural Coding, pp. 269–298 (2006)
28. Todorov, E., Erez, T., Tassa, Y.: MuJoCo: a physics engine for model-based control. In: IEEE International Conference on Intelligent Robots and Systems, pp. 5026–5033. IEEE (2012). https://doi.org/10.1109/IROS.2012.6386109, http://ieeexplore.ieee.org/document/6386109/
29. Virtanen, P., et al.: SciPy 1.0: fundamental algorithms for scientific computing in Python. Nat. Meth. 17, 261–272 (2020). https://doi.org/10.1038/s41592-019-0686-2

Intelligent Agents for Observation and Containment of Malicious Targets Organizations

Thayanne França da Silva[1(✉)], Matheus Santos Araújo[1],
Raimundo Juracy Campos Ferro Junior[1], Leonardo Ferreira da Costa[2],
João Pedro Bernardino Andrade[2], and Gustavo Augusto Lima de Campos[1]

[1] State University of Ceara, Fortaleza, Brazil
{thayanne.silva,math.araujo,junior.ferro}@aluno.uece.br,
gustavo.campos@uece.br
[2] Federal University of Ceara, Fortaleza, Brazil
{leonardo.costa,jpandrade}@alu.ufc.br

Abstract. The problem addressed in this work is an extension of the Cooperative Multi-Robot Observation of Multiple Moving Targets Problem (CMOMMT). The scenario remains the same, but the targets are structured as an organization to achieve the highest possible percentage of exploration of the environment and avoid robots. Targets can be organized as hierarchy, holarchy, team, and coalition, but they can also be unorganized. Our work seeks to apply computer vision to assist robots in classifying the target team's organizational structure faced with a group of malicious target agents. Thus, robots can select the most appropriate strategy among the containment strategies implemented for each organizational structure or continue with the method proposed by literature for cases where the targets are not organized. The results showed that our approach had satisfactory results since, in luck, robots have a 20% chance of hitting the structure (hierarchy, holarchy, team, coalition, or random). Our approach had an accuracy of 63.28%. The containment strategies obtained satisfactory results in the robots' performance regarding the depreciation of the Percentage of the Environment Explored by the Targets (PEET) compared to the previous approach for robots. However, for the Average Number of Observed Targets (ANOT), the previous strategy was better. The new organizational approach to targets in CMOMMT was better than random in the desired exploration of the desired environment.

Keywords: Multi-agent system · Cooperative Multi-Robot Observation of Multiple Moving Targets · Organization · Computational vision · Containment strategies

1 Introduction

The observation of moving targets is an essential multi-robot application in MAS that still presents numerous open challenges, including the effective coordination

© Springer Nature Switzerland AG 2021
A. Britto and K. Valdivia Delgado (Eds.): BRACIS 2021, LNAI 13073, pp. 48–63, 2021.
https://doi.org/10.1007/978-3-030-91702-9_4

of robots [11]. The CMOMMT addresses the scenario where a group of agents, called robots, seek to retain within an "observation band" a maximum number of agents from the opposite group, called targets [14, 16].

In one of the reformulations of CMOMMT, Cooperative Target Observation (CTO) [13, 18] proposed targeting strategies based on structures to improve the performance of this team in the CTO. The results showed that the strategies of [18] were better for target performance compared to the proposed strategy [13]. Thus, in our approach, we implement four organizational paradigms for multi-agent systems [10], in the target team in CMOMMT, in a scenario where targets seek not only to walk randomly but explore the environment while avoiding robots.

Research has shown that it is possible to recognize patterns in images captured by satellites [21], infrared cameras [6], security cameras [3,4], etc. Furthermore, there are researches in the area of multi-agent simulation that made use of computer vision in the simulation image to locate agents in the environment [15]. Thus, this research applied computer vision to help classify the four organizational paradigms modeled for the targets (hierarchy, holarchy, team, and coalition) and the random model through images captured from the environment.

Thus, if we model the organizational structures in the targets so that communication between them depends on the location of these agents with their coordinators, sub-coordinators, or group partners, we can apply computer vision to classify the organizational paradigms present in the targets. With this information, robots can now select an appropriate containment approach for each of the four organizations developed or remain in the strategy proposed by [14] if the targets are walking randomly.

This article aims to develop a strategy for the decision-making of artificial robot agents, aiming to maximize the observation metric and minimize the exploration of the environment by target agents organized to explore and perform malicious actions in the background.

This paper is organized into four more sections. Section 2 presents the literature review. Section 3 describes the approaches used by targets and robots, materials, and methods. Section 4 shows the experiment and the results. Finally, Sect. 5 concludes the paper with final observations and future research.

2   A Literature Review

The CMOMMT Problem, initially described in [14], is defined in a simple, two-dimensional polyhedral spatial region, with inputs/outputs containing two teams of agents, the targets and the robots. The team of robots has 360° observation sensors. This team's objective is to maximize the collective time during which each target in the environment is monitored by at least one robot during the simulation time. We say that a robot is tracking a target when the target is within the robot's sensory field of observation.

Some reformulations of this problem have been created, such as FCMOMMT [8], P-CMOMMT [2,9]. The CTO is another reformulation of the CMOMMT [13]. The main difference between these two problems is that in the CTO, the targets provide information about their location. The observers' objectives remain the same as that of the robots. The targets were modeled with random movements, the target team being superior in numbers to the observer team, but the observers are faster.

In [1], it was observed that observers at the CTO work in a hierarchical structure with the K-means and Fuzzy C-means algorithm. Thus, [18] proposed three strategies so that the targets were as organized as the observers, hierarchy with K-means, the hierarchy with Fuzzy C-means, and Holarchy, and a strategy with neural networks. The results showed that the method based on the organizational paradigms was superior to the CTO literature approach.

[19] proposed an approach to classify the organizational structure of a group of target mobile agents that are continuously monitored by a smaller group of mobile observer agents in the CTO problem, a reformulation of the CMOMMT. The approach considers that the group of target agents can be organized according to eight different paradigms. These agents communicate through the exchange of messages whose contents are performative of the speech act.

This approach proved effective in comparison with the *Dummy* classifier that simulated human logic based on the frequency strategy. However, it was observed in this approach that the organizational strategy adopted for the targets, targets, and observers provided their location information to each other in the CTO. This type of scenario, where opposing teams exchange information when their goals neutralize each other, is not a realistic scenario. Thus, in the CMOMMT scenario, where the environment is partially observable, the targets and robots (observers) do not contribute with their locations, the simulation becomes more realistic.

In [5,15], computer vision was applied to the images captured by agents to help achieve their goals. [5] presented the development of a robotic multi-agent system, called SMART, in which there are two groups of agents, hardware and software agents, that work cooperatively. The hardware agents are robots with three and four legs and an IP camera that captures images of the scene where the cooperative task takes place. [15] presented a behavior-based approach for maintaining robot formation. The robot's objective is to circulate through the environment, keeping a relative position between them and avoiding shock.

Based on these concepts of an image capture agent, the use of the computer vision technique, and the limitation of target communication to maintain the organization in CMOMMT, we can assist robots in classifying target structures through the application of computer vision in the simulation scenario, such as [5,15].

3   The Approach for Observing and Containment of Malicious Targets Organizations

In this article, we present an extension to the CMOMMT problem. The scenario remains the same, but the targets are structured as an organization aiming to achieve the highest possible percentage of scenario exploration while avoiding robots. Robots continue with the surveillance objective, seeking to maximize the vision on the targets and minimize the effect of the organized targets on the environment through strategies to contain the exploration of the targets.

In the original CMOMMT problem, robots were only concerned with observing targets. However, this approach is not sufficient for robots in this extension of the CMOMMT problem. The targets are intended to explore the environment, and they are organized as hierarchy, holarchy, team, or coalition. Therefore, the proposed approach for robots in this new CMOMMT problem is using image classification to select the containment strategy to minimize the exploration of targets organized in the scenario.

Considering the assumption that there is a drone in the scenario capable of capturing images of both the target team and the robots, it is possible to generate a set of examples, label them and train a classification system that can recognize patterns in movement in each organizational paradigm present in the targets to classify the structure. Thus, through classification, the work proposes four associated containment strategies to solve the problem.

3.1 Target Strategy

Targets, in this extension of the CMOMMT problem, see the environment in quadrants. There are four quadrants: upper right, upper left, lower left, and lower right.

Among the paradigms raised by [10], hierarchy, holarchy, team, and coalition were selected for this work. Because, according to [10], starting from these four, it is possible to generate the others. The Subsections below detail each organizational structure modeled on the target team.

Hierarchy. In this organizational structure, only the two-level hierarchy was considered. At the level above, there is a target responsible for calculating the quadrant that contains the smallest percentage of exploration closest to it, based on the information obtained by the targets, to request the movement of this team to this particular quadrant at each time interval, called of coordinator. The targets that inform the area explored by them and perform the action requested by the coordinator are called subordinates. Therefore, subordinates must remain within the message range to receive the information, as communication is carried out through the speech act.

The state machine that demonstrates target team communication is shown in Fig. 1. In the initial state (q_0), the coordinator (**c**), belonging to the target team (**T**), requests that all subordinates (**s**) tell you the coordinates of your current state. In state q_1, subordinates report their status to the coordinator. Finally, in the state (q_2), the coordinator requests all targets to move towards the goal calculated by it. As the communication is not continuous but occurs every period after reaching the final state, the state machine will only restart again in the initial state at the time determined to exchange messages from the target team.

Fig. 1. Hierarchy communication state machine.

Holarchy. In holarchy, targets have been separated into two *holon* structured as a simple hierarchy, each containing a sub-coordinator who has authority over the subordinates of his holon. Our approach has only been tested with this simpler holarchy. There is a general coordinator that performs the same calculation process as the hierarchy coordinator. However, this one only has access to the sub-coordinators. Therefore, the coordinator transmits the message to the sub-coordinators, and these send the message to their subordinates. Hierarchy Communication State Machine.

At each given time, the coordinator informs the sub-coordinators of their environment analysis based on information obtained by all members of this paradigm (coordinator, sub-coordinator, and subordinates). The sub-coordinators request an action from their subordinates, who are within the sub-coordinator's speech act range, based on the analysis of the general coordinator. The coordinator's message is only received if the sub-coordinators are within the coordinator's message transmission range. Likewise, the subordinates of each holon must be within range of their sub-coordinator's message transmission.

Figure 2 shows the state machine of the target team communication in the holarchy. In the initial state (q_0), the coordinator (**c**), belonging to the target team (**T**), requests that all sub-coordinators (**sc**) ask all their subordinates to inform them of the coordinates of their current status. In the q_1 state, sub-coordinators request this information from subordinates. In the state q_2 and q_3, the subordinates report their status to their sub-coordinators, and these report the status of their subordinates to the coordinator, respectively. In the next state, the coordinator asks the sub-coordinators to ask their subordinates to go towards the goal calculated by the coordinator. Finally, in the q_5 state, the subordinates perform the action forwarded by their sub-coordinator. Finally, in the final state (q_0), the machine is shut down until the next communication period between the target team.

Team. In this organizational paradigm, all members are at the same level and divided into four groups containing the same number of members. Each group is sent to a region of the environment (quadrant) to accomplish its objective. Each sub-team member must remain within a certain radius to maintain communication and must stay in the area assigned to their sub-team. As shown in Fig. 3, in the initial state, (q_0), the targets of each sub-team report the exploration rate around them to their teammates. In the final state, (q_1), each sub-team will

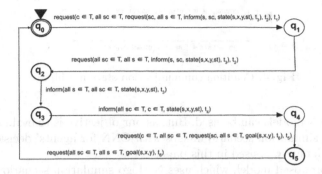

Fig. 2. Holarchy communication state machine.

ask its members to go to the coordinates of the target that obtained the lowest exploration rate around it in that sub-team.

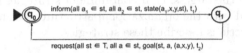

Fig. 3. Team communication state machine.

Coalition. On our problem, there are two coalitions, the members located on the right in the scene form a coalition, and those found on the left form another coalition. As targets are placed randomly in the environment, each coalition can have a different number of members. Communication, just like in the team, takes place at a certain point in time, and the members of each coalition must stay within a certain radius to enable communication.

The two coalitions tend to be separated from each other, as each is allocated to a specific area. However, the targets of each coalition must be close to each other to carry out decision-making.

In Fig. 4, the communication state machine is presented. In the initial state (q_0), each target of each coalition informs the exploration state of the region around them. In the q_1 state, coalitions request that targets belonging to their coalition go to the target coordinates of each coalition that has the lowest exploitation rate in its surroundings.

3.2 Robots Strategy

Computer Vision. Initially, a survey of classic CNN models was carried out to evaluate each model for classification of organizational paradigms and the random model. MobileNetsv2 was selected because it is a simple but efficient convolutional neural network. In order to carry out the classification in real-time simulation, a light network is needed for this task.

Fig. 4. Coalition communication state machine.

Other CNN models can be used. But, as our objective is to evaluate the benefits of using simulation scenario images with CNN for agents' decision-making, only MobileNetsv2 was used in this paper.

For our proposed model, which uses Netlogo simulation scenario images, we train our dataset to 100 epochs using the *Adam* optimizer and an initial learning rate of $1e - 3$. In this work, we test the lot size valued at 32.

80% from our image bank was used for the training step and 20% for the validation step. In the training stage, we obtained 74.56% accuracy. In the validation step, the accuracy of 72.14% was obtained.

During the simulation, images of the scenario are sent to classify the organizational structure of the targets. Four containment strategies for the robot team are proposed to deal with the organizational structure that targets may adopt. The following subsections describe these strategies.

Strategy Against Hierarchy. In this type of organization, in which the targets are submitted to a higher-level agent and where decision-making depends on the latter's endorsement, the containment strategy adopted was to disable the coordinator's communication with the targets. As only the coordinator requests an action and the agents' act of speaking represents in the real world the human speech itself, as soon as the hierarchical structure is detected, the robots move between the targets in order to see which one is communicating an action. Thus, the robots will disable this agent's communication, and the targets will be incapacitated, as they will not be able to reach their goals because the agent in charge of transmitting information about their goals has been disabled. However, robots continue to look for more coordinators, as they do not have the details on how the hierarchy is structured, two levels or more.

Strategy Against Holarchy. In holarchy, it is necessary to disable the general coordinator and the sub-coordinators. For however much the sub-coordinators are submissive to the general coordinator, they have authority over the targets of their group. If so, the holarchy could turn into coalitions, and targets could still achieve their goals. Therefore, the containment strategy adopted for this structure is to disable the communication of all targets that have authority over other targets.

Strategy Against the Team. In the team's case, the targets are independent of each other but cooperate to achieve their goals; that is, they explore the area allocated to their team to maximize the percentage of the region explored by the targets.

As there is no essential agent in this structure that its suppression disables, as all targets are at the same level, the strategy adopted was an extension of the approach proposed by [14]. In this extension, robots have two behaviors, that of seeking sub-times and that of disrupting the communication of targets under observation. As the targets of each subteam are close to each other, if the force field proposed by [14] were used to repel all other robots, only four robots would be trying to disable communication, while the other eight would be idle. Thus, a conditional was added to this force field. It now allows a maximum of two robots and repels the others when it exceeds this margin. So, instead of just one robot trying to disable the communication of six targets, there are now two for this purpose in each sub-team. The four remaining robots are responsible for looking for more sub-teams, as the robot team does not know how many sub-teams there are in the simulation scenario.

Strategy Against the Coalition. As the number of members in each coalition can be unequal and there are no agents with authority over others, the containment strategy was an extension of the method proposed by [14] as well. They are seeking, as well as in the team, to break communication between the coalition targets, as there are not enough robots to disable all targets to minimize the exploration of the scenario.

Strategy Overview. Thus, the robot approach consists of them, every 200 steps of time on the Netlogo platform, sending an image of the simulation scenario to the Jupyter Notebook. Then, the classification of the organizational paradigm by Mobilenetv2 is performed, and the result is returned to the robots team. Thus, from the returned response, the most suitable containment strategy is selected. The simulation only returns to processing when the *Jupyter Notebook* returns the classification value performed by CNN.

3.3   Materials and Methods

The Netlogo [20] platform was selected for scenario simulation, as [1,18,19] used to simulate a reformulation of the CMOMMT Problem, the CTO. This platform was also chosen for its integrality with the Jupyter Notebook [12] platform used in this research. MobileNetV2 [17] was loaded by Keras [7], an open-source neural network library written in Python.

In the generation of the image bank, the Data Augmentation technique was used to supplement our dataset. In addition, the use of this technique simulates a drone flying over the scene and capturing images of the robot team and targets from various angles and positions.

Our image bank was generated from the simulation scenario images. The starting position of each target and robot is random, and the choice of the following position is based on the strategy adopted by each team.

Seeking to diversify the image bank, the target communication range sensor can vary between 5, 10, 15, 20, and 25 Netlogo distance units. Thus, targets can be further away or closer depending on the communication range setting.

Initially, the image bank contains 750 images of each organizational paradigm and the random model for [14] targets, in which there are 150 images for each communication range sensor configuration, totaling 3750 images. The photos were captured manually in the most diverse positions, rotations, and transitions from one quadrant to another for better learning of the model.

After capturing the 3750 images of the scenery, the Data Augmentation technique was applied, which generated our final image bank with 41, 250 images[1].

Figure 5 shows the simulation environment with the targets structured as a hierarchy, where the "arrows" agents are the robots and the "people" agents are the targets. According to the robot closest to them, the colors of the target agents are responsible for their observation. There is a quadrant division to aid in viewing target movements and image processing.

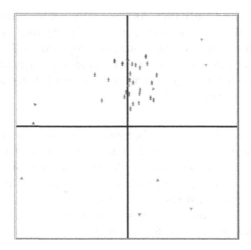

Fig. 5. Netlogo simulation example.

4   Experiments and Results

4.1   Test Settings

The configured parameters were based on the settings used by [13].

- Targets and robots are in a rectangular field with dimensions 150 by 150 units;
- There are 1500 interaction steps per simulation;

---

[1] The dataset is available at the following link: https://drive.google.com/drive/folders/1PwjDRzP23sT4qZSZF_wnYQUEhOT9qcDQ?usp=sharing.

- Observer speed is 1 step at each interaction step;
- Target speed is 0.9 steps at each interaction step;
- Sensor range is 25 units;
- Communication range varies between 5, 10, 15, 20 and 25 units;
- 24 targets;
- 12 robots;
- In the case of Hierarchy there is an extra target, called coordinator;
- In the case of Holarchy there are three extra targets, called general coordinator and sub-coordinators.

In order to evaluate the performance of the robots, it was configured for targets with the highest speed and range of vision defined by [13]. Well, this is the most challenging scenario for the team of configured robots today. Thus, if satisfactory results were obtained in this scenario, in better scenes for robots, the performance tends to be acceptable as well.

4.2 Result of the Classification of Organizational Structures and the Random Model

For our proposed model, we train our dataset for 1000 epochs with the lot size value of 32. Our model was tested to classify images that were not included in training or validation sets. Our test suite consists of 3750 images, with 750 from each organizational paradigm and random model.

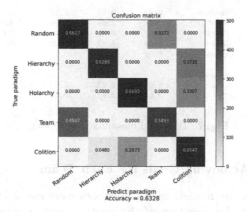

Fig. 6. Confusion matrix.

As we can see in the confusion matrix presented in Fig. 6, our model obtained an accuracy of 0.6328. If we were to consider robots drawing lots to predict the paradigms, the probability of getting it right would be 20%, while our model gets a 63.28% hit odds rate.

Analyzing the classification of each paradigm, we can see that the random model and the team obtained a considerable rate of false positives between them.

This could be due to the team paradigm spreading the four subteams across the four quadrants of the scenario. For high team communication range values (20 or 25), where targets can move further between their sub-team members, the team structure begins to resemble the random model.

Some images of the organizational paradigm of the hierarchy were classified as the coalition. As in the coalition, the number of members in each coalition can be unbalanced, to the point where there are 22 targets in one coalition and only two in another. Thus, in these cases, the hierarchy can resemble the coalition and vice versa, especially if the communication range is high, as the targets spread further across the environment.

Holarchy was also classified as a coalition due to the fact that they have a common characteristic; there are two target groups in the scenario. However, while one has the same amount of members, the other may contain unbalanced values. Therefore, for scenarios where there were the same amount of members or approximate, the model could misclassify these two paradigms.

Figure 7 presents the test step rating report, which contains the accuracy and *recall* for each organizational and random model. The model had the best accuracy for holarchy; that is, when it predicts that targets are arranged as a holarchy, it is correct at 93% of the time. In the case of *recall*, the results were close, with the exception of the coalition. However, all were above 50%. That is, our model correctly identifies around 60% most of the target-structured models.

	precision	recall	f1-score	support
Random	0.60	0.66	0.63	750
Hierarchy	0.48	0.65	0.56	750
Holarchy	0.93	0.63	0.75	750
Team	0.69	0.67	0.68	750
Colition	0.62	0.55	0.58	750
accuracy			0.63	3750
macro avg	0.66	0.63	0.64	3750
weighted avg	0.66	0.63	0.64	3750

Fig. 7. Test step classification report.

4.3   Results of Achieving Goals for Each Team

One hundred times, the models with the robots were run using or not the classification approaches for selection of the containment strategy. After that, the average of their values was calculated. Tables 1, 2, 3, 4, and 5 present the results of the Percentage of Environment Explored by Targets (PEET) and the Average Number of Observed Targets (ANOT) with robots using or not the classifications with the containment strategies for each range of target communication.

Table 1. Result with communication range equal to 5.

Structures	PEET without the approach	PEET with the approach	ANOT without the approach	ANOT with the approach
Random	33,574	–	23,683	–
Hierarchy	30,437	16,512	23,995	23,996
Holarchy	32,867	17,977	23,994	23,995
Team	52,764	33,163	23,982	23,981
Coalition	55,835	34,051	23,898	23,893

Table 2. Result with communication range equal to 10.

Structures	PEET without the approach	PEET with the approach	ANOT without the approach	ANOT with the approach
Random	34,137	–	23,813	–
Hierarchy	33,255	20,304	23,995	23,995
Holarchy	37,578	22,891	23,995	23,996
Team	55,196	35,285	23,988	23,981
Coalition	57,664	37,224	23,924	23,920

Table 3. Result with communication range equal to 15.

Structures	PEET without the approach	PEET with the approach	ANOT without the approach	ANOT with the approach
Random	35,266	–	23,868	–
Hierarchy	35,704	21,049	23,987	23,991
Holarchy	40,476	25,117	23,994	23,991
Team	58,527	39,581	23,972	23,970
Coalition	59,556	41,033	23,951	23,948

Table 4. Result with communication range equal to 20.

Structures	PEET without the approach	PEET with the approach	ANOT without the approach	ANOT with the approach
Random	34,691	–	23,794	–
Hierarchy	38,348	25,722	23,984	23,989
Holarchy	42,167	27,564	23,984	23,989
Team	60,631	42,813	23,969	23,963
Coalition	60,511	43,097	23,964	23,961

Table 5. Result with communication range equal to 25.

Structures	PEET without the approach	PEET with the approach	ANOT without the approach	ANOT with the approach
Random	35,482	–	23,762	–
Hierarchy	46,013	29,248	23,981	23,985
Holarchy	57,352	32,173	23,992	23,989
Team	62,409	48,386	23,966	23,962
Coalition	62,393	49,033	23,973	23,969

Note that in scenarios where robots used these strategies, targets had the lowest PEET for any communication range when compared to scenarios where robots did not use them. This means that robots that used classification approaches with containment strategies were more efficient in preventing the progress of exploration of the environment by the targets.

In hierarchy and holarchy, it was observed in the behavior of the targets that when agents with authority over them were disqualified, the targets remained around the last location passed by their coordinators. In the team and coalition, the behavior of the agents who departed from communication with their team or coalition was also to walk around the last location where they had communication with their sub-team or coalition. In some cases, it was possible for this isolated agent to meet again with his group, while the rest continued with the objective of exploring the environment.

The results for the ANOT for each strategy are not significantly different. But, the hierarchy and holarchy were observed because the targets are only circling in the last location passed by their coordinators or sub-coordinators when these are disabled by robots, which facilitates the observation of the robot team. However, in the case of team and coalition, while some robots seek to break communication, others seek to locate other teams and coalitions across the scene. Thus, the number of robots for observation of targets is lower than the scenario where robots did not use classification approaches.

A containment strategy for the random structure was not proposed, as the movement of targets in this structure is unpredictable. So the robots remained in the approach proposed by [14]. However, it was evaluated along with the paradigms in order to compare the achievement of the goals with the targets being organized or not. Regarding the avoidance of observers, the random strategy was the best since the Average Number of Observed Targets (ANOT) was lower for this strategy; that is, the targets were a little better at avoiding the robots since in this strategy, the targets were more spread out, while the organizational targets are closer to communicate.

For the purpose of exploring the environment, organizational strategies were better, as they reached higher percentages of exploration, except for hierarchy and holarchy that obtained better results from a communication reach equal to or greater than 15 and 10, respectively. Because the targets in these strategies are organized for this purpose, as they focus on regions that have not been explored so far, while the random one can repeat areas already explored more than once. In addition, team and coalition obtained the best results for exploration, as they do not have an organizational structure as restricted as hierarchy and holarchy. Thus, they have more freedom to spread out across the environment in order to reach their goal. The random model is not affected by the change in communication range, as this strategy does not use communication; the targets just walk randomly.

5 Conclusion

The contribution given by this research is to show how the use of computer vision enables the classification of organizational structures in multi-agent simulations, which until now had not been proposed. In addition to introducing new strategies and objectives to targets and robots in the CMOMMT problem.

Our classification approach showed satisfactory results when compared to drawing lots since the model obtained an accuracy of 63.28%. Our containment approaches also showed promising results for the robots team in terms of exploration of the environment, allowing them to minimize the percentage of the territory explored by the targets and thus avoid further damage to the scenarios that the targets could cause. In the ANOT issue, hierarchy and holarchy had the best results, while the team and coalition containment strategies performed less than the performance obtained by robots that did not use the containment strategies but used the method proposed by [14]. Furthermore, in the case of targets, our distance-to-target-based organizational approach for communication was adequate for the targets' objective of exploring the environment compared to the random strategy. However, it is not suitable for the targets' other goal, avoiding robots.

For future work, we intend to implement other simulation scenarios, for example, with obstacles, in order to evaluate our approach, even in the real world. In addition to implementing more robust frameworks to analyze the performance of our model in complex environments. We intend to evaluate the approach with other CNN models to select the best model to be used in terms of accuracy and processing time. Finally, we want to examine the impact of the wrong choice of containment approaches, in addition to implementing the containment strategies for each new organizational paradigm implemented and improving those already implemented.

References

1. Andrade, J.P., Oliveira, R., da Silva, T.F., Maia, J.E.B., de Campos, G.A.: Organization/fuzzy approach to the CTO problem. In: 2018 7th Brazilian Conference on Intelligent Systems (BRACIS), pp. 444–449. IEEE (2018)
2. Araújo, M.S., et al.: Cooperative observation of smart target agents. In: Cerri, R., Prati, R.C. (eds.) BRACIS 2020. LNCS (LNAI), vol. 12320, pp. 77–92. Springer, Cham (2020). https://doi.org/10.1007/978-3-030-61380-8_6

3. Arroyo, R., Yebes, J.J., Bergasa, L.M., Daza, I.G., Almazán, J.: Expert video-surveillance system for real-time detection of suspicious behaviors in shopping malls. Expert Syst. Appl. **42**(21), 7991–8005 (2015)
4. Bennett, M.K., Younes, N., Joyce, K.: Automating drone image processing to map coral reef substrates using google earth engine. Drones **4**(3), 50 (2020)
5. Cena, C.G., Cardenas, P.F., Pazmino, R.S., Puglisi, L., Santonja, R.A.: A cooperative multi-agent robotics system: design and modelling. Expert Syst. Appl. **40**(12), 4737–4748 (2013)
6. Cho, C., Kim, J., Kim, J., Lee, S.-J., Kim, K.J.: Detecting for high speed flying object using image processing on target place. Clust. Comput. **19**(1), 285–292 (2016). https://doi.org/10.1007/s10586-015-0525-x
7. Chollet, F., et al.: Keras (2015). https://github.com/fchollet/keras
8. Ding, Y., He, Y.: Flexible formation of the multi-robot system and its application on CMOMMT problem. In: 2010 2nd International Asia Conference on Informatics in Control, Automation and Robotics (CAR 2010), vol. 1, pp. 377–382. IEEE (2010)
9. Ding, Y., Zhu, M., He, Y., Jiang, J.: P-CMOMMT algorithm for the cooperative multi-robot observation of multiple moving targets. In: 2006 6th World Congress on Intelligent Control and Automation, vol. 2, pp. 9267–9271. IEEE (2006)
10. Horling, B., Lesser, V.: A survey of multi-agent organizational paradigms. Knowl. Eng. Rev. **19**(4), 281–316 (2004)
11. Khan, A., Rinner, B., Cavallaro, A.: Cooperative robots to observe moving targets. IEEE Trans. Cybern. **48**(1), 187–198 (2016)
12. Kluyver, T., et al.: Jupyter Notebooks - a publishing format for reproducible computational workflows. In: Loizides, F., Schmidt, B. (eds.) Positioning and Power in Academic Publishing: Players, Agents and Agendas, pp. 87–90. IOS Press (2016)
13. Luke, S., Sullivan, K., Panait, L., Balan, G.: Tunably decentralized algorithms for cooperative target observation. In: Proceedings of the Fourth International Joint Conference on Autonomous Agents and Multiagent Systems, pp. 911–917 (2005)
14. Parker, L.E.: Cooperative motion control for multi-target observation. In: Proceedings of the 1997 IEEE/RSJ International Conference on Intelligent Robot and Systems. Innovative Robotics for Real-World Applications. IROS 1997, vol. 3, pp. 1591–1597. IEEE (1997)
15. Payá, L., Gil, A., Reinoso, O., Ballesta, M., Ñeco, R.: Behaviour-based multi-robot formations using computer vision. In: Proceedings of the 6th IASTED International Conference on Visualization, Imaging and Image Processing (2006)
16. Robin, C., Lacroix, S.: Multi-robot target detection and tracking: taxonomy and survey. Auton. Robot. **40**(4), 729–760 (2015). https://doi.org/10.1007/s10514-015-9491-7
17. Sandler, M., Howard, A., Zhu, M., Zhmoginov, A., Chen, L.C.: MobileNetv 2: inverted residuals and linear bottlenecks. In: Proceedings of the IEEE Conference on Computer Vision and Pattern Recognition, pp. 4510–4520 (2018)
18. França da Silva, T., et al.: Smart targets to avoid observation in CTO problem. In: Proceedings of the 18th International Conference on Autonomous Agents and MultiAgent Systems, pp. 1958–1960 (2019)
19. Silva, T., Araújo, M., Junior, R.F., Costa, L., Andrade, J., Campos, G.: Classifying organizational structures on targets in the cooperative target observation. In: Anais do XVII Encontro Nacional de Inteligência Artificial e Computacional, pp. 718–729. SBC (2020)

20. Tisue, S., Wilensky, U.: NetLogo: a simple environment for modeling complexity. In: International Conference on Complex Systems, vol. 21, pp. 16–21. Boston, MA (2004)
21. Vargas-Cuentas, N.I., Roman-Gonzalez, A., Mantari, A.A., Muñoz, L.A.: Chagas disease study using satellite image processing: a Bolivian case. Acta Astronaut. **144**, 216–224 (2018)

MAS4GC: Multi-agent System for Glycemic Control of Intensive Care Unit Patients

Tiago Henrique Faccio Segato[1,2]([✉]) [iD], Rafael Moura da Silva Serafim[1] [iD],
Sérgio Eduardo Soares Fernandes[3] [iD], and Célia Ghedini Ralha[2] [iD]

[1] Instituto Federal de Brasília (IFB), Brasília, Brazil
tiago.segato@ifb.edu.br, rafael.serafim@estudante.ifb.edu.br
[2] Universidade de Brasília (UnB), Brasília, Brazil
ghedini@unb.br
[3] Escola Superior de Ciências da Saúde (ESCS), Brasília, Brazil
sergio.fernandes@escs.edu.br

Abstract. The coronavirus disease (COVID-19) pandemic has brought significant challenges worldwide through the consequences of increasing demand for the Intensive Care Unit (ICU) resources. This work presents the Multi-Agent System for Glycemic Control (MAS4GC) to assist health professionals leading with critical patients in the ICU. More specifically, the MAS4GC manages patients' blood glucose through glycemic predictions, treatment, and monitoring recommendations to health professionals. Prediction models are applied to monitor patients' blood glucose allowing health professionals to carry out preventive treatments. The glycemic control is included in the FAST HUG mnemonic to remember the key issues in the supportive care of critically ill patients. The MAS4GC methodological development process is presented with Tropos modeling, architectural design, and implementation with the PADE framework. Agents' inference mechanisms are based on production rules defined by intensive care physician specialists applying their knowledge to indicate treatments for patients. Two experiments using patients with synthetic data were conducted to evaluate the results of the MAS4GC: (1) the prediction model achieved 90% accuracy in blood glucose predictions for the next 4 h, (2) 84% similarity of treatment recommendations compared to a human specialist, and 78% in recommendations for monitoring glycemic of critical patients.

Keywords: Agent-based system · Blood glucose · Health care · ICU · Rule-based reasoning

1 Introduction

The Intensive Care Unit (ICU) is considered a high-risk care setting where medical carelessness or errors can cause deaths or complications to patients' health [1]. In pandemic times, such as that of the coronavirus disease (COVID-19), the ICU's importance in the treatment of critically ill patients became even more

© Springer Nature Switzerland AG 2021
A. Britto and K. Valdivia Delgado (Eds.): BRACIS 2021, LNAI 13073, pp. 64–78, 2021.
https://doi.org/10.1007/978-3-030-91702-9_5

evident. Since patients need special care varying from basic requirements to the need for equipment for patient monitoring and life support, such as respirators and mechanical ventilators [2,3]. Artificial Intelligence (AI) applications currently are used in the management of such complex tasks. AI can assist in the monitoring and treatment of patients with chronic diseases and in critical conditions hospitalized in the ICU, where these cases occur more frequently [4].

The ICU patient monitoring must rigorously take place. The FAST HUG - Feeding, Analgesia, Sedation, Thromboembolic prevention, Head of the bed elevated, stress Ulcer prophylaxis, and Glucose control - is a simple and significant mnemonic to highlight seven of the main aspects that must be monitored by health professionals for each patient in the ICU to minimize possible problems [1].

Focusing in the FAST HUG, glycemic control has the function of checking the patient's glucose level in the blood, keeping as long as possible in the target range [1]. Regardless of which is the ideal target range, it is vital to keep it monitored. Effective glycemic control in the ICU environment has the potential to decrease mortality rates and the patient's length of stay in the ICU, optimizing hospital resources [5]. The overload of health professionals in pandemic times is crucial, and factors such as this point out that automated systems to monitor and assist patients' treatment can bring benefits. Thus, the FAST HUG, or at least some of its items, can be improved with process automation combined with AI techniques.

In the literature review, AI-based works for glycemic control of ICU patients are presented [5–7], as well as the application of Multi-Agent System (MAS) for patients glycemic control [8], and MAS in the ICU context [9,10]. However, none of the works applies MAS for patients' glycemic control admitted to the ICU. Some work points to prediction models as good solutions for glycemic control [11–13].

Considering the cited scenario, the objective of this work is to present a system to track and monitor the glycemic control of critical patients in the ICU through a MAS approach. In a previous work [14], the MAS development process was presented without implementation results. More specifically, the following hypothesis has to be proven: a MAS can suggest patients' treatment recommendations using prediction models and a knowledge base with inference rules similar to specialist intensive care physicians.

The rest of the manuscript is organized as follows. In Sect. 2 some works found in the literature are presented. In Sect. 3, the materials and methods are presented. In Sect. 4 the experiments were carried out with results. Finally, in Sect. 5 the final considerations and possible future work are indicated.

2   Literature Review

The literature review used the *Portal de Periódicos Capes*, seeking articles in English published from 2015 to 2021, with the following combined keywords: artificial intelligence, multiagent systems, intensive care unit, glucose control.

The works of DeJournett et al. [6,7] present an autonomous glucose control system (artificial pancreas) to reduce problems resulting from the glycemia of patients in the ICU. In the first work [6], the authors perform an insulin test with simulated patients, using an AI-based glucose controller. As an AI technique, they used rule-based reasoning (RBR). In the second work [7], the same system was used, but the objective was to evaluate the system's safety and performance by applying simulated tests with swine in a clinical setting.

In Jemal et al. [10], a model was proposed, and a specialized decision support system was implemented and validated to detect the degree of risk of patients in the ICU. A MAS was used as the main technology combined with a knowledge base and Intuitionistic Logic Fuzzy (IFS). In Malak et al. [9] an architecture based on agents with decision support and in real-time for the management of high-risk newborns admitted to the ICU-N was presented. Both studies showed that MAS is a good solution to be used in healthcare systems.

When it comes to glycemic control, studies such as those presented in Vehi et al. [11], Bertachi et al. [12], and Kim et al. [13] point to prediction models as interesting solutions, where the prevention of hypo or hyperglycemic events tends to be more efficient than the correction of these episodes. The solutions presented in these works include (1) prediction and prevention of hypoglycemic events in diabetics [11]; (2) a prediction model for episodes of nocturnal hypoglycemia in diabetics [12]; and (3) a glucose prediction model for hospitalized type 2 diabetic patients [13].

Table 1 summarizes the qualitative aspects of the related work, limited to the application context (glycemic control, ICU patients) and technologies used (agent-based, prediction model). Note that no work presents a solution with MAS to manage glycemic control with predictions in the ICU setting.

Table 1. Related work overview.

Reference	Glycemic control	ICU patients	Agent-based	Prediction model
Dejournett et al. (2016, 2020) [6,7]	✓	✓	–	–
Malak et al. (2018) [9]	–	✓	✓	–
Jemal et al. (2019) [10]	–	✓	✓	–
Vehí et al. (2020) [11]	✓	–	–	✓
Bertachi et al. (2020) [12]	✓	–	–	✓
Kim et al. (2020) [13]	✓	–	–	✓
This work (2021)	✓	✓	✓	✓

3   Materials and Methods

The methodological process used was based on previous work [14] and consists basically of four steps as presented in Fig. 1 and described in the sequence.

3.1 Problem Definition

The problem definition is based on the literature review (Sect. 2). The theoretical foundation includes FAST HUG, glycemic control, and MAS concepts.

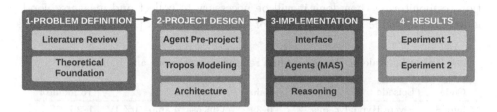

Fig. 1. Methodological workflow.

FAST HUG in ICU Management. According to the National Health Surveillance Agency (Anvisa) [15], ICU is considered a complex unit to be managed in a hospital being the place where critical condition patients deserving greater care are concentrated. Such patients are very difficult to manage safely and effectively due to their complex, nonlinear, and highly variable physiology. Therefore, improving patients care and treatment are the main current challenges for ICU settings, where personalization and automation of care offer opportunities to cause significant impacts [16]. Mnemonics are commonly used in medical procedures as cognitive aids to guide doctors around the world. Thus, Vincent [1] suggested the simple mnemonic FAST HUG that contains seven essential aspects to be verified during the care of critical patients in an ICU during medical rounds. Such checks should take place at least once a day and ideally, whenever any of the professionals assist the patient.

Glycemic Control. The glycemic control concerns the amount of glucose (sugar) that the patient has in the blood. Both, high glucose level (hyperglycemia), and low blood glucose level (hypoglycemia) are frequent problems in ICU patients causing damage to health [17]. One of the studies in Vincent's [1] highlights the importance of glycemic control in the ICU. The author demonstrates that maintaining blood glucose levels at approximately $140mg/dL$ results in a 29.3% decrease in-hospital mortality rates, and a 10.8% reduction in the ICU stay.

Hyperglycemia is prevalent in ICU being a good example that the strict glycemic control of these patients can have a great impact. Such an episode occurs due to the non-suppressed production of glucose by the body, medications, suppression of the body's insulin secretion, among others. All these factors effectively damage the body's normal feedback control mechanisms resulting in less insulin-mediated glucose uptake. Therefore, some type of supplementary glycemic control is necessary, possibly in a closed circuit and highly personalized in intensive care [16].

Table 2 was prepared in conjunction with a specialist doctor and presents, in addition to the values, what must be done for each glycemic episode. It synthesizes the information necessary to identify the glycemic episode, using a code (in the system implementation) and blood glucose values. It also presents treatment suggestions (applying glucose or insulin with the respective dosage), in addition to monitoring (how long it will be necessary to collect and measure blood glucose). This table is an adaptation of the previous work [14].

Table 2. Scale of glycemic values related to treatment and monitoring.

Code	Episode	Values	Treatment	To monitor
hypoS	Severe Hypoglycemia	0–39	Glucose: 4 amp–50% IV	1–2 h
hypoM	Mild Hypoglycemia	40–69	Glucose: 2 amp–50% IV	2–4 h
normoG	Normoglycemia	70–140	Keep watching	6–8 h
hyperM	Mild Hyperglycemia	141–180	Regular Insulin: 2 un SC	4–6 h
hyperS	Severe Hyperglycemia	181–250	Regular Insulin: 4 un SC	2–4 h
hyperVS	Very Severe Hyperglycemia	>251	Regular Insulin: 6 un SC	1 h

MAS Aspects. A MAS is composed of two or more intelligent agents capable of perceiving events in the environment through sensors, reasoning, and acting in the environment through actuators [18]. According to [8], a MAS has an intelligent distributed approach suitable for modular, changeable, and complex applications, with characteristics such as autonomy, integration, reactivity, and flexibility, becoming an interesting solution for modeling large-scale health systems.

Regarding the agents' reasoning, different AI techniques can be used including the combination of them. In this work, agents should make predictions according to the patient's blood glucose and data. Also, to make treatment and monitoring suggestions for the patients. Regarding predictions, predictive models can be highlighted, which can assist in decision-making [11–13]. In the case of monitoring and suggesting treatments, a good solution would be to represent the knowledge of medical specialists through RBR [6].

– Predictive model: A regression model is based on the correlation between two (Simple Linear Regression - SLR) or more variables (Multiple Linear Regression - MLR), where one depends on another or others [19]. The use of regression models obtained satisfactory results in previous works of [20, 21] in the health area, more specifically, the prediction of new cases of COVID-19 and prediction of glucose levels in critically ill patients, respectively.
  An MLR is expressed by the Eq. 1, where y is the dependent variable or the value to be predicted. The β_0 is the constant that represents the intercept of the line on the y axis, and the independent variables $\beta_1 x_i + \beta_2 x_2 + ... + \beta_k x_k$ represent the slope of the line. The x is the independent variable or predictor variable, this has the power to influence the variable to be found, and ε the variable that represents the residual factors of the measurement errors [20].

$$y = \beta_0 + \beta_1 x_i + \beta_2 x_2 + ... + \beta_k x_k + \varepsilon \qquad (1)$$

- RBR: Dejournett [6] associates systems based on rules or knowledge with AI controllers that seek to capture the human thought process, creating rules that mimic the exact reason used by human beings. Such a system is created when a domain expert joins a knowledge engineer and explains his lines of reasoning to perform certain functions when trying to control the system in question. The engineer in turn transforms the lines of reasoning into a series of if-then rules that mimic the thinking of experts in the field.

Assuming the situation where a patient is in the ICU and the blood glucose collected value is 60mg/dL, the rule compatible with this case would be:

IF glycemia = hypoM THEN "Glucose: 2 amp - 50% IV"

3.2 Project Design

This step includes the agents' identification with their respective objectives and the construction of diagrams that will assist in the implementation step.

Agents Pre-project. A MAS project includes the identification of the perceptions, actions, objectives, performance, and environment of each agent in the system. The pre-project includes these definitions through the acronym PEAS (Performance, Environment, Actuators, Sensors). The MAS pre-project serves to identify in which environment the agent will act and its respective characteristics. Based on the objectives of each agent, it is also possible to describe what are the mechanisms by which they will perceive the information and how they will act in response to such stimuli [18].

Tropos Modeling. The MAS modeling can use Tropos software development methodology for agent-oriented software systems. Tropos is based on the i* framework (ISTAR - Intentional STrategic Actor Relationships modeling), modeling the functionality of an application based on objectives using diagrams [22]. Although there are other methodologies for agent modeling, the Tropos was chosen since it encompasses all five phases of software development supporting the analysis of initial requirements to implementation with the diagrams: initial requirements, final requirements, architectural design, detailed design, and implementation. The five Tropos diagrams of MAS4CG can be found in [14].

Figure 2 presents the diagram of the late requirements of the proposed system. Red circles represent MAS external *Actors* (e.g., people, systems), Yellow circles with a top straight line illustrate MAS *Agents*. The green rounded rectangles represent the agent *Goals* that can be understood as the system requirements. The purple rectangles represent *Resources*, a physical or informational entity needed by the actor or agent to perform a task. The MAS agents are Patient Analyzing Agent (PAA), Proposed Treatment Agent (PTA), and Adjustment Monitoring Agent (AMA) with functionalities detailed Fig. 3.

Other diagrams, such as the architecture and UML (Unified Modeling Language), can be developed to complement the understanding of the proposed solution representing details aimed at implementing the solution. In the implementation stage, details of how all technologies, systems, and agents were implemented are presented.

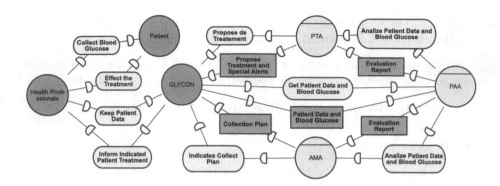

Fig. 2. Tropos late requirements diagram.

Architecture. Figure 3 presents the MAS4GC architecture. The PAA's main objective is to collect the patient's data and respective glycemia, whenever new data is inserted or updated in the Glycon Database. Glycon is a web system used as an interface by healthcare professionals, both for entering information about the patient and their blood glucose, and for displaying the recommendations made by agents. This agent should analyze such data and make a report assessing the patient's situation in comparison with previous data. This will allow the agent to calculate and make predictions of how the patient's next blood glucose will be. This report will be sent to the PTA and AMA agents, who in turn will analyze it and propose a treatment (what should be applied, glucose or insulin, and how much) or adjust the blood glucose monitoring (indicate the appropriate frequency of blood collection). It will be possible to send an alert containing the recommendation to health professionals through the Glycon interface.

3.3   MAS4GC Implementation

In this work, we present a MAS developed using objective-oriented agent modeling for glycemic control of patients admitted to the ICU. The three agents PAA, PTA, and AMA interact with each other and with a Web system called Glycon, used as an interface for collecting and presenting patient data, including blood glucose levels. The MAS4GC consists of three integrated systems, one of which is an interface that serves for the entry and visualization of data through health professionals. The other is the MAS itself, where, through a framework, the three agents were implemented. Finally, the agents' reasoning was built in the form of a rule base, also with the help of a framework. All source code,

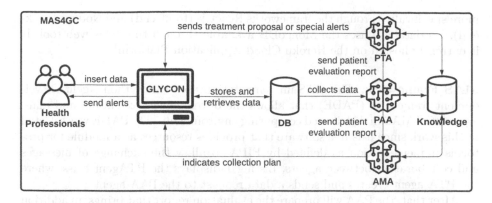

Fig. 3. MAS4GC architecture.

diagrams, dataset and frameworks used are available in the repository: https://github.com/tiagosegato/mas4gc.

Fig. 4. Glycon interface

Interface. The interface system called Glycemic Control On-line (Glycon) [14] is a web system whose main objective is to receive patient data and blood glucose. Health professionals can maintain this data by inserting, updating, and listing. However, the two most important tasks of the system are to record blood glucose levels and the possible applications of glucose or insulin that the patient may have received. Such information is listed and presented using graphics in a kind of patient's dashboard and alerts regarding monitoring (time of next blood collect) and treatment (glucose, keep watching, regular insulin) suggestions are displayed on the interface's initial screen, serving as treatment suggestions to health professionals, as shown in Fig. 4. Glycon was developed in Javascript, more specifically through the frameworks React.js (front-end) and Node.js (back-end). Its database uses the MongoDB available through the Atlas web tool. It is currently hosted on the Heroku Cloud Application Platform[1].

MAS Framework. The MAS implementation used the Python Agent DEvelopment framework (PADE) that allows the development, execution, and management of MAS in distributed computing environments [23]. PADE was selected in this work since it is free software that provides resources as a module for protocols implementation as defined by FIPA to allow the exchange of messages and collaboration between agents. Listing 1 displays the PTAgent class, where the PTA agent creates and sends a data request to the PAA agent.

After that, the PAA will prepare the evaluation report that brings, in addition to patient and blood glucose data, a forecast of the patient's next blood glucose after 4 h. The prediction was calculated using an MLR and LinearRegression(). The functions $Fit(x, y)$ and $predict()$ from the Python library scikit-learn were used [24].

```
class PTAgent(Agent):
    def __init__(self, aid, paa_name):
        super(PTAgent, self).__init__(aid=aid)

        message = ACLMessage(ACLMessage.REQUEST)
        message.set_protocol(ACLMessage.FIPA_REQUEST_PROTOCOL)
        message.add_receiver(AID(name=paa_name))
        message.set_content('Do you have a New Collection?')

        self.comport_request = CompRequest(self,message)
        self.comport_temp = ComportTemporal(self,10.0,message)
        self.behaviours.append(self.comport_request)
        self.behaviours.append(self.comport_temp)
```

Listing 1. The PTAgent class.

Agents' RBR. After the PAA agent generates the assessment report containing the patients' situation, it sends the report to the PTA and AMA agents, who in turn consult the knowledge base coded using Experta. Experta is a Python library that can be used in the development of rule-based systems. A system developed with Experta can pair a set of facts with a set of rules for those facts and perform some actions based on the rules of correspondence [25]. With Experta the specialist's knowledge was transcribed in a knowledge base to be used by the MAS4GC agents. At the moment, the rule base has 48 initial rules that indicate both treatment recommendations and blood glucose collections' frequency. The rules were defined based on the knowledge of an intensive care specialist. In addition, new rules may be manually introduced in the rule base as tests take place in conjunction with specialists.

---

[1] Available at http://glycon.herokuapp.com/.

Listing 2 presents a rule where a hypoS (Severe Hypoglycemia in Table 2) situation is received and indicates that the treatment should be "Glucose: 4 amp–50% IV". Afterward, this information is inserted in the database shared with Glycon and it will appear on the screen of that interface.

```
@Rule(AND( BloodGlucose ( glycemia='hipoS ') ,
    BloodGlucose ( idPatient=MATCH. idPatient )))
    def bg_hipoG ( self , idPatient ):
        tratamento = "Glucose : _4_AMP_−_50%_IV"
        response = connection . collection . update_one (
            {" _id " : idPatient } ,
            {" $set " : {" treatment " : treatment }})
```

Listing 2. Treatment verification rule.

The Experta syntax does not use $if - then$ for condition and action as other RBR tools. The condition applies to the @Rule expression and the result of the action is presented in a specific function in def.

4 Experiments and Results

To evaluate the results of MAS4GC, two experiments were carried out. The first aimed to identify the regression model to be used to predict patient events, as well as its accuracy, and the second to compare MAS treatment suggestions with human physicians.

4.1 Experiment 1

This experiment aims to identify which type of regression is the most appropriate one (SLR or MLR) to predict a patient's next glycemic event, as well as to gauge whether such a model presents satisfactory results. The experiment systematization is illustrated in Fig. 5 and described in the sequence.

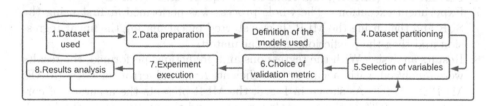

Fig. 5. Experiment 1 and 2 pipeline.

1. Dataset used - A dataset was created by an intensive care physician based on his experience to simulate synthetic data from 50 patients. Each patient has 30 blood glucose collections according to the initial collection plan defined by the physician, assuming patients are not in treatment. The information present in the dataset include patient, gender, age, height, weight, BMI[2], diabetes, time, time of day, food, and blood glucose.
2. Data preparation - Some of the values available in the dataset are categorical and have undergone adjustments. From the patient, a sequential numeric code started at 1 was used, for sex 0 was defined for female and 1 for male. The values 0, 1, and 2 were used to represent those who do not have diabetes, had this information ignored, or have diabetes, respectively. The time field received a scale of values according to the difference in hours in which the collections took place and the time of day field received only the value that represents the hours (discarded minutes, seconds). The rest of the values are numeric and have been maintained, except in the case of numbers with decimal places that have been approximated for their correspondents in integer.
3. Definition of the models used - Simple and multivariate linear regressions were used (SLR and MLR).
4. Dataset partitioning - In the SLR and MLR, the regressions were subjected to cross-validation, whose dataset was divided into 10 groups of 5 patients each, where the accuracy will be measured 10 times between the training and test sets.
5. Selection of variables - In the case of SLR, only the time variables will be used, as the independent variable and blood glucose with the dependent one. In MLR, glycemia is also the dependent variable, as this is the value that is intended to be predicted and all other variables were used as an independent. After some significance tests, the p-value was analyzed and only the variables patient, sex, BMI, diabetes, time, time of day, food (how many hours did the patient receive food), and the last blood glucose were selected for use.
6. Choice of validation metric - The K-fold method was applied by creating 10 subgroups from the current base. The results will be analyzed using the following measures: coefficient R^2, Mean Absolute Error (MAE), Mean Absolute Percent Error (MAPE), and Root Mean Square Error (RMSE).
7. Experiment execution - The whole process was performed at first using the SLR and then using the MLR. It was started by calculating the linear regression and then the predictions were calculated.
8. Results analysis - With the regressions and predictions performed, the results were analyzed with accuracy (coefficient R^2), and error metrics (MAE, MAPE, RMSE). As shown in Fig. 6, the MLR presents the accuracy of correct answers 90% of the times that recommend the treatment to the health professional, against 13% of the SLR. The MLR presents errors inferior compared to SLR. Thus, the MLR is more adequate in the applied context in accordance with the specialist opinion.

---

[2] BMI is a person's weight in kilograms divided by the square of height in meters, it is the adult body mass index.

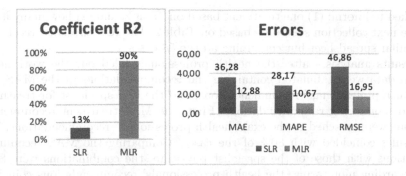

Fig. 6. The Experiment 1 result analysis with the R^2 coefficiente, and MAE, MAPE, RMSE error metrics per prediction of next action.

Experiment 1 was useful to assess the accuracy of blood glucose predictions made by the PAA and consequently enable health professionals to provide preventive treatment for patients. In Experiment 2, the idea is to verify whether the treatment and monitoring recommendations made by the PTA and AMA are similar to real physicians.

4.2 Experiment 2

The objective of Experiment 2 is to compare the treatment and monitoring recommendations made by the MAS4GC with real health professionals. Six health professionals contributed to the experiment. Five professionals were from different specialties, such as endocrinology, neurology, urology, gastroenterology, and physiotherapy. In addition, an IUC intensivist contributed with his knowledge in treating ICU patients. The experiment systematization contains the same eight steps as in Experiment 1. Steps 1, 2, and 5, that is, dataset, data preparation, and selection of variables are the same. The other steps will be presented in the sequence:

- Definition of the models used - the agents' knowledge used to make the recommendations were RBR, while human health professionals used their knowledge based on their studies and experience.
- Dataset partitioning - the $29^t h$ collection from each of the 50 patients were used, both by the intensive care physician, who analyzed the 50 cases and by health professionals from different areas, where each of the five professionals analyzed 10 cases each.
- Choice of validation metric - the MAS responses compared to the health professionals and the specialist ones.
- Experiment execution - five spreadsheets were created containing an explanation of the experiment and the 10 cases. It is understood by case the patient's data that include blood glucose among other information, described in the Selection of variables step. For each case, the healthcare professional was

asked to inform: (1) one treatment based on Table 2; and (2) how many hours the next collection should be based on Table 2. The ICU specialist received a similar spreadsheet but containing all the 50 cases.

– Results analysis - after the health professionals filled out the worksheets, the data were tabulated containing the recommendations of the MAS, the health professionals' recommendations, and the ICU specialist for treatment and monitoring recommendations. First, the MAS treatment recommendations were matched to the exact health professionals' recommendations. The results coincided with 80% of the cases. Comparing the MAS recommendations with those of the specialist physician, the combinations were 84%. Regarding monitoring, the health professionals' recommendations coincided with MAS in 48% of times. Concerning the MAS' recommendations with the specialist, the combinations were 78%. The results are presented in Table 3.

Table 3. Comparison of treatment and monitoring recommendations.

Comparison	Treatment	Monitoring
Health Professionals x MAS	80%	48%
ICU Specialist x MAS	84%	78%

According to the ICU specialist physician, an acceptable hit rate would be close to 80% of the cases. Experiment 2 shows that the MAS recommendations were satisfactory in most cases, except for the health professionals' monitoring recommendation. Considering Table 2, the divergence of the MAS and health professionals' monitoring recommendations was because some professionals use their frequency parameters to measure blood glucose based on their ICU knowledge or hospital rules (e.g., hourly collections or continuous monitoring standards). The ICU specialist confirmed this observation.

Although the presented results are promising, other experiments should better investigate the parameters used by the software agents' decisions, performance, and usability aspects of the MAS4CG.

5 Conclusion

This work presents a system based on agents capable of performing glycemic control in ICU patients. For this, predictive models and a knowledge base with inference rules were used to compose the agents' intellectual capacity. According to the literature, such methods are highlighted as viable solutions in healthcare applications including glycemic control. However, it was not found the combination of these techniques with MAS to solve the problem in question.

Two experiments were conducted to validate MAS4CG. Experiment 1 shows that the prediction model has a 90% accuracy rate of blood glucose for the next

4 h, allowing health professionals to anticipate and perform preventive treatment on their patients. In Experiment 2, the MAS treatment recommendations coincided in 84% of the cases with those of the specialist and 78% about the recommendations for monitoring (collection of blood glucose) of the patients, proving to be a viable solution.

As future work, the rule base might improve the agents' capacity about the recommendation, mainly in terms of glycemic monitoring of patients in different contexts, such as in continuous monitoring environments. The qualitative analysis on the discrepancy in time predicted for next collection is interesting and can be quantified using diagnostic test validation. Considering the results as a whole, the MAS4GC had a good capacity for predictions and recommendations. The promising results indicate that the system can be tested in a real clinical environment.

Acknowledgement. Prof. C. G. Ralha thanks the financial support from the Brazilian National Council for Scientific and Technological Development (CNPq) under grant number 311301/2018-5.

References

1. Vincent, J.L.: Give your patient a fast hug (at least) once a day. Crit. Care Med. **33**(6), 1225–1229 (2005)
2. Noronha, K.V.M.d.S., et al.: The COVID-19 pandemic in Brazil: analysis of supply and demand of hospital and ICU beds and mechanical ventilators under different scenarios. Cadernos de Saúde Pública **36**(6), e00115320 (2020)
3. Socolovithc, R.L., et al.: Epidemiology, outcomes, and the use of intensive care unit resources of critically ill patients diagnosed with COVID-19 in Sao Paulo, Brazil: a cohort study. PLoS ONE **15**(12), e0243269 (2020)
4. Contreras, I., Vehi, J.: Artificial intelligence for diabetes management and decision support: literature review. J. Med. Internet Res. **20**(5), 1–24 (2018)
5. DeJournett, J., DeJournett, L.: Comparative simulation study of glucose control methods designed for use in the ICU setting via a novel controller scoring metric. J. Diab. Sci. Technol. **11**(6), 1207–1217 (2017)
6. DeJournett, L., DeJournett, J.: In silico testing of an artificial-intelligence-based artificial pancreas designed for use in the intensive care unit setting. J. Diab. Sci. Technol. **10**(6), 1360–1371 (2016)
7. DeJournett, J., Nekludov, M., DeJournett, L., Wallin, M.: Performance of a closed-loop glucose control system, comprising a continuous glucose monitoring system and an AI-based controller in swine during severe hypo and hyperglycemic provocations. J. Clin. Monit. Comput. **35**, 1–9 (2020)
8. Darabi, Z., Zarandi, M.H., Solgi, S.S., Turksen, I.: An intelligent multi-agent system architecture for enhancing self-management of type 2 diabetic patients. In: IEEE Conference on Computational Intelligence in Bioinformatics and Computational Biology (CIBCB), pp. 1–8. IEEE, Niagara Falls (2015)
9. Malak, J.S., Safdari, R., Zeraati, H., Nayeri, F.S., Mohammadzadeh, N., Seied Farajollah, S.S.: An agent-based architecture for high-risk neonate management at neonatal intensive care unit. Electron. Phys. **10**(1), 6193–6200 (2018)

10. Jemal, H., Kechaou, Z., Ben Ayed, M.: Multi-agent based intuitionistic fuzzy logic health care decision support system. J. Intell. Fuzzy Syst. **37**(2), 2697–2712 (2019)
11. Vehí, J., Contreras, I., Oviedo, S., Biagi, L., Bertachi, A.: Prediction and prevention of hypoglycaemic events in type-1 diabetic patients using machine learning. Health Informatics J. **26**(1), 703–718 (2020)
12. Bertachi, A., et al.: Prediction of nocturnal hypoglycemia in adults with type 1 diabetes under multiple daily injections using continuous glucose monitoring and physical activity monitor. Sensors (Basel) **20**(6), 1705 (2020)
13. Kim, D.-Y., et al.: Developing an individual glucose prediction model using recurrent neural network. Sensors (Basel) **20**(11), 6460 (2020)
14. Segato, T.H.F., Ralha, C.G., Fernandes, S.E.S.: Development process of multiagent system for glycemic control of intensive care unit patients. Artif. Intell. Res. **10**(1), 43–56 (2021)
15. Anvisa. https://www.gov.br/anvisa/pt-br. Accessed 15 May 2021
16. Chase, J.G., Benyo, B., Desaive, T.: Glycemic control in the intensive care unit: a control systems perspective. Annu. Rev. Control. **48**, 359–368 (2019)
17. Braga, A.A., Fernandes, M.C.C., Madeira, M.P., Júnior, A.A.P.: Associação entre Hiperglicemia e Morbidade em Pacientes Críticos na Unidade de Terapia Intensiva de um Hospital Terciário de Fortaleza - CE. J. Health Biolog. Sci. **3**(30), 132–136 (2015)
18. Russell, S., Norvig, P.: Artificial Intelligence A Modern Approach, 3rd edn. Pearson Education (2010)
19. Barbettam, P.A., Bornia, A.C., Reis, M.M.: Estatística Para Cursos De Engenharia E Informática. 3rd edn. Atlas (2010)
20. Rath, S., Tripathy, A., Tripathy, A.R.: Prediction of new active cases of coronavirus disease (Covid-19) pandemic using multiple linear regression model. Diab. Metab. Syndr. Clin. Res. Rev. **14**(1), 1467–1474 (2020)
21. Zhang, Z.: A mathematical model for predicting glucose levels in critically-ill patients: the PIGnOLI model. Peer J., 1–11 (2016)
22. Pimentel, J., Castro, J.: piStar tool - a pluggable online tool for goal modeling. In: Proceedings IEEE 26th International Requirements Engineering Conference, pp. 1–2 (2018)
23. PADE - Python agent development framework. https://pade.readthedocs.io/en/latest/. Accessed 17 May 2021
24. Scikit-learn homepage. https://scikit-learn.org/stable/. Accessed 14 June 2020
25. Experta. https://experta.readthedocs.io/en/latest/index.html. Accessed 17 May 2021

On the Impact of MDP Design
for Reinforcement Learning Agents
in Resource Management

Renato Luiz de Freitas Cunha$^{(\boxtimes)}$ and Luiz Chaimowicz

Programa de Pós Graduação em Ciência da Computação, Universidade Federal
de Minas Gerais (PPGCC-UFMG), Belo Horizonte, MG, Brazil
{renatoc,chaimo}@dcc.ufmg.br

Abstract. The recent progress in Reinforcement Learning applications
to Resource Management presents Markov Decision Processes (MDPs)
without a deeper analysis of the impacts of design decisions on agent performance. In this paper, we compare and contrast four different MDP
variations, discussing their computational requirements and impacts on
agent performance by means of an empirical analysis. We conclude by
showing that, in our experiments, when using Multi-Layer Perceptrons as
approximation function, a compact state representation allows transfer
of agents between environments, and that transferred agents have good
performance and outperform specialized agents in 80% of the tested scenarios, even without retraining.

Keywords: Reinforcement Learning · Resource management · Markov
Decision Processes

1 Introduction

Deep Reinforcement Learning (DRL) has the potential of finding novel solutions
to complex problems, as outlined by recent progress in diverse areas such as
Control of Gene Regulatory Networks [9], adaptive video acceleration [11], and
management of computational resources [7].

Resource management, the process by which we map computational resources
to the tasks and jobs (programs) that require them, in particular, is an area in
which recent learning approaches have demonstrated superior performance over
classical algorithms and optimization techniques. Still, we see that, in recent
work, each approach defines their own MDP formulations, with different design
decisions. Thus, the literature lacks an analysis of the impact of certain decisions
on agent performance.

In this paper, we investigate what happens to agent performance as we modify an MDP, observing the impacts when we change the state representation,
the transition function, and when we shape the reward signal, performing an
empirical investigation using open-source software from the deep learning and
Reinforcement Learning (RL) communities.

© Springer Nature Switzerland AG 2021
A. Britto and K. Valdivia Delgado (Eds.): BRACIS 2021, LNAI 13073, pp. 79–93, 2021.
https://doi.org/10.1007/978-3-030-91702-9_6

Our main contribution is the formulation and analysis of a set of MDPs derived from one in the literature [7] that allows agents to learn faster, and to perform transfer learning with various function approximation methods. We also show that doing so does not degrade performance in the task.

The rest of this paper is organized as follows: In Sect. 2, we describe the papers that influenced this one, together with other deep RL work for resource management. In Sect. 3, we describe the theoretical background with an ongoing example applied to a resource management problem. In Sect. 4, we describe our methods and the proposed extensions to a base MDP. In Sect. 5, we describe our experimental framework, along with the experiments designed to evaluate our MDPs. In Sect. 6 we present our concluding remarks and a brief discussion of consequences of the work described here.

2 Related Work

In recent years, interest in Deep Reinforcement Learning (RL) applied to executing the scheduling of computing jobs was probably inspired by the Deep Resource Management (DEEPRM) [7] agent and environment. DEEPRM presented an approach of using Policy Gradients to schedule jobs based on CPU and memory requirements. DEEPRM's approach uses images to represent jobs, and a window of jobs from which it can choose which job to schedule next. It was shown that DEEPRM can learn to schedule based on different metrics. Domeniconi et al. [3] proposed CuSH, a system that built on DEEPRM to schedule for CPUs and GPUs, but proposed a hierarchical agent by introducing an additional Convolutional Neural Network (CNN) that chooses which job is going to be scheduled next, and then uses a policy network to choose the scheduling policy to use with the previously selected job. It is important to highlight a major difference between DEEPRM and CuSH: whereas DEEPRM learns the scheduling *policy* itself, CuSH is essentially a classifier, which chooses between two existing policies. This means that, even without training, CuSH's behavior is more stable than that of DEEPRM, since DEEPRM-style schedulers might get stuck in local minima, as reported by de Freitas Cunha and Chaimowicz [2], who investigated the behavior of DEEPRM-style agents when trained with state-of-the-art RL algorithms such as Advantage Actor-Critic (A2C) and Proximal Policy Optimization (PPO), and proposed an OpenAI Gym environment for easier evaluation of RL agents for job scheduling.

Another agent that has been proposed recently, and that learns the scheduling policy itself, is RLSCHEDULER [16]. RLSCHEDULER is a PPO-based agent with a fully convolutional neural network for scoring jobs in a fixed window of size 128. The major innovation in RLSCHEDULER is in the training setting, in which the authors combine synthetic workload traces with real workload traces to present the agent with ever more difficult settings, similar to learning a curriculum of tasks.

Similarly to CuSH, other agents that use a classification approach have been proposed. A recent one is the Deep Reinforcement agent for Scheduling in HPC (DRAS) [4], which classifies jobs in three categories: *ready, reserved,* and *backfilled.* After this classification step, the cluster scheduler takes the output of this classification and allocates jobs accordingly (for example, by reserving slots in the future for reserved jobs, scheduling immediately ready jobs, and finding "holes" in the schedule for backfilled jobs). DRAS uses a five-layer CNN that works in two levels, with the first level selecting jobs for immediate and reserved execution, and the second layer for backfilled execution.

None of the papers mentioned above discuss the impact of their design decisions, resorting to only comparing their results with existing algorithms. In this paper, we aim to analyze how decisions in MDP design impact DEEPRM-style algorithms, and how they impact both computational performance and scheduling performance.

3 Background

In this section, we describe the background needed to understand the techniques and methodology presented in this paper. To help in understanding, we will use a resource management problem as running example throughout this section.

3.1 Batch Job Scheduling

The primary goal of a job scheduler is to manage the job queue and coordinate execution of jobs in High Performance Computing (HPC) clusters, while matching jobs to resources in an efficient way. In a discrete time setting, at each time step, zero or more jobs may arrive in the queue for processing, and the scheduler's job is to allocate jobs to resources while satisfying their resource requirements. The job scheduler guarantees jobs execute when requested resources are available, and usually guarantee there won't be oversubscription of resources[1]. Given this primary goal, secondary goals vary between schedulers and HPC facilities, depending on whether the hosting institution prefers to satisfy the needs of individuals submitting jobs, or the whole group of users [5].

When optimization of response time is a subgoal, it is usually modeled as the minimization of the average response time, with response time used as a synonym to turnaround time: the difference between the time a job was submitted to the time it *completed* execution. A metric commonly used to evaluate this is the *slowdown* of a job, which, for job j is defined as

$$\text{slowdown}(j) = \frac{(t_f(j) - t_s(j))}{t_e(j)} = \frac{t_w(j) + t_e(j)}{t_e(j)} = \frac{1}{t_e(j)} \left(\sum_{i=1}^{t_w(j)} 1 + \sum_{i=1}^{t_e(j)} 1 \right), \quad (1)$$

where $t_s(j)$ is the time job j was submitted, $t_e(j)$ is the time it took to execute job j, and $t_f(j)$ is the finish time of job j. The equality in the middle holds because the wait time, t_w, of a job j is defined as $t_w(j) = t_f(j) - (t_e(j) + t_s(j))$.

---

[1] Some schedulers allow for oversubscription of memory resources in their default configuration, inspired by the fact that jobs don't use peak memory during their complete lifetimes.

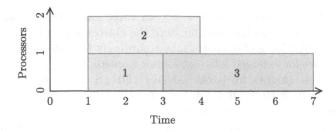

Fig. 1. A possible schedule when three jobs arrive in a scheduling system at discrete time step 1 and no more jobs are submitted to the system at least until time step 7, the last one shown in the figure.

Consider the case of three batch jobs, $j_1 = \square$, $j_2 = \square$, and $j_3 = \square$, submitted to a scheduling system with two processors, and that the three jobs were submitted "between" time step 0 and 1, such that, when transitioning from the first time step to the second, now there are three jobs waiting. Also consider that, for these jobs, the generated schedule is the one displayed in Fig. 1. As shown in the figure, the jobs execute for two, three and four time steps respectively, and all of them use a single processor.

The reader should observe that different schedules can yield substantially different values of (average) slowdown. For example, the schedule shown in Fig. 1 has an average slowdown equal to $1/3 \sum_{i=1}^{3}$ slowdown$(j_i) = 1/3(1 + 1 + 3/2) = 7/6$, whereas, if we swapped j_3 with j_1, and started j_1 soon after j_2 finished, the slowdown would be $1/3(\frac{3+2}{2} + 1 + 1) = 9/6 = 3/2$, a $\approx 29\%$ increase. Therefore, a scheduler should choose job sequences wisely, otherwise its performance can be degraded.

In this paper, we will focus our discussion on what happens when an RL system tries to minimize the average slowdown, but our conclusions are general and apply to other metrics and problems as well.

3.2 Deep Reinforcement Learning and Job Scheduling

In a Reinforcement Learning (RL) problem, an agent interacts with an unknown environment in which it attempts to optimize a reward signal by sequentially observing the environment's state and taking actions according to its perception. For each action, the agent receives a reward. Thus, in the end, we want to find the sequence of actions that maximizes the total reward, as we will detail in the next paragraphs.

RL formalizes the problem as a Markov Decision Process (MDP) represented by a tuple $\mathcal{M} = \langle \mathcal{S}, \mathcal{A}, \mathcal{R}, \mathcal{T}, \rho, \gamma \rangle$[2]. At each discrete time step t the agent is in state $S_t \in \mathcal{S}$. From S_t, the agent takes an action $A_t \in \mathcal{A}$, receives reward

---

[2] Some authors leave the γ component out of the definition of the MDP. Leaving it in the definition yields a more general formulation, since it allows one to model continuous (non-ending) learning settings.

$R_{t+1} \in \mathcal{R}$ and ends up in state $S_{t+1} \in \mathcal{S}$. Therefore, when we assume the first time step is 0, the interaction between agent and environment create a sequence $S_0, A_0, R_1, S_1, A_1, R_2, \ldots$ of states, actions and rewards. To a specific sequence $S_0, A_0, R_1, S_1, A_1, R_2, \ldots$ of states, actions, and rewards we give the name of trajectory, and will denote such sequences by τ. The transition from state S_t to S_{t+1} follows the probability distribution defined by $\mathcal{T} : \mathcal{S} \times \mathcal{A} \rightarrow \mathcal{S}$ or, in an equivalent way, $\mathcal{T}$ gives the probability of reaching any new state s' when taking action a when in state s: $p(s'|s, a) = p(S_{t+1} = s'|S_t = s, A_t = a)$. ρ is a distribution of initial states, and γ is a parameter $0 \leq \gamma \leq 1$, called the discount rate. The discount rate models the present value of future rewards. For example, a reward received k steps in the future is worth only γ^k now. This discount factor is added due to the uncertainty in receiving rewards and is useful for modeling stochastic environments. In such cases, there is no guarantee an anticipated reward will actually be received and the discount rate models this uncertainty.

To map our presentation of RL into our problem of job scheduling, we consider $\rho(\llcorner\lrcorner) = 1$ (the only possible initial state is the empty cluster), with the first state consisting of the empty cluster, with no jobs in the system, $S_0 = \langle\llcorner\lrcorner\rangle$ and $A_0 = \emptyset$, since there is no job to schedule.

Recall our discussion about classifying jobs to be processed by different policies, *versus* choosing the next job to enter the system. In this paper, we are modeling an MDP in which the next job is chosen by the agent, so the agent is learning a scheduling policy. In our example, one can obtain a reward function by using the sequential version of slowdown, shown in the rightmost equality of (1), such that the reward at each time step is given by the sum of the current slowdown for all jobs in the system: $\mathcal{R} = -\sum_{j \in \mathcal{J}} 1/t_e(j)$. When the reward function is such that it computes the online version of slowdown for *all jobs in the system*, if $A_1 = \emptyset$, $R_2 = 1/2 + 1/3 + 1/4$. Moreover, if jobs j_1, j_2, and j_3 are chosen in sequence, the next state, shown in Fig. 1, will be given by sequentially applying the transition function $\mathcal{T}$ as $\mathcal{T}(\llcorner\lrcorner, \square)\mathcal{T}(\llcorner\llcorner, \square)\mathcal{T}(\sqsubseteq, \square)$. If the episode finished immediately after the state shown in Fig. 1, the trajectory τ_1 would be given by $\tau_1 = \langle S_0 = \llcorner\lrcorner, A_0 = \square, R_1 = 0, S_1 = \llcorner\llcorner, A_1 = \square, R_2 = 0, S_2 = \sqsubseteq, \ldots\rangle^3$.

The reward signal encodes all of the agent's goals and purposes, and the agent's sole objective is to find a policy π_θ parameterized by θ that maximizes the expected return

$$G(\tau) = R_1 + \gamma R_2 + \cdots + \gamma^{T-1} R_T = \sum_{t=0}^{T-1} \gamma^t R(S_t, A_t \sim \pi_\theta(S_t)), \qquad (2)$$

which is the sum of discounted rewards encountered by the agent. When T is unbounded, $\gamma < 1$. Otherwise, Eq. (2) would diverge. In our example, a deterministic policy that always chose the smallest job first would yield $\pi_\theta(\langle\llcorner\lrcorner, \square, \square, \square\rangle) = \square$, while a stochastic policy would assign a probability to each job, and either choose the one with highest probability or sample from the

---

[3] The value shown for R_2 might contradict the previous discussion, but the MDP is set in a way that, *when jobs are scheduled successfully*, $R_{t+1} = 0$.

jobs according to that distribution. In practice, when neural networks are used for approximation, the last layer of the neural network is usually a softmax, so that each action gets a number that can be interpreted as a probability. As mentioned before, in the example in this section, rewards are based on the negative online slowdown and, therefore, returns will also depend on the slowdown. The policy π is a mapping from states and actions to a probability of taking an action A when in state S, and the parameters θ relate to the approximation method used by the policy[4]. Popular function approximators include linear combinations of features [6] and neural networks [13,15].

3.3 Policy Gradients

In this section we present the main optimization method we use to find policies: policy gradients. As implied by the name, we compute gradients of policy approximations, and use them to find better parameters for those functions.

Formally, we generalize policies to define distributions over trajectories with

$$\phi_\theta(\tau) = \rho(S_0) \prod_t \pi_\theta(A_t|S_t) \underbrace{\mathcal{T}(S_{t+1}|S_t, A_t)}_{\text{Environment}}, \tag{3}$$

in which π_θ is being optimized by the agent, and ρ and $\mathcal{T}$ are provided by the environment. What (3) says is that we can assign probabilities to any trajectory, since we know the distribution of initial states ρ, and we know that the policy will assign probabilities to actions given states, and that, when such actions are taken, the environment will sample a new state for the agent. When we do so, we can define an optimization objective to find the optimal set of parameters

$$\theta^* = \arg\max_\theta J(\theta) = \arg\max_\theta \int_\tau G(\tau)\phi_\theta(\tau)d\tau, \tag{4}$$

where $J(\theta)$ is the performance measure given by the expected return of a trajectory, which can be approximated by a Monte Carlo estimate $\widehat{J(\theta)} = 1/N \sum_i G(\tau_i)\phi_\theta(\tau_i)$[5]. If we construct $\widehat{J(\theta)}$ such that it is differentiable, we can approximate θ^* by gradient ascent in θ, such that $\theta_{j+1} \leftarrow \theta_j + \alpha\nabla\widehat{J(\theta)}$, with $\alpha > 0$, yielding

$$\nabla_\theta J(\theta) \approx \frac{1}{N} \sum_i G(\tau_i)\nabla_\theta \log \phi_\theta(\tau_i); \tau_i \sim \pi_\theta. \tag{5}$$

By expanding (2) by one time-step, we get the update $G(\tau) = R_1 + \gamma G(\tau_1)$ or, more generally, $G(\tau_t) = R_{t+1} + \gamma G(\tau_{t+1})$, where τ_i indicates the trajectory τ

---

[4] In our example, for each job j_i, in time step 1, π would give the probabilities of choosing each job given an empty cluster: $\pi(\square|\sqcup)$, $\pi(\square|\sqcup)$, and $\pi(\blacksquare|\sqcup)$ such that, by total probability, $\pi(\square|\sqcup) + \pi(\square|\sqcup) + \pi(\blacksquare|\sqcup) = 1$.

[5] Normalization is needed to approximate the average value of $\widehat{J(\theta)}$. Otherwise, $\widehat{J(\theta)} \to \infty$ as $N \to \infty$.

starting from offset i. This update is usually written as $G_t = R_{t+1} + \gamma v_\pi(S_{t+1})$, where G_t is shorthand notation for $G(\tau_t)$, and $v_\pi(S_{t+1})$ is the return when starting at state S_{t+1} and following policy π (which generated trajectory τ). Another function related to $v_\pi(S_t)$ is $q_\pi(S_t, A_t)$, which gives the return when starting at state S_t and taking action A_t then following policy π. With these two functions, we can define a third one, which gives the relative *advantage* of taking action A_t when in state S_t, defined as $a_\pi(S_t, A_t) = q_\pi(S_t, A_t) - v_\pi(S_t)$, and called the advantage function. As with the policy, q_π, v_π, and a_π can also be approximated and, thus, learned. When such an approximation is used, the update (5) becomes

$$\nabla_\theta J(\theta) \approx \frac{1}{N} \sum_i \nabla_\theta \log \pi_\theta(A_t|S_t) \hat{a}_\pi(S_t, A_t); S_t, A_t \sim \pi_\theta, \qquad (6)$$

where $\hat{a}_\pi$ is an approximation of a_π, and which can be further split into two estimators as $\hat{a}_\pi = \hat{q}_\pi - \hat{v}_\pi$, with the arguments S_t and A_t dropped for better readability. In this setting, the π_θ approximator is called an actor, and the $\hat{a}_\pi$ approximator is called a critic.

In the literature, we find techniques that regularize updates [8,12], but as presented, Eq. (6) is sufficient for understanding of the techniques discussed in this paper.

4 Methodology

Although the discussion in the previous section is helpful for conceptualizing the problem we are interested in, it is not enough to help us *implement* a solution, since it does not specify when the agent is invoked for learning, how a state is actually represented, nor how rewards are computed for each action. In this section, we will detail our design decisions, and will elaborate on what changes are required to assess the impact of said decisions in RL performance. We begin by describing the base MDP, and then we will describe incremental changes that can be made to the environment so that it may become faster to compute, and easier to learn, leading to faster convergence.

We implemented the base MDP and each incremental change discussed in this section. Then, we evaluated all implementations, observing both convergence performance and final agent performance in the task of scheduling HPC jobs.

4.1 The Base, Image-Like MDP

We begin by following the design of DeepRM [7], summarized here, and exemplified in Fig. 2. We start by representing states as images whose height corresponds to a look to a time horizon of H time-steps "into" the future, and the width comprising: the number of processors in the system and their occupancy state, a window of configurable size W (in the Figure, $W = 2$) containing the first W jobs in the wait queue times the number of processors in the system, and a column vector indicating jobs in a "backlog"[6]. If there are more jobs in the system that can fit the window and the backlog, they are omitted from the state representation[7].

For the action, the agent can either choose to schedule a job from one of the W slots, or it can refuse to schedule a job, totalling three possible actions in the example of Fig. 2. Regarding *when* actions are taken, the MDP was built in such a way that agents see every simulation time step and "intermediate" time-steps as well: when a job is scheduled, there is a state change in the MDP, with the job moving to the in-use processors, and the queue being re-organized so that all slots in window W are filled.

In the base MDP, whenever a job is scheduled, the agent receives a reward of zero. In all other cases, the reward is given by the negative online slow-

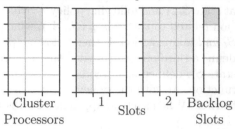

Cluster   1   Slots   2   Backlog
Processors   Slots

Fig. 2. Dense state representation with images representing state. In the figure, there is one job in execution (with two processors for the next two time steps), three waiting jobs in total, two of them within window $W = 2$, one using one CPU for at least five time steps, and another using three CPUs for four time steps. Details for the third job, in the backlog are omitted.

down (1): $\mathcal{R} = -\sum_{j \in \mathcal{J}} \frac{1}{t_e(j)}$. For a detailed description of the environment, we direct the reader to Mao et al. [7], who first introduced it.

4.2 Compact State Representation

The first realization we had was that the state representation in the base MDP is wasteful, in the sense that one can reduce the size of the state without losing information. Particularly when working with larger clusters, or with a larger number of job slots, it may be the case that full trajectories take too much space in memory, reducing the computational performance of learning agents. Due to that, and based on a set of features found in the literature for machine learning with HPC jobs [1], we devised a set of features that can represent states in a compact way. In our new state representation, jobs in the queue are represented by the features shown in Table 1, where "work" is computed by multiplying the number of processors a job requires by the time it is expected to run, and cluster features are a pair that indicates the number of processors in use, and the number of free processors. The features related to the cluster state still use a time horizon H but instead of using a matrix, we used a pair of integers representing how many processors are in use, and how many processors are free in a given time-step. As an example, assuming the job in the cluster was submitted at time 1,

---

[6] Jobs in the wait queue that the agent *cannot* choose to schedule.

[7] Truncating the list of jobs violates the Markov property, since once it overflows, the agent cannot know how many jobs are in the system.

the job in slot 1 was submitted at time 2, and the job in slot 2 was submitted in time 3, the state shown in Fig. 2 can be fully described by the concatenation of vectors with cluster state $\langle (2,1), (2,1), (0,3), (0,3), (0,3) \rangle$, jobs in window W $\langle (1,5,1,0,0,1,6,0), (2,4,3,1,5,1,4,0) \rangle$ and backlog $\langle 1 \rangle$[8]. The features in the jobs slots are presented in the same order as the ones shown in Table 1.

Table 1. Job features in a compact state representation.

Feature	Description
Submission time	Time at which the job was submitted
Requested time	Amount of time requested to execute the job
Requested processors	Number of processors requested at the submission time
Queue size	Number of jobs in the wait queue at job submission time
Queued work	Amount of work that was in the queue at job submission time
Free processors	Amount of free processors when the job was submitted
Remaining work	Amount of work remaining to be executed at job submission time
Backlog	The number of jobs waiting outside window W

A side-effect of using this new compact state representation is that, when H and W are fixed between different cluster configurations, learned features are directly transferable between clusters even when using function approximation methods that depend on a fixed number of features.

4.3 Sparse State Transitions

Another deficiency we've identified in the base MDP is that the agent sees *all* time-steps in the simulation, but this causes the agent to have to take an action even when there is no good action to take. Consider, for example, the case in which all resources are in use (there are no free resources). In cases such as this, any action the agent takes will lead to the same outcome: increasing the simulation clock, receiving negative rewards related to the slowdown of the jobs, and having *no* new jobs scheduled. This will be repeated for all time steps between the start of the last job that exhausted resources until the finish of the first job that frees them, causing non-negative rewards to be more sparse, making the reinforcement signal noisier and, therefore, harder to learn. The opposite is also true: if there are no jobs waiting to be scheduled, no matter what the agent chooses, the outcome will be the same: no jobs will be scheduled.

Due to that, we updated the environment to only call the agent and, therefore, to only add states, actions and rewards to a trajectory, when it was possible for the agent to take an action that could result in a job being scheduled. In short, we

---

[8] Parentheses group elements. In the first vector, there are five parenthesized pairs to indicate the time horizon of 5, and two parenthesized elements to represent job slows in window W.

change the transition function $\mathcal{T}(S_{t+1} \mid S_t, A_t)$ so that all state transitions from S_t to S_{t+1} will always have at least one job that may be scheduled by the agent in state S_{t+1}. We did not change the initial state, though, so $\rho = \{\llcorner\}$ still holds. This essentially turns the MDP into a semi-MDP [14]. To make our formulation compatible with a semi-MDP, we extend the reward function to return zero in all intermediate states after successfully scheduling a job.

4.4 Reducing the Noise of the Reward Signal

Based on the idea of only showing the agent what it can use to learn and act, we noticed that the reward signal could be further improved by, instead of computing the online slowdown of all jobs in the system $\mathcal{J}$, considering only the jobs that are in the waiting queue, and within the job slots window W: the jobs that can be directly influenced by the agent's actions. Therefore, we defined the set $\mathcal{W}$ that contains the subset of jobs from $\mathcal{J}$ that are within the window W, and the reward function became $\mathcal{R} = -\sum_{j \in \mathcal{W}} \frac{1}{t_e(j_i)}$ when the action taken doesn't schedule a job, and 0 otherwise.

We evaluate the impact of the various MDPs on agent performance by performing two sets of experiments, one in which we observe the impact of the changes proposed in Sects. 4.2 through 4.4 (called *Compact*, *Sparse*, and *Reduced* respectively in the experiments) as opposed to the dense MDP, and another in which we observe the impact of using an event-based simulation and bounded rewards both in dense and compact MDPs.

5   Experiments

In order to evaluate our methodology, we used open-source libraries to implement both our agents and environment, with `stable-baselines3` [10] providing the PPO agent and its training loop, and `sched-rl-gym` [2] providing the simulator and environment implementation.

All our experiments consisted of training a PPO agent in the different formulations of the previous section. We also fixed the neural network architecture used for function approximation, consisting of a two-layer neural network with 64 units in each layer, and with parameter sharing between policy and value networks. The fixed number of units implies the image-like representation will use more parameters, as it contains more data than the compact representation. The hyper-parameters used for training the agent are summarized in Table 2. We performed no hyper-parameter optimization, and used values found in the literature when training the image-like agent. For a full description of PPO, we direct the reader to Schulman et al. [12].

Table 2. List of hyper-parameters used when training agents.

Hyper-parameter	Value
Learning rate	10^{-4}
n steps	50
Batch size	64
Entropy coefficient	10^{-2}
GAE λ	0.95
Clipping ϵ	0.2
Surrogate epochs	10
γ	0.99
Value coefficient	0.5

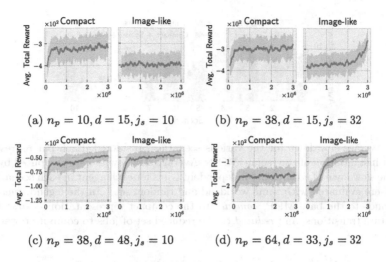

(a) $n_p = 10, d = 15, j_s = 10$ (b) $n_p = 38, d = 15, j_s = 32$

(c) $n_p = 38, d = 48, j_s = 10$ (d) $n_p = 64, d = 33, j_s = 32$

Fig. 3. Learning curves for various scenarios with $H = 20$ contrasting learning using a compact representation with learning with an image-like representation. Curves are an average of six agents, with shaded areas representing one standard deviation, and show a moving average of total rewards received by the agents during training.

We also maintained the environment specification fixed for all agent evaluations and used $W = 10$ job slots, with simulations of length $T = 100$ time-steps and time horizon $H \in \{20, 60\}$. These two horizon values enable us to contrast cases in which agents can see when jobs will complete, or not. Regarding the workload, we used a workload generator from the literature [2,7], which submitted a new job with 30% chance on each time step. Of these, a job had 80% chance of being a "small" job, and "large" otherwise. The number of processors n_p was chosen in the set $\{10, 32, 64\}$, while the maximum job length (duration) d varied from $\{15, 33, 48\}$ and the size of the largest job (number of processors) j_s came from the set $\{10, 32, 64\}$. In the workload generator, the length of small jobs was sampled uniformly from $[1, {}^d/_5]$, and the length of large jobs was sampled uniformly from $[{}^{2d}/_3, d]$. The number of processors used by any job was sampled from $[{}^{n_p}/_2, n_p]$.

All agents were trained for three million time-steps as perceived by the agent. This means that all agents will see the same number of states, and will take the same number of actions, but the number of time steps in the underlying simulation will vary, due to the event-based case becoming a semi-MDP. We evaluated agents with a thousand independent trials, reporting average values.

In Fig. 3 we show a sampling of learning curves comparing the learning performance of agents that were trained using the image-like representation and the compact representation with rewards computed from all jobs. The compact representation converges faster than the image-like representation, probably due to its smaller number of parameters. We also notice that although convergence is faster, the compact representation is not necessarily better (Fig. 3c, 3d). There

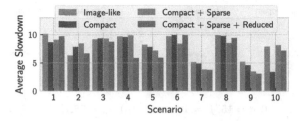

Fig. 4. Average slowdown for the various scenarios considered. Each bar represents a different instantiation of the (semi-)MDPs. Average slowdowns were computed by averaging the slowdown of a thousand independent trials for each agent in each scenario. All agents were evaluated with same workload and random seed. In the legend, *image-like* corresponds to the base MDP, *compact* to the compact representation, *sparse* to the sparse state transitions, and *reduced* to the reduced set of jobs to compute rewards.

doesn't seem to be a general rule, but we noticed that when jobs are shorter (the d parameter is smaller), the compact representation dominates (Fig. 3a, 3b). When d increases and most jobs use few processors ($j_s \ll n_p$), the compact representation tends to have comparable performance with the image-like representation (Fig. 3c), whereas when jobs use many processors *and* have a longer duration, agents using the image-like representation learn the environment better (Fig. 3d). For this set of experiments, the size of the time horizon (H) doesn't impact the learning performance, as curves obtained with $H = 60$ (not shown) are indistinguishable from visual inspection to the ones obtained with $H = 20$. When evaluating agents, we performed t-tests to check whether there was a difference in agent performance when using these different H values. In other words, the null hypothesis was that performance was equal, and the alternative hypothesis was that agent performance varied. In this setting, the null hypothesis was rejected only 36.6% of the time when considering p-values $\leq 1\%$.

When evaluating agents after one million iterations, scheduling performance was similar between agents when the maximum number of processors used by jobs was smaller (which implies less parallelism). Given job submission rates in all environments was the same, clusters were less busy in these situations: as long as jobs are scheduled, there shouldn't be significant differences in average slowdown, due to smaller queues.

In Fig. 4, with key to scenarios shown in Table 3, we show average slowdown of the agents for the scenarios in which there was some variability in per-

Table 3. Key to the scenarios presented in Fig. 4. *Procs.* refers to the number of processors in the cluster, *Max Length* refers to the maximum job length, and *Max Size* refers to the maximum number of processors used by jobs.

Scenario	Procs.	Max length	Max size
1	10	15	10
2	10	48	10
3	38	15	32
4	38	33	32
5	38	48	32
6	64	15	64
7	64	33	32
8	64	33	64
9	64	48	32
10	64	48	64

formance between agents. From the figure, we see that, apart from scenarios 2 and 6, agent performance in the "Compact + Sparse + Reduced" MDP is not worse than that of the image-like MDP. Of these two, only the difference for scenario 6 is statistically significant, with p-value $\leq 5\%$ when performing a t-test with alternative hypothesis of different distributions. For the cases where "Compact + Sparse + Reduced" agents are better, the results are statistically significant (p-value $\leq 5\%$) in scenarios 3, 4, 5, 7, and 9. Scenarios 2 and 6 are interesting, since they were configured to have shorter jobs of at most 15 time-steps, with scenario 1 having 10 processors, and scenario 6 having 64, both with jobs with the potential of using all cluster resources.

In Fig. 5 we contrast the training times for the various agents. As can be seen, training times for agents based on the image-like MDP are highly variable, due to the fact that different MDP configurations result in different sizes of state representations, which impacts training performance. As an example, the image-like agent requires 301068, 1089548, and 1821708 parameters for the scenarios with 10, 38, and 64 processors, while all compact agents require a fixed number of parameters: 24332. Times were measured in a Linux 5.10.42 desktop with an NVIDIA GTX 1070 GPU and an i7–8700K processor using the *performance* CPU frequency-scaling governor.

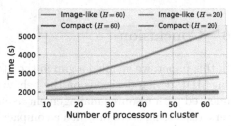

Fig. 5. Time needed to train agents for three million iterations. The shaded area represents one standard deviation. Increasing the time horizon increases the training time of compact agents by a constant factor, while it adds a linear factor to the training times of agents that use an image-like representation.

The compact MDPs proposed in this paper all have the characteristic of having a state representation with a fixed size, which allows for *transfer* of learned weights between MDPs. Here, we consider transfer the ability to change cluster configuration without the need for retraining an agent from scratch, which is simply not possible when using the image-like representation. In Fig. 6, for example, we show the performance of an agent trained in the bounded reward, event-based, compact MDP with 64 processors and with jobs of length 33 (the best agent in Fig. 4, corresponding to scenario 9) evaluated in a compact environment *without* event-based updates. With this same agent, we were able to evaluate its performance in all different scenarios, without the need for retraining. We see that, for the most part, slowdown is kept low, and not only that: this agent outperformed other agents in 80% of scenarios (differences are statistically significant, with p-value $\leq 1\%$, except for scenario 9, since this is the same agent, and scenario 5, where the test has low power to reject the null hypothesis). This highlights the advantage of using a representation that allows for easy transfer between agents, enabling good performance in a variety of cluster settings.

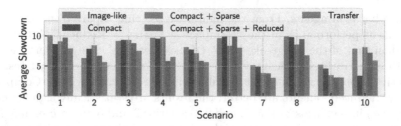

Fig. 6. Bar chart contrasting the performance of a transferred agent to agents trained specifically in their environments.

6   Conclusion

In this paper, we've filled a gap in the literature by analyzing the effects of different MDP design decisions on the behavior of RL agents. In particular, we experimented with resource management agents for job scheduling in computing clusters, discussing cases in which a compact representation outperforms a dense one, and vice-versa. We proposed a new state representation, a transition function, and a reward function for an MDP studied in the literature, and we saw that these environments support transferring agents between different cluster settings, while also keeping agent memory consumption constant, and processing requirements stable. We also saw that these compact representations are no worse than image-like ones, and, thus, might be preferable when constant memory usage is a requirement. Moreover, our results indicate that transferred agents may outperform specialized agents in 80% of the tested scenarios without the need for retraining.

References

1. Cunha, R.L., Rodrigues, E.R., Tizzei, L.P., Netto, M.A.S.: Job placement advisor based on turnaround predictions for HPC hybrid clouds. Future Gener. Comput. Syst. **67**, 35–46 (2017). ISSN 0167–739X
2. de Freitas Cunha, R.L., Chaimowicz, L.: Towards a common environment for learning scheduling algorithms. In: 2020 28th International Symposium on Modeling, Analysis, and Simulation of Computer and Telecommunication Systems (MASCOTS), pp. 1–8 (2020)
3. Domeniconi, G., Lee, E.K., Morari, A.: CuSH: cognitive scheduler for heterogeneous high performance computing system. In: Proceedings of DRL4KDD 19: Workshop on Deep Reinforcement Learning for Knowledge Discovery (DRL4KDD), vol. 12 (2019)
4. Fan, Y., Lan, Z., Childers, T., Rich, P., Allcock, W., Papka, M.E.: Deep reinforcement agent for scheduling in HPC. arXiv preprint arXiv:2102.06243 (2021)
5. Feitelson, D.G., Rudolph, L.: Toward convergence in job schedulers for parallel supercomputers. In: Feitelson, D.G., Rudolph, L. (eds.) JSSPP 1996. LNCS, vol. 1162, pp. 1–26. Springer, Heidelberg (1996). https://doi.org/10.1007/BFb0022284

6. Liang, Y., Machado, M.C., Talvitie, E., Bowling, M.: State of the art control of atari games using shallow reinforcement learning. In AAMAS (2016)
7. Mao, H., Alizadeh, M., Menache, I., Kandula, S.: Resource management with deep reinforcement learning. In: Proceedings of the 15th ACM Workshop on Hot Topics in Networks, pp. 50–56 (2016)
8. Mnih, V., et al.: Asynchronous methods for deep reinforcement learning. In: International Conference on Machine Learning, pp. 1928–1937 (2016)
9. Nishida, C.E.H., Costa, A.H.R., da Costa Bianchi, R.A.: Control of gene regulatory networks basin of attractions with batch reinforcement learning. In: 2018 7th Brazilian Conference on Intelligent Systems (BRACIS), pp. 127–132 (2018)
10. Raffin, A., Hill, A., Ernestus, M., Gleave, A., Kanervisto, A., Dormann, N.: Stable baselines3 (2019). https://github.com/DLR-RM/stable-baselines3
11. Ramos, W., Silva, M., Araujo, E., Marcolino, L.S., Nascimento, E.: Straight to the point: fast-forwarding videos via reinforcement learning using textual data. In: Proceedings of the IEEE/CVF Conference on Computer Vision and Pattern Recognition, pp. 10931–10940 (2020)
12. Schulman, J., Wolski, F., Dhariwal, P., Radford, A., Klimov, O.: Proximal policy optimization algorithms. arXiv preprint arXiv:1707.06347 (2017)
13. Silver, D., et al.: A general reinforcement learning algorithm that masters chess, shogi, and Go through self-play. Science 362(6419), 1140–1144 (2018)
14. Sutton, R.S., Precup, D., Singh, S.: Between MDPs and semi-MDPs: a framework for temporal abstraction in reinforcement learning. Artif. Intell. 112(1), 181–211 (1999). https://doi.org/10.1016/S0004-3702(99)00052-1. ISSN 0004-3702
15. Tesauro, G.: TD-Gammon, a self-teaching backgammon program, achieves master-level play. Neural Comput. 6(2), 215–219 (1994)
16. Zhang, D., Dai, D., He, Y., Bao, F.S., Xie, B.: RLScheduler: an automated HPC batch job scheduler using reinforcement learning. In: SC20: International Conference for High Performance Computing, Networking, Storage and Analysis, pp. 1–15. IEEE (2020)

Slot Sharing Mechanism in Multi-domain Dialogue Systems

Bruno Eidi Nishimoto[1,2](✉)(iD) and Anna Helena Reali Costa[1](iD)

[1] Universidade de São Paulo, São Paulo, Brazil
{bruno.nishimoto,anna.reali}@usp.br
[2] Itaú Unibanco, São Paulo, Brazil

Abstract. Task-oriented dialogue systems are very important due to their wide range of applications. In particular, those tasks that involve multiple domains have gained increasing attention in recent years, as the actual tasks performed by virtual assistants often span multiple domains. In this work, we propose the Divide-and-Conquer Distributed Architecture with Slot Sharing Mechanism (DCDA-S2M) system, which includes not only a distributed system aimed at managing dialogues through the pipeline architecture, but also a slot sharing mechanism that allows the system to obtain information during a conversation in one domain and reuse it in another domain, in order to avoid redundant interactions and make the dialogue more efficient. Results show that the distributed architecture outperforms the centralized one by 21.51% and that the slot sharing mechanism improves the system performance both in the success rate and in the number of turns during the dialogue, demonstrating that it can prevent the agent from requesting redundant information.

Keywords: Dialogue systems · Reinforcement learning · Transfer

1 Introduction

Dialogue systems aim to interact with humans, employing conversations in natural language. They can be broadly divided into three categories: socialbots, question & answering, and task-oriented systems. Socialbots aim to have an entertaining conversation to keep the user engaged, without having a specific goal other than being friendly and keeping company. Question & answering systems aim to provide a concise and straightforward answer to the user's question, possibly using information stored in knowledge bases. Finally, task-oriented systems help users to complete a specific task [6]. These tasks range from simple tasks, such as setting an alarm and making an appointment, to more complex tasks, such as finding a tourist attraction, booking a restaurant or taking a taxi.

Due to their wide range of applications, task-oriented systems have been showing great relevance in recent years, with both academia and industry drawing their attention to them. There are roughly two types of architectures to model a task-oriented agent: end-to-end and pipeline (or modular) [24].

© Springer Nature Switzerland AG 2021
A. Britto and K. Valdivia Delgado (Eds.): BRACIS 2021, LNAI 13073, pp. 94–108, 2021.
https://doi.org/10.1007/978-3-030-91702-9_7

End-to-end approaches consider the system composed of a single component that maps user input in natural language directly to the system output also in natural language (see Fig. 1). On the other hand, pipeline architectures comprise three components: Natural Language Understanding (NLU), Dialogue Management (DM) and Natural Language Generation (NLG) (see Fig. 2). NLU extracts the major information from the user utterance and transforms it in a structured data known as dialogue act. The dialogue act shows relevant information for the dialogue comprehension and it is defined by the tuple *[domain, intent, slot, value]*, encoding the *domain* of this particular act, the *intent*, i.e., its broad objective (*inform*, *request*, or *thanks*, for instance), and the *slot* and *value* representing specific pieces of information in this domain. The example illustrated in Fig. 2 shows that utterance *I want a restaurant located in the centre of town* would have the following dialogue act: *[restaurant, inform, area, centre]*. DM contains two sub-modules, the Dialogue State Tracking (DST) which keeps track of the dialogue state and the Policy (POL) which decides the best response to give to the user. Finally, the NLG component transforms the DM output, which is also a dialogue act, into natural language to present to the user.

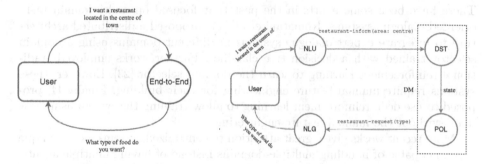

Fig. 1. Illustration of end-to-end architecture.

Fig. 2. Illustration of pipeline architecture.

Early works on task-oriented dialogue systems [9,12,13] focus on problems with a single domain. However, it is common that real systems such as the popular virtual assistants span their applications over multiple domains. This means that a single system must be able to act and complete the user task in more than one domain, for example, the user first asks to find a tourist attraction for sightseeing and then, in the same conversation, she requests a restaurant to have lunch. Dealing with multi-domain dialogue systems is a problem much harder since the complexity of user goals and conversations increases a lot. Besides the complexity of the problem, which is already challenging, another issue faced by multi-domain systems is related to the redundancy of information given by the user during a conversation [15]. For instance, suppose that the user first reserves a restaurant table for two people and then she requests to reserve a hotel room for that night. It is usual that a single-domain dialogue system asks again the number of people although this kind of information—that can be shared among the domains—was already given by the user. These redundant turns make the conversation longer and are not nice for the user experience [14].

In this paper, we use the pipeline architecture with a focus on the DM component and adopt the divide-and-conquer approach, that is, several agents are trained independently, each one in a specific domain, and aggregate them to act in the multi-domain scenario. We show that this approach leads to better results when compared to the same algorithm trained in all domains at once. Furthermore, we propose the use of a mechanism that enables the system to reuse some shareable slots among the domains, avoiding the agent to ask for redundant information.

The remainder of this paper is organized as follows: Sect. 2 shows some related work, Sect. 3 describes our methods and proposal, Sect. 4 shows our experiments and results obtained, and finally Sect. 5 highlights our conclusions and directions for further work.

2 Related Work

There have been some efforts in the past that focused on multi-domain task-oriented dialogue systems. Komatani et al. [11] proposed a distributed architecture to integrate expert dialogue systems in different domains using a domain selector trained with a decision tree classifier. Further works employed traditional reinforcement learning to learn the domain selector [23]. However, these systems require manual feature engineering for their building. Finally [4] proposed to use deep reinforcement learning to allow training the system using raw data, without the manual feature engineering.

More recent works give a great attention to centralized systems, i.e., a unique system capable of handling multiple-domains instead of having multiple agents, each one specialized for each domain. One reason for this is the increase in the power processing in modern computers. Traditional works using the end-to-end architectures rely on the use of recurrent neural networks (RNN) with a sequence-to-sequence approach [2,19]. Recurrent Embedding Dialogue Policy (REDP) [21] learns vector embeddings for dialogue acts and states showing it can adapt better to new domains than the usual RNNs. Vlasov et al. [22] proposed to use a transformer architecture [20] with the self-attention mechanism operating over the sequence of dialogue turns outperforming previous models based on RNNs.

Another line of research focus on the DM module using the pipeline architecture. There are some attempts to use supervised learning to learn a policy for the DM, but as it can be seen as a sequential decision making problem, RL is more used [5]. However, RL algorithms are too slow in general when trained from scratch. Many works attempt to include some expert knowledge either by supervised pre-training or by warm-up, i.e., pre-filling the replay buffer with rule agents in DQN algorithms. DQfD (Deep Q-learning from Demonstrations) uses the expert demonstrations to guide the learning and encourages the agent to explore in high reward areas, avoiding random exploration [7,8]. Redundancy with respect to the overlapping slots between domains is another issue in dialogue systems. Chen et al. [3] address this problem by implementing the policy with a graph neural network where the nodes can communicate to each other to share information but they assume the adjacency matrix for this communication is known.

Although recent work on multi-domain settings does not consider a distributed architecture, it is relevant for two reasons: it can reuse well-established algorithms for a single domain and, in the need to add a new domain to the system – which is common for real applications such as virtual assistants – it is unnecessary to retrain the entire system. Therefore, we adopted a distributed architecture using the divide-and-conquer approach. Furthermore, to the best of our knowledge, it is the first work that focuses on learning this slot-sharing mechanism.

3 Proposal

In this work we propose the Divide-and-Conquer Distributed Architecture with Slot Sharing Mechanism (DCDA-S2M) showed in Fig. 3, which uses the pipeline architecture (Fig. 2) primarily focusing on the DM component.

The dialogue state must encode all useful information collected during interactions. The MultiWOZ[1] annotated states basically comprise the slots informed, and the ones that are required to complete the task in each domain. For example, the hotel domain requires, among others, the area of the city where the hotel is located, the number of stars it has been rated, as well as its price range.

Our proposal to train POL is divided into two steps. The first step is to use a distributed architecture in a divide-and-conquer approach to build a system capable of interacting in a multi-domain environment. The second stage consists of a mechanism that shares slots between domains. In the following sections we detail each component of DCDA-S2M, the Divide-and-Conquer Distributed Architecture (DCDA) and the Slot Sharing Mechanism (S2M).

3.1 Divide-and-Conquer Distributed Architecture

We implemented seven agents, each one specifically trained for a domain of the MultiWOZ dataset (attraction, hospital, hotel, police, restaurant, taxi, and train), as illustrated in Fig. 3. The idea is that, using a simple reinforcement learning algorithm, we can have a multi-domain system with better performance than if we had a single agent trained in all domains at once.

However, just having multiple agents, each one for each domain, is not enough. We need a controller capable of perceiving when there is a domain change and selecting the right agent to collect the right response. The controller

---

[1] MultiWOZ is a fully-labeled collection of human-human written conversations spanning over multiple domains and topics – https://github.com/budzianowski/multiwoz.

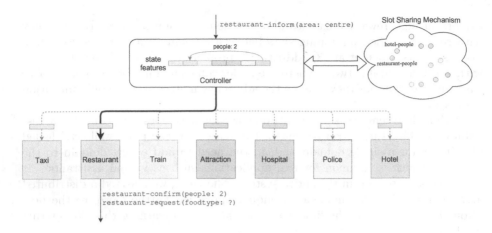

Fig. 3. Illustrative figure of our proposed architecture.

keeps track of all state features from all agents allowing it to know the past and current domains.

As mentioned before, in multi-domain systems the dialogue act is defined by the tuple *[domain, intent, slot, value]*, i.e., it already includes the *domain* of the conversation [10]. Therefore, the controller simply observes the *domain* element of the dialogue act. If it is a new domain (different from the current one) it checks with S2M (detailed in Sect. 3.2) the slots from past domains that can share values with the new domain, and it copies the values of these shareable slots. Given this, the controller sends the state features to the corresponding agent of the current domain. Each agent is trained using reinforcement learning to learn an optimal policy (POL component of DM) in each domain.

Reinforcement Learning. In reinforcement learning problems there is an agent that learns by interacting with an environment through the Markov Decision Process (MDP) framework defined as a tuple $(\mathcal{S}, \mathcal{A}, \mathcal{R}, \mathcal{T}, \gamma)$, where $\mathcal{S}$ is the set of possible states, $\mathcal{A}$ is the set of actions, $\mathcal{R}$ is the reward function, $\mathcal{T}$ is the state transition function, and $\gamma \in [0, 1]$ is the discount factor that balances the trade-off between immediate rewards and future rewards. In the context of dialogue systems, the state represents the dialogue state. The actions are the set of dialogue acts of the agent, and the reward is $+R$ if the dialogue succeeds, i.e., the agent achieve the user goal, $-R$ if the dialogue fails and -1 for each turn to encourage the agent lasts the minimum number of turns.

The agent's objective is to maximize the cumulative discounted rewards, $\sum_{t=0}^{T} \gamma^t r_t$, $r_t \in \mathcal{R}$, received during interactions in order to find its optimal policy that maps state $s_t \in \mathcal{S}$ to the best action $\pi(s_t) = a_t$, $a_t \in \mathcal{A}$.

In this work the Proximal Policy Optimization (PPO) algorithm [18] was employed which is a policy gradient method which aim to directly optimize the parameterized policy to maximize the expected reward. The intuition of PPO is to make the greatest improvement in policy without stepping so far from current policy to avoid a performance collapse. Formally, it optimizes the loss function, $\mathcal{L}(\theta) = \min\left(\rho(\theta)\hat{A}, \ clip(\rho(\theta), 1 - \epsilon, 1 + \epsilon)\hat{A}\right)$, where $\rho(\theta) = \frac{\pi_\theta(a|s)}{\pi_{\theta_{old}}(a|s)}$ denotes the probability ratio, $\hat{A}$ is the estimated advantage function and ϵ is a hyperparameter that indicates how far can we go from old policy.

For the advantage function estimation, we use the Generalized Advantage Estimation (GAE) [17], $\hat{A}(s_t, a_t) = \delta(s_t) + \gamma\lambda\hat{A}(s_{t+1}, a_{t+1})$, with $\delta(s_t) = r_t + \gamma\hat{V}(s_{t+1}) - \hat{V}(s_t)$, where γ is the discount factor, λ is a hyperparameter to adjust the bias-variance tradeoff and $\hat{V}(s)$ is the estimate of value function, i.e., the expected reward the agent receives being at state s.

However reinforcement learning algorithms often suffer in sparse reward environments. One possible option to deal with this issue is to use some pre-training method, such as imitation learning, where the agent tries to clone expert behaviors. We use the Vanilla Maximum Likelihood Estimation (VMLE) algorithm [25] to pre-train the agents. VMLE employs a multiclass classification with data extracted from the MultiWOZ dataset. The algorithm optimizes its policy trying to mimic the behavior presented by the agent in the dataset.

User Simulator. Furthermore, training the agent with real users is impracticable since it requires a great number of interactions. Therefore there is a need of an user simulator. The used user simulator follows an agenda-based approach [16]. First, it generates a user goal which comprises all information needed to complete the task. Then, the user simulator generates an agenda in a stack-like structure with all actions it needs to take (informing its constraints and/or requesting information). During the conversation, as the agent requests or informs something, the user can reschedule the agenda accordingly. For example, if the agent requests the type of restaurant, than the user can move the action "inform the type of restaurant" to the top of the stack. The conversation lasts until the stack is empty or the maximum number of turns is reached.

We made a small modification to the user simulator to handle the confirmation actions provided by the agent using the slot sharing mechanism. The simulator checks if the values of the confirmation act are correct and if they are wrong it informs the correct value; otherwise, it just removes the action regarding that slot-value pair from the agenda and continues with its policy.

Finally, after all the seven agents are trained using reinforcement learning we plug them with the controller resulting in the DCDA system.

3.2   Slot Sharing Mechanism

Some domains contain overlapping slots, i.e., slots that can share their values in a conversation. For instance, if the user is looking for a restaurant and a hotel, it is likely that they are for the same day and in the same price range. More complex relationships can be represented, such as the time of reservation for a restaurant and the time when the taxi must arrive at its destination (which would be the restaurant). However, not all slots are shareable between two domains.

Therefore, we need to know which slots whose contents can be transferred from one domain to another.

In the following subsections we show how we learned these relationships. But for now, suppose we already know what such slots are. Therefore, during the conversation, when the controller notices a domain change, the information-sharing mechanism first gets all informed slots in previous domains and then checks whether any of those slots can share their values with any slots in the current domain. In this case, the controller copies the value from that slot to the slot of the new domain. For example, suppose the system has already interacted with the user in the hotel domain and knows that the demand is for two people. When the conversation changes to the restaurant domain, the slot sharing mechanism will see that *hotel-people* slot can be shared with *restaurant-people* slot and the controller will transfer this information (two people) to the restaurant domain. The agent then asks for confirmation and acts considering this transferred slot. This can speed up the dialogue and improve the user experience during interactions by avoiding asking redundant information. In the worst case, if the transferred value is wrong, the user informs the correct value and continues the interaction normally.

Learning Shareable Slots. Our proposal to learn which slots can be shared, named Node Embedding for Slots Sharing Mechanism, uses the node embedding technique in which each node represents a *domain-slot* pair. The similarity of two nodes indicates whether they can share the same value in a conversation. It is defined by a simple scalar product, that is, given nodes $u = [u_1, u_2, \ldots, u_d]$ and $v = [v_1, v_2, \ldots, v_d]$, we have: $similarity(u, v) = \langle u, v \rangle = \sum_{i=1}^{d} u_i \cdot v_i$, where d is the embedding dimension. For instance, the nodes *restaurant-day* and *hotel-day* must be similar, i.e., have a high scalar product, while *restaurant-name* and *hotel-name* must have a low scalar product.

Before learning the node embedding, we need to build a similarity matrix $A \in \mathbb{R}^{n \times n}$, a $n \times n$ matrix where n is the number of nodes, i.e., number of *domain-slot* pairs and the cell A_{uv} represents the similarity between nodes u and v normalized to fall between 0 and 1, that is, $similarity(u, v) \in [0, 1]$. This is done using the dialogues from dataset $\mathcal{D}$ as shown in Algorithm 1. At the end of each conversation, we observe the final state (the state of the dialogue in the last interaction) and check if each node pair presents the same value (a node represents a *domain-slot* pair). If they do, the weight between these two nodes is increased by one. In the end, all weights are normalized to the number of times each pair of nodes appeared in the dialogues.

However, keeping a matrix with $\mathcal{O}(n^2)$ of space complexity does not scale with the number of domains and slots. For this reason, we trained a node embedding representation for each *domain-slot* pair. The learning of node embedding uses the similarity matrix $A \in \mathbb{R}^{n \times n}$ and follows Algorithm 2 proposed by Ahmed et al. [1]. In each step, for each node pair $(u, v) \in E$, where E is the set containing all node pairs, it performs an update to minimize the following error $L(A, Z, \lambda) = \frac{1}{2} \sum_{(u,v) \in E} (A_{uv} - \langle Z_u, Z_v \rangle)^2 + \frac{\lambda}{2} \sum_u \|Z_u\|^2$, where Z represents the embedding

---

Algorithm 1. Similarity Matrix

Require: Dataset $\mathcal{D}$
1: Initialize similarity matrix $A \in \mathbb{R}^{n \times n}$ with zeros
2: Initialize pair-counter matrix $C \in \mathbb{R}^{n \times n}$ with zeros
3: **for all** dialogue $d \in \mathcal{D}$ **do**
4:      Set $goal \leftarrow$ final state of dialogue d
5:      **for all** $(u, v) \in goal$, where u and v form a *domain-slot* pair **do**
6:          **if** $value(u) = value(v)$ **then**
7:              $A_{uv} \leftarrow A_{uv} + 1$
8:          **end if**
9:          $C_{uv} \leftarrow C_{uv} + 1$
10:     **end for**
11: **end for**
12: $A \leftarrow A/C$ ▷ normalize similarity matrix
13: **return** A

---

space and Z_u is the vector for node u. The t in Algorithm 2 can be thought as a learning rate for each update.

This S2M is independent of the agent trained, so it can be plugged with any agent we want. Gathering the DCDA and S2M we get the DCDA-S2M system (Fig. 3), where the controller asks for the S2M for slots that can share their values when there is a domain shift and then send the state features to the respective agent to get the response to the user.

4   Experiments and Results

To evaluate our proposal we did three experiments: the first was to train the embedding of nodes and to do a qualitative analysis of the learned embedding. We then evaluate the divide-and-conquer approach with the information-sharing mechanism, and finally we evaluate the same approach without the mechanism.

Algorithm 2. Node Embedding

Require: Matrix $A \subset \mathbb{R}^{n \times n}$, embedding dimension d, regularization factor λ, set of all node pairs E
1: Initialize $Z \in \mathbb{R}^{n \times d}$ at random
2: $t \leftarrow 1$
3: **repeat**
4:      **for all** $(u, v) \in E$ **do**
5:          $\eta \leftarrow \frac{1}{\sqrt{t}}$
6:          $t \leftarrow t + 1$
7:          $Z_u \leftarrow Z_u + \eta \left[(A_{uv} - \langle Z_u, Z_v \rangle) Z_v \right] + \lambda Z_u$
8:      **end for**
9: **until** no more epochs
10: **return** Z

4.1 Experimental Setup

We used the ConvLab-2[2] platform to run our experiments. It provides a platform with a user simulator and some implementations of all dialogue systems components (NLU, DST, POL and NLG). In this way, it is easy to assess new algorithms for each of the components. We trained the agents for each domain using the PPO algorithm with the standard parameters of the ConvLab-2 platform: discount factor $\gamma = 0.99$, clipping factor $\epsilon = 0.2$ and $\lambda = 0.95$ (for advantage value estimation). The training lasted 200 epochs and in each epoch we collected around 100 turns and sampled a batch of size 32 for optimization. The agent contains two separate networks, one for policy estimation with two hidden layers of size 100 and other for value estimation with two hidden layers of size 50. The optmizers used for policy and value networks are RMSProp and Adam and learning rate $lr_p = 10^{-4}$ and $lr_v = 5 \cdot 10^{-5}$, respectively. The reward function is -1 for each turn (to encourage the agent complete the task more quickly), 40 for success dialogue and 20 for fail dialogue. For pre-training we employed the VMLE algorithm using RMSProp optimizer with $lr_{vmle} = 10^{-3}$ and binary cross entropy with logits as loss function. We also used the available PPO model trained with all domains at once to compare with our results.

For our node embedding, we built the adjacency matrix using the dialogue corpus available at ConvLab-2 platform. For hyperparameters, we used an embedding dimension $d = 50$, regularization factor $\lambda = 0.3$, and 1000 epochs for training.

4.2 Node Embedding

For visualization of the learned node embedding we used a t-distributed stochastic neighbor embedding (t-SNE) model with perplexity 5 using the scalar product as similarity function. Figure 4 shows this visualization.

Figure 4 clearly shows some groups of nodes that are related to each other. For example, *hotel-area*, *attraction-area*, and *restaurant-area* forms a group, indicating that users generally request places in the same area. It also happens for the price range (*restaurant-pricerange* and *hotel-pricerange*), day (*restaurant-day*, *hotel-day*, and *train-day*) and people (*restaurant-people*, *hotel-people*, and *people*) slots. Although *hotel-stay* and *hotel-stars* looks close to the group with slot "people", computing their similarity with *restaurant-people* we got 0.118 and 0.088, respectively. Thus they are not similar and should not share values. On the other hand, the similarity between *restaurant-area* and *attraction-area* is 0.91 showing that they are similar and must share their values inside a conversation. Here we used a similarity of 0.8 as a threshold for sharing the slots values.

An interesting observation is that *attraction-name* and *hotel-name* are quite close to *taxi-departure*, with similarity 0.57 and 0.668, respectively, but they are not close to each other, i.e., the similarity between them is 0.011. This is expected since it is not common an attraction with the same name as a hotel.

---

[2] https://github.com/thu-coai/ConvLab-2.

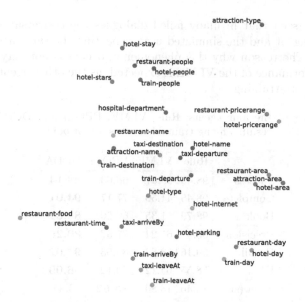

Fig. 4. Visual representation of the learned node embedding.

4.3   DCDA Evaluation

To evaluate our proposal we assessed four models: the baseline Rule-based policy available in ConvLab-2, the VLME policy (obtained from the VLME algorithm), and the PPO algorithm trained in both approaches: a centralized system with a unique agent trained to handle all domains at once (PPO_{all}) and our proposal (DCDA). We also evaluated the effects of using or not the S2M in the rule and DCDA agents. The metrics are automatically computed by the evaluator presented in ConvLab-2 and encompasses the complete rate, success rate, book rate, precision, recall, F1-score for the informed slots, and average number of turns for both the dialogues that were successful and the total set of dialogues. The complete rate indicates the rate of dialogues that could finish (either with success or fail) before achieving the maximum number of turns. The precision, recall and F1-score indicate the ability of the agent to fulfill the slots of the user goal, i.e., leads to the correct slot. Tests were performed over 2000 dialogues.

Table 1 shows the evaluation results for all the four models trained in the pipeline setting, i.e., without the NLU and NLG modules. As expected the rule policy performs almost "perfectly" succeeding in 98.45% of the dialogues and it can serve as a baseline. Among the trainable agents, DCDA performs better in almost all aspects achieving 88.14% of success rate, 94.01% of completion rate, and 88.01% of book rate. It beats the PPO_{all} by almost 21.51% in the success rate, showing a much better performance and efficiency as it can solve user tasks using less number of turns. The average number of turns in all dialogues, 14.92, is very close to the baseline rule policy (13.48) showing it could learn a very good policy in solving tasks. The large increase in the average number of turns for all dialogues can be explained by analysing the failed dialogues during test. It can be seen that in many failed dialogues the conversation went in a loop with the agent and the simulated user, repeating the same act of dialogue consecutively. The reason why this phenomenon occurs is not very clear to us. The worst performance of the VLME is expected, as the other agents depend on the VLME for pre-training.

Table 1. Results of the four agents: Rule, VLME, PPO_{all} and DCDA tested in a pipeline setting. Best results among trainable agents are in bold.

	Rule	VLME	PPO_{all}	DCDA
Success	98.45	39.57	66.63	**88.14**
Complete	98.45	41.50	77.17	**94.01**
Book	98.79	1.35	60.20	**88.01**
Precision	83.47	65.24	77.57	**80.26**
Recall	99.16	68.82	86.53	**97.02**
F1	88.55	64.12	79.12	**86.00**
Turn (suc)	13.40	13.80	**13.62**	13.84
Turns (all)	13.48	22.39	19.82	**14.92**

The results of the second experiment regarding the use of the slot sharing mechanism are presented in Table 2. We evaluated both the Rule policy and our proposed model DCDA. Results show that for the Rule policy the sharing mechanism also helped the agent to have a slightly better performance. Although the success rate for DCDA did not change much, the sharing mechanism also helped it to have a better complete and book rate. Another enhancement was in the average number of turns. The average number of turns required in successful dialogues for the Rule and DCDA policies decreased from 13.40 to 13.20 and from 13.84 to 13.43, respectively, when the sharing mechanism was incorporated. Thus we can see that the sharing mechanism makes the agent to complete dialogues faster than without this mechanism for both agents.

Table 2. Evaluation of the use of S2M in the Rule policy and DCDA with the goal generator generating random goals. Best results are in bold.

	Rule		DCDA	
	With S2M	No S2M	With S2M	No S2M
Success	**98.60**	98.45	**88.24**	88.14
Complete	**98.65**	98.45	**95.94**	94.01
Book	**99.25**	98.79	**90.50**	88.01
Precision	83.32	**83.46**	**80.31**	80.26
Recall	99.14	**99.16**	**97.04**	97.02
F1	88.46	**88.55**	**86.02**	85.99
Turn (suc)	**13.20**	13.40	**13.43**	13.84
Turns (all)	**13.23**	13.48	**14.77**	14.92

An interesting fact is that besides the slightly better performance with the sharing mechanism, the precision, recall and F1-score did not followed the same behavior, i.e., they had better results or very close (less than 0.05%) results as those without the sharing mechanism. This result is not very surprising because as the agent with the sharing mechanism tries to "guess" the slots of new domains within the conversation, it ends up reporting more wrong slots of the user goal causing worse precision, recall, and F1-score.

All theses experiments was assessed with the user simulator generating random goals based on a distribution of the goal model extracted from the dataset. So this can include simple goals within only one domain and/or goals that span to more than one domain but do not have any slot with the same value. Indeed, among all 2000 goals generated during testing, only about 400 contain common values between slots. With that in mind, we ran another test of the sharing mechanism that restricts the user simulator to only generating goals that contain common slots. Therefore, the generated goals end up being more complex in general than those generated in the first test.

Table 3 shows the results. There is an expected significant decrease in the general performance due to the increase in user goals complexity. However, here we can clearly observe the great advantage of the sharing mechanism in this setting.

Table 3. Evaluation of the use of S2M in the Rule policy and DCDA with the goal generator generating slots with common values. Best results are in bold.

	Rule		DCDA	
	With S2M	No S2M	With S2M	No S2M
Success	**92.60**	80.35	**78.41**	68.42
Complete	**92.50**	80.45	**92.13**	85.49
Book	**96.11**	95.88	**83.58**	83.10
Precision	**79.89**	79.34	**76.82**	74.88
Recall	**95.43**	86.80	**95.15**	90.01
F1	**83.99**	77.38	**82.92**	78.77
Turn (suc)	**17.15**	18.24	**19.97**	20.69
Turns (all)	**17.10**	17.60	**21.28**	21.57

There is a 12.25% and 9.99% success rate difference with the Rule and DCDA policies, respectively. We also see a bigger impact on the average number of turns. It affects mostly the successful dialogues because the number of turns is affected only when the transferred slots values are correct – otherwise the user would still need to inform these slots – and chances of a successful dialogue increase when it happens. Finally, we also see a better precision, recall and F1-score for the agent with the sharing mechanism. Since all goals in this tests have at least one common value among the slots, the agent "guesses" are more likely to be correct.

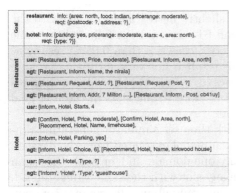 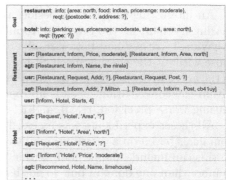

Fig. 5. Example of dialogue using the slot sharing mechanism, resulting in a dialogue length 8.

Fig. 6. Example of dialogue that does not use slot sharing mechanism, resulting in a dialogue length 11.

Figures 5 and 6 show examples of system generated dialogues using and not using the slot sharing mechanism, respectively. Observe that when the domain switched to the hotel domain, the agent in Fig. 5 asked for confirmation if the price is moderate and area is north and recommended a hotel with these constraints. In natural language we could think in this dialogue act as: "Do you want a hotel in north with a moderate price, right? There is the hotel Limehouse". In this way, the user did not need to inform these slots again, saving some turns until task completion. While in Fig. 6 the agent needed to ask again the area and price for the user, resulting in a redundant dialogue which takes more turns to be completed (11 turns against 8 turns).

One drawback for the DCDA-S2M is the training time required for training all the agents. Table 4 shows the average training time for each agent. As we can see, the total amount of time required to train all seven agents is 291.29 minutes, which is approximately 16% more than centralized system training. However, it is worth noting that agents could be trained in parallel, which would require greater computational power.

Table 4. Training time in minutes for each agent.

	All	Attraction	Hospital	Hotel	Police	Restaurant	Taxi	Train
Training time (min)	188.97	32.01	28.87	29.97	26.85	32.01	29.77	111.81

5   Conclusions

In this work we show that the use of a distributed architecture, with multiple agents trained separately for each domain, can leverage the system performance compared to the same algorithm used to train a single agent for all domains at once. This is because each agent can specialize in solving its own problem well, which is much simpler than solving tasks well in all domains, as with the centralized approach in a single agent. Furthermore, distributed systems can add new domains without the need to retrain the entire system.

The use of the slot sharing mechanism also proved to enhance system performance, especially for tasks where the goal has some common slot across domains. Besides improving the system's success rate, it also decreases the average number of turns, showing that the system avoided asking for redundant information.

A major disadvantage of DCDA-S2M is the need to train several agents separately and this can be time and energy consuming. In this sense, for future work we intend to explore transfer learning techniques in reinforcement learning to accelerate the training of new agents.

Acknowledgments. The authors acknowledge the support from the Conselho Nacional de Desenvolvimento Científico e Tecnológico (CNPq grant 310085/2020-9), the Coordenação de Aperfeiçoamento de Pessoal de Nível Superior (CAPES Finance Code 001), and from Itaú Unibanco, through the scholarship program of *Programa de Bolsas Itaú* (PBI).

References

1. Ahmed, A., Shervashidze, N., Narayanamurthy, S., Josifovski, V., Smola, A.J.: Distributed large-scale natural graph factorization. In: Proceedings of the 22nd International Conference on World Wide Web, pp. 37–48 (2013)
2. Bordes, A., Weston, J.: Learning end-to-end goal-oriented dialog. In: 2017 International Conference on Learning Representations (ICLR) (2017)
3. Chen, L., Chen, Z., Tan, B., Long, S., Gasic, M., Yu, K.: AgentGraph: towards universal dialogue management with structured deep reinforcement learning. ArXiv:abs/1905.11259 (2019)
4. Cuayáhuitl, H., Yu, S., Williamson, A., Carse, J.: Scaling up deep reinforcement learning for multi-domain dialogue systems. In: 2017 International Joint Conference on Neural Networks (IJCNN), pp. 3339–3346 (2017)
5. Dai, Y., Yu, H., Jiang, Y., Tang, C., Li, Y., Sun, J.: A survey on dialog management: recent advances and challenges. ArXiv:abs/2005.02233 (2020)
6. Gao, J., Galley, M., Li, L.: Neural approaches to conversational AI. Found. Trends® Inf. Retr. **13**, 127–298 (2019)
7. Gordon-Hall, G., Gorinski, P.J., Cohen, S.B.: Learning dialog policies from weak demonstrations. In: Proceedings of the 58th Annual Meeting of the Association for Computational Linguistics, pp. 1394–1405 (July 2020)
8. Gordon-Hall, G., Gorinski, P.J., Lampouras, G., Iacobacci, I.: Show us the way: learning to manage dialog from demonstrations. ArXiv:abs/2004.08114 (2020)
9. Ilievski, V., Musat, C., Hossmann, A., Baeriswyl, M.: Goal-oriented chatbot dialog management bootstrapping with transfer learning. In: Proceedings of the Twenty-Seventh International Joint Conference on Artificial Intelligence (2018)
10. Kim, Y., Lee, S., Stratos, K.: ONENET: joint domain, intent, slot prediction for spoken language understanding. In: 2017 IEEE Automatic Speech Recognition and Understanding Workshop (ASRU), pp. 547–553 (2017)

11. Komatani, K., et al.: Multi-domain spoken dialogue system with extensibility and robustness against speech recognition errors. In: Proceedings of the 7th SIGdial Workshop on Discourse and Dialogue, pp. 9–17 (July 2006)
12. Li, X., Chen, Y.N., Li, L.: End-to-end task-completion neural dialogue system. In: Proceedings of the The 8th International Joint Conference on Natural Language Processing, pp. 733–743 (2017)
13. Peng, B., Li, X., Gao, J., Liu, J., Wong, K.F., Su, S.Y.: Deep Dyna-Q: integrating planning for task-completion dialogue policy learning. In: ACL (2018)
14. Peng, B., et al.: Composite task-completion dialogue policy learning via hierarchical deep reinforcement learning. In: Proceedings of the 2017 Conference on Empirical Methods in Natural Language Processing, pp. 2231–2240 (2017)
15. Saha, T., Gupta, D., Saha, S., Bhattacharyya, P.: Towards integrated dialogue policy learning for multiple domains and intents using hierarchical deep reinforcement learning. Expert Syst. Appl. **162**, 113650 (2020)
16. Schatzmann, J., Young, S.: The hidden agenda user simulation model. IEEE Trans. Audio Speech Lang. Process. **17**, 733–747 (2009)
17. Schulman, J., Moritz, P., Levine, S., Jordan, M., Abbeel, P.: High-dimensional continuous control using generalized advantage estimation. ArXiv:abs/1506.02438 (2015)
18. Schulman, J., Wolski, F., Dhariwal, P., Radford, A., Klimov, O.: Proximal policy optimization algorithms. ArXiv:abs/1707.06347 (2017)
19. Serban, I.V., Sordoni, A., Benggio, Y., Courville, A., Pineau, J.: Building end-to-end dialogue systems using generative hierarchical neural network models. In: Proceedings of the Thirtieth AAAI Conference on Artificial Intelligence, pp. 3776–3783 (2016)
20. Vaswani, A., et al.: Attention is all you need. In: Advances in Neural Information Processing Systems, pp. 5998–6008 (2017)
21. Vlasov, V., Drissner-Schmid, A., Nichol, A.: Few-shot generalization across dialogue tasks. CoRR:abs/1811.11707 (2018)
22. Vlasov, V., Mosig, J.E.M., Nichol, A.: Dialogue transformers. ArXiv:abs/1910.00486 (2019)
23. Wang, Z., Chen, H., Wang, G., Tian, H., Wu, H., Wang, H.: Policy learning for domain selection in an extensible multi-domain spoken dialogue system. In: Proceedings of the 2014 Conference on Empirical Methods in Natural Language Processing (EMNLP), pp. 57–67, Doha, Qatar (October 2014)
24. Zhang, Z., Takanobu, R., Zhu, Q., Huang, M., Zhu, X.: Recent advances and challenges in task-oriented dialog system. Sci. China Technol. Sci. **63**, 2011–2027 (2020)
25. Zhu, Q., et al.: ConvLab-2: an open-source toolkit for building, evaluating, and diagnosing dialogue systems. In: Proceedings of the 58th Annual Meeting of the Association for Computational Linguistics (2020)

Evolutionary Computation, Metaheuristics, Constrains and Search, Combinatorial and Numerical Optimization

A Graph-Based Crossover
and Soft-Repair Operators for the Steiner
Tree Problem

Giliard Almeida de Godoi[1]([✉]) [iD], Renato Tinós[2] [iD],
and Danilo Sipoli Sanches[1] [iD]

[1] Universidade Tecnológica Federal do Paraná – Cornélio Procópio, Paraná, Brazil
giliardgodoi@alunos.utfpr.edu.br, danilosanches@utfpr.edu.br
[2] Universidade de São Paulo (USP) – Ribeirão Preto, São Paulo, Brazil
rtinos@ffclrp.usp.br

Abstract. In the Steiner Tree Problem in Graphs, a subset of nodes, called terminals, must be efficiently interconnected. The graph is undirected and weighted, and candidate solutions can include additional nodes called Steiner vertices. The problem is NP-complete, and optimization algorithms based on population metaheuristics, e.g., Genetic Algorithms (GAs), have been proposed. However, traditional recombination operators may produce inefficient solutions for graph-based problems. We propose a new recombination operator for the Steiner Tree Problem based on the graph representation. The new operator is a kind of partition crossover: it breaks the graph formed by the union of two parents solutions and decomposes the evaluation function by finding connected subgraphs. We also investigate two soft-repair operators that produce small changes in candidate solutions: an MST transformation and a pruning repair. The GA with the proposed crossover operator was able to find the global optimum solution for all tested instances and has a success rate of 28% in the worst case. The experiments with the proposed crossover presented a quick convergence to optimal solutions, and an average solution cost lower in most cases than other approaches.

Keywords: Steiner Tree Problem in Graphs · Partition based crossover · Genetic algorithms

1 Introduction

Steiner Tree Problems (STP) are relevant in the combinatorial optimization field [9]. In STP, a minimum special tree should be found in a weighted graph $G(V, E)$, where an edge $e_{ij} \in E$ represents the relationship between vertices v_i and $v_j \in V$, and a function $c : E \to \mathbb{R}_*^+$ maps each edge to a positive value. The graph is also undirected, i.e., we can reach a node v_j from v_i and vice-versa by the same edge and cost. Moreover, suppose there is a subset of requested vertices $Z \subset V$, also called terminals.

© Springer Nature Switzerland AG 2021
A. Britto and K. Valdivia Delgado (Eds.): BRACIS 2021, LNAI 13073, pp. 111–125, 2021.
https://doi.org/10.1007/978-3-030-91702-9_8

The Steiner Tree Problem in Graphs (STPG) seeks the Steiner Minimum Tree (SMT), a subgraph of G that connects all the terminals nodes in Z and minimize Eq. 1, i.e., has the minimum cost.

$$f(T) = \sum_{i=1} c(e_i) \quad \forall e_i \in E' \subset E \tag{1}$$

A Steiner tree (ST) is a candidate solution for the problem with $Z \subset V'$ and $E' \subset E$ forming a tree $T(V', E')$. Some non-terminal vertices from the set $V - Z$ might appear in V' to assure connectivity - they are Steiner vertices. An ST must follow these constraints: 1) The terminal nodes must be connected; 2) Any cycle must be broken to reduce the solution cost; 3) Leaf nodes that are not terminals must be removed, reducing the solution cost without lost connectivity.

Hamiki (1971) [4] first described the SPTG for a finite graph with vertices and well-defined edges. The STPG is a well-known NP-complete problem [6] in the combinatorial optimization field. It means no known algorithm assures that deterministically can find the best possible solution in a polynomial time for the general case. As a consequence, heuristics and metaheuristics have been proposed for the STPG.

The STPG can model a comprehensive range of practical problems. Several studies describe applications in the telecommunication and related fields as Multicast Routing Problems and Wireless Sensor Networks. However, it is also applicable in supply chain management, VSLI circuit projects, genetic phylogeny, and drug repositioning. Due to its applicability, several studies attempted to find better solutions for the STPG.

Ljubic (2021) [9] describes different approaches for special cases of the STPG: Lagrangian relaxations, dual-ascent methods, heuristics, approximate and exact algorithms. The author in [9] pointed out that metaheuristics also has been developed, like genetic algorithms (GA), ant colony optimization, and particle swarm optimization. However, metaheuristics are not competitive in many practical Steiner Tree benchmark problems where traditional approaches have been applied. The worst results, when compared to conventional approaches, can be partially credited to the use of transformation operators that end up producing inefficient or unfeasible solutions in graph-based representations of the problem.

In addition, the interaction between decision variables is ignored in black-box optimization. Alternatively, in gray-box optimization, the structure of the problems is efficiently explored by transformation operators to produce better solutions [14]. Partition Crossover (PX) operators explore the interaction between decision variables in problems with graph-based representation to recombine solutions efficiently. PX operators break the union graph formed by two parents' solutions and decompose the problem's evaluation function.

For instance, the Generalized Partition Crossover (GPX) improved the solution cost for the Traveling Salesman Problem (TSP) when it was combined with the state-of-art Lin-Kernighan-Helsgaum heuristic [13]. This operator is respectful because the offspring inherits all common edges from its parents. It also transmits alleles, which means all edges present in the offspring can be found in the parents. Moreover, when two local optima are recombined, the offspring is also local optima.

Accordingly to Raidl et al. (2003) [11], an efficient crossover operator for tree-graph problems is expected to be respectful and transmit alleles. In this paper, we propose a new recombination operator for the STPG based on partition crossover. It is important to observe that PX operators were not previously proposed for Steiner Tree Problems. In addition, we use two soft-repair operations to repair unfeasible solutions eventually generated by recombination. The proposed operator is compared to a Prim's random-based crossover. The GA with the proposed recombination and repair operators is also compared to a classical GA using binary chromosome representation.

This paper is organized as follows: Sect. 2 describes GA approaches for the STPG and some expected features for an efficient crossover operator. Section 3 explains our partition crossover for Steiner Trees. Section 4 describes the experiments, Subsect. 4.1 gives more detail about GA tested, and Subsect. 4.2 discusses results. Finally, a conclusion is given in Sect. 5.

2   Related Works

In two early studies, Kapsalis et al. (1993) [5] and Esbensen (1995) [3] proposed a GA with a binary chromosome to encode an ST. The binary chromosome indicates if a non-terminal node (from the set $V - Z$) belongs to an ST. If it belongs, it is set the value one to the corresponding node. Otherwise, it will set a value of zero if the node does not belong. The algorithms in [5] and [3] differ by the process of decoding the ST from the chromosome.

The decoding procedure in [5] first identifies the Steiner vertices from the chromosome and the terminals nodes that form the set of nodes $V' \subset V$ for the ST. Then, it adds an edge $e_{ij} = (v_i, v_j) \in E$ to the subgraph G' if both incidents vertices v_i and v_j belongs to V'. If so, the edge will also belong to G'. Finally, the ST is the Minimum Spanning Tree (MST) from the subgraph G'.

The decoding procedure adopted by [5] can lead to disconnected individuals. However, their fitness evaluation receives a penalization proportionally to the number of disconnected components. They also used traditional GA operators such as 1 point crossover and bit-flip mutation. The GA found the best solution for all OR-Library B class instances in at least one algorithm run. However, the authors applied a diverse range of parameters configuration to achieve the results, and the GA required a long running time.

Esbensen (1995) [3] embodied the Distance Network Heuristic (DNH) [7] to avert disconnected solutions in his decoding procedure. The heuristic uses the distance graph (representing all paths and distances between every pair of vertices) obtained by the Floyd-Wharsall algorithm [2]. Even though it is computed once, its time complexity of $O(n^3)$ might prevent its use for large graphs. The decoding procedure computes the ST by applying the heuristic over the subgraph G' induced by the chromosome subset of vertices V'.

The author in [3] applied his GA in C, D, and E instance classes from OR-Library [1]. The author also used reduction techniques to reduce the size of the instance graphs. Reduction techniques try to identify edges and vertices that always will belong to an ST—contracting those pairs of vertices and edges—or identifying those that will never be part of an ST and remove them. However, the GA achieved a near-optimum solution and took a long time running.

Raidl et al. (2003) [11] presented an Edge Set encoding for tree-based optimization problems. In this approach, just the set of edges represents the tree. They tested it considering different tree-based problems, but the adaptation to STPG is simple. They also proposed three randomized strategies for population initialization, based on Prim [10], Kruskal [8], and random walk algorithms.

The random walk starts at an arbitrary node v_{start} and moves a particle to some adjacent node v_k. The edge (v_{start}, v_k) is selected to compose the resulting tree. Then the particle moves to another adjacent node of v_k and inserts a new edge into the resulting tree if the selected edge does not insert a cycle in the tree. This procedure is repeated until all terminal nodes belong to the individual. The authors in [11] proved that this procedure does not favor a specific tree topology.

Prim (1957) [10] and Kruskal (1956) [8] have proposed two well-known algorithms to compute the MST from a graph. These algorithms greedily choose the lowest cost edges to form the MST, but they differ in selecting such edges. Kruskal's algorithm looks at the entire edges' set and selects one even though it does not connect with the previously chosen. Therefore, this algorithm uses a disjoint set data structure to ensure that the selected edges will not form a cycle. Instead, Prim's algorithm starts from a node v_0, considers its incident edges as eligible, and chooses some edge (v_0, v_j) with the lowest cost. The new incident edges from v_j will be eligible too. The set of eligible edges keeps growing until all graph edges are reached.

As a procedure to population initialization, the authors in [11] proposed a variation of Prim and Kruskal algorithms to select a new edge randomly instead of in a greedy way. Furthermore, the authors claimed that these procedures favor a star topology in the individuals. Hence, they named these variations as *PrimRST* and *KruskalRST*.

Also, in [11] the initialization procedures are adapted as recombination procedures by simply applying them to the union graph of the two candidate solutions. In addition, they stated that other recombination strategies are possible depending on what to do with the shared edges, transmitting them directly to the offspring or not. However, they conclude that recombination operators for spanning trees should always include the common edges in the offspring.

The authors in [11] stated that these operators offer locality (small changes in the genotype imply small changes in the phenotype) and heritability (the offspring consist mainly of alleles of its parents). These are two of many properties expected for proper encoding. However, to the best of our knowledge, those recombination strategies were not evaluated for the STPG case. Additionally, there is no formal demonstration that *PrimRST* always includes all the common edges between two individuals chosen for recombination.

In this paper, we propose a crossover operator for ST based on PX. It generates two offspring, which will have: the shared edges from both parents, the select edges from recombinant partitions, and the remaining edges from non-recombinant partitions, respectively, from each parent.

The principles behind our operator come from the PX for the Traveling Salesman Problem (TSP) proposed in [12,13,15], and [16]. The original Generalized Partition Crossover (GPX) was designed to recombine two Hamiltonian cycles. Under these circumstances, a new recombination operator is necessary for the Steiner Tree, i.e., recombine tree graph solutions.

The GPX takes two individuals (Hamiltonian cycles) for recombination. Then, it joins the edges from each individual in a union graph G_u. The vertices with a degree equal to 2 are removed, and their incident edges are transferred directly to the offspring. The remaining edges form candidate partitions for recombination. One can think each partition is the union of two paths, each one from a parent.

The recombinant partition must ensure that it can ensemble a cycle in the offspring if either inner path is chosen. It is made by checking the portal and simplified graph for each parent graph. A simplified graph reduces a path to a single edge. A portal vertex connects a partition to another through common edges. So it will be easy to verify if the path from both parents connects to the same portal vertex. Nevertheless, vertices with a degree equal to 4 are split by inserting a ghost vertex. The edges are then divided between each vertex–the original and the ghost one.

3 Proposed Operator

As mentioned before, the original GPX is not suitable for the STPG case due to differences in candidate solutions' topology. Algorithm 1 shows the proposed recombination operator based on partition crossover ($PXST$).

Both operators (the original and the proposed one) keep some resemblance. The offspring will inherit all the common edges between the parents. Uncommon edges will form partitions, and some tests will identify the comparable ones, also called recombinant partitions. For such partitions, those with the lowest cost will compose the offspring. Moreover, if there are k recombining components, the operator will generate the best of 2^k potential offspring.

Furthermore, the PXST's operation can be divided into four phases: identifying common elements (vertices and edges); identifying partitions candidates to recombination; matching, selecting, and updating the recombinant partitions; handle partitions unfeasible to recombination and return the offspring. Each phase will be detailed as follows.

a) **Identifying Common Elements.** The common vertices and edges between two ST, play an essential role in the PXST. They work as a cutting point, increasing the number of partitions and will help to identify recombinant partitions.

The vertices from the union graph between two Hamiltonian cycles will have degrees equal to 2, 3, or 4, and GPX explores it. First, the operator removes the vertices with a degree equal to 2. The vertices with a degree equal to 3 belong to

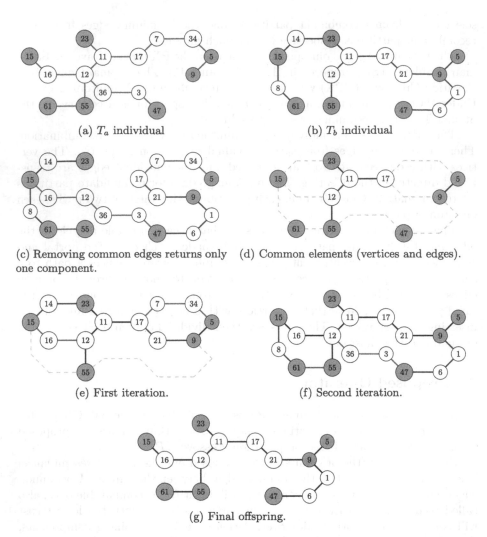

(a) T_a individual

(b) T_b individual

(c) Removing common edges returns only one component.

(d) Common elements (vertices and edges).

(e) First iteration.

(f) Second iteration.

(g) Final offspring.

Fig. 1. An example of $PXST$ crossover between two ST, T_a and T_b, showed in Figs. 1a and 1b. Figure 1c shows the resulting graph after removing the common edges from the union $T_a \cup T_b$; the resulting graph has only one connected component. Figure 1d shows the common elements (vertices and edges) between T_a and T_b. They will be used to explore more partitions. $PXST$ represents common elements as a disjoint set, where vertices are linked if they share a common edge (steps 1, 2, and 3, of Algorithm 1). Figure 1e shows two cycles identified in the first iteration. The distinct paths of non-common edges are connected to the same components of common elements. Then, we can choose the best path. The selected path is added to the common elements, and two more cycles are then identified in the second iteration (Fig. 1f). Again, the respective paths for each parent are compared, and the best paths are chosen. Figure 1g shows the offspring. All the partitions are feasible for recombination in this example, but this may not be the case for some pairs of solutions (ST).

some partition. Moreover, the insertion of a ghost vertex will separate the edges adjacent to vertices with a degree equal to 4, leading to more partitions.

Unlike, the ST inner vertices do not have such degree constraints. An inner node can have any number of adjacent edges, and only the leaves must have a degree equal to one. Consequently, the union graph between two ST solutions might have more connections within its vertices, leading to fewer independent components.

Figures 1a and 1b show two examples of ST, T_a and T_b. Figure 1c shows why the strategy adopted by GPX is not suitable for the STPG. Figure 1c the resulting graph obtained by removing the common edges from the union between T_a and T_b has only one connected component. In this simple case, choosing either one or another parent's edges did not result in a new individual.

Figure 1d shows the common edges and vertices between T_a and T_b. Notice that the vertices 15, 61, and 47 will act like cutting points, breaking the component shown in Fig. 1c in smaller partitions.

Algorithm 1: Partition based crossover for Steiner Tree (PXST)

Input: trees $T_a(V_a, E_a)$ and $T_b(V_b, E_b)$
Output: offspring $T_y(V_y, E_y)$ and $T_z(V_z, E_z)$

1 Define the subset $S_{ab} := V_a \cap V_b$ of common nodes between the two solutions. Include leaf nodes that are not terminal and belong to just one solution.
2 For each node in S_{ab} define a unity set in a disjoint set DS.
3 For each common edge $(v_i, v_j) \in E_a \cap E_b$, apply a union operation in DS between the vertices v_i and v_j.
4 Define the subgraph T_a^* and T_b^* by removing the common edges from T_a and T_b, respectively. The common edges will compose the two resulting offspring.
5 Proceeds a depth-first search in T_a^*, starting with a common vertex $v_s \in S_{ab}$ to find its partitions, and then save them in L_a. A partition is a set of non-common edges and vertices, except by the portals vertices. The portal vertices are common vertices defined in step 1. Repeat it for T_b^* and save in a list L_b.
6 Consulting DS maps each partition's portal vertices to a key representing which component (of common elements) the partition is connected to. The key is necessary to identify related partitions in L_a and L_b.
7 While there is a key-matching between partitions in L_a and L_b, select the best subtree partition (lower cost). If two subtrees partition has an equal cost, select it randomly.
8 Update DS by applying a union operation with the portal vertices from the selected partition.
9 Repeat steps 6, 7, and 8 until there are not more related subtrees partitions in L_a and L_b. The remaining partitions are unfeasible for recombination.
10 Return two offspring (T_y and T_z) formed by the common edges identify in step 3, the partition edges selected in steps 7, 8, and 9, and the respective remaining edges identified in step 9.

Moreover, there are two sets of common edges: one formed by the edge $\{(5,9)\}$ and the other formed by the edges $\{(55,12),(12,11),(23,11),(17,11)\}$. Also, notice how candidate partitions connect to these sets by different vertices. One first test to identify a recombinant partition might be to check when two distinct partitions (one from each parent) connect to the same components of common elements, i.e., a set of connected edges or an isolated vertex.

To keep track of such common elements, PXST uses a disjoint set (DS) data structure. It allows three operations: the "make a set" initializes the disjoint set; the "find" operation returns the same representative element for any other element in the set; the "union" operation joins two sets and defines a new element to represent the new set. An efficient implementation of the union operation (using union by rank and path compression) has $O(m \log n)$ time complexity [2].

Step 1 from Algorithm 1 firstly identifies the common vertices between the parent's solutions. Then, step 2 initializes DS by creating a unitary set for all common vertices. Finally, DS's initial unitary sets are jointed when a common edge connects two distinct vertices (step 3). Notice that the set of common edges (showed in Fig. 1d) will be represented as a disjoint set, i.e., unique key-vertex.

Another detail is that some individuals might have leaf nodes that are not terminals. Some of these leaf nodes might belong to just one of the parent's solutions. Moreover, such leaf nodes will become a portal vertex for one partition. Since checking portal vertices will allow identifying recombinant partition, such leaves will be included in the DS, although they may not be common (step 1).

b) Identifying Candidate Partitions to Recombination. The $PXST$ identifies partitions directly on the graphs T_a^* e T_b^* (step 4). They are the resulting graphs from deleting the common edges from T_a and T_b, respectively. Removing an edge from a tree results in two subtrees. Therefore, T_a^* and T_b^* are compound by many subtrees. However, those subtrees are not partitions yet. They still will be divided into smaller parts taking the common vertices as cutting points.

A depth-first search looks for partitions in T_a^* e T_b^*, separately (step 5). A partition is formed mainly by non-common elements, where all its internal vertices and edges belong to one of the solutions. Since the common vertices work as cutting points, where the partitions are breaking, these vertices are called portals. Figure 1 shows partitions only with two portal vertices. However, a partition with more than two portal vertices is possible when an internal node has a degree greater than two, i.e., the partition has a subtree topology.

c) Matching, Selecting, and Updating the Recombinant Partitions. For instance, consider two STs with common vertices $u, v \in T_a \cap T_b$. Pretend that T_a has a partition with a path-like topology P_a and T_b has a P_b partition. Both P_a and P_b connect u and v by different edges. The union between P_a and P_b will form a cycle closing in the vertices u and v. Moreover, u and v are the portal vertices for these partitions. Alternatively, the portal vertices for the partitions are coincident, and one can select the partition with the lower cost.

However, Figs. 1e and 1f show that partitions can connect to the same component of common edges by different vertices. Then, the partition will not have coincident portal vertices. So, a more refined way to find recombinant partitions is needed.

One can argue that if we shrink all common edges to a single vertex, we might get the well-formed cycle as mentioned earlier. For example, in Fig. 1e, vertices 12 and 23 will become coincident, and the partitions will form a cycle between nodes 15 and 12 (or 23). In the same way, shrinking the edge $(5, 9)$ will close another cycle with the path-like partitions up to node 17. Then, we will be able to compare such distinct paths.

Instead of shrinking common edges and checking the exact correspondence between portal vertices from distinct partitions, we check if they are mapped to the same representative nodes by the disjoint set (step 6). That indicates the partitions are connected to the same components of common elements (Fig. 1d), and they are interchangeable. Then, the operator can choose between the partitions without losing the offspring connectivity or adding a loop into it.

When a partition is liable for recombination, the disjoint set from each portal vertex can be joined, expanding the original disjoint set (common elements set). This update is helpful to discover new recombination opportunities, and it will be performed until there are no new recombinant components (step 9).

For instance, Fig. 1f shows other opportunities for recombination. In the first one, we can choose the path from node 61 up to 15 or from 61 to 55. However, to compare these paths, it is necessary to resolve the connectivity to node 15 first (shown in Fig. 1e). When we resolve the best connectivity up to node 15, this node will belong to the same disjoint set with nodes 12, 23, and 55. The portal vertices 15 and 55 will be mapped to the same arbitrary representative element, and the operator can decide the best path to node 61. A similar situation occurs to the partitions connecting node 47. Firstly, we must decide the best connectivity up to node 9 to include this node to the same disjoint set as portal vertex 12.

Some partitions might have equal cost even though they do not have the same edges. In such cases, the selection occurs randomly, with equal probability for each partition. When their total cost is different, $PXST$ chooses the one with a lower cost. Thus, the partition's selected edges will belong to both offspring.

d) Handle Unfeasible Partitions to Recombination and Return Offspring. By the end of this process, some non-recombinant partitions might persist for both parents. One alternative is to choose the remaining partitions from one of the parents with the lowest sum cost. In preliminary tests, this strategy leads to an early population stagnation and higher solution cost outcomes. Another option is to return two offspring, one for each parent's remaining partitions (step 10). Therefore, we chose this last alternative as it showed to lead to better outcomes.

New ways for identifying recombinant partitions can further be developed. It may improve the operator's efficiency and lead to better solutions.

Table 1. Average and standard deviation of the best solution found along with the runs of each experiment. The symbols '=', '+' or '−' indicate if the average cost is equal, better, or worse than GA with PXST. The letter s indicates if the differences are statistically significant. Properties of each instance are also presented (number of vertices $|V|$, number of edges $|E|$, number of terminals vertices $|Z|$ and the optimum solution cost (Opt.).

Instances	$|V|$	$|E|$	$|Z|$	Opt.	PXST + R(mst, pruner)	PrimRST + R(mst, pruner)	Binary GA
B1	50	63	9	82	**82.00 ± 0.00**	82.10 ± 0.30 (−)	84.88 ± 3.15 (s−)
B2	50	63	13	83	**83.00 ± 0.00**	83.00 ± 0.00 (=)	84.34 ± 2.23 (s−)
B3	50	63	25	138	**138.00 ± 0.00**	138.00 ± 0.00 (=)	**138.00 ± 0.00** (=)
B4	50	100	9	59	60.22 ± 1.53	**59.30 ± 0.84** (s+)	63.02 ± 2.25 (s−)
B5	50	100	13	61	61.10 ± 0.46	**61.00 ± 0.00** (+)	62.10 ± 0.68 (s−)
B6	50	100	25	122	122.84 ± 1.31	123.44 ± 0.86 (s−)	**122.12 ± 0.44** (s+)
B7	75	94	13	111	111.02 ± 0.14	**111.00 ± 0.00** (+)	124.96 ± 8.23 (s−)
B8	75	94	19	104	**104.00 ± 0.00**	104.06 ± 0.42 (−)	107.96 ± 3.48 (s−)
B9	75	94	38	220	**220.00 ± 0.00**	220.00 ± 0.00 (=)	220.80 ± 1.03 (s−)
B10	75	150	13	86	87.26 ± 1.99	**86.00 ± 0.00** (s+)	97.00 ± 4.98 (s−)
B11	75	150	19	88	**88.06 ± 0.31**	88.12 ± 0.48 (−)	97.46 ± 3.81 (s−)
B12	75	150	38	174	174.18 ± 0.39	176.26 ± 1.07 (s−)	**174.16 ± 0.37** (+)
B13	100	125	17	165	**165.64 ± 1.48**	167.70 ± 2.53 (s−)	190.98 ± 9.35 (s−)
B14	100	125	25	235	**235.84 ± 0.82**	238.04 ± 0.20 (s−)	244.68 ± 6.00 (s−)
B15	100	125	50	318	**319.46 ± 0.93**	325.20 ± 1.12 (s−)	322.26 ± 2.42 (s−)
B16	100	200	17	127	**130.98 ± 3.66**	138.78 ± 3.22 (s−)	150.32 ± 6.72 (s−)
B17	100	200	25	131	132.52 ± 1.52	**131.00 ± 0.00** (s+)	137.90 ± 4.57 (s−)
B18	100	200	50	218	**218.46 ± 0.65**	219.02 ± 0.94 (s−)	218.88 ± 0.87 (s−)

4   Experiments

This section describes the experiments comparing the proposed operator ($PXST$), the $PrimRST$ crossover operator [11] and a binary GA based on [5]. The section also describes soft-repair operators used in our algorithm. In the experiments, three algorithms were compared:

- **$PXST$+Repair(pruner, mst)**: GA with the proposed operator $PXST$ (Algorithm 1), mutation that replaces edges randomly, and the pruning and MST repair strategies.
- **$PrimRST$+Repair(pruner, mst)**: GA with the $PrimRST$ crossover operator [11] and pruning and MST repair strategies.
- **Binary GA**: GA with binary chromosome [5]. It used a traditional 2 point crossover operator and bit-flip mutation.

4.1   Experimental Design

The algorithms were tested over 18 instances of class B from the OR-Library [1]. Table 1 presents the properties for each instance and statistics about the solution cost discussed later. In all three experiments, the same stop condition was adopted. The execution stops if the algorithm finds the global optimum solution, if the best solution is not improved within an interval of 500 fitness evaluations, or if the number of iterations reaches 4.000. However, this last condition was never reached.

The experiments with $PXST$ and $PrimRST$ differ only by the crossover operator. They have the same parameters and underlying mechanisms for selection (roulette) and mutation. An adjacency list data structure represents the ST. The population size is 100 individuals, initialized by a heuristic based on the random walk procedure presented in [11]. This initialization strategy generates only connected individuals.

The mutation by replacing a random edge (RRE) is presented in [11]. The operator removes a random edge from the tree producing two disconnected subtrees. Then, it reconnects them by inserting a new edge selected randomly from the problem instance graph. The mutation rate is fixed to 0.3. This operator produces a slight change in the ST since it replaces just one edge when possible.

The $PrimRST$ crossover takes two individuals and joins them in a graph G_u. Like the original Prim's algorithm, the operator computes a spanning tree from G_u (see Sect. 2). Instead of taking the lowest cost edge, the operators insert a new edge randomly – of course, if the new edge does not form a cycle with the previously selected edges. The algorithm stops when all terminal vertices are inserted in the offspring or when all edges from G_u were processed.

The offspring's edges generated by $PrimRST$ came only from G_u, i.e., the operator does not insert a new edge that does not belong to one of the parents. Some common edges will appear in the offspring, although there is no proof that all common edges will be transmitted every time. In that regard, both $PrimRST$ and $PXST$ transmit alleles (edges) from the parent to the offspring. However, $PXST$ transmits all of the common edges to the offspring, and for $PrimRST$, this is not guaranteed. Moreover, $PrimRST$ favors the edge exchange between the parents, while $PXST$ looks for connected components.

$PrimRST$ and $PXST$ generate connected individuals. Still, some offspring might not be completely ST because some of them might have leaf nodes that are not terminal. Thereby, a pruning repair iteratively seeks and removes leaf nodes that are not terminal. It reduces the solution cost without losing the offspring's connectivity. Thus, the pruning repair was constantly applied for all individuals over the generations (fitness evaluations).

In [5] and [3] a solution is obtained computing the MST of the induced subgraph from Steiner vertices represented in the chromosome. Similarly, an ST can be improved by replacing some of its edges with lowest cost edges that span the solution vertices, computed from the induced MST subgraph. For that reason, an MST-based repair is applied.

The MST-based repair operator takes one individual $T_k(V_k, E_k)$ and determines a subgraph induced from its vertices V_k. The induced subgraph has all edges from the individual plus the edges from the problem instance graph when both incident vertices are in the individual's vertex set. Then, it applies Prim's

Table 2. Average and standard deviation of the execution time (in seconds) and the success rate (SR), i.e., the percentage of execution where the best-known solution was found. The symbols =, +, and − indicate if the average result was equal, better or worse than the GA with PXST. The letter s indicates if the differences in the samples are statistically significant.

Instances	PXST + R(mst, pruner)		PrimRST + R(mst, pruner)		Binary GA	
	Exec. time (s)	SR (%)	Exec. time (s)	SR (%)	Exec. time (s)	SR (%)
B1	**0.14 ± 0.05**	100.0	2.04 ± 4.64 $(s-)$	90.0	23.65 ± 6.66 $(s-)$	18.0
B2	**0.21 ± 0.09**	100.0	0.52 ± 0.32 $(s-)$	100.0	17.20 ± 8.73 $(s-)$	68.0
B3	**0.24 ± 0.09**	100.0	0.26 ± 0.08 $(-)$	100.0	2.30 ± 0.93 $(s-)$	100.0
B4	11.11 ± 11.47	52.0	**3.39 ± 6.01** $(s+)$	86.0	26.21 ± 8.25 $(s-)$	6.0
B5	**2.15 ± 6.98**	94.0	2.34 ± 2.98 $(s-)$	100.0	20.92 ± 5.68 $(s-)$	18.0
B6	15.37 ± 21.31	68.0	35.03 ± 12.51 $(s-)$	24.0	**9.45 ± 6.47** $(+)$	92.0
B7	0.78 ± 4.10	98.0	**0.32 ± 0.07** $(s+)$	100.0	42.57 ± 11.70 $(s-)$	0.0
B8	**0.20 ± 0.05**	100.0	1.03 ± 3.48 $(s-)$	98.0	34.60 ± 11.36 $(s-)$	22.0
B9	**0.32 ± 0.07**	100.0	0.44 ± 0.15 $(s-)$	100.0	31.30 ± 13.59 $(s-)$	48.0
B10	12.54 ± 15.08	60.0	**0.80 ± 0.41** $(+)$	100.0	44.87 ± 12.34 $(s-)$	0.0
B11	**1.88 ± 6.59**	96.0	3.64 ± 7.06 $(s-)$	94.0	42.81 ± 11.21 $(s-)$	0.0
B12	**13.32 ± 25.93**	82.0	70.48 ± 20.67 $(s-)$	2.0	23.33 ± 11.73 $(s-)$	84.0
B13	**6.71 ± 14.57**	84.0	24.39 ± 15.91 $(s-)$	40.0	60.08 ± 14.92 $(s-)$	0.0
B14	**37.56 ± 25.51**	34.0	65.24 ± 14.26 $(s-)$	0.0	70.69 ± 17.36 $(s-)$	0.0
B15	**58.87 ± 36.28**	28.0	92.33 ± 27.59 $(s-)$	0.0	67.04 ± 19.47 $(-)$	2.0
B16	**25.52 ± 17.99**	34.0	53.59 ± 17.32 $(s-)$	0.0	72.10 ± 24.21 $(s-)$	0.0
B17	34.48 ± 25.47	36.0	**7.78 ± 6.16** $(s+)$	100.0	75.61 ± 17.86 $(s-)$	0.0
B18	**34.23 ± 42.30**	62.0	74.83 ± 36.45 $(s-)$	30.0	45.99 ± 18.31 $(-)$	36.0

algorithm to compute the MST from the induced subgraph. This repair operator is used with a fixed rate of 0.3.

Since $PXST$ and $PrimRST$ operators do not produce disconnected individuals, repair operations to reconnect the candidate solution or a penalty function (as in [5]) are useless.

The binary GA is based on [5] (see Sect. 2). A binary chromosome represents the Steiner vertices that belong to some ST. A decoding procedure based on computing the MST tree from the induced graph does not assure the candidate solution's connectivity. Therefore, a penalty function increases the fitness of infeasible solutions (disconnected ones). The GA used a population size of 100 individuals. The binary chromosome is randomly initialized, and the genetic operators are roulette selection, a traditional crossover operator with 2 points cut, and a mutation by bit flip by a rate of 0,2.

The operators and GA implementations were in Python language, version 3.7.9, and the code is available publicly.[1] Experiments were conducted in a virtual machine set up in *Google Cloud*[TM] platform, using Linux Ubuntu OS (version 18.4), 32 GB of RAM, and CPU with 8 cores *Intel Cascade Lake*. Each experiment runs 50 times, with a seed randomly defined by the system.

---

[1] https://github.com/GiliardGodoi/ppgi-stpg-gpx.

4.2   Experimental Results and Analysis

Table 1 presents the instances graphs features alongside the average and standard deviation for the best solution cost found in all executions. The values in bold indicate the lower average cost. Table 2 shows the average and standard deviation of execution time (in seconds) and the success rate (SR), i.e., the proportion of runs where the algorithm found the global optimum solution.

For the average cost and execution time results, we applied the Wilcoxon signed-rank test to verify if the differences in the samples are statistically significant. We adopted a significance level of $\alpha = 0.05$.

The GA with $PXST$ found the global optimum solution for all instances in at least one run. This GA found the global optimum solution for all execution in five instances (B1, B2, B3, B8, and B9). For the other six cases (B11, B13, B14, B15, B16, and B18), the GA obtained the lowest average solution cost, as showed in Table 1. Moreover, GA with $PXST$ has an SR in the worst case of 28% (instance B15), as shown in Table 2. In all other cases, it has an SR above 30%, indicating certain robustness of the algorithm.

The GA with $PrimRST$ found the global optimum solution in all runs for seven instances: B2, B3, B5, B7, B9, B10, and B17. It obtained the lowest average solution cost for the B4 instance. Significant improvements are observed for B4, B10, and B17. This GA could not find the optimum solution in any run for B14, B15, and B16 and obtained a low SR for instances B6 (24%) and B12 (2%) (Table 2).

The statistical test does not indicate significant differences for the best cost sample to B1, B2, B3, B5, B7, B8, B9, B11, and B18 between $PXST$ and $PrimRST$ experiments. For example, GA with $PrimRST$ found the global optimum solution, for instance, B5 in all runs (SR 100%) – note that the average best cost is equal to the global optimum, and the standard variation is zero. In contrast, GA with $PXST$ obtained 94% of the SR and an average best cost of 61.10 and a standard deviation of 0.46. However, the test indicates that the difference observed in the best cost sample was not significant for the B5 instance.

Binary GA obtained a lower best cost average for instances B6 and B12. However, the differences for best cost are statistically significant just for B6. All of the three GA solved the instance B3 in all runs. For all other 15 instances, the best cost average for binary GA was worse than the GA with $PXST$. This result can be credited to the use of specific operators for tree-graph problems rather than traditional operators that produce infeasible (disconnected) solutions.

The GA with $PXST$ was faster for fourteen instances (see Table 2). Moreover, GA with $PXST$ was, in general, faster than $PrimRST$ for those instances where the average best cost was similar, i.e., differences were not statistically significant. That is the case for B1, B2, B3, B5, B8, B9, and B11 (except B7). This result is because the crossover $PXST$ is greedy and converges quickly to a local optimum solution.

The GA with $PrimRST$ was faster when it found the global optimum solution in all runs (B7, B10, and B17) and in B4, where the SR is 86%. $PXST$ and

PrimRST GA took less than a second on average to solve some instances (see B2, B3, B7, B8, B9, and B10).

As expected, binary GA usually took a long time to run on average, as showed in Table 2. For B6, binary GA was faster than the other two GAs. Compared with GA with *PXST*, the execution time is similar (not statistically significant) for B6, B15 and B18, and worse for all other instances. Both *PXST* and *PrimRST* operators do not rely on high complexity cost heuristic as [3]. Hence, they might be suitable for large instance graphs in an appropriate execution time.

5   Conclusions

In this paper, we presented a graph-based crossover for the STPG. *PXST* is based on partition crossover, firstly proposed to the TSP problem. The main adaptation is identifying partitions since the candidate solutions are tree-graphs, in this particular case.

Two soft-repair operators were applied in the GA using a graph representation of the individuals. The first repair relates to the property that the STPG can be solved by determining the MST of an appropriate set of Steiner vertices. An MST transformation calculates the MST spanning the vertices for a given ST. In this case, many edges can be replaced by lower-cost edges. Then, the GA would not necessarily explore all possible ST.

A second repair operation pruned non-terminal leaves, reducing the solution cost without losing the tree connectivity. The repair operator is applied since *PXST* and *PrimRST* might generate individuals with such leaves. In this first study, we do not analyze the impact of varying the numeric parameters (mutation or repair rate, for example).

Compared with two other GA for the STPG, the one using the *PXST* was faster for almost all instances. It also reached the global optimum solution for all instance problems with a success rate above 30% in general, considering 50 executions of the experiment.

PXST and *PrimRST* are similar considering the offspring are constituted only with edges of their parents, and common edges are transmitted to the offspring, most of them in the *PrimRST* case. The differences are that *PXST* does not delete edges or nodes present in the parents and exchanges more significant structures such as paths or subtrees. Because of their similarities, both crossover operators are at least competitive in their results.

The partition crossover operators, which *PXST* was based on, have the property of tunneling between optima [13]. It produces the best of 2^k potential offspring, where k is the number of feasible partitions. Thus, if two parents are local optima, the offspring will be a local optimum with a lower cost. Considering that, *PXST* also can be used combined with other heuristics as in [13].

Furthermore, the GA using *PXST* is greedy and quickly converges to local or global optima. A mechanism to control the population diversity might contribute to GA with *PXST* performance.

In the $PXST$, some partitions remain unfeasible for recombination. Further studies can improve the partition discovery or fusion procedures to find more opportunities for recombination and improve the operator's efficiency.

References

1. Beasley, J.E.: OR-library: distributing test problems by electronic mail. J. Oper. Res. Soc. **41**(11), 1069–1072 (1990)
2. Cormen, T.H., Leiserson, C.E., Rivest, R.L., Stein, C.: Algoritmos: Teoria e Prática. Editora Elsevier (2012)
3. Esbensen, H.: Computing near-optimal solutions to the Steiner problem in a graph using a genetic algorithm. Networks **26**(4), 173–185 (1995). https://doi.org/10.1002/net.3230260403
4. Hakimi, S.L.: Steiner's problem in graphs and its implications. Networks **1**(2), 113–133 (1971). https://doi.org/10.1002/net.3230010203
5. Kapsalis, A., Rayward-smith, V.J., Smith, G.D.: Solving the graphical Steiner tree problem using genetic algorithms. J. Oper. Res. Soc. **44**(4), 397–406 (1993)
6. Karp, R.M.: Reducibility among Combinatorial Problems, pp. 85–103. Springer, Boston (1972). https://doi.org/10.1007/978-1-4684-2001-2_9
7. Kou, L., Markowsky, G., Berman, L.: A fast algorithm for Steiner trees. Acta Inf. **15**(2), 141–145 (1981). https://doi.org/10.1007/BF00288961
8. Kruskal, J.B.: On the shortest spanning subtree of a graph and the traveling salesman problem. Proc. Am. Math. Soc. **7**(1), 48–50 (1956). http://www.jstor.org/stable/2033241
9. Ljubić, I.: Solving Steiner trees: recent advances, challenges, and perspectives. Networks **77**(2), 177–204 (2021)
10. Prim, R.C.: Shortest connection networks and some generalizations. Bell Syst. Tech. J. **36**(6), 1389–1401 (1957)
11. Raidl, G.R., Julstrom, B.A.: Edge sets: an effective evolutionary coding of spanning trees. IEEE Trans. Evol. Comput. **7**(3), 225–239 (2003). https://doi.org/10.1109/TEVC.2002.807275
12. Tinós, R., Whitley, D., Ochoa, G.: Generalized Asymmetric Partition Crossover (GAPX) for the asymmetric TSP. In: Proceedings of the 2014 Annual Conference on Genetic and Evolutionary Computation, GECCO 2014, pp. 501–508. ACM, New York (2014)
13. Tinós, R., Whitley, D., Ochoa, G.: A new generalized partition crossover for the traveling salesman problem: tunneling between local optima. Evol. Comput. **28**(2), 255–288 (2020). https://doi.org/10.1162/evco_a_00254, pMID: 30900928
14. Whitley, D.: Next generation genetic algorithms: a user's guide and tutorial. In: Gendreau, M., Potvin, J.-Y. (eds.) Handbook of Metaheuristics. ISORMS, vol. 272, pp. 245–274. Springer, Cham (2019). https://doi.org/10.1007/978-3-319-91086-4_8
15. Whitley, D., Hains, D., Howe, A.: Tunneling between optima: partition crossover for the traveling salesman problem. In: Proceedings of the 11th Annual Conference on Genetic and Evolutionary Computation, GECCO 2009, pp. 915–922. ACM, New York (2009)
16. Whitley, D., Hains, D., Howe, A.: A hybrid genetic algorithm for the traveling salesman problem using generalized partition crossover. In: Schaefer, R., Cotta, C., Kołodziej, J., Rudolph, G. (eds.) PPSN 2010. LNCS, vol. 6238, pp. 566–575. Springer, Heidelberg (2010). https://doi.org/10.1007/978-3-642-15844-5_57

A Modified NSGA-DO for Solving Multiobjective Optimization Problems

Jussara Gomes Machado[1], Matheus Giovanni Pires[1(✉)],
Fabiana Cristina Bertoni[1], Adinovam Henriques de Macedo Pimenta[2],
and Heloisa de Arruda Camargo[3]

[1] State University of Feira de Santana, Feira de Santana 44036-900, BA, Brazil
`jussara@ecomp.uefs.br`, {`mgpires,fcbertoni`}`@uefs.br`
[2] Adventist University Centre of São Paulo, São Paulo 05858-001, SP, Brazil
`adinovan.pimenta@unasp.edu.br`
[3] Federal University of São Carlos, São Carlos 13565-905, SP, Brazil
`heloisacamargo@ufscar.br`

Abstract. This paper presents a novel Multiobjective Genetic Algorithm, named Modified Non-Dominated Sorting Genetic Algorithm Distance Oriented (MNSGA-DO), which aims to adjust the NSGA-DO selection operator to improve its diversity when applied to continuous multiobjective optimization problems. In order to validate this new Genetic Algorithm, we carried out a performance comparison among it and the genetic algorithms NSGA-II and NSGA-DO, regarding continuous multiobjective optimization problems. To this aim, a set of standard benchmark problems, the so-called ZDT functions, was applied considering the quality indicators *Generational Distance*, *Inverted Generational Distance* and *Hypervolume* as well as a time evaluation. The results demonstrate that MNSGA-DO overcomes NSGA-II and NSGA-DO in almost all benchmarks, obtaining more accurate solutions and diversity.

Keywords: Multiobjective genetic algorithm · Multiobjective optimization · NSGA-II · NSGA-DO

1 Introduction

Multiobjective optimization is an area of multiple-criteria decision making, concerning mathematical optimization problems involving more than one objective function to be optimized simultaneously [12]. Multiobjective optimization has been used to many fields of science and engineering, which provides multiple solutions representing the trade-offs among objectives in conflict. Genetic Algorithms have been applied to several classes of Multiobjective Optimization Problems (MOP) and have been shown to be promising for solving such problems efficiently [2,8–10,13,16], along with the advantage of evaluating multiple potential solutions in a single iteration because they deal with a population of solutions.

© Springer Nature Switzerland AG 2021
A. Britto and K. Valdivia Delgado (Eds.): BRACIS 2021, LNAI 13073, pp. 126–139, 2021.
https://doi.org/10.1007/978-3-030-91702-9_9

Given the diversity of existing genetic algorithms for solving MOP, named Multiobjective Genetic Algorithms (MOGA), it is necessary to be aware of their benefits and drawbacks. Thereby, this work performs a comparative study among the well-known MOGA NSGA-II [6], a modification of it, the NSGA-DO [15] and the novel MNSGA-DO (Modified Non-Dominated Sorting Genetic Algorithm Distance Oriented), designed as a adjustment of the NSGA-DO. The use of crowding distance by NSGA-II as the selection criterion can prioritize individuals that are farther from the optimal front. A selection process which guide the solutions to converge towards the ideal points along the Pareto front, as proposed by NSGA-DO, have been developed as an alternative to enhance a diversity of solutions. However, the NSGA-DO was developed based on problems with discrete search space. So, when applied to continuous problems, it does not guarantee the assignment of solutions to all the ideal points what might concentrate its set of solutions in specific regions. Another disadvantage is the amount of ideal points, which is proportional to the size of Pareto front. This imbalance between the quantity of ideal points and the number of solutions that need to be selected makes the convergence time-consuming.

In order to overcome these failures, we proposed the Modified Non-Dominated Sorting Genetic Algorithm Distance Oriented (MNSGA-DO). This MOGA calculates the length of Pareto front and then estimates a partition considering the number of solutions to be found, setting the coordinates of the ideal points. In addition, MNSGA-DO assigns one solution to each ideal point, what ensures a diversity of solutions. To perform the proposed comparative analysis, the ZDT family of functions [20] was selected, because it is a broad and popular set of test functions for benchmarking the performance of multiobjective optimization methods. For each NSGA, a study of its convergence and distribution of solutions along the Pareto Front was performed by applying the quality indicators Generational Distance, Inverted Generational Distance and Hypervolume, according to the literature. In order to complement the analysis, the optimal Pareto front were visually compared to the boundaries computed by the three MOGA. Finally, we carried out an analysis of runtime and a statistical evaluation of the results achieved.

2 Multiobjective Optimization Problems

According to [12], a multiobjective optimization problem (MOP) can be defined as follows:

$$
\begin{aligned}
Minimize\ \ &F(x) = (f_1(x), ..., f_m(x))^T \\
subject\ to\ \ &g_j(x) \geq 0, j = 1, ..., J \\
&h_k(x) = 0, k = 1, ..., K \\
&x \in \Omega
\end{aligned}
\tag{1}
$$

where J and K are the numbers of inequality and equality constraints, respectively. $\Omega = \Gamma_{i=1}^n [a_i, b_i] \subseteq \mathbb{R}^n$ is the space, $x = (x_1, ..., x_n)^T \in \Omega$ is a candidate solution. $F : \Omega \longrightarrow \mathbb{R}^m$ constitutes m conflicting objective functions and $\mathbb{R}^m$ is called the objective space. The attainable objective set is defined as $\Theta = \{F(x) \mid x \in \Omega, g_j(x) \geq 0, h_k(x) = 0\}$, for $j \in \{1, ..., J\}$ and $k \in \{1, ...K\}$.

x^1 is said to dominate x^2 (denoted as $x^1 \preceq x^2$) if and only if $f_i(x^1) \leq f_i(x^2)$ for every $i \in \{1, ..., m\}$ and $f_l(x^1) < f_l(x^2)$ for at least one index $l \in \{1, ..., m\}$. A solution x^* is Pareto-optimal to (1) if there is no other solution $x \in \Omega$ such that $x \preceq x^*$. $F(x^*)$ is then called a Pareto-optimal (objective) vector. The set of all Pareto-optimal solutions is called the Pareto-optimal set (PS). Therefore, the set of all Pareto-optimal vectors, $EF = \{F(x) \in \mathbb{R}^m \mid x \in PS\}$, is called the efficient front (EF) [14].

3   Multiobjective Genetic Optimization

Genetic algorithms are suitable to solving multiobjective optimization problems because they deal simultaneously with a set of possible solutions (or a population). This allows to find several members of the Pareto optimal set in a single run of the algorithm, instead of having to perform a series of separate runs. Additionally, genetic algorithms are less susceptible to the shape or continuity of the Pareto front [3]. MOGA are usually designed to meet two often conflicting goals: convergence, viewed as minimizing the distances between solutions and the EF, and diversity, which means maximize the spread of solutions along the EF. Balancing convergence and diversity becomes much more difficult in many-objective optimization [12]. MOGA are population-based approaches which initiate with a randomly created population of individuals. Then, the algorithm starts an iterative process that creates a new population at each generation, by the use of operators which simulate the process of natural evolution: selection, crossover and mutation.

Among of all MOGA approaches, the literature shows that the NSGA-II is one of the most used for solving multiobjective optimization problems [9,10, 16]. Recently, a modification on the NSGA-II was proposed by [15], seeking to improve the diversity of the set of non-dominated solutions. This new MOGA was called Non-Dominated Sorting Genetic Algorithm Distance Oriented (NSGA-DO). In order to refine the NSGA-DO algorithm, we proposed an adjustment on it. A brief description of these approaches, which are evaluated in this work, is presented in Sects. 3.1, 3.2 and 3.3, depicting only the procedures used by the selection operator, that is the point at which the three algorithms differ. A detailed mathematical formulation is left to the references cited.

3.1   NSGA-II

The NSGA-II algorithm is based on an elitist dominance sorting. For each solution i, contained in the population of solutions, two values are calculated: nd_i, the number of solutions that dominate solution i; and U_i, the set of solutions that are dominated by solution i. Solutions with $nd_i = 0$ are contained in the F_1 front (Pareto front). Then, for each solution j in U_i, the nd_j is decremented for each $i \prec j$, where $i \in F_1$. If $nd_j = 0$, then solution j belongs to the next front, in this case, F_2. This procedure is repeated until all solutions are classified in a front. This procedure consists of classifying the solutions of a set M in different fronts $F_1, F_2, ..., F_f$ according to the dominance degree of such solutions. To guarantee diversity at the front, NSGA-II employs an estimate of the density of solutions that surround each individual in the population. Thus, the average distance of the two solutions adjacent to each individual is calculated for all objectives by Crowding Distance selection operator. The suitability of each solution (individual) i is determined by the following values: $rank_i = f$, the $rank_i$ value is equal to the number of the F_f front to which it belongs; and $crowdist_i$, the crowding distance value of i. Thus, in the dominance sorting process, a solution i is more suitable than a solution j if: i has a ranking lower than j, that is, $rank_i < rank_j$; or if both solutions have the same ranking and $rank_i$ has a higher crowding distance value.

Offspring population and current generation population are combined and the individuals of the next generation are set by dominance sorting process. The new generation is filled by each front subsequently until the number of individuals reach the current population size.

3.2 NSGA-DO

As NSGA-DO is based on NSGA-II, the way it works is similar. The difference between them is due to the selection operator that begins with the estimation of the ideal points in the Pareto front F_1. In order to find the ideal points, the algorithm calculates the length of the Pareto Front and then estimates an uniform partition, setting the coordinates of the ideal points. Thereafter, the selection of solutions to be inserted in the next generation instead of considering the crowding distance as in NSGA-II, considers the tournament distance between each solution of a certain front F and the calculated ideal points, with the aim to enhance the diversity of the solutions.

In order to improve understanding of the difference selection criterion of the closest solution to an ideal point (NSGA-DO) instead of the crowding distance (NSGA-II), consider the following example, presented by [15] and illustrated in Fig. 1. It is possible to observe two representations of the Pareto front, formed by the solutions belonging to the fronts F_1 and F_2.

Solutions belonging to the front F_1 dominate the solutions belonging to the front F_2 and are not dominated by any other solution, composing the Pareto-optimal front. The black dots (I_1, I_2, I_3, I_4 and I_5) represent the ideal points and the gray dots (F_1S_1, F_1S_2, F_1S_3, F_1S_4, F_1S_5, F_2S_1, F_2S_2, F_2S_3 and F_2S_4) represent the solutions found by a MOGA. The f_1 and f_2 are the conflicting objectives to be optimized.

Assuming that eight solutions would be selected for the next generation of the MOGA, NSGA-II and NSGA-DO would form different solution sets. Initially, all solutions belonging to the front F_1 (F_1S_1, F_1S_2, F_1S_3, F_1S_4, F_1S_5) would be selected, because the number of solutions on this front is less than the size

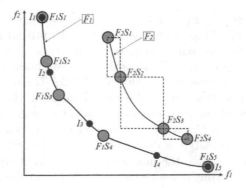

Fig. 1. Difference between NSGA-II and NSGA-DO selection operator [15].

of population (F_1 SIZE $< N$). Up to this point, the NSGA-II solution set is the same as the NSGA-DO solution set, both of which are:

$$S_{\text{NSGA-II}} = S_{\text{NSGA-DO}} = \{F_1S_1, F_1S_2, F_1S_3, F_1S_4, F_1S_5\}$$

Three solutions will be selected from F_2 front (N - F_1 SIZE). The crowding distance operator from NSGA-II choose solutions with greater distance from its neighbors, in this case, solutions S_1 and S_4, because they are located at the extremity of front. Solution S_2 would be the last one to be chosen because its crowding distance is greater than the one of S_3. On the other hand, the solutions chosen by NSGA-DO are those closest to the ideal points. Among the solutions of front F_2 the ranking of choice would be S_3, S_4 and S_2, because they have the lowest distance to an ideal point. Therefore, the final solution sets are:

$$S_{\text{NSGA-II}} = \{F_1S_1, F_1S_2, F_1S_3, F_1S_4, F_1S_5, F_2S_1, F_2S_4, F_2S_2\}$$
$$S_{\text{NSGA-DO}} = \{F_1S_1, F_1S_2, F_1S_3, F_1S_4, F_1S_5, F_2S_3, F_2S_4, F_2S_2\}$$

3.3   MNSGA-DO

Modified Non-Dominated Sorting Genetic Algorithm Distance Oriented (MNSGA-DO) is an extension of NSGA-DO, with the purpose to improve its performance. In the way NSGA-DO works, it estimates ideal points in the Pareto front based on an uniform partition, proportional to its size, and for each solution of a certain front F, the distance from the ideal points is calculated. Then, these distances are sorted, selecting the S solutions associated with the smallest distances, where S is the number of individuals to reach the current population size. In this way, if the S smallest distances are associated with a single ideal point, the selected solutions will be concentrated in a single region of the front, harming diversity.

Instead of estimating ideal points based on an uniform partition proportional to the size of Pareto front, MNSGA-DO calculates an uniform partition based on the number of solutions to be found, setting the coordinates of the ideal points in the Pareto front F_1, what decrease convergence time by reducing the number of points when calculating distances. To improve the approximation to the Pareto front, MNSGA-DO uses the number of ideal points equal to twice the number of solutions to be selected ($2S$). Then, it calculates the distances from solutions in F to ideal points, starting from the ends to the center of the Pareto front, switching the selection between the two ends. Moreover, MNSGA-DO does not allow the repetition of the same solution for different points. When a solution has already been associated with a point, MNSGA-DO do not considers this solution to the other ideal points. Thus, MNSGA-DO guarantee the assignment of one different solution to each ideal point, avoiding the concentration of solutions in some of them and ensuring the diversity of the solutions. MNSGA-DO Selection Operator is illustrated in Algorithm 1.

Algorithm 1: MNSGA-DO Selection Operator

Input : Pareto Front (F_1), front in which the solutions are selected (F) and
            the amount of solutions to be found (S)
Output: Selected solutions set ($SelectedSolutions$)
1 $IdealPoints \leftarrow GenerateIdealPoints(F_1, 2S)$;
2 $SelectedSolutions \leftarrow \emptyset$;
3 $alternate \leftarrow true$;
4 $Solution \leftarrow SearchNearestSolution(IdealPoints[0], F)$;
5 $SelectedSolutions \leftarrow SelectedSolutions \cup Solution$;
6 $Solution \leftarrow SearchNearestSolution(IdealPoints[|IdealPoints| - 1], F)$;
7 $SelectedSolutions \leftarrow SelectedSolutions \cup Solution$;
8 **for** $i \leftarrow 2$ **to** $|IdealPoints|$ **do**
9   **if** $alternate = true$ **then**
10       $index \leftarrow i$;
11   **else**
12       $index \leftarrow |IdealPoints| - i$;
13   **end**
14    $alternate \leftarrow$ **not** $alternate$;
15    $Solution \leftarrow SearchNearestSolution(IdealPoints[index], F)$;
16    $SelectedSolutions \leftarrow SelectedSolutions \cup Solution$;
17   **if** $|SelectedSolutions| = S$ **then**
18       $break$;
19   **end**
20 **end**

---

4 Experimental Setup

For our experiments, the ZDT family of functions [20] was selected, because it is a widely used set of test functions for benchmarking the performance of multi-objective optimization methods. All of the ZDT functions contain two objectives and a particular feature that is representative of a real world optimization problem that could cause difficulty in converging to the Pareto-optimal front [18], as showed in Table 1.

Table 1. ZDT family characteristics.

Test instance	Characteristics
ZDT1	Convex front
ZDT2	Non convex front
ZDT3	Discontinuous convex front
ZDT4	Convex front, multimodal
ZDT6	Convex front non uniform

In this study, 5 out of 6 ZDT functions were considered (ZDT{1-4} and ZDT6) and for each one, 30 runs were conducted using each MOGA, in order to ensure the results were not biased based upon the initial population. We performed our experiments on 2.40 GHz PC with 8 GB RAM and operating system 64 bits. The implementations of MOGAs were done in the Framework JMetal 5 [7] and assumed the following parameters:

Codification: The chromosomes $Ci = (ci_1, ci_2, ...ci_n)$ are encoded as a vector of floating point numbers in which each component of the vector is a variable of the problem.

Initial Population and *Stopping Criterion* were defined following a grid search procedure [4]. The values tested for the population size were 50, 100, 150 and 200. For the number of fitness function evaluations were considered values 10000, 20000, 25000 and 30000. After these tests, initial population was generated randomly considering population size of 100 individuals and stopping criterion was defined as 25000 fitness function evaluations, in all problems.

Fitness Function: The fitness function for MOP is the objective function $F(x)$ to be minimized. The goal of solving a MOP is to find the Pareto-optimal set or at least a set of solutions close to Pareto-optimal set.

Intermediate Population: Based on [5,11,17], we have used Binary Tournament selection, Simulated Binary Crossover (SBX) and Polynomial Mutation. The crossover probability and distribution index were respectively defined as 0.9 and 30. Similarly, the mutation probability was set as $1/n$, where n is the number of problem variables, and the mutation distribution index was 20.

The study of convergence and distribution of solutions along the Pareto Front was performed to the MOGAs by applying three well known quality indicators, Generational Distance (GD), Inverted Generational Distance (IGD) and Hypervolume (HV) [1]. The first one measure the convergence and the other indicators can simultaneously measure the convergence and diversity of obtained solutions.

In the following section, the performance of the NSGA-II, NSGA-DO and MNSGA-DO on ZDT test problems is investigated and the results are presented based on the mentioned quality indicators.

5 Experimental Results

As stated in Sect. 4, we have selected five functions from ZDT family to evaluate NSGA-II, NSGA-DO and MNSGA-DO algorithms: ZDT{1–4} and ZDT6. According to the MOGA configurations presented in Sect. 4 and regarding ZDT functions mentioned, we applied the quality indicators GD, IGD and HV, which make a comparison between the set of points sampled from the Pareto-optimal (set mathematically calculated) and the set of points found by the MOGA to be evaluated.

Thirty design variables x_i were chosen to ZDT{1–4} and ten to ZDT6. Each design variable ranged in value from 0 to 1, except to ZDT4, which the variables ranged from −5 to 5. The numerical results of GD, IGD and Hypervolume are showed in Tables 4, 5 and 6, respectively. The values highlighted in gray means the best for each ZDT function and **SD** column shows standard deviation.

According to the results, for the quality indicator GD, which only evaluates convergence, NSGA-DO found the best results. For both quality indicators IGD and HV, which simultaneously measure the convergence and diversity, the MNSGA-DO had the best results, except ZDT3 problem, in which the NSGA-II was slightly better. This happens because both NSGA-DO and MNSGA-DO consider the entire extension of Pareto front for the definition of ideal points, not taking into account the gaps of the discontinuous ZDT3 problem. Thereby, ideal points are assigned in infeasible regions of the search space, thus compromising their effectiveness.

In Table 2, the execution time of the algorithms is presented in seconds. NSGA-II performed better than MNSGA-DO. NSGA-DO achieved the worst results, with execution time much higher than the other two multiobjective genetic algorithms.

Table 2. Execution time results.

Problem	NSGA-II				NSGA-DO				MNSGA-DO			
	Best	Average	Worst	SD	Best	Average	Worst	SD	Best	Average	Worst	SD
ZDT1	1.60	4.92	8.66	2.17	10.75	13.33	20.45	2.29	2.22	4.69	8.07	1.62
ZDT2	0.84	4.88	8.38	2.63	9.63	14.86	18.86	1.97	1.96	4.85	8.87	1.98
ZDT3	0.60	3.72	7.02	1.89	4.33	13.35	16.75	3.59	0.68	4.31	8.07	2.11
ZDT4	0.38	0.78	0.94	0.13	4.71	6.01	7.47	0.69	0.78	1.05	1.24	0.12
ZDT6	0.73	0.83	0.91	0.05	4.55	5.91	6.71	0.69	0.58	1.06	1.21	0.15

As the result of a two-objective Pareto optimization study is a set of points on a curve (the Pareto front), we plotted the last fronts from each algorithm with the aim of visualising the results variation. Hence, a visual comparison between the algorithms is performed based on the Pareto fronts. Figures 2, 3, 4, 5 and 6 illustrate the Pareto-optimal, NSGA-II, MNSGA-DO and NSGA-DO fronts of each ZDT function. The x and y axes of all figures represent objectives f_1 and f_2, respectively.

After analysing the fronts, we can observe that the NSGA-II and the MNSGA-DO fronts are very similar comparing to Pareto-optimal front. NSGA-II fronts shows some gaps, what demonstrates that the NSGA-II fronts diversity is worst than MNSGA-DO. On the other side, the NSGA-DO front is very different from Pareto-optimal front. This problem can be attributed to the process of individuals selection to reach the current population size, which consider the increasing sorting of the distances from the solutions to the ideal points. Such sorting can add individuals from a single region of the front in the new population. These individuals, combined with others, generate new individuals with characteristics similar to theirs, leading to agglomeration of solutions in regions of the objective space. In addition, as the NSGA-DO algorithm does not balance the number of solutions selected by each ideal point, one same point can select various solutions and another none.

In order to verify if there is statistical difference among the MOGAs results, we have applied the Wilcoxon signed-rank test [19], with level of significance = 0.05. The statistical results are shown in Table 3, in which one − symbol means the null hypothesis was accepted, and one ▲ or ▽ symbol means the null hypothesis was rejected. The ▲ symbol indicate that the algorithm from the line was significantly better than the algorithm from the column, and the ▽ symbol indicate the opposite. Each −, ▲ or ▽ symbols refers to a function, that is, the first symbol refers to ZDT1 function, the second symbol refers to ZDT2 function, and so on.

Table 3. Statistical comparison on quality indicators GD, IGD and HV.

		MNSGA-DO	NSGA-II
GD	NSGA-DO	▲ ▲ ▲ ▲ −	▲ ▲ ▲ ▲ −
	MNSGA-DO		▲ ▲ ▽ ▽ −
IGD	NSGA-DO	▽ ▽ ▽ ▽ ▽	▽ ▽ ▽ ▽ ▽
	MNSGA-DO		▲ ▲ ▽ ▲ ▲
HV	NSGA-DO	▽ ▽ ▽ ▽ ▽	▽ ▽ ▽ ▽ ▽
	MNSGA-DO		▲ ▲ − ▲ ▲
Time	NSGA-DO	▽ ▽ ▽ ▽ ▽	▽ ▽ ▽ ▽ ▽
	MNSGA-DO		− − − ▽ ▽

Based on Table 3, when considering quality indicator GD, the NSGA-DO was significantly better than MNSGA-DO and NSGA-II in almost all functions, only to the ZDT6 function there was not statistical difference among them. Between

MNSGA-DO and NSGA-II, MNSGA-DO was better than NSGA-II to ZDT1 and ZDT2 functions, and NSGA-II was better in ZDT3 and ZDT4 functions. Regarding quality indicator IGD, MNSGA-DO and NSGA-II were significantly better than NSGA-DO in all functions. Take into account only MNSGA-DO and NSGA-II, MNSGA-DO was better in almost all functions, except to the ZDT3 function. Considering quality indicator HV, MNSGA-DO and NSGA-II were significantly better than NSGA-DO in all functions, once more. Comparing MNSGA-DO and NSGA-II, they are similar to ZDT3 function, but MNSGA-DO was markedly better in all others. Finally, evaluating execution time, MNSGA-DO and NSGA-II were significantly better than NSGA-DO in all the problems. Regarding MNSGA-DO and NSGA-II, they are similar to ZDT1-3, but MNSGA-DO was worse than NSGA-II to ZDT4 and ZDT6.

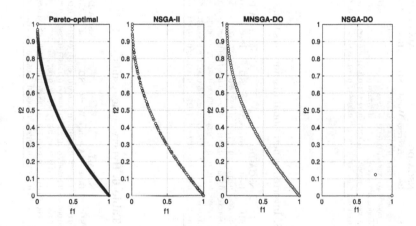

Fig. 2. Pareto-optimal, NSGA-II, MNSGA-DO and NSGA-DO from ZDT1 function.

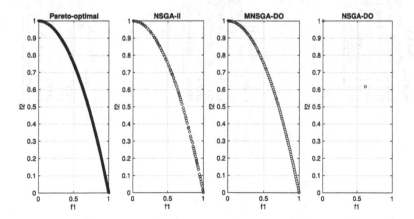

Fig. 3. Pareto-optimal, NSGA-II, MNSGA-DO and NSGA-DO from ZDT2 function.

Table 4. Generational distance results.

Problem	NSGA-II				NSGA-DO				MNSGA-DO			
	Best	Average	Worst	SD	Best	Average	Worst	SD	Best	Average	Worst	SD
ZDT1	0.000107	0.000191	0.000280	0.000036	0.000000	0.000009	0.000092	0.000022	0.000142	0.000164	0.000234	0.000019
ZDT2	0.000063	0.000117	0.000172	0.000027	0.000000	0.000002	0.000033	0.000006	0.000045	0.000063	0.000106	0.000019
ZDT3	0.000103	0.000121	0.000142	0.000009	0.000000	0.000700	0.007449	0.001884	0.000132	0.000138	0.000181	0.000009
ZDT4	0.000051	0.000073	0.000109	0.000012	0.000001	0.001255	0.024971	0.004746	0.000103	0.000104	0.000105	0.000000
ZDT6	0.000038	0.000054	0.000075	0.000011	0.000000	0.000358	0.008617	0.001568	0.000040	0.000048	0.000064	0.000009

Table 5. Inverted generational distance results.

Problem	NSGA-II				NSGA-DO				MNSGA-DO			
	Best	Average	Worst	SD	Best	Average	Worst	SD	Best	Average	Worst	SD
ZDT1	0.000172	0.000185	0.000201	0.000007	0.015840	0.021189	0.028856	0.002522	0.000140	0.000147	0.000163	0.000006
ZDT2	0.000173	0.000191	0.000207	0.000008	0.009518	0.024154	0.029420	0.004149	0.000138	0.000142	0.000168	0.000005
ZDT3	0.000120	0.000133	0.000144	0.000006	0.019166	0.027923	0.030006	0.003872	0.000134	0.000136	0.000146	0.000003
ZDT4	0.000171	0.000181	0.000202	0.000007	0.002492	0.016385	0.028865	0.007478	0.000137	0.000137	0.000137	0.000000
ZDT6	0.000194	0.000217	0.000253	0.000011	0.011906	0.023008	0.027756	0.004219	0.000134	0.000134	0.000134	0.000000

Table 6. Hypervolume results.

Problem	NSGA-II				NSGA-DO				MNSGA-DO			
	Best	Average	Worst	SD	Best	Average	Worst	SD	Best	Average	Worst	SD
ZDT1	0.660834	0.660349	0.659756	0.000248	0.203235	0.043911	0.000000	0.065461	0.661833	0.661629	0.661028	0.000214
ZDT2	0.327608	0.327242	0.326590	0.000229	0.145905	0.015223	0.000000	0.040851	0.326668	0.328511	0.327918	0.000142
ZDT3	0.515473	0.515327	0.515174	0.000075	0.033111	0.002495	0.000000	0.007440	0.515381	0.515332	0.515222	0.000038
ZDT4	0.661023	0.660764	0.660062	0.000199	0.615101	0.080119	0.000000	0.167728	0.662000	0.661953	0.661839	0.000045
ZDT6	0.399493	0.398467	0.397643	0.000455	0.028007	0.001206	0.000000	0.005276	0.401490	0.401259	0.400846	0.000253

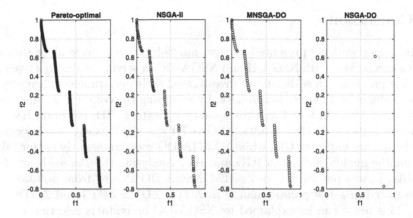

Fig. 4. Pareto-optimal, NSGA-II, MNSGA-DO and NSGA-DO from ZDT3 function.

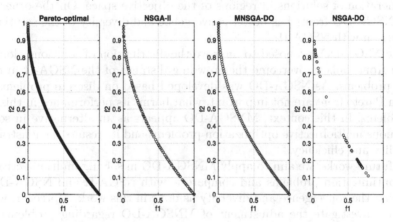

Fig. 5. Pareto-optimal, NSGA-II, MNSGA-DO and NSGA-DO from ZDT4 function.

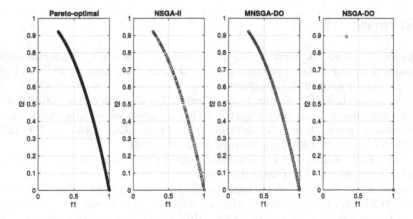

Fig. 6. Pareto-optimal, NSGA-II, MNSGA-DO and NSGA-DO from ZDT6 function.

6   Conclusion

In this paper we have presented a novel multiobjective genetic algorithm and made a comparison with NSGA-II and NSGA-DO in solving multiobjective optimization problems. In addition, we investigated their performance when applied to the popular ZDT family benchmark, by analysing the convergence and diversity of these MOGAs based on three quality indicators, GD, IGD and HV.

According to statistical results, NSGA-DO was significantly better considering the quality indicator GD, while MNSGA-DO was significantly better taking account the quality indicators IGD and HV. Analysing the Pareto fronts from MOGAs, it was possible to verify that NSGA-DO Pareto front solutions are concentrated in a few points, mainly in ZDT1, ZDT2, ZDT3 and ZDT6 functions. This result can be explained by NSGA-DO individuals selection process, which allow one ideal point select various solutions and another none, leading to agglomeration of solutions in regions of the objective space. On the other side, the MNSGA-DO fronts has better convergence and diversity of the solutions in comparison with NSGA-II.

The NSGA-DO, proposed to improve the distribution of solutions along the Pareto front, fails to overcome the crowding distance of the NSGA-II in continuous problems. As NSGA-DO was developed based on discrete problems, the gaps in Pareto front are not important, what harms its performance in this class of problems. In this context, MNSGA-DO appears as an alternative in solving continuous multiobjective optimization problems and its results demonstrate its feasibility and efficiency.

In future works, we aim to apply MNSGA-DO in solving multiobjective discrete optimization problems and compare it with NSGA-II and NSGA-DO by analysing the convergence and diversity as done in this work. Moreover, we are going to investigate the adjustment of MNSGA-DO regarding problems with discontinuous fronts.

References

1. Audet, C., Bigeon, J., Cartier, D., Le Digabel, S., Salomon, L.: Performance indicators in multiobjective optimization. Eur. J. Oper. Res. **292**(2), 397–422 (2021). https://doi.org/10.1016/j.ejor.2020.11.016
2. Bora, T.C., Mariani, V.C., dos Santos Coelho, L.: Multi-objective optimization of the environmental-economic dispatch with reinforcement learning based on non-dominated sorting genetic algorithm. Appl. Therm. Eng. **146**, 688–700 (2019). https://doi.org/10.1016/j.applthermaleng.2018.10.020
3. Chang, K.H.: e-design: computer-aided engineering design. Elsevier Sci. (2015). https://doi.org/10.1016/C2009-0-63076-2
4. Coroiu, A.M.: Tuning model parameters through a genetic algorithm approach. In: 2016 IEEE 12th International Conference on Intelligent Computer Communication and Processing (ICCP), pp. 135–140 (2016). https://doi.org/10.1109/ICCP.2016.7737135

5. Deb, K., Jain, H.: An evolutionary many-objective optimization algorithm using reference-point-based nondominated sorting approach, part i: solving problems with box constraints. IEEE Trans. Evol. Comput. **18**(4), 577–601 (2014)
6. Deb, K., Agrawal, S., Pratab, A., Meyarivan, T.: A fast elitist non-dominated sorting genetic algorithm for multi-objective optimization: NSGA-II. In: International Conference Parallel Problem Solving from Nature, vol. 1917 (2000)
7. Durillo, J.J., Nebro, A.J.: jMetal: a Java framework for multi-objective optimization. Adv. Eng. Softw. **42**(10), 760–771 (2011)
8. George, T., Amudha, T.: Genetic algorithm based multi-objective optimization framework to solve traveling salesman problem. In: Sharma, H., Govindan, K., Poonia, R.C., Kumar, S., El-Medany, W.M. (eds.) Advances in Computing and Intelligent Systems. AIS, pp. 141–151. Springer, Singapore (2020). https://doi. org/10.1007/978-981-15-0222-4_12
9. Guerrero, C., Lera, I., Juiz, C.: Genetic algorithm for multi-objective optimization of container allocation in cloud architecture. J. Grid Comput. **16**(1), 113–135 (2017). https://doi.org/10.1007/s10723-017-9419-x
10. Hamdy, M., Nguyen, A.T., Hensen, J.L.M.: A performance comparison of multi-objective optimization algorithms for solving nearly-zero-energy-building design problems. Energy Build. **121**, 57–71 (2016). https://doi.org/10.1016/j.enbuild. 2016.03.035
11. Ishibuchi, H., Imada, R., Setoguchi, Y., Nojima, Y.: Performance comparison of NSGA-II and NSGA-III on various many-objective test problems. In: 2016 IEEE Congress on Evolutionary Computation, pp. 3045–3052. IEEE (2016)
12. Li, K., Deb, K., Zhang, Q., Kwong, S.: An evolutionary many-objective optimization algorithm based on dominance and decomposition. IEEE Trans. Evol. Comput. **19**(5), 694–716 (2015)
13. Maghawry, A., Hodhod, R., Omar, Y., Kholief, M.: An approach for optimizing multi-objective problems using hybrid genetic algorithms. Soft. Comput. **25**(1), 389–405 (2020). https://doi.org/10.1007/s00500-020-05149-3
14. Miettinen, K.: Nonlinear Multiobjective Optimization. Kluwer Academic Publishers (1999)
15. Pimenta, A., Camargo, H.: NSGA-DO: non-dominated sorting genetic algorithm distance oriented. In: IEEE International Conference Fuzzy System, pp. 1–8 (2015). https://doi.org/10.1109/FUZZ-IEEE.2015.7338080
16. Saborido, R., Ruiz, A.B., Bermúdez, J.D., Vercher, E., Luque, M.: Evolutionary multi-objective optimization algorithms for fuzzy portfolio selection. Appl. Soft Comput. **39**, 48–63 (2016). https://doi.org/10.1016/j.asoc.2015.11.005
17. Seada, H., Deb, K.: U-NSGA-III: a unified evolutionary optimization procedure for single, multiple, and many objectives: proof-of-principle results. In: Gaspar-Cunha, A., Henggeler Antunes, C., Coello, C.C. (eds.) EMO 2015. LNCS, vol. 9019, pp. 34–49. Springer, Cham (2015). https://doi.org/10.1007/978-3-319-15892-1_3
18. Priya, V., Umamaheswari, K.: Enhanced continuous and discrete multi objective particle swarm optimization for text summarization. Clust. Comput. **22**(1), 229–240 (2018). https://doi.org/10.1007/s10586-018-2674-1
19. Wilcoxon, F.: Individual comparisons by ranking methods. Biometrics Bull. **1**(6), 80–83 (1945). https://doi.org/10.2307/3001968
20. Zitzler, E., Deb, K., Thiele, L.: Comparison of multiobjective evolutionary algorithms: empirical results. Evol. Comput. **8**(2), 173–195 (2000). https://doi.org/10. 1162/106365600568202

An Enhanced TSP-Based Approach for Active Debris Removal Mission Planning

João Batista Rodrigues Neto[(✉)] and Gabriel de Oliveira Ramos[(✉)]

Graduate Program in Applied Computing, Universidade do Vale do Rio dos Sinos,
São Leopoldo, RS, Brazil
jrneto@edu.unisinos.br, gdoramos@unisinos.br

Abstract. The extensive exploration of the Low Earth Orbit (LEO) has
created a dangerous spacial environment, where space debris has threatened the feasibility of future operations. In this sense, Active Debris
Removal (ADR) missions are required to clean up the space, deorbiting
the debris with a spacecraft. ADR mission planning has been investigated
in the literature by means of metaheuristic approaches, focused on maximizing the amount of removed debris given the constraints of the spacecraft. The state-of-the-art approach uses an inver-over and maximal open
walk algorithms to solve this problem. However, that approach fails to deal
with large instances and duration constraints. This work extends the state
of the art, increasing its performance and modeling all the constraints.
Experimental results evidence the improvements over the original approach, including the ability to run for scenarios with thousands of debris.

Keywords: Time-dependent traveling salesman problem · Space
debris · Active debris removal · Genetic algorithm

1 Introduction

In the 1960 s, space exploration began to boost the development of new technologies feasible through the use of satellites distributed in orbits. Among these
orbits, Low Earth Orbit (LEO) was widely used for satellite networks with a
large number of objects per service in orbit [20]. Over time, these satellites
became depreciated, lose communication or got out of control, thus becoming
space debris. A high population of debris represents a hazard to the operating
structures in orbit, since they are objects out of control and at high speed [12].

According to some predictive models, a sufficiently large population of debris
will increase the probability of collisions and, therefore, increase the debris population again, thus making this population increase recursively for many years.
This phenomenon is known as Kessler syndrome and may cause the collapse of
the LEO, rendering it useless for years [12].

In fact, the literature already pointed out that the debris population in LEO has
already reached a critical point [16], and now measures to mitigate this situation

© Springer Nature Switzerland AG 2021
A. Britto and K. Valdivia Delgado (Eds.): BRACIS 2021, LNAI 13073, pp. 140–154, 2021.
https://doi.org/10.1007/978-3-030-91702-9_10

are needed. While there are documented techniques that would stabilize the debris population, for now, the only approach capable of reducing it consists in Active Debris Removal (ADR) missions [16]. These missions aim to clean up orbit space by forcing the re-entry of certain debris as performed by specific spacecraft. Due to the limited resources and time to perform the rendezvous maneuvers and the debris population size, the selection of the debris to be removed has became a hard combinatorial problem [3, 4, 7].

A basic ADR should consider the bounds of cost and duration, dependent position values of the moving debris along time [3, 4], and also prioritizing the removal of the most threatener debris first, thus increasing the benefits of such missions [17]. Various approaches to this problem have been proposed in the literature [7, 9, 13]. However, most of these works share the same limitations: small instance sizes, unbounded approaches, and non-time-dependent modeling.

In this work, we propose an enhanced genetic algorithm to optimize ADR mission planning. Our approach builds upon the work by [9], improving its performance through a novel combination of genetic operators. The final algorithm resembles the original *inver-over* genetic operator, with modifications to its reversing strategy. Moreover, a new variant of the *k-opt* algorithm is implemented using a stochastic approach. Finally, the open-walk algorithm is improved with one additional constraint. Through extensive experiments, our approach yielded better solutions with instances larger than the previous largest ones [9].

2   Literature Review

In order to solve the ADR optimization problem, a few methods have been approached and documented in the last 10 years. Exact solutions were used by Braun et al. [2] with brute force, and branch and bound variations by [3, 14, 19]. However, in both classes of methods can only be applied to small instances. Approximate solutions were implemented using simulated annealing [4, 7], reinforcement learning [25], and genetic algorithms [18, 24]. Nonetheless, all of these works were also tested only on small instances.

On the other hand, a few approximate methods considered bigger instances. Barea et al. [1] used a linear programming method, which has a high complexity as drawback. Yang et al. [26] used a greedy heuristic, but requires instance-dependent parameters. Ant Colony Optimization was used in [13, 21, 27], but ignoring mission constraints, strongly simplifying the cost dynamics to reduce the complexity, or even leaving mission duration unbounded. Finally, Izzo et al. [9] and Kanazaki et al. [11] used genetic algorithms, though both did not model all the necessary mission constraints.

Generally speaking, the majority of the works do not prioritize debris by hazard, consider bigger instances, model the time-dependence, or bound the cost or the mission duration, meaning that most works fail to fully meet the ADR mission requirements. Building against this background, in this work we introduce the enhanced inver-over operator to deal with large instances, and an enhanced maximal open walk algorithm to model the cost and duration constraints while prioritizing the most threatener debris.

3 Problem Formulation

An ADR problem is the combinatorial problem of finding the correct sequence to rendezvous maneuvers towards debris in order to maximize the mission profit given some constraints. In this sense, ADR can be seen as a complex variant of the Travelling Salesman Problem (TSP), where one wants to find the minimum weight path in a dynamic complete graph, where the debris are the cities and the dynamic transference trajectories are the edges. The dynamicity is due to the time-dependent cost of the transference, so the correct generalized version of the TSP will be a time-dependent TSP (TDTSP). This work will make use of the integer linear TDTSP problem formulation by [7].

Hereafter, we will follow the notation typically used in the literature [7]. Let $V = \{1, ..., n\}$ be the set of n debris. The distance tensor is represented by $C = (c_{ijtm})$, where c_{ijtm} is the cost of transfer from debris i at epoch t to debris j at epoch $t+m$. Also, let $X = (x_{ijtm})$ be a binary tensor, where $X_{ijtm} = 1$ indicates that this transference is part of the solution and $X_{ijtm} = 0$ otherwise. The n_t possible epochs of departure and arrival are discretized following $n_t \geq n + 1$ and $M \leq n_t - (n-1)$. Usually, in order to grant some freedom at the mission planning, n_t is far larger than n, while M limits the maximum duration of the transfers.

Along these lines, the problem of finding the optimized route can be modelled as finding the X matrix that minimizes the total cost with due respect to the constraints, which can be formulated as follows.

$$\min_x \sum_{i=1}^{n} \sum_{j=1}^{n} \sum_{t=1}^{n_t} \sum_{m=1}^{M} c_{ijtm} x_{ijtm} \tag{1}$$

$$\text{s.t.} \sum_{i=1}^{n} \sum_{j=1}^{n} \sum_{t=1}^{n_t} \sum_{m=1}^{M} x_{ijtm} = n \tag{2}$$

$$\sum_{i=1}^{n} \sum_{t=1}^{n_t} \sum_{m=1}^{M} x_{ijtm} = 1 \qquad j = 1, ..., n \tag{3}$$

$$\sum_{j=1}^{n} \sum_{t=1}^{n_t} \sum_{m=1}^{M} x_{ijtm} = 1 \qquad i = 1, ..., n \tag{4}$$

$$\sum_{j=1}^{n} \sum_{t=2}^{n_t} \sum_{m=1}^{M} t x_{ijtm} - \sum_{j=1}^{n} \sum_{t=1}^{n_t} \sum_{m=1}^{M} t x_{ijtm} = \sum_{j=1}^{n} \sum_{t=1}^{n_t} \sum_{m=1}^{M} m x_{ijtm} \qquad i = 2, ..., n \tag{5}$$

$$\sum_{i=1}^{n} \sum_{j=1}^{n} \sum_{t=1}^{n_t} \sum_{m=1}^{M} m x_{ijtm} \leq n_t \tag{6}$$

In the above formulation Eq. (1) represents the objective function and Eqs. (2)–(6) represent the problem constraints, to which we will refer simply as constraints (2)–(6) hereafter. Constraint (2) guarantees that the solution tensor has all the n debris. Constraints (3) and (4) ensure that there is no loops in the solution (one departure and one arrival transfer for each debris). Constraint (5) enforces the transfer duration. Finally, constraint (6) limits the total mission duration. As the result, this formulation has a $\mathcal{O}(2^n)$ search space, using n^4 binary decision variables and $3n - 1$ constraints.

3.1   Orbit Transfer

The trip between one debris and another require impulses (Δv) of the thrusters to change the orbit of the spaceship. Low thrust propulsion systems can perform this maneuvers efficiently. However, they require the optimization of the trajectories to make the mission time available [5]. Determining a minimal fuel transfer trajectory between two debris is a complex optimization problem in general case. Thus, major works simplify this task by using a generic transfer strategy [4].

The major used transfer strategies are the Hohmann and Lambert transfers. Since this work's scope does not focus on the orbital transfer optimization problem, the mechanics of the transfers will be briefly described. In [6], Hohmann is described as a minimum two-impulse elliptic maneuver to transfer from coplanar orbits. Hohmann transfer is a high thrust transfer. Since debris are not always in co-planar orbits it is also presented a variation of the Hohmann transfer, namely the Edelbaum transfer, which is a three-impulse bi-elliptic transfer that allows transferences between non-coplanar orbits.

In [5], Lambert is described as two-impulse trajectory to transfer from coplanar orbits given a certain transference duration. Also, it is possible to make use of the J_2 gravitational earth perturbation to wait for the natural alignment of orbital planes, saving some fuel but increasing the mission duration [5,7].

Finally, the cost of a transference between two debris relies on the mass of the spacecraft, since the thrusters consume propellant mass at each impulse, as the mission goes on the cost of the transfers became lower due to the mass lost in the previous maneuvers. So, it is also possible to optimize the removal sequence taking in account the resultant masses of the objects [3].

4   Proposed Approach

In this work, the ADR problem is approached with an improved heuristic solution, similar to the method used by [9] with the inver-over operator in a Genetic Algorithm and the maximal open walk algorithm. The inver-over operator optimizes the total cost of the mission with a local search strategy through genetic operations on the individuals. The maximal open walk algorithm constraints the path. This work enhances this solution with a greedier implementation of the algorithms. Moreover, to avoid the local optimum, a stochastic 2-opt is proposed to improve the solution, creating new connections. Finally, the most rewardable open walk is extracted from the best individual as the final solution.

As this is a bi-optimization problem, the complexity was divided in two stages. In the first part, the effectiveness of a solution will be given by the total cost of the removal sequence, calculated with the Edelbaum transfer [6] (Eqs. (7) and (8)), a consistent and reliable cost approximation for cheap orbital maneuvers. In the second part, the effectiveness of a solution became the removed threat of the LEO, calculated with the sum of radar cross section (RCS) area of the removed debris. The RCS area is an abstraction of the size of the debris and is widely used for the threat calculation in the literature, being a measure about how much detectable an object is for a ground radar.

$$\Delta v = \sqrt{v_0^2 - 2V_0 V_f \cos \frac{\pi}{2} \Delta i + v_f^2} \tag{7}$$

$$\cos \Delta i = \cos i_1 \cos i_2 + \sin i_1 \sin i_2 (\cos \Omega_1 \cos \Omega_2 + \sin \Omega_1 \sin \Omega_2) \tag{8}$$

$$T = \frac{1}{2} \sqrt{\frac{4\pi^2 a^3}{\mu}} \tag{9}$$

Also, in order to minimize the complexity of the problem, major works in the literature have assumed a few dynamics simplifications. In this work, the following assumptions were made with the same purpose:

- The time dependence of the problem is relaxed by the correlation explored in [9], where an optimal solution can remain optimal up to 50 d, depending on the size of the instance.
- Since the Edelbaum transfer is an optimized variation of the Hohmann transfer [6], the duration of the transference arcs can be computed using Kepler's third law of planetary motion (Eq. (9)), which measures the orbital transference duration of an object (spacecraft) between two orbits (debris).
- The transfer cost also depends on mass of the spacecraft, that will decrease during the mission, where the fuel mass will be consumed. Moreover, the gravitational effects of the earth on the spacecraft also influence the transfer cost. In this work, the masses of the objects and the orbital perturbation effects are neglected in the cost transfer dynamics.
- There are more steps of the rendezvous process to remove a debris, and each step take some time to be performed [15]. In this work the duration of the mission will be given by the sum of the duration of the transferences, the other stages will be neglected.

4.1 Inver-Over

The inver-over operator [22] is a unary genetic operator that resembles characteristics of mutation and crossover at the same time. The evolution of an individual is based on simple population-driven inversions and recombinations of genes. This is a well established operator, know by its good performance with larger instances [22].

Algorithm 1. Inver-over

```
 1: generate random population P;
 2: while not satisfied termination-condition do
 3:     for each individual $S_i \in P$ do
 4:         $S' \leftarrow S_i$
 5:         $c \leftarrow$ select random gene $\in S'$
 6:         repeat
 7:             $S'' \leftarrow$ select random individual $\in P$ that $S'' \neq S'$
 8:             $c' \leftarrow$ gene after $c$ in $S''$
 9:             if $c'$ is the next or previous gene to $c$ in $S'$ then
10:                 exit from repeat loop;
11:             end if
12:             if $c' > c$ then
13:                 $c_{temp} \leftarrow$ gene after $c'$ in $S'$
14:             else
15:                 $c_{temp} \leftarrow$ gene before $c'$ in $S'$
16:             end if
17:             inverse the section from the gene after $c$ until $c'$ in $S'$
18:             $c \leftarrow c_{temp}$
19:         until
20:         if $eval(S') \leq eval(S_i)$ then
21:             $S_i \leftarrow S'$
22:         end if
23:     end for
24: end while
```

In this work a different version on the algorithm is used, mixing the original implementation with the [9] implementation. This work allows array cyclic inversions to happen, inversions that include the section of the last to the first gene. However, the next gene pick depends of the previous order of the selected genes for inversion. This inversion process is analogous to a crossover operator.

Furthermore, it is stated that to avoid the local optima, a process analogous to the mutation operator has to be used to create new connections that do not exist in population [22]. However, these mutations are not greedy and usually delay the convergence process, so for this implementation it was removed. The implemented algorithm pseudo-code is sketched in Algorithm 1.

4.2   Stochastic 2-Opt

The 2-opt optimization [8] is a TSP local search algorithm that adjusts the routing sequence greedily with simple inversions. The main idea is to break the route in two paths and reconnect it invertedly, if it improves the fitness so the inversion is kept in the solution. Unfortunately, this is an exact algorithm with a complexity of $\mathcal{O}(n^2)$, so a lot of solutions evaluations need to be performed in order to improve the solution. Nonetheless, there are other methods to improve the 2-opt performance, such as search parallelism and the Lin and Kernighan technique [10]. The present work made use of a stochastic approach for the algorithm.

The stochastic implementation relies on the observations that even with a N^2 search space, the actual number of improvements performed by the algorithm is roughly N [10]. So, with a random exploration of the space, there could be more

Algorithm 2. Stochastic 2-opt

```
1:  given an individual S_i ∈ P;
2:  S' ← S_i
3:  moves ← {}
4:  while not satisfied termination-condition do
5:      repeat
6:          c ← select random gene ∈ S'
7:          c' ← select random gene ∈ S'
8:      until (c, c') ∉ moves
9:      moves ← moves + (c, c')
10:     if c' < c then
11:         tmp ← c
12:         c ← c'
13:         c' ← tmp
14:     end if
15:     S'' ← inverse the section from the gene c until c' of S'
16:     if eval(S'') ≤ eval(S') then
17:         S' ← S''
18:     end if
19: end while
20: if eval(S') ≤ eval(S_i) then
21:     S_i ← S'
22: end if
```

chances of finding a profitable move. In order to keep control of the algorithm run time, a termination condition is used to limit the exploration. Also, a set of explored moves prevents duplicated evaluations. The implemented algorithm pseudo-code is sketched in Algorithm 2.

4.3 Maximal Open Walk

The maximal open walk proposed by [9] as "City Selection", is a separated algorithm that searches for the contiguous part of a Hamiltonian path with the maximal cumulative value limited to some constraint. The path is the optimal solution found, while the value and constraint are respectively, the threat, given by the RCS area, and the total cost. In this work, another constraint is added to this problem, the duration of the open walk, calculated with Kepler's third law. The implemented algorithm pseudo-code is sketched in Algorithm 3.

5 Experimental Evaluation

The objective of these experiments is to evaluate the performance of the approach, understand the improvement gain by each technique and find some optimal parameters. To preserve the comparability, all the runs used 20000 fixed iterations as the termination-condition of the inver-over algorithm, a 100 individuals population, the original method runs used 0.05 as ri (mutation probability), the constrained runs were performed with a cost constraint of 1000 m/s and a time constraint of 1 year.

Algorithm 3. Maximal Open Walk

1: given an individual $S_i \in P$;
2: $arcs \leftarrow \{\}$
3: **for** each pair of contiguous genes $(g', g'') \in S_i$ **do**
4:     $arcs \leftarrow arcs + (cost(g', g''), duration(g', g''))$
5: **end for**
6: $walks \leftarrow \{\}$
7: **for** each of arc $a' \in arcs$ **do**
8:     $walk \leftarrow \{a'\}$
9:     $cost \leftarrow a'_{cost}$
10:     $duration \leftarrow a'_{duration}$
11:     **for** each of arc a'' after $a' \in arcs$ **do**
12:         **if** $cost + a''_{cost} > max_{cost}$ **or** $duration + a''_{duration} > max_{duration}$ **then**
13:             exit from second for loop;
14:         **else**
15:             $walk \leftarrow walk + a''$
16:             $cost \leftarrow cost + a''_{cost}$
17:             $duration \leftarrow duration + a''_{duration}$
18:         **end if**
19:     **end for**
20:     $genes \leftarrow$ the genes of the arcs in $walk$
21:     $reward \leftarrow threat(genes)$
22:     $walks \leftarrow walks + (genes, cost, duration, reward)$
23: **end for**
24: **return** $walk \in walks$ with the biggest $walk_{reward}$

The data about the debris were extracted from the Satellite Catalog (SAT-CAT), a catalogue of all the objects on the Earth orbit, maintained by the United States Space Command (USSPACECOM). The following instances of the problem were extracted: Iridium-33, Cosmos-2251 and Fengyun-1C, with respectively 331, 1048 and 2653 debris. Data was collected at respectively 11-Jun-2021 00:06 UTC, 13-Jun-2021 22:06 UTC and 13-Jun-2021 22:06 UTC.

To preserve the comparability of some results, back-propagated instances were generated inputting the actual instances in a SGPD4 orbital propagator [23] that backtracked the debris positions back to 01-Jan-2015 at 00:00 UTC, the same epoch of the instances used by [9]. Unfortunately this process is not very precise, though still feasible. All the debris in the clouds were considered, including the ones that will decay during the mission time.

For the sake of clearness, the *Time (min)* values on the experiments are concerned to the computation time taken for the run, and *Std. dev.* is the abbreviation for Standard Deviation. All experiments were conducted on a public online machine with a 2.30GHz CPU and 12.69 GB of RAM. Also, for all the experiments, the statistical data results out of 10 independent runs.

5.1 Back-Propagated Instances

The experiments performed with the back-propagate debris are intended to provide comparative results to the work of [9] and guide the definition of the parameters. The inver-over algorithm implemented in this work differs from the original

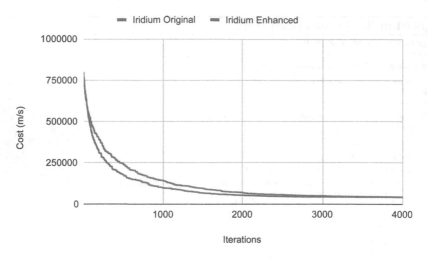

Fig. 1. Iridium-33 convergence results

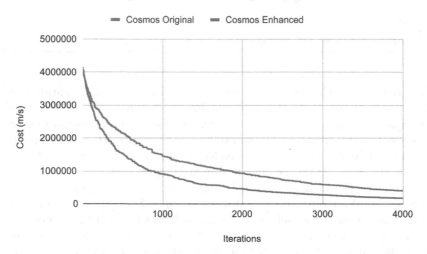

Fig. 2. Cosmos-2251 convergence results

implementation on two points, each point will be tested separately to show the advantages and justify its usage in this approach.

The changes made on the inversion implementation in this work aims to improve the convergence of the algorithm, to do so, this work approach is more population driven and less random mutated. To analyse the performance of the changes, the instances Iridium-33 and Cosmos-2251 were each run twice for 10 times, the first runs with the [9] implementation, and the other ones with this work implementation. The Figs. 1 and 2 demonstrate the improvement of the convergence, for a better visualization, just the first 4000 iterations were drawn.

Table 1. Inversion results

Instance	Method	Mean (min)	Std. dev. (min)
Iridium-33	Original	2.14	0.02
Iridium-33	Enhanced	**1.25**	0.02
Cosmos-2251	Original	12.09	0.07
Cosmos-2251	Enhanced	**7.30**	0.08

It is possible to notice a considerably change on the shape of the curve and the taken computation time in Table 1, making this works implementation convergence better. For the record, the majority of runs in each instance got the same final result, indicating that, for small and medium sized instances, the fast convergence does not deteriorate the result.

To deviate from the local optima, the stochastic 2-opt (S2opt) is used in this work. Parametric tests were conducted to analyse its performance when it matters to time and achieved result. Iridium-33 and Cosmos-2251 were submitted to 9 different runs, running 10 times each, with a different combination of two parameters: how often does the S2opt runs and with how many iterations at each time. All individuals of the population were processed at each S2opt iteration.

Since the search area of the S2opt is large, a range of possible attempts should be chosen, being neither too small, so no improvement move is found, or too big, so almost the whole search space is tested, turning it into a exact solution. In this work, the chosen range is from 10,000 to 1,000,000 attempts, while the parameters are equally spaced discrete values where its configuration do not underflow or overflow the range. The parameters per run are the following:

- Run 1: At each 100 main iterations, run S2opt with 100 iterations.
- Run 2: At each 100 main iterations, run S2opt with 500 iterations.
- Run 3: At each 100 main iterations, run S2opt with 1000 iterations.
- Run 4: At each 500 main iterations, run S2opt with 100 iterations.
- Run 5: At each 500 main iterations, run S2opt with 500 iterations.
- Run 6: At each 500 main iterations, run S2opt with 1000 iterations.
- Run 7: At each 1000 main iterations, run S2opt with 100 iterations.
- Run 8: At each 1000 main iterations, run S2opt with 500 iterations.
- Run 9: At each 1000 main iterations, run S2opt with 1000 iterations.

Analysing the results in Figs. 3 and 4 it is possible to state that due to the small size of the Iridium-33 debris, all the runs achieved the optimal solution. Also, the number of S2opt iterations is directly proportional to the computation time. And finally, the Runs 4, 7 an 8 have the better performances ratios, with low values of cost and time, among these, Run 7 is the best one.

Also, to preserve the idea of a competitive evolution, the following experiments use S2opt with an elitist strategy. This time, instead of running the S2opt for the whole population at each S2opt iteration, just the better individuals will be improved. Parametric tests were performed with 5 different sizes of elites, running 10 times each. To preserve the elitist characteristic, the elite group should not be greater than 30% of the population.

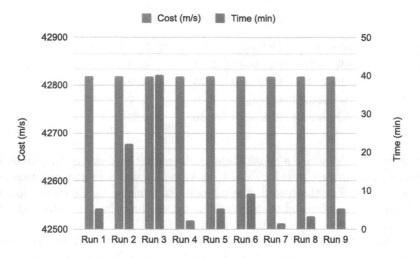

Fig. 3. Iridium-33 Stochastic 2-opt results

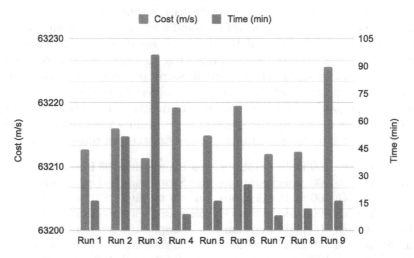

Fig. 4. Cosmos-2251 Stochastic 2-opt results

In these experiments, Iridium-33 will be discarded, since its small size does not help to fully analyse the performance of the algorithm. Here, the tests are set with the same parameters of the previous Run 7 (best run), at each 1000 main iterations, run S2opt with 100 iterations.

Analysing the results in Table 2 it is possible to state that the elitist improvement of the best 5 individuals at each iteration is the wise strategy to follow, having a lower cost with a little bit more computational time taken.

To emphasises the importance of each technique, ablation experiments were conducted for this solution. It is important to state that the Elitism is applied on the S2Opt, so there is no possible scenario using Elitism without S2Opt. Table 3,

Table 2. Elitist results

Run	Elite size	Cost (m/s)		Time (min)	
		Mean	Std. dev.	Mean	Std. dev.
Run 1	1	63,226	29.75	8.01	0.12
Run 2	5	**63,211**	0.10	7.54	0.05
Run 3	10	63,216	16.92	**7.52**	0.11
Run 4	20	63,217	22.01	8.02	0.13
Run 5	30	63,220	26.56	8.07	0.10

Table 3. Ablation results

Technique			Cost (m/s)		Time (min)	
Inver-over	S2Opt	Elitism	Mean	Std. dev.	Mean	Std. dev.
Y	Y	Y	**63,211**	0.10	**7.54**	0.05
Y	Y	N	63,211	2.14	8.27	0.31
Y	N	N	63,214	8.23	7.49	0.07
N	Y	Y	63,225	21.01	12.57	0.12
N	Y	N	63,252	36.87	13.34	0.14
N	N	N	63,256	42.70	12.50	0.08

summarizes the results of each combination. The experiments were performed on instance Cosmos-2251, with 10 runs each. The S2opt parameters are 100 runs at each 1000 main iterations, and the elitist parameter is 5.

With the ablation experiments, it is possible to understand how each technique of our approach affects the final results. It is clear that the conjunction of the techniques improves the found solutions.

5.2 Actual Instances

The experiments performed with the actual instances are intended to provide results for the present status of the debris clouds. The runs were performed using the best parameters found in the previous experiments. Also, for a comparative result, the original implementation was used with the actual instances, and its solutions, inputted to the enhanced maximal open walk of this work. For each instance were performed 10 runs, for the sake of clarity, the settings are: Enhanced inversion implementation, with S2opt running 100 iterations for the best 5 individuals at each 1000 main iterations, with a cost constraint of 1000 m/s and a time constraint of 1 year.

The results on both sections of Table 4 are from the same runs, the results at the bottom are given by the maximal open walk applied to the optimized path at the top. Being the missions objective: clear the maximum possible area are under the cost and time constraints [17], our approach focused on making an optimized use of the mission resources to outperforms other approaches. Cleaning more area, even if the mission duration and cost are bigger, bounded to time and cost constraints, of course.

Table 4. Final results

Instance	Method	Cost (m/s)		Duration (min)		Time (min)	
		Mean	Std. dev.	Mean	Std. dev.	Mean	Std. dev.
Iridium-33	Original	**55,788**	0.0	**16,345**	0.57	4.27	0.04
Iridium-33	Enhanced	55,810	12.69	16,346	0.80	**2.56**	0.03
Cosmos-2251	Original	186,896	4,512	52,069	1.23	46.52	1.29
Cosmos-2251	Enhanced	**156,067**	8,022	52,069	1.41	**31.00**	0.14
Fengyun-1C	Original	2,438,068	24,731	**134,904**	0.71	188.29	6.50
Fengyun-1C	Enhanced	**899,179**	11,468	134,905	0.82	**115.37**	5.24

Constrained results

Instance	Method	Cost (m/s)		Duration (min)		Area (m^2)	
		Mean	Std. dev.	Mean	Std. dev.	Mean	Std. dev.
Iridium-33	Original	971.16	26.17	34,722	3,785	3,984	196.0
Iridium-33	Enhanced	**952.42**	61.79	**34,720**	3,789	**3,993**	167.0
Cosmos-2251	Original	974.77	19.55	**26,675**	6,705	1,666	86.0
Cosmos-2251	Enhanced	**967.90**	19.62	33,163	5,009	**1,703**	41.0
Fengyun-1C	Original	**859.63**	270.0	**148.0**	84.52	671.4	45.4
Fengyun-1C	Enhanced	951.67	67.63	403.5	60.21	**905.5**	221.6

Summarizing, in the average of the runs, the enhanced approach decreased the cost by **26.51%** and the computation time by **35.56%**. Also, when constrained, the solutions of the enhanced approach produced paths that performs significantly better than the original approach solutions, cleaning **12.82%** more area under the same constraints. This is possible by a better usage of the constrained resources, increasing the cost by **5.28%** and the mission duration by **1.06%**. Finally, it is possible to notice that the most profitable mission is the enhanced Iridium-33, cleaning way more area with the same constraints.

6   Conclusions

The exploration of TSP approaches when dealing with space debris have considerably evolved the TDTSP problem modeling and its available solutions. However, it still lacks from approaches that fulfill the ADR mission requirements with a feasible performance and big instance sizes. This work proposed an enhanced method as a strong candidate to future approaches on TSP based ADR mission plannings. Using the inver-over as a fast convergence algorithm to deal with the greedy search throught fast inversions, the S2opt as a solution diversity creator to deviate the search from local optima, our method found optimized solutions in a feasible time out of large datasets. Real world instances were used to evaluate this approach performance and execute parametric tests, the retrieved results were considerably better than the original approach by [9].

However, this approach does not model the time dependence of the problem, meaning that the produced solutions may not be good solutions in a real scenario, since it does not consider the moving dynamics of the debris. Also, this approach first optimizes the cost of the solution, and later chooses the most rewardable sub-path, so it is not a fully bi-optimization algorithm. Meaning that solutions with a good cost versus reward ratio could be missed.

As future work, we would like to implement a time-dependent removal sequence to produce a complete solution for the ADR mission planning problem. We also plan to implement Lin and Kernighan's algorithm to improve the convergence of local search heuristic. The optimization of the transference cost with the consideration of body masses and J_2 effect on the transfers represents another interesting direction. Finally, we would like to improve the approximation of the mission duration with a more suitable equation for the Edelbaum (rather than Hohmann) transfer duration.

Acknowledgements. We thank the anonymous reviewers for their constructive feedback. This work was partially supported by FAPERGS (grant 19/2551-0001277-2).

References

1. Barea, A., Urrutxua, H., Cadarso, L.: Large-scale object selection and trajectory planning for multi-target space debris removal missions. Acta Astronautica **170**, 289–301 (2020). https://doi.org/10.1016/j.actaastro.2020.01.032
2. Braun, V., et al.: Active debris removal of multiple priority targets. Adv. Space Res. **51**(9), 1638–1648 (2013)
3. Cerf, M.: Multiple space debris collecting mission-debris selection and trajectory optimization. J. Optim. Theor. Appl. **156**, 761–796 (2013). https://doi.org/10.1007/s10957-012-0130-6
4. Cerf, M.: Multiple space debris collecting mission: optimal mission planning. J. Optim. Theor. Appl. **167**(1), 195–218 (2015). https://doi.org/10.1007/s10957-015-0705-0
5. Koblick, D., et al.: Low thrust minimum time orbit transfer nonlinear optimization using impulse discretization via the modified picard-chebyshev method. Comput. Model. Eng. Sci. **111**(1), 1–27 (2016). https://doi.org/10.3970/cmes.2016.111.001
6. Edelbaum, T.N.: Propulsion requirements for controllable satellites. ARS J. **31**(8), 1079–1089 (1961). https://doi.org/10.2514/8.5723
7. Federici, L., Zavoli, A., Colasurdo, G.: A time-dependent tsp formulation for the design of an active debris removal mission using simulated annealing (2019)
8. Flood, M.M.: The traveling-salesman problem. Oper. Res. **4**(1), 61–75 (1956). https://doi.org/10.1287/opre.4.1.61
9. Izzo, D., Getzner, I., Hennes, D., Simões, L.F.: Evolving solutions to tsp variants for active space debris removal. In: Proceedings of the 2015 Annual Conference on Genetic and Evolutionary Computation, pp. 1207–1214. GECCO '15, ACM (2015)
10. Johnson, D., McGeoch, L.A.: The traveling salesman problem: A case study in local optimization (2008)
11. Kanazaki, M., Yamada, Y., Nakamiya, M.: Trajectory optimization of a satellite for multiple active space debris removal based on a method for the traveling serviceman problem. In: 2017 21st Asia Pacific Symposium on Intelligent and Evolutionary Systems (IES), pp. 61–66 (2017). https://doi.org/10.1109/IESYS.2017.8233562

12. Kessler, D.J., Cour-Palais, B.G.: Collision frequency of artificial satellites: the creation of a debris belt. J. Geophys. Res. Space Phys. **83**(A6), 2637–2646 (1978). https://doi.org/10.1029/JA083iA06p02637

13. Li, H., Baoyin, H.: Optimization of multiple debris removal missions using an evolving elitist club algorithm. IEEE Trans. Aerospace Electron. Syst. **56**(1), 773–784 (2020). https://doi.org/10.1109/TAES.2019.2934373

14. Li, H.Y., Baoyin, H.X.: Sequence optimization for multiple asteroids rendezvous via cluster analysis and probability-based beam search. Sci. Chin. Technol. Sci. **64**(1), 122–130 (2020). https://doi.org/10.1007/s11431-020-1560-9

15. Liou, J.C.: An active debris removal parametric study for leo environment remediation. Adv. Space Res. **47**(11), 1865–1876 (2011). https://doi.org/10.1016/j.asr.2011.02.003

16. Liou, J.C., Johnson, N.: Instability of the present leo satellite populations. Adv. Space Res. **41**(7), 1046–1053 (2008). https://doi.org/10.1016/j.asr.2007.04.081

17. Liou, J.C., Johnson, N., Hill, N.: Controlling the growth of future leo debris populations with active debris removal. Acta Astronautica **66**(5), 648–653 (2010). https://doi.org/10.1016/j.actaastro.2009.08.005

18. Liu, Y., Yang, J., Wang, Y., Pan, Q., Yuan, J.: Multi-objective optimal preliminary planning of multi-debris active removal mission in LEO. Sci. Chin. Inform. Sci. **60**(7), 1–10 (2017). https://doi.org/10.1007/s11432-016-0566-7

19. Madakat, D., Morio, J., Vanderpooten, D.: Biobjective planning of an active debris removal mission. Acta Astronautica **84**, 182–188 (2013). https://doi.org/10.1016/j.actaastro.2012.10.038

20. Seumahu, E.S.: Exploration of the Equatorial LEO Orbit for Communication and Other Applications, pp. 217–228. Springer, Netherlands, Dordrecht (1996). https://doi.org/10.1007/978-94-011-5692-9_25

21. Stuart, J., Howell, K., Wilson, R.: Application of multi-agent coordination methods to the design of space debris mitigation tours. Adv. Space Res. **57**(8), 1680–1697 (2016). https://doi.org/10.1016/j.asr.2015.05.002

22. Tao, G., Michalewicz, Z.: Inver-over operator for the TSP. In: Eiben, A.E., Bäck, T., Schoenauer, M., Schwefel, H.-P. (eds.) PPSN 1998. LNCS, vol. 1498, pp. 803–812. Springer, Heidelberg (1998). https://doi.org/10.1007/BFb0056922

23. Vallado, D., Crawford, P.: SGP4 Orbit Determination, pp. 18–21 (2008). https://doi.org/10.2514/6.2008-6770

24. Wang, D., Li, L., Chen, L.: An efficient genetic algorithm for active space debris removal planning. In: 2019 IEEE Congress on Evolutionary Computation (CEC), pp. 514–521 (2019). https://doi.org/10.1109/CEC.2019.8790081

25. Yang, J., Hou, X., Hu, Y.H., Liu, Y., Pan, Q.: A reinforcement learning scheme for active multi-debris removal mission planning with modified upper confidence bound tree search. IEEE Access **8**, 108461–108473 (2020)

26. Yang, J., Hu, Y.H., Liu, Y., Pan, Q.: A maximal-reward preliminary planning for multi-debris active removal mission in leo with a greedy heuristic method. Acta Astronautica **149**, 123–142 (2018). https://doi.org/10.1016/j.actaastro.2018.05.040

27. Zhang, T., Shen, H., Li, H., Li, J.: Ant Colony Optimization based design of multiple-target active debris removal mission (2018). https://doi.org/10.2514/6.2018-2412

Dynamic Learning in Hyper-Heuristics to Solve Flowshop Problems

Lucas Marcondes Pavelski[1]([⊠])[ID], Marie-Éléonore Kessaci[2][ID],
and Myriam Delgado[1][ID]

[1] CPGEI - Universidade Tecnológica Federal do Paraná, Curitiba, Brazil
lpavelski@alunos.utfpr.edu.br, myriamdelg@utfpr.edu.br
[2] CRIStAL - Univ. Lille, CNRS - Centrale Lille - UMR 9189, 59000 Lille, France
marie-eleonore.kessaci@univ-lille.fr

Abstract. Hyper-heuristics (HHs) are algorithms suitable for designing
heuristics. HHs perform the search divided in two levels: they look for
heuristic components in the high level and the heuristic is used, in the
low level, to solve a set of instances of one or more problems. Different
from offline HHs, hyper-heuristics with dynamic learning select or generate heuristics during the search. This paper proposes a hyper-heuristic
associated with a dynamic learning strategy for selecting Iterated Greedy
(IG) components. The proposal is capable of selecting appropriate values for six IG components: local search, perturbation, destruction size,
neighborhood size, destruction position and local search focus. The proposed HH is tested with six dynamic adaptation strategies: random, ϵ-
greedy, probability matching, multi-armed bandit, LinUCB, and Thompson Sampling (TS). The hyper-parameters of each strategy are tuned by
irace. As a testbed, we use several instances with four different sizes (20,
50, 100 and 200 jobs) of three different formulations of flowshop problems
(permutation, no-wait, no-idle), two distinct objectives (makespan, flowtime), and four processing time distributions (uniform, exponential and
job or machine correlated). The results show that different strategies are
most suitable for adapting different IG components, TS performs quite
well for all components and, except for local search, using adaptation
is always beneficial when compared with the IG running with standard
parameters.

Keywords: Hyper-heuristic · Iterated Greedy algorithm · Adaptive
Components · Flowshop Variants

1 Introduction

An automated methodology for selecting or generating heuristics to solve optimization problems is the focus of hyper-heuristics (HHs) [4]. HHs can be further
classified according to the learning phase feedback, where they possibly have no
learning, offline learning, or online learning. The later is also known, and referred

© Springer Nature Switzerland AG 2021
A. Britto and K. Valdivia Delgado (Eds.): BRACIS 2021, LNAI 13073, pp. 155–169, 2021.
https://doi.org/10.1007/978-3-030-91702-9_11

in this text, as dynamic learning, that selects or generates heuristics during the search process.

As occurs in a simpler level for meta-learning approaches, in a broader view, hyper-heuristics make algorithm design more adaptable to different problems [14]. Another face on the same problem is given by the algorithm selection, particularly for dynamic schedules of algorithms, which generalizes static and per-instance algorithm selection approaches [10]. Hyper-heuristics can contribute therefore to the development of algorithms for solving a wide range of optimization problems.

Flowshop problems (FSP) involve deciding how J jobs will be processed on M machines in series [2]. This paper investigates three FSP formulations: permutation (with no schedule constraints), and no-wait, no-idle which include constraints (there are no waiting jobs, and no idle machines, respectively).

Different proposals of parameter adaptation and HHs exist in the context of FSPs and scheduling problems in general [3,16,25,26]. In this context, one of the first works proposed in the literature uses an adaptive genetic algorithm [26], with online selection of four types of crossover and three mutations for the permutation FSP with makespan objective. The algorithm produces new offspring using different operators proportional to their contributions on previous generations. Results show that the adaptive genetic algorithm presents a good performance when compared with an algorithm with static parameters and uniform selection of operators.

A HH based on Variable Neighborhood Search (VNS) is proposed in [16]. The VNS strategy adapts the shaking mechanism and local search, providing different low-level heuristics. The shaking is adapted by maintaining a tabu-list of non-improving heuristics, while different local searches are chosen greedily on a rank metric based on improving moves. The proposal performs well on four different combinatorial optimization problems, including permutation FSPs.

Another related work is presented in [25], using the Iterated Local Search (ILS) with different neighborhood types. A greedy strategy selects the best neighborhood based on the fitness improvement, number of times each operator is used and the time to perform the local search. Results show advantages on problems considering makespan, as well as flowtime objective.

A recent work [3] proposes an Iterated Greedy (IG) algorithm enhanced by hyper-heuristics to solve hybrid flexible FSPs with setup times. IG is a metaheuristic with excellent results for some FSP variants. It is based on initialization-destruction-construction phases, followed by a local search, which provides at the end a solution that can be accepted or discarded depending on an acceptance criterion. In [3], the neighborhood types used by the local search (swap, insert and inverse) are considered the low-level heuristics to be selected during the search. The enhanced IG is competitive while solving real-world instances.

Inspired by the fact that IG has been adapted, presents good performance on several combinatorial optimization problems, and performs particularly well on permutation FSP [18], in the present paper we propose and analyze different dynamic strategies used by a hyper-heuristic for selecting IG components. By adapting different components like destruction size and position, neighborhood size, perturbation, local search and its focus, the proposed HH is tested with distinct dynamic adaptation strategies: random, ϵ-greedy, probability matching, multi-armed bandit, LinUCB, and Thompson sampling. The most suitable hyper-parameters of each strategy are set in the tuning phase performed by irace. The proposal's performance is evaluated on three formulations (constraints) of FSPs, with four different sizes, two objectives, and four processing time distributions. This way, we intent to contribute to the FSP understanding and to find general strategies for solving different formulations of the problem.

The main contributions of the paper can be summarized as: (i) adapting six different IG components; (ii) testing six different dynamic learning strategies in the proposed HH; (iii) tuning main HH hyper-parameters; (iv) addressing several FSP variants; (v) providing a high performance adaptive IG capable of outperforming the standard IG [18] in many FSP variants. As far as we know, there is no other previous work considering different HHs with dynamic learning to FSP. Moreover, no other previous work considers the simultaneously adaptation of multiple IG components. Finally, there is no reported results with HH outperforming standard IG on different FSP types.

The paper is organized as follows. Section 2 discusses the basic concepts necessary to understand the proposal. Section 3 details the adaptive IG that is being proposed here. The methodology adopted in the experiments is described in Sect. 4. Results are presented and analyzed in Sect. 5. Finally, Sect. 6 concludes the paper and discusses future perspectives.

2   Background

This section presents the basic concepts regarding the application context (Sect. 2.1 details the FSP) and the proposal (Sect. 2.2 describes the dynamic adaptive strategies used by the proposed hyper-heuristic).

2.1   Flowshop Problems (FSPs)

Flowshop is a combinatorial optimization problem of scheduling. The problem involves deciding how J jobs will be processed on M machines in series. Given the processing times on each machine, a permutation $x = (x_0, \ldots, x_J)$ informs the order jobs will be executed on all machines. The most common formulation considers that jobs and machines are available any time, with processing times known in advance, and machine processes are sequence-independent and occur without interruptions [2].

In permutation FSPs, the completion time of a job x_j in the m-th machine can be determined by:

$$C_{x_j,m} = \max(C_{x_j,m}, C_{x_{j-1},m}) + p_{x_j,m} \tag{1}$$

where $p_{x_j,m}$ is the processing time of job x_j on machine m.

Two common objectives in FSPs are makespan and flowtime. Makespan is the time required to complete all jobs, i.e., $\max_j C_{x_j,M}$, and flowtime is the sum of all completion times, $\sum_j C_{x_j,M}$.

Besides the permutation FSPs formulations, other variants like the no-wait and no-idle include constraints on the schedules. The no-wait FSP variant only considers schedules where there are no waiting times between job operations. The no-wait completion times are given by:

$$C_{x_j,m} = d_{x_{j-1},j} + \sum_{m=0}^{M} p_{x_j,m} \tag{2}$$

where d are the precomputed delay times [20].

No-idle schedules completion times are computed using [23]:

$$F(x_1, m, m+1) = p_{x_j,m+1} \tag{3}$$

$$F(x_j, m, m+1) = \max\left\{F(x_{j-1}, m, m+1) - p_{x_j,m}, 0\right\} + p_{x_j,m+1} \tag{4}$$

$$C_{x_j,m} = \sum_{m=1}^{M-1} F(x_j, m, m+1) + \sum_{k=1}^{j} p_{x_k,1}. \tag{5}$$

where $F(x_j, m, m+1)$ is the minimum difference between the completion of processing up to job x_j, on machines $k+1$ and k, restricted by the no-idle constraint.

In addition to objectives and constrains, a FSP formulation includes the definition of processing times, which can be correlated or non-correlated and whose distributions can be uniform or exponential. Large number of jobs and uniform processing times usually make the problem harder to solve. Alternatively, simple heuristics perform well when there are correlations between processing time [24]. Also, in this paper, we investigate processing times with exponential distributions, and job or machine correlated processing times.

2.2   Hyper-Heuristics and Their Adaptation Strategies

According to [15], a hyper-heuristic (HH) works with a two-level structure, at the high-level it looks for heuristic configurations $h \in H$, considering H the heuristic space. At the low-level, each solution $x \in X$ of the target optimization problem p is generated by the heuristic $h \in H$. There are two evaluation functions: in the first level, the HH's success is measured by function $F : h \to \Re$ and in the second level, each solution $x \in X$ is evaluated by an objective function $f : x \to \Re$.

From a mapping function $M : f(x) \to F(h)$, it is possible to define the purpose of a selection HH. The HH must optimize $F(h)$ by means of the search for the optimal heuristic configuration h^*, in H, thus h^* generates the optimal solution(s) x^* [15]. The formal HH definition, in a minimization optimization problem is summarized in Eq. 6.

$$F(h^* \mid h^* \to x^*, h^* \in H) \leftarrow f(x^*, x^* \in X) = min\{f(x), x \in X\} \tag{6}$$

In this paper, h relies on different choices for each IG component, f is associated with makespan or flowtime objectives, and F is measured by a reward function detailed in Sect. 3.

Hyper-heuristics aim therefore at providing more generalized solutions for optimization problems by producing good results when dealing with a set of instances or a set of problems. For this, HHs work in the heuristic space rather than the solution space. Based on specific strategies, they adapt low-level heuristics, which are used to solve the target problem(s).

We investigate six adaptation strategies commonly used in the HH literature: Random, ϵ-greedy, Probability Matching, Multi-armed Bandit, LinUCB and Thompson sampling.

Random parameter selection is the simplest strategy and in most cases, it serves as a baseline for comparison between static and dynamic parameter selection. It might be beneficial depending on the chance of selecting the best parameter combination [5].

ϵ-*greedy* is a simple strategy often referenced on the exploration-exploitation dilemma [22]. With probability ϵ, the parameter with the best average reward is chosen, otherwise, with probability $1 - \epsilon$, a random one is selected.

Probability Matching (PM) [7] works as a roulette wheel selection biased towards the operators with best quality. The probability of selecting the k-th operator from a set of K operators at iteration t is given by:

$$P_{k,t} = P_{min} + (1 - K \times P_{min}) \frac{q_{k,t}}{\sum_{j=1}^{K} q_{j,t}} \tag{7}$$

where $q_{k,t}$ is the quality of k-th operator and $0 < P_{min} < 1/K$ is used to guarantee that every operator will have a minimum chance of being chosen. The quality values are updated according to the rewards:

$$q_{k,t} = q_{k,t-1} + \alpha \times (r_{k,t}^{acc} - q_{k,t-1}) \tag{8}$$

where α is the learning rate parameter and $r_{k,t}^{acc}$ is the accumulated reward of operator k during a given update window of size W. The accumulation in $r_{k,t}^{acc}$ considers either the average or extreme and, optionally, normalized reward values.

Multi-armed bandit (MAB) [1] algorithms are based on the Upper Confidence Bound for exploitation-exploration trade-off. In particular, the Fitness-Rate-Rank-Based Multi-armed Bandit [11] considers the dynamic search behavior with a sliding window of size W to store the rewards of each operator. The selected operator maximizes the expression:

$$FRR_{k,t} + C_s \left(\frac{2 \ln \sum_{l=1}^{K} n_l^t}{n_k^t} \right) \tag{9}$$

where C_s is a scaling parameter, n_k^t is the number of times operator k is applied during the window of size W, and $FRR_{k,t}$ is the k-th operator credit value given by:

$$FRR_{k,t} = \frac{D^{rank_k} \times r_k}{\sum_{l=1}^{K} D^{rank_l} \times r_l} \tag{10}$$

in which D is the best operator influence decay parameter, r_k^{acc} is the k-th operator accumulated reward, and $rank_k$ is the k-th operator reward sum rank.

Linear Upper Confidence Bound (LinUCB) [12] works by assuming that the reward for a given operator is linearly proportional to the values of contextual features, i.e., $E[r_{k,t}|\phi_{k,t}] = \phi_{k,t}^T \theta^*_k$, where $\phi_{k,t}$ is the feature vector and θ^*_k the unknown coefficients. We consider four fitness landscape metrics as the context for a local search procedure, calculated online during the local search step [17]:

- Adaptive walk length: the total number of steps of the local search;
- Autocorrelation: correlation between the fitness values observed with the fitness values of the previous solutions;
- Fitness-distance correlation: correlation between fitness and insertion distance considering the initial and final solutions;
- Neutrality: proportion of neighbors with equal fitness values.

Using a ridge regression formulation, the coefficients can be found efficiently with the following steps:

$$\begin{aligned} \theta^*_{k,t} &= A_{k,t}^{-1} b_{k,t} \\ P_{k,t} &= \theta^*_{k,t} \phi_{k,t} + \alpha \sqrt{\phi_{k,t} A_{k,t}^{-1} \phi_{k,t}} \end{aligned} \tag{11}$$

where α is a learning rate parameter. The operator with maximum $P_{k,t}$ is chosen, yielding the reward value $r_{k,t}$, and the model update follows:

$$\begin{aligned} A_{k,t} &= A_{k,t-1} + \phi_{k,t} \phi_{k,t}^T \\ b_{k,t} &= b_{k,t-1} + r_{k,t} \phi_{k,t}. \end{aligned} \tag{12}$$

Thompson Sampling (TS) [19] strategy starts with a prior distribution, chooses the best operator by sampling, observes the output and updates the distribution. The Beta distribution $Beta(S_{k,t}, F_{k,t})$ models Bernoulli trials where operator k has $S_{k,t}$ successes (rewards $r_k > 0$) and $F_{k,t}$ fails (rewards $r_k \leq 0$). Therefore we choose the operator with:

$$op = \arg \max_k Sample[Beta(S_{k,t}, F_{k,t})] \tag{13}$$

and update the distribution after the reward:

$$\begin{aligned} S_{k,t} &= S_{k,t-1} + 1_{r_{k,t}>0} \\ F_{k,t} &= F_{k,t-1} + 1_{r_{k,t}\leq0}. \end{aligned} \tag{14}$$

Alternatively, the Dynamic TS [8] introduces a window size parameter W and a modified update rule (after iteration W) as follows:

$$\begin{aligned} S_{k,t} &= (S_{k,t-1} + 1_{r_{k,t}>0})\frac{W}{W+1} \\ F_{k,t} &= (F_{k,t-1} + 1_{r_{k,t}\leq0})\frac{W}{W+1}. \end{aligned} \tag{15}$$

Table 1. Hyper-parameters of the addressed adaptation strategies.

Strategy	Hyper-Parameter	Domain
ϵ-greedy	Greedy choice probability (ϵ)	$[0.0, 1.0]$
Probability matching	Min. probability (P_{min})	$[0.05, 0.2]$
	Learning rate (α)	$[0.1, 0.9]$
	Update window (W)	$[1, \ldots, 500]$
	Reward accumulation	{average, extreme}
	Reward normalization	{yes, no}
Multi-armed bandit	Scale (C_s)	$[0.01, 100.0]$
	Decay (D)	$[0.25, 1.0]$
	Update window (W)	$[10, \ldots, 500]$
LinUCB	Learning rate (α)	$[0.01, 1.5]$
TS	Update type	{$static, dynamic$}
	Update window (W)	$[1, \ldots, 500]$
Common	Warm up period	$[0, 1000, 2000]$
	Reward type[a]	$[bLaL, bLaI, bIaL, bIaI]$

[a] Reward types are detailed in Sect. 3.

Table 1 shows a summary of all hyper-parameters used by the adaptation strategies, as well as two hyper-parameters common to all strategies: reward type and warm-up period. The warm-up period is considered at the beginning of the iterations where strategies are chosen randomly. Reward type is detailed in the next section.

3   Adaptive IG Proposal

Considered the state of the art for some FSP variants, the Iterated Greedy (IG) algorithm [18] is a successful iterative metaheuristic that encompasses five main steps: (1) the incumbent solution x is initialized, (2) a destruction phase randomly removes d jobs, (3) a construction procedure inserts each job at the best position, (4) a local search generates a new solution by exploiting the solution resulted from construction and (5) the new solution replaces the incumbent x according to an acceptance criterion. The last step accepts the new solution x'' based on the following probability:

$$P_{accep}(x'') = \begin{cases} 1.0 & \text{if } f(x'') < f(x) \\ \exp\left(\frac{-(f(x'')-f(x))}{Temp}\right) & \text{otherwise} \end{cases} \tag{16}$$

where $f(.)$ is the cost function and $Temp$ is the temperature defined by:

$$Temp = T \times \frac{\sum_{j=1}^{J} \sum_{m=1}^{M} p_{j,m}}{J * M * 10}. \tag{17}$$

IG has two main parameters: the destruction size d and the temperature factor T, whose values IG authors [18] recommend to be set as $d = 4$ and $T = 0.5$.

Other recommended configurations are Nawaz–Enscore–Ham (NEH) construction heuristic as the initialization and iterative improvement as local search. The local search iteratively inserts a job, chosen randomly without replacement, on the best position until there are no improvements [21]. This version is referred from now on as Standard IG.

The hyper-heuristic proposed in the present paper considers an adaptive strategy (one fixed among the six possible strategies described in Sect. 2.2) to update some components of an IG algorithm. An adaptive strategy requires a set H of possible choices and a reward function indicating how well a particular choice performs.

An important issue for defining an adaptive strategy is the reward function. It returns a real number representing the quality of a given choice. Good choices have positive rewards and bad choices receive negative ones. In the case of using IG to solve FSPs, reward of kth parameter at iteration t can be calculated based on the relative decrease in cost function value (e.g. makespan or any other objective function being considered):

$$r_{k,t} = (f(x_{before}) - f(x_{after}))/f(x_{before}). \tag{18}$$

Here x_{before} and x_{after} represent solutions before and after the reward evaluation period (whose reference can be either local search or iteration). We have four possible periods during which a solution has its performance evaluated: before the iteration (bI) or the local search (bL) and after the iteration (aI) or local search (aL), giving rise to four types of reward:

- bLaL: $(f(x_{\text{before local search}}) - f(x_{\text{after local search}}))/f(x_{\text{before local search}})$;
- bIaL: $(f(x_{\text{before iteration}}) - f(x_{\text{after local search}}))/f(x_{\text{before iteration}})$;
- bLaI: $(f(x_{\text{before local search}}) - f(x_{\text{after iteration}}))/f(x_{\text{before local search}})$;
- bIaI: $(f(x_{\text{before iteration}}) - f(x_{\text{after iteration}}))/f(x_{\text{before iteration}})$.

As Fig. 1 shows, each reward type considers a different period to get the feedback for each solution considering its quality increase/decrease.

```
x <- neh_initialization()
while stopping criteria is not met:
    removed_jobs <- destruction(x)
    x' <- construction(x, removed_jobs)
    x'' <- local_search(x')
    if acceptance_criterion(x, x''):
        x <- x''
```

bLaL:
x' - x''

bIaL:
x - x''

bLaI:
x' - x

bIaI:
x - x

Fig. 1. IG algorithm and reward types: bLaL, bIaL, bLaI and bIaI.

Using the standard IG [18] as basis, with a budget computation time as the stopping criteria, we identify some components that can be adapted dynamically: type of the local search, type of perturbation, destruction size, neighborhood size, destruction position, local search focus. A summary of these adaptive components and their values are presented in Table 2.

In our case, different choices are possible for each IG component, for example, the number of deconstructions. We refer to the parameter values as discrete choices (also referred to arms in multi-armed bandit literature) indexed by $\{1, \ldots, K\}$, where $K = |H|$ is the total number of choices. Therefore, according to Eq. 6, we have each choice representing a heuristic $h \in H$, where the dynamic adaptation strategy of the HH chooses h as the k-th choice, $k = 1, \ldots, |H|$.

The pool regarding the first choice sets encompasses the most usual options for Local search and Perturbation in the context of combinatorial optimization. The options for Destruction size are the ones often chosen in IG implementations for FSP. Finally, the pools for the last three choice sets are defined aiming to produce different granularities, starting from a coarse, ending to a fine search.

Table 2. IG adaptive parameters.

Parameter	Choice set(s) H
Local search	{Iterative improvement, first improvement, First-best improvement, random best improvement}
Perturbation	{Destruction-construction, Random swaps +transposition, Destruction-construction with local search}
Destruction size	$\{2,4\}$, $\{4,6\}$, $\{2,4,6\}$ or $\{4,8\}$
Neighborhood size	$\{1/2, 2/2\}$, $\{1/3, 2/3, 3/3\}$, $\{1/5, \ldots, 5/5\}$, $\{1/10, \ldots, 10/10\}$
Destruction position	$\{[0, 1/3], \ldots, [2/3, 3/3]\}$, $\{[0, 1/10], \ldots, [9/10, 10/10]\}$ or $\{1, \ldots, J\}$
Local search focus	$\{[0, 1/3], \ldots, [2/3, 3/3]\}$, $\{[0, 1/10], \ldots, [9/10, 10/10]\}$ or $\{1, \ldots, J\}$

The HH based on dynamic learning proposed in this paper is capable of adapting different components (e.g. local search and perturbation) for each iteration of the adaptive IG. In local search for example, it can choose between the original IG's iterative improvement, first improvement, first-best improvement, random best improvement. The last three local search options consider all possible insertions. The perturbation adaptation considers three possibilities: (1) the IG destruction-construction steps; (2) two random swaps and a transposition, as used in [21] for ILS on FSP, and (3) destruction-construction with iterative improvement local search between destructions, recently proposed in [6].

The neighborhood size adaptation considers the percentage of the neighborhood explored during the local search step. For example, exploring only half of the neighborhood at the beginning of the search could save time for exploitation at the end of the search. Similarly, choosing the destruction size parameter dynamically might improve the search during exploration-exploitation phases.

As mentioned, there are two mechanisms in IG that randomly choose jobs to be re-inserted: destruction and best-insertion local search. An adaptive mechanism can be applied in the last two IG components shown in Table 2 to focus these operators on parts of the solution that have a better chance of improvement. For that, we propose partitioning the solution into chunks and adaptively selecting from each chunk the job that will be sampled.

As shown in Table 2, for some parameters there are multiple possible choice pools $\{H_1, H_2, ...\}$. For example, the destruction size can be chosen from the set $H_1 = \{2,4\}$, $H_2 = \{4,6\}$, $H_3 = \{2,4,6\}$ or $H_4 = \{4,8\}$. Also on destruction position and local search focus, the solution can be partitioned into 3, 10 or J chunks (the last considering one arm for each job). Some HH hyper-parameters like the pool H_i for a component, reward type and update window are determined in a parameter tuning phase described in Sect. 4.

4   Tuning and Testing Phases

Based on the hyper-heuristic with dynamic learning and its six possible adaptive strategies described in Sect. 2.2, Sect. 3 has detailed the proposed adaptive IG. This section presents the hyper-parameter configuration performed by irace (tuning) using part of available data and the test performed on the tuned HH using the remaining one. The configurations are evaluated with a budget of $J \times (M/2) \times 30$ ms on tuning and test phases. Algorithms are implemented in C++ using the Paradiseo library [9]. The experiments have been executed on a server with 8-core AMD EPYC 7542 processors and 16 GB of RAM. For the results analysis, we used the R language and relevant packages[1].

Tuning phase Before running the experiments, we perform a tuning phase aiming to determine which strategy configuration (shown in Table 1) works best for each IG component (shown in Table 2). The random strategy is not considered for destruction position and local search focus, because a random choice in these cases is equivalent to the standard IG behaviour.

The irace algorithm [13] with default parameters and 5000 configurations evaluations is used to tune each combination of the six adaptive components and the six dynamic strategies. There are multiple choice pools $\{H_1, H_2, ...\}$ for destruction size, neighborhood size, destruction position and local search focus components. For example, the destruction size can be chosen from $H_1 = \{2,4\}$ or $H_2 = \{4,8\}$. In these cases, the pool choice is considered as an additional categorical parameter for irace during the tuning phase.

The instance set for the tuning phase is composed of 48 ($3 \times 2 \times 3 \times 4$) instances resulted from the combinations of: 3 sizes (20 or 50 jobs and 10 machines), 2 objectives (makespan or flowtime), 3 types (permutation, no-wait or no-idle FSPs), and 4 processing times distributions: (exponential, uniform, job-correlated or machine-correlated processing times).

Testing phase After tuning, the best configurations are tested with 10 restarts on a set of unseen instances with the same features, but sampled with different random seeds. Aiming to better evaluate algorithms' generalization

---

[1] Tuning and Testing instances, as well as the code used in the paper, are available at https://github.com/lucasmpavelski/Adaptive-IG.

capabilities, we also include larger instances of $J = 100$ and $J = 200$ jobs and $M = 20$ in the testing phase.

The evaluation for each algorithm is done using the Average Relative Percentage Deviation (ARPD) given by:

$$ARPD_{alg} = \frac{1}{R} \sum_{r=1}^{R} 100 \times \frac{f(x_r^{alg}) - f(x^{best})}{f(x^{best})} \tag{19}$$

where R is the number of runs and x_r^{alg} is the best solution found by algorithm alg. The reference solution x^{best} is given by the best solution found by a standard IG (the same used by the authors in [18]) with a higher budget of $J \times (M/2) \times 120$ ms and 30 restarts.

Finally, the configurations are also compared with the standard IG configuration to evaluate the effectiveness of the adaptive components in the presented scenario. In all comparisons, we highlight the lowest ARPDs and perform a Friedman rank sum test with Nemenyi post hoc to verify if the differences are statistically significant with p-value threshold of 0.05: results with no statistically significant differences from the best one are highlighted with gray background.

5 Results

Table 3 shows some hyper-parameter values (reward type, choice sets and update window) tuned by irace for each IG component addressed in the paper.

Different reward types have been tuned for the different adaptive components and HHs, with no dominance between the different options. The choice sets for the destruction size parameter enable more exploration with values higher than the default $d = 4$, which is present in all options. The choice sets for neighborhood size are small (2 or 3), with preference for a coarse search, while the local search focus uses partitions with 10 or more choices in three among five cases. Finally, the update window is short for MAB and TS on perturbation component, indicating that its value (4 or 8 destructions) is often switched during the search. However, in most cases the strategies prefer less frequent changes, since large window sizes are the tuned parameters.

The ARPD values for each adaptive component and strategy are shown in Table 4, and they are calculated using Eq. 19, with $R = 10$ and considering all the different testing instances. The values for standard IG configuration [18] are computed with the same budget ($J \times (M/2) \times 30$) as the HHs proposals on the testing instances. Notice that this budget is lower than the one used to compute the reference $f(x_{best})$ values for ARPD.

The adaptive components are able to improve the static standard IG configuration for all (perturbation, destruction size, destruction position and local search focus) but one IG component. Local search adaptation is not effective independently of the strategy used by the HH, meaning that the iterative improvement performed by the standard IG is quite effective compared with the other choices. TS adaptation achieves the lowest ARPD for perturbation

Table 3. Tuned hyper-parameters.

Adaptive component	Random	ϵ-greedy	PM	MAB	LinUCB	TS
Reward type						
Local search	-	bLaI	bIaI	bLaI	bLaI	bIaL
Perturbation	-	bLaL	bIaI	bIaL	bLaL	bIaL
Destruction size	-	bIaI	bIaL	bIaL	bLaI	bLaI
Destruction position	-	bLaI	bIaI	bLaL	bLaL	bIaL
Neighborhood size	-	bIaL	bLaL	bIaI	bIaI	bLaI
Choice sets						
Destruction size	$\{4, 8\}$	$\{2, 4, 6\}$	$\{4, 6\}$	$\{4, 8\}$	$\{4, 8\}$	$\{4, 8\}$
Neighborhood size	$\{1/2, 1\}$	$\{1/3, 2/3, 1\}$	$\{1/3, 2/3, 1\}$	$\{1/2, 1\}$	$\{1/2, 1\}$	$\{1/2, 1\}$
Destruction position*	-	3	3	3	3	10
Local search focus*	-	3	J	10	3	10
Update window						
Local search	-	-	283	188	-	414
Perturbation	-	-	283	59	-	21
Destruction size	-	-	128	406	-	static
Destruction position	-	-	357	350	-	static
Neighborhood size	-	-	340	32	-	358
Local search focus	-	-	125	271	-	299

* number of partitions.

and destruction position components, while ϵ-greedy performs well on the local search focus task. As (biased) adaptation might not be the best option in some cases [5], the simple random strategy performs well for selecting the destruction size and neighborhood size. In all cases, the adaptive strategies TS and MAB are among the strategies with the lowest or statistically equivalent to the lowest ARPD values.

We see from Table 4 that TS is robust for different components but provides the lowest among all results when adapting IG perturbation. However, it is equivalent to most other HH proposals.

We compare TS configuring only perturbation with two others that try to adapt multiple components simultaneously: *Adapt all* components, *Adapt all* components *except local search* (for which no adaptation strategy was capable of improving the performance). The *Adapt all* approaches use strategies with the best ARPD values for each component from Table 4, which means, TS for perturbation and destruction position, Random for destruction and neighborhood sizes and ϵ-greedy for Local search focus. The results in Table 5 show the metrics separated by objective, FSP type, processing times distribution and size. Overall, IG with adaptive perturbation obtains the lowest ARPDs. Adapting all components at the same time does not provide benefits, but when we eliminate the local search it performs like the best approach.

Table 4. Adaptation strategies ARPDs (and standard deviation) for each adaptive component and strategy. Lowest mean values are highlighted in bold, statistically equivalent values are highlighted with gray background.

Adaptive component	Random	ε-greedy	PM	MAB	LinUCB	TS	Standard IG
Local search	$.694_{1.0}$	$.686_{1.0}$	$.686_{1.0}$	$.666_{.94}$	$.696_{1.0}$	$.662_{.96}$	$\mathbf{.595}_{.80}$
Perturbation	$.348_{.63}$	$.377_{.62}$	$.351_{.63}$	$.353_{.61}$	$.434_{.70}$	$\mathbf{.343}_{.63}$	$.595_{.80}$
Destruction size	$\mathbf{.416}_{.60}$	$.491_{.67}$	$.466_{.64}$	$.454_{.69}$	$.441_{.61}$	$.438_{.67}$	$.595_{.80}$
Neighborhood Size	$\mathbf{.567}_{.75}$	$.599_{.80}$	$.593_{.79}$	$.589_{.80}$	$.589_{.80}$	$.598_{.81}$	$.595_{.80}$
Destruction position	-	$.575_{.75}$	$.563_{.73}$	$.553_{.73}$	$.672_{.79}$	$\mathbf{.548}_{.73}$	$.595_{.80}$
Local search focus	-	$\mathbf{.546}_{.74}$	$.583_{.80}$	$.554_{.76}$	$.570_{.76}$	$.556_{.76}$	$.595_{.80}$

Table 5. Adaptation strategies ARPDs (and standard deviation) for all adaptive components, perturbation and destruction size adaptation and standard (static) IG. Best mean and statistically equivalent values are in bold and gray background, respectively.

Problem set All problems		Adapt all	Adapt all w/o local search	Adapt only perturbation	Standard IG
		$.455_{.77}$	$.380_{.64}$	$\mathbf{.343}_{.63}$	$.595_{.80}$
Objective	Flowtime	$.592_{.95}$	$.466_{.80}$	$\mathbf{.436}_{.78}$	$.763_{.95}$
	Makespan	$.319_{.49}$	$.294_{.42}$	$\mathbf{.251}_{.40}$	$.426_{.56}$
Type	No-idle	$.888_{1.1}$	$.759_{.90}$	$\mathbf{.708}_{.90}$	$.793_{.92}$
	No-wait	$.144_{.26}$	$.136_{.25}$	$\mathbf{.106}_{.22}$	$.592_{.76}$
	Permutation	$.334_{.50}$	$.245_{.39}$	$\mathbf{.216}_{.35}$	$.400_{.63}$
Distribution	Exponential	$.805_{1.0}$	$\mathbf{.614}_{.82}$	$.614_{.84}$	$1.09_{.99}$
	Uniform	$.658_{.78}$	$.589_{.72}$	$\mathbf{.461}_{.67}$	$.814_{.81}$
	Job-correlated	$.317_{.52}$	$.286_{.50}$	$\mathbf{.269}_{.48}$	$.403_{.53}$
	Machine-correlated	$.042_{.08}$	$.031_{.06}$	$\mathbf{.030}_{.07}$	$.070_{.13}$
J	$\{20, 50\}$	$.214_{.52}$	$.206_{.45}$	$\mathbf{.172}_{.39}$	$.251_{.50}$
	$\{100, 200\}$	$.697_{.89}$	$.554_{.76}$	$\mathbf{.515}_{.75}$	$.938_{.88}$

6 Conclusions

This paper proposed and analyzed the use of hyper-heuristic with dynamic strategies to adapt different components of the Iterated Greedy algorithm. Six different adapting strategies (random, ε-greedy, probability matching, multi-armed bandit, LinUCB, and Thompson sampling) were tested to adapt six IG components (local search, perturbation, destruction size, neighborhood size, destruction position and local search focus). After a tuning phase performed by irace to set the best strategy hyper-parameters for each component being adapted, the proposal was tested in different variants of the flowshop problems.

Results show that, in most cases, the adaptation is able to improve the performance over the static standard IG configuration, especially the perturbation operator adapted using dynamic Thompson Sampling. Also, using multiple adaptive components did not seem to be beneficial, unless we fix the local search as iterated improvement, a fact that deserves further investigation.

The work can be expanded by including different flowshop problems (objectives and constraints), adaptation strategies, and alternative operators. In addtion, we intend to propose modifications to improve the performance of the *Adapt all* approach.

Acknowledgment. M. Delgado acknowledges CNPq (grants 439226/2018-0, 314699/2020-1) for her partial financial support.

References

1. Auer, P.: Using confidence bounds for exploitation-exploration trade-offs. J. Mach. Learn. Res. **3**(9), 397–422 (2002)
2. Baker, K.R., Trietsch, D.: Principles of Sequencing and Scheduling. Wiley Publishing, New Jersey USA (2009)
3. Burcin, O.F., Sagir, M.: Iterated greedy algorithms enhanced by hyper-heuristic based learning for hybrid flexible flowshop scheduling problem with sequence dependent setup times: a case study at a manufacturing plant. Comput. Oper. Res. **125**, 105044 (2021). https://doi.org/10.1016/j.cor.2020.105044
4. Burke, E.K., Hyde, M.R., Kendall, G., Ochoa, G., Özcan, E., Woodward, J.R.: A classification of hyper-heuristic approaches: revisited. In: Gendreau, M., Potvin, J.-Y. (eds.) Handbook of Metaheuristics. ISORMS, vol. 272, pp. 453–477. Springer, Cham (2019). https://doi.org/10.1007/978-3-319-91086-4_14
5. Chakhlevitch, K., Cowling, P.: Hyperheuristics: recent developments. In: Studies in Computational Intelligence, vol. 136, pp. 3–29. Springer, Berlin, Heidelberg (2008). https://doi.org/10.1007/978-3-540-79438-7_1
6. Dubois-Lacoste, J., Pagnozzi, F., Stützle, T.: An iterated greedy algorithm with optimization of partial solutions for the makespan permutation flowshop problem. Comput. Oper. Res. **81**, 160–166 (2017). https://doi.org/10.1016/j.cor.2016.12.021
7. Goldberg, D.E.: Probability matching, the magnitude of reinforcement, and classifier system Bidding. Mach. Learn. **5**(4), 407–425 (1990). https://doi.org/10.1023/A:1022681708029
8. Gupta, N., Granmo, O.C., Agrawala, A.: Thompson sampling for dynamic multi-armed bandits. In: 2011 10th International Conference on Machine Learning and Applications and Workshops, pp. 484–489. IEEE, Honolulu, HI, USA (2011). https://doi.org/10.1109/ICMLA.2011.144
9. Humeau, J., Liefooghe, A., Talbi, E.G., Verel, S.: ParadisEO-MO: From Fitness Landscape Analysis to Efficient Local Search Algorithms. Research Report RR-7871, INRIA (2013)
10. Kerschke, P., Hoos, H.H., Neumann, F., Trautmann, H.: Automated algorithm selection: survey and perspectives. Evol. Comput. **27**(1), 3–45 (2019)
11. Li, K., Fialho, A., Kwong, S., Zhang, Q.: Adaptive operator selection with bandits for a multiobjective evolutionary algorithm based on decomposition. IEEE Trans. Evol. Comput. **18**(1), 114–130 (2014). https://doi.org/10.1109/TEVC.2013.2239648
12. Li, L., Chu, W., Langford, J., Schapire, R.E.: A contextual-bandit approach to personalized news article recommendation. In: Proceedings of the 19th International Conference on World Wide Web - WWW '10, p. 661. ACM Press, Raleigh, North Carolina, USA (2010). https://doi.org/10.1145/1772690.1772758

13. López-Ibáñez, M., Dubois-Lacoste, J., Cáceres, L.P., Stützle, T., Birattari, M.: The irace package: iterated racing for automatic algorithm configuration. Oper. Res. Perspect. **3**, 43–58 (2016). https://doi.org/10.1016/j.orp.2016.09.002
14. Pappa, G.L., Ochoa, G., Hyde, M.R., Freitas, A.A., Woodward, J., Swan, J.: Contrasting meta-learning and hyper-heuristic research: the role of evolutionary algorithms. Genetic Program. Evolvable Mach. **15**(1), 3–35 (2013). https://doi.org/10.1007/s10710-013-9186-9
15. Pillay, N., Qu, R.: Hyper-heuristics: theory and applications. Springer Nature, 1 edn. (2018). https://doi.org/10.1007/978-3-319-96514-7
16. Hsiao, P.-C., Chiang, T.-C., Fu, L.-C.: A VNS-based hyper-heuristic with adaptive computational budget of local search. In: 2012 IEEE Congress on Evolutionary Computation, pp. 1–8. IEEE, Brisbane, Australia (2012). https://doi.org/10.1109/CEC.2012.6252969
17. Pitzer, E., Affenzeller, M.: Recent Advances in Intelligent Engineering Systems. Studies in Computational Intelligence, vol. 378, chap. A Comprehensive Survey on Fitness Landscape Analysis, pp. 161–186. Springer, Berlin, Heidelberg (2012). https://doi.org/10.1007/978-3-642-23229-9
18. Ruiz, R., Stützle, T.: A simple and effective iterated greedy algorithm for the permutation flowshop scheduling problem. Eur. J. Oper. Res. **177**(3), 2033–2049 (2007). https://doi.org/10.1016/j.ejor.2005.12.009
19. Russo, D., Van Roy, B., Kazerouni, A., Osband, I., Wen, Z.: A Tutorial on Thompson Sampling. arXiv:1707.02038 [cs] (2020)
20. Sapkal, S.U., Laha, D.: A heuristic for no-wait flow shop scheduling. Int. J. Adv. Manuf. Technol. **1**, 1327–1338 (2013). https://doi.org/10.1007/s00170-013-4924-y
21. Stützle, T.: Applying iterated local search to the permutation flow shop problem. Technical report, FG Intellektik, TU Darmstadt, Darmstadt, Germany (1998)
22. Sutton, R.S., Barto, A.G.: Reinforcement Learning: An Introduction. Adaptive Computation and Machine Learning Series. 2nd edn. The MIT Press, Cambridge, Massachusetts (2018)
23. Tasgetiren, M.F., Pan, Q.K., Suganthan, P.N., Liang, Y.C.: A discrete differential evolution algorithm for the no-wait flowshop scheduling problem with total flowtime criterion. In: 2007 IEEE Symposium on Computational Intelligence in Scheduling, pp. 251–258. IEEE, Honolulu, USA (2007)
24. Watson, J.P., Barbulescu, L., Howe, A.E., Whitley, L.D.: Algorithm performance and problem structure for flow-shop scheduling. In: AAAI/IAAI, pp. 688–695. American Association for Artificial Intelligence, Menlo Park, CA, USA (1999)
25. Yahyaoui, H., Krichen, S., Derbel, B., Talbi, E.G.: A hybrid ILS-VND based hyper-heuristic for permutation flowshop scheduling problem. Procedia Comput. Sci. **60**, 632–641 (2015). https://doi.org/10.1016/j.procs.2015.08.199
26. Zhang, L., Wang, L., Zheng, D.Z.: An adaptive genetic algorithm with multiple operators for flowshop scheduling. Int. J. Adv. Manuf. Technol. **27**(5–6), 580–587 (2006). https://doi.org/10.1007/s00170-004-2223-3

Hyper-Heuristic Based NSGA-III for the Many-Objective Quadratic Assignment Problem

Bianca N. K. Senzaki$^{(\boxtimes)}$ ⓘ, Sandra M. Venske ⓘ, and Carolina P. Almeida ⓘ

Midwestern Paraná State University - UNICENTRO, Paraná, Brazil
{ssvenske,carol}@unicentro.br

Abstract. The Quadratic Assignment Problem (QAP) can be subdivided into different versions, being present in several real-world applications. In this work, it is used a version that considers many objectives. QAP is an NP-hard problem, so approximate algorithms are used to address it. This work analyzes a Hyper-Heuristic (HH) that selects genetic operators to be applied during the evolutionary process. HH is based on the NSGA-III framework and on the Thompson Sampling approach. Our main contribution is the analysis of the use of a many objective algorithm using HH for QAP, as this problem was still underexplored in the context of many objective optimization. Furthermore, we analyze the behavior of operators forward the changes related to HH (TS). The proposal was tested considering 42 instances with 5, 7 and 10 objectives. The results, interpreted using the Friedman statistical test, were satisfactory when compared to the original algorithm (without HH), as well as when compared to algorithms in the literature: MOEA/DD, MOEA/D, SPEA2, NSGA-II and MOEA/D-DRA.

Keywords: NSGA-III · Hyper-Heuristic · Thompson sampling · MaQAP

1 Introduction

The Quadratic Assignment Problem (QAP) is an NP-Hard problem even for small instances [1]. The QAP was initially introduced to model a plant location problem, with the objective of placing facilities in places so that the sum of the product between distances and flows is minimal [2]. Depending on the needs, several variations of the problem arose, such as the linear allocation problem, quadratic bottleneck allocation problem, the quadratic multi-objective allocation problem, among others. Each variation works to solve different real-world problems [1]. The multi-objective QAP (mQAP) [3] and the many objective QAP (MaQAP) are versions of the problem where a distance matrix and several flow matrices (objectives) are used, with two or more functions configuring the mQAP and with more than three functions the MaQAP. The greater the number of objective functions to be optimized, the amount of non-dominated

© Springer Nature Switzerland AG 2021
A. Britto and K. Valdivia Delgado (Eds.): BRACIS 2021, LNAI 13073, pp. 170–185, 2021.
https://doi.org/10.1007/978-3-030-91702-9_12

solutions and the computational cost for calculating the fitness of the solutions and the operations of the algorithm increases [4]. These versions are used to solve disposition problems, for example, the allocation of clinics and departments in hospitals to reduce the distance between doctors, patients, nurses and medical equipment [1].

NSGA-III [5] is an Evolutionary Algorithm for many objective problems that uses a non-dominance classification approach based on reference points, being a hybrid strategy. Different classifications are used according to the method that the problem is solved [4]. Such an approach has been showing good results in the literature of the area, so it will be investigated in this work.

Hyper Heuristics (HHs) [6] are high-level methods that generate or select heuristics, used to free the user from choosing a heuristic or its configuration. The Thompson Sampling (TS) [7] approach is a HH that selects a heuristic based on information accumulated during the execution of the algorithm to improve their performance.

In this work, a study was carried out on the methods already used to solve MaQAP, implementing and testing the NSGA-III with the Thompson Sampling, which makes the selection of the crossover operator combined with a mutation operator. We tested different versions by modifying the way of Thompson Sampling action. TS uses success and failure factors to determine the best operator choices to apply. Different ways of handling these factors and the frequency of resetting them are proposed. The changes of success and failures factors are based on the dominance relation and the average of the objectives. In addition to the comparison between the previously mentioned versions, the proposal, named NSGA-III$_{TS}$, is also evaluated in terms of effectiveness by comparison with the original algorithm that does not use HH. Finally, the performance of the NSGA-III$_{TS}$ is compared with five other evolutionary multi-objective algorithms: MOEA/DD [8], MOEA/D [9], SPEA2 [10], NSGA-II [11] and MOEA/D-DRA [12]. The main contribution of this work is to analyze the behavior of a many objective algorithm based on hyper-heuristic applied to MaQAP. MaQAP is an under-explored problem in the literature of the area, which motivated its choice for use in this work due to several fronts of scientific contribution. In addition, to analyze how the choice of operators proceeds when important factors related to the HH used (TS) are modified.

This paper is organized as follows. Section 2 presents concepts related to work, such as the QAP, the used algorithm (NSGA-III) and the HHs. The section ends with some related works. Section 3 presents the proposed methodology for the development of the work. The results are shown and discussed in Sect. 4. Section 5 ends the work with the conclusions and future directions.

2   Background

2.1   Many-Objective Quadratic Assignment Problem - MaQAP

QAP's objective is to associate a set of n facilities to n different locations, in order to minimize the cost of transporting the flow between the facilities [2]. Over time the problem has been adapted to solve different real-world problems

[13]. One of its existing variations is the QAP with many objectives, which is an extension of the multi-objective QAP, but which more than 3 flow matrices.

The MaQAP is formalized by the Eq. 1, with the condition of Eq. 2 being satisfied. π is a solution that must belong to a set of feasible solutions.

$$Minimize\ \overrightarrow{C}(\pi) = \{C^1(\pi), C^2(\pi), \ldots, C^m(\pi)\} \\ \pi \in P(n) \tag{1}$$

$$C^k(\pi) = \sum_{i=1}^{n}\sum_{j=1}^{n} a_{ij} b^k_{\pi_i \pi_j}, k \in 1..m \tag{2}$$

The calculation of each objective is shown in Eq. 2, and two matrices of the same size are used, the matrix a_{ij} has the distance between the locations, $b^k_{\pi_i \pi_j}$ is a matrix of the k-th objective flow, with π_i and π_j being the position of i and j in π; n is the number of locations and k is an iterator. The objective is to find a permutation π, belonging to the set of all possible permutations ($P(n)$), which simultaneously minimizes m objective functions each represented by the k-th C. In order to represent problems with many objectives, the constraint $m > 3$ must be added.

2.2   Hyper-Heuristics (HH)

Hyper heuristics are high-level methodologies developed for the optimization of complex problems [6], looking for the best option to fulfill your objective by generating heuristics through components (generation HH) or by selecting from a set of low-level heuristics (selection HH). The basic composition of a selection HH is a high-level strategy, which uses a mechanism to decide which low-level heuristic to use, and an acceptance criterion that can accept or reject a solution. The other element of HH is a set of low-level heuristics, which must be appropriate for the tested problem [14]. Low-level heuristics work on the set of solutions, they can be of two types, constructive or perturbative.

Among the ways that the HHs can learn are: online, where the learning takes place during the execution of the search process, or offline where the learning already comes with the rules defined before the execution of the algorithm. The use of HHs is due to their ability to find good results for NP-hard problems, seeking the best way to achieve a good result and removing the need for the user to choose a single heuristic or its configuration [15]. But as the HHs have evolved, updated information is needed to better organize the classifications to reflect the challenges found in the current world, [16] presents an extended version of the categorization of selection heuristics.

In the Fig. 1, a diagram showing the elements of the HH used in this work is shown. In which, the high level strategy is formed by the TS as a heuristic selection mechanism, and the NSGA-III acceptance criterion is used. The low and high level heuristics are separated because they work in different search spaces, the high level ones works searching in the space of low level heuristics, and the low level ones work in the space of solutions. The heuristic selection mechanism chooses within a predefined set of low-level heuristics based on its own evaluation criteria. In this case, Thompson Sampling uses the success and failure factors associated to each operator who performs a count using different methods to define whether a low-level heuristic was efficient or not. Finally, the solution is sent to the acceptance criterion, which can choose whether or not to accept the new solution (NSGA-III acceptance criterion).

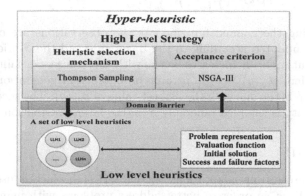

Fig. 1. Hyper-Heuristic elements. Adapted from [14].

Low-Level Heuristics - Genetic Operators: The set of genetic operators is from the permutational type, so they don't generate infactible solutions. These operators use two solutions, to create two new solutions, and they are: **i)** Partially Matched Crossover (PMX) [17]: 2 cutoff points are selected, so both parents are divided into 3 parts. When creating an offspring, the elements between the points are copied in the same absolute position, then the elements of the other parent are taken. If they have not been used, they are copied in the same position, and if there is a conflict, they are exchanged according to their position in the offspring. The parents order to raise a second offspring is reversed. **ii)** Cyclic Crossover (CX) [18]: sequences of connected positions are created by the values and positions. It starts at the first value on the left, takes the position at parent 2, checks and looks for the value at parent 1 and repeats the process until it finds cycles. Each cycle is passed alternately to the offspring. **iii)** Permutational Two Points Crossover (2P) [17]: two points are chosen at random, separating the individual into three parts. The initial and final parts of a parent are passed on to the offspring. Then, the parent is changed and the offspring receive values according to the order in the parent of the missing elements.

In this work the Swap mutation [19] was used, then two positions of the solution are chosen at random and they are exchanged with each other.

High Level Heuristic - Thompson Sampling: Thompson Sampling was proposed by [7] and is used for online decision problems, where information is accumulated during the execution of the algorithm to performance optimization. As shown in [20], TS can incorporate a Bayesian model to represent the uncertainty associated to decision making on K operators (actions). When used, a k

operator produces, in the context of Bernoulli, a reward of 1 with probability θ_k and a reward of 0 with probability $1 - \theta_k$. In the present work, for each k action, the previous probability density function of θ_k is given by a β-distribution with parameters $\boldsymbol{\alpha} = (\alpha_1, ..., \alpha_K)$ e $\boldsymbol{\beta} = (\beta_1, ..., \beta_K)$:

$$p(\theta_k) = \frac{\Gamma(\alpha_k + \beta_k)}{\Gamma(\alpha_k)\Gamma(\beta_k)} \theta_k^{\alpha_k - 1}(1 - \theta_k)^{\beta_k - 1} \tag{3}$$

with Γ denoting the gamma function. As the observations are collected, the distribution is updated according to the Bayes rule and the β-prior conjugate. This means that α_k or β_k increases by one with each success or failure observed, respectively. Here $p_{op}(g) = \theta_k$ (operator probability in generation g) and the operator chosen is the one with the highest value of $p_{op}(g)$. The success estimate of the probability is sampled randomly from the later distribution.

2.3 NSGA-III

NSGA-III [5] arose from adaptations of NSGA-II [11] which is used for multi-objective problems, in order to better address problems with many objectives. NSGA-III is an algorithm based on Pareto dominance and Decomposition concepts. Pareto dominance happens when an individual has at least one of the objectives better than the others and all the others must be equivalent or better [21]. As decomposition-based algorithms, the NSGA-III uses an adaptive reference point mechanism, when the population for the next generation is formed, to preserve the diversity and convergence of the algorithm.

Any structure of reference points can be used, but in this work, as well as in many decomposition based algorithms, is considered the technique applied in [5]. In this technique, in order to avoid generating a very large number of reference points, the reference points are generated twice and then joined.

2.4 Related Works

When analyzing the papers on QAP [1], mQAP [22–24] and MaQAP [25, 26], it is possible to observe that with the increasing objectives and, consequently, on the problem complexity, there is less research being carried out on the theme. In researching different sources of scientific work, few correlated works are found for the QAP with many objectives. In [25] the Pareto stochastic local search is used. The authors compare different versions using algorithms that make use of the Cartesian product of scaling functions to reduce the number of objectives in the search space. The benchmark considered for testing has problems with 3 objectives, and the quality indicator used was the hypervolume. In [26] a comparison research was carried out between 18 state-of-art evolutionary algorithm to solve QAP with many objectives. The instances used in the tests have 3, 5 and 10 objectives. The performance metric used was IGD (Inverted Generational Distance). In the work [27] two frameworks for multi-objective optimization, MOEA/D and NSGA-II, were hybridized with Transgenetic Algorithms for the multi- and many-objective QAP solution. The benchmark instances considered

2, 3, 5, 7 and 10 objectives. The results were analyzed by different quality indicators of the Pareto frontier and statistical tests. The work shows the performance gain of the inclusion of operators based on computational transgenetics in the analyzed frameworks.

The many objective HHs have already been used for some problems in the literature [28–32], especially for continuous optimization problems with benchmark instances, although they are still under-explored. However, there is a gap for MaQAP, where the use of HH has not yet been explored. This is one of the contributions that we intend to obtain with this work.

3 Proposed Approach

The algorithm proposed in this work, named NSGA-III$_{TS}$, is presented in Algorithm 1, in which the functions marked in gray are part of the TS-based HH. The NSGA-III$_{TS}$ algorithm starts with the random generation of the initial population, of size NP, and the evaluation of its solutions. The NP reference points (Z) are initialized as an uniformly distributed hyperplane intersecting the axis of each objective (Step 1). Then, a main loop (Steps 3–27) begins , each iteration is a new generation (g). Internally to the main repetition, another loop (Steps 5–12) is used to create a new population who starts by selecting two solutions (parents), carried out at random (Step 6). The Thompson Sampling chooses among three different crossover operators (Step 7), the PMX, the CX or the 2P, combined with the swap mutation, using the beta distribution. After the application of the chosen crossover and, followed by the Swap mutation (Step 8), the new solutions are evaluated (Step 9) and inserted in a NP size population formed only by the offspring. The next action of the algorithm (Step 10) is related to the different versions of the algorithm which have been created. This step returns the obtained reward for each operator, after applying the swap mutation.

When the population of offspring is full, respectively in Steps 13 and 14, an union with the population of the parents is made, and the joined population is organized by the non-dominance ordering method. The first set of solutions to be part of next population is formed by the fronts of non-dominance that fully fit into the new population (Steps 16–19). If individuals are missing, the algorithm normalizes the objectives (Step 23) to determine the ideal and extreme points and adjust the reference points accordingly. The projected distance of each solution on the lines formed by the ideal and reference points are used to associate each solution with a reference point (Step 24) and a niche procedure is performed to select the solutions remaining, prioritizing clusters with fewer solutions (Step 25). Unused solutions are discarded. The algorithm ends when it reaches the maximum number of evaluations performed.

The different versions tested in Step 10 represent contributions of this work. What differentiates one version from another is the success and failure factor and the frequency of resetting successes and failures count, used by TS. The success and failure factor concerns the decision on which operator applications will contribute to the update of the TS counters (success - α_k and failure - β_k in Eq. 3).

The NSGA-III$_{TS}$'s versions that change the success and failures factors are based on the dominance relation (ND) and the average of the objectives (AVG) and work as follows: **i) ND**: after applying the crossover, a two-phase method is used. In the first, parents and offspring are ordered by level of non-dominance. If all parents and offspring are at the same level, the best average of the objectives is used as success. On the other hand, if at least one offspring succeeds in dominating a parent, it is considered success, otherwise it is failure; **ii) AVG**:

Algorithm 1: NSGA-III$_{TS}$ Procedure

Require: **f**: a many-objective problem;

$MaxEv$: the maximum number of evaluations;

P_0: the initial population of size NP;

Ensure: A Pareto front approximation of **f**

1 Assign to Z the NP reference points evenly distributed

2 $t \leftarrow NP$

3 **while** $t < MaxEv$ **do**

4 | $a \leftarrow 0$

5 | **while** $a < NP/2$ *and* $t < MaxEv$ **do**

6 | | PARENTSELECTION()

7 | | TSOPERATORSELECTION()

8 | | $Q_g \leftarrow$ CROSSOVERANDMUTATION(P_g)

9 | | OFFSPRINGEVALUATION()

10 | | OPERATORREWARDCALCULATION()

11 | | $a \leftarrow a + 1; t \leftarrow t + 2$

12 | **end**

13 | $R_g \leftarrow P_g \bigcup Q_g$

14 | $(F_1, F_2, \dots) \leftarrow$ NON-DOMINATEDSORTING(R_g)

15 | $i \leftarrow 1$

16 | **while** P_{g+1} *have space to include* F_i **do**

17 | | $P_{g+1} \leftarrow P_{g+1} \bigcup F_i$

18 | | $i \leftarrow i + 1$

19 | **end**

20 | **if** $|P_{g+1}| = NP$ **then**

21 | | Go to the next generation

22 | **else**

23 | | Normalize objectives and adapt Z

24 | | Associate each solution in P_{g+1} to a reference point and count the number of solutions in each cluster.

25 | | Choose $NP - |P_{g+1}|$ solutions from F_i using niching information and complete P_{g+1}

26 | **end**

27 **end**

28 **return** P_{g+1}

after applying the crossover, the mean of the objectives is used. Comparing the average of the parents with that of the offspring, if there is improvement it is considered success, otherwise failure;

The different ways of treating the success and failure counting of operators are presented below: **i) V1**: the success and failure counter variables are not reset at any point in the search; **ii) V2**: the variables success and failure counters are reset when creating a new population, that is, at each iteration; **iii) V3**: the variables success and failure counters are reset during the execution, at the moment when they reach 1/3 and 2/3 of the evaluations performed.

4   Experiments and Discussion

For the tests, 42 MaQAP instances were used with 50 locations and 5, 7 and 10 objectives (flows), these instances were proposed in [27] using the generators presented in [33]. Each instance is named following the format: mqap$X$-Yfl-TW, with X is the number of locations, Y is the number of objectives (or flow types), T means that the correlation is positive (p) or negative (n) and W shows whether the problem is uniform (uni) or real-like (rl) [27].

All experiments are evaluated by the IGD+ (Inverted Generational Distance plus) quality indicator [34]. This metric evaluates different properties of a Pareto front approximation and provides a single performance value for the set. IGD+ indicates the distance from the set found by the algorithm in relation to a reference set. The reference set used in this work is composed of the non-dominated solutions resulting from the union of the approximation front of all the algorithms involved in the comparison.

The experiments were organized in four stages. Firstly, the different strategies for resetting the count of successes and failures are compared with each other, for the two success and failure factors: by the dominance relation criterion (ND) and by the average criterion (AVG). Next, considering the best counting strategies, AVG and ND are confronted. In the third stage, the best version of the proposed algorithm is compared with the separately applied operators. In the last stage, the best proposed HH is compared with some approaches from literature.

The proposed approaches, as well as the other considered algorithms, are executed 30 times with different seeds. The parameters for all experiments in this study are in Table 1. TS does not have any input parameters to be defined, which represents a positive point for the technique. All algorithms, quality indicator and statistical test were developed in jMetal framework [35].

Table 1. Parameters considered in the proposed and in the literature approaches.

	Values	Description
CrR	100%	Crossover rate (percentage)
MutR	100%	Mutation rate (percentage)
NP	105/294/275	Population size (5/7/10 objectives)
MaxEv	300000	Maximum number of function evaluations

Tables 2, 3, 4, 5 and 6 show a comparison between the algorithms, involving all instances for 5, 7 and 10 objectives, using the test statistical analysis of Friedman test [36]. The tables show the rank of the algorithms and the hypothesis that the first place approach is equivalent to the others. Whenever the hypothesis is rejected, the first ranked approach is statistically superior to those in the rejected hypothesis line.

4.1 Stage 1: Selecting the Success and Failure Counting Strategy

The number of succeed and fail operator applications is the main element for the operation of the Thompson Sampling heuristic. Therefore, experiments are presented below for the analysis of different strategies for resetting these counts throughout the search process. This analysis is performed for the two proposed success factor - dominance relation (ND) and the average of the objectives (AVG), detailed in Sect. 3.

Table 2 analyzes the different ways of counting on the ND success factor (Algorithm column) and shows the Friedman test on the IGD+ indicator (Ranking and Hypothesis columns), considering all 42 instances for all number of objectives. The percentage of usage of the different operators used is also shown (% PMX, % 2P, % CX columns. From the table it is possible to observe that the NSGA-III$_{TS-ND}$V3, which resets the count in two moments of the search process, is the best algorithm, followed by NSGA-III$_{TS-ND}$V1 (version without reset). The algorithm that resets the success and failure counts at each iteration (NSGA-III$_{TS-ND}$V2) has the worst performance. This table also shows the average percentage usage of operators throughout the search process. It is worth noting that the algorithms that selected the CX operator more often have better performance. In addition, the version that resets success and failure counts at each iteration tends to distribute the use of operators more evenly, as well as in a random selection. The same analysis is performed in Table 3, considering the success and failure factors by the average of the objectives (AVG). According to the IGD+ quality indicator and the Friedman test, the NSGA-III$_{TS-AVG}$V1, which does not reset the count during the search, is the best algorithm, while the NSGA-III$_{TS-AVG}$V3 algorithm got second place. Again, the algorithm that resets the success and failure counts at each iteration (NSGA-III$_{TS-AVG}$V2) has the worst performance. The two algorithms that have the highest choice frequency for the CX operator have the best performance.

Table 2. Ranking and average percentage usage of operators - comparison between the three ways of restarting the counting of successes and failures (V1, V2 and V3), for success factor based on the dominance criterion (ND), according to Friedman's test.

Algorithm	Ranking	% PMX	% 2P	% CX	Hypothesis
NSGA-III$_{TS-ND}$V3	1.500	7.05	8.2	84.75	-
NSGA-III$_{TS-ND}$V1	1.619	15.00	8.31	76.68	Rejected
NSGA-III$_{TS-ND}$V2	2.881	30.27	17.31	52.41	Rejected

Table 3. Ranking and average percentage usage of operators - comparison between the three ways of restarting the counting of successes and failures (V1, V2 and V3), for success factor based on the objectives average (AVG), according to Friedman's test.

Algorithm	Ranking	% PMX	% 2P	% CX	Hypothesis
NSGA-III$_{TS-AVG}$V1	1.500	0.91	0.64	98.44	-
NSGA-III$_{TS-AVG}$V3	1.667	0.67	0.43	98.88	Rejected
NSGA-III$_{TS-AVG}$V2	2.833	16.51	9.86	73.63	Rejected

In view of these analyzes, it is possible to observe that the best count reset strategy for the ND success and failure factors is that which reset the variables in 1/3 and 2/3 of the maximum number of performed evaluations (V3). As for the AVG success and failure factors, is the best strategy that accumulates the scores throughout all evaluations, without resetting them at any time (V1). The next section compares the two success and failure factor (ND and AVG), taking into account your best success and failure count reset strategies, NSGA-III$_{TS-AVG}$V1 and NSGA-III$_{TS-ND}$V3. In order to understand how the different forms of counting and factors of success and failure behave in the operators choice, the frequencies of use for each operator during the search were analyzed. Figures 2 and 3 show how often each of the three available operators is selected by HH during the search, taking as example three instances, they are: $mqap50 - 5fl - n25uni$, $mqap50 - 7fl - p00rl$, $mqap50 - 10fl - n75rl$. The choice of instances aimed to vary their characteristics such as the number of objectives, the correlation and the form of generation (rl or uni). These figures show the percentage of use of the operator (y axis) over the generations (x axis), considering the average of 5 independent executions. Each point on the curve represents the frequency with which each operator is used in a given generation. Figure 2 shows, for the AVG success and failure factor, that the version with the best performance (NSGA-III$_{TS-AVG}$V1) selects the CX more frequently than the others operators. However, all three restart strategies favor the choice of the CX operator. V2, which resets the counts at each iteration, presents an oscillation in the choice of operators throughout the search. V3, which resets the count in two moments, behaves similarly to the version that does not reset the counts (V1). Considering ND, Fig. 3 also shows that the version with the best performance (NSGA-III$_{TS-ND}$V3) selects the CX operator more frequently than the others. The oscillation in the choice of operators in the version that reinitializes the count at each iteration (V2) is even greater in this success and failure factors (ND). In order to make a fair comparison, the same stopping criterion was applied in all experiments ($MaxEv = 300000$). Thus, as can be seen in the figures, the number of generations for each instance is equal to 2832 for 5 objectives, 1022 for 7 objectives and 1088 for 10 objectives.

4.2   Stage 2: Selecting the Success and Failure Factor

The tests carried out in this stage answer the research question related to which the most appropriate success and failure factor in the proposed scenario. It is

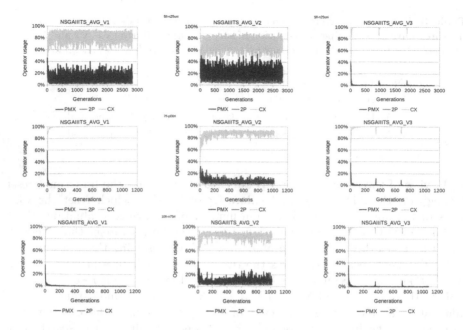

Fig. 2. Operators usage over the generations for the NSGA-III$_{TS-AVG}$V1 approaches (left), NSGA-III$_{TS-AVG}$V2 (in the middle) and NSGA-III$_{TS-AVG}$V3 (right), for 5, 7 and 10 objectives.

based on the best strategy selected in the previous simulation stage, that is, performs the comparison of the NSGA-III$_{TS-AVG}$V1 and NSGA-III$_{TS-ND}$V3 algorithms.

Friedman's test, in Table 4, makes clear the supremacy of the version that considers the average of the objectives when deciding whether the application of a given operator is a success or a failure (AVG). NSGA-III$_{TS-AVG}$V1 represents the best HH version proposed in the scope of this work and will be compared with the operators applied in isolation in the next section.

Table 4. Average ranking of algorithms considering the comparison between NSGA-III$_{TS-AVG}$V1 and NSGA-III$_{TS-ND}$V3, according to Friedman test.

Algorithm	Ranking	Hypothesis
NSGA-III$_{TS-AVG}$V1	1.357	-
NSGA-III$_{TS-ND}$V3	1.643	Rejected

4.3 Stage 3: Effect of Operator Selection by Hyper-Heuristic

The use of a hyper-heuristic in the selection of different operators throughout the search process releases the user from choosing the best operator to be applied

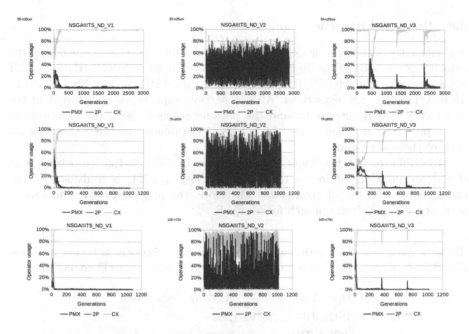

Fig. 3. Operators usage over the generations for the NSGA-III$_{TS-ND}$V1 (left), NSGA-III$_{TS-ND}$V2 (in the middle) and NSGA-III$_{TS-ND}$V3 (right), for 5, 7 and 10 objectives.

Table 5. Average ranking of algorithms considering comparison between the best HH (NSGA-III$_{TS-AVG}$V1) and operators applied in isolation, according to Friedman's test.

Algorithm	Ranking	Hypothesis
NSGA-III$_{CX}$	1.476	-
NSGA-III$_{TS-AVG}$V1	1.524	Accepted
NSGA-III$_{PMX}$	3.048	Rejected
NSGA-III$_{2P}$	3.952	Rejected

in the solution of a given problem. This avoids the necessary simulations and even a more in-depth knowledge of the technique and the problem in question. The simulations of this stage aim to analyze the performance of the automatic selection made by Thompson Sampling in relation to the performance of the operators applied in isolation. Table 5 shows that the performance of HH and the CX operator are similar to each other, while the PMX and 2P operators perform worse, considering the IGD + quality indicator. Friedman's test indicates statistical equivalence between NSGA-III$_{TS-AVG}$V1 and NSGA-III$_{CX}$, which shows that the automation of the choice of operators is done properly.

4.4   Stage 4: Literature Comparison

The efficiency of the best proposed HH is measured at this stage in the face of the algorithms reported in the literature. Table 6 compares the NSGA-III$_{TS-AVG}$V1 algorithm with the MOEA/DD [8], MOEA/D [9], SPEA2 [10], NSGA-II [11], MOEA/D-DRA [12] algorithms. All comparison approaches are present in the framework jMetal version 4.5 and the parameters used, with the exception of those presented in Table 1, are the standard parameters of the framework for each of these algorithms.

According to the Friedman test, presented in Table 6, the hypothesis that the algorithm with the lowest rank (NSGA-III$_{TS-AVG}$V1) is equivalent to the others is rejected for all algorithms. Therefore, the proposed algorithm can be seen as a competitive approach for MaQAP.

Table 6. Average ranking of the algorithms considering a comparison between the best HH (NSGA-III$_{TS-AVG}$V1) and literature algorithms, according to Friedman's test.

Algorithm	Ranking	Hypothesis
NSGA-III$_{TS-AVG}$V1	2.048	-
MOEA/DD	2.143	Rejected
MOEA/D	2.881	Rejected
SPEA2	3.976	Rejected
NSGA-II	4.024	Rejected
MOEA/D-DRA	5.929	Rejected

5   Conclusions

In this work, it was proposed an approach combining selection HH with the NSGA-III framework. The selection HH considered was Thompson Sampling. The HH was used to generate each offspring through the selected operator from a set of LLH, according to an updated probability based on their previous performance. Three candidates for operators were considered: PMX, 2P and CX. A set of 42 benchmarks instances was considered with 5, 7 and 10 objectives.

Two success and failure factors have been proposed with different ways of resetting the counters. The results were analyzed with the IGD+ quality indicator and Friedman's statistical test. The IGD+ points out that the best way to restart the success and failure counts varies according to the success and failure factor adopted. The best proposed HH (NSGA-III$_{TS-AVG}$V1) presented equivalent, or even better, performance to operators applied in isolation. This shows the benefit of automatic selection made by TS, as it releases the user from determining the best operator to be applied in the problem solution. In comparison with the literature, the NSGA-III$_{TS-AVG}$V1 outperforms all the considered algorithms.

It should be noted that, in addition to its competitive results, Thompson Sampling is a simple technique in relation to other high-level heuristics implemented for HHs and without parameters to be adjusted. As future work we can consider the adaptive control of the size of the set of operators and the test in different optimization problems with multiple and many objectives.

References

1. Abdel-Basset, M., Manogaran, G., Rashad, H., Zaied, A.N.: A comprehensive review of quadratic assignment problem: variants, hybrids and applications. In: Journal of Ambient Intelligence and Humanized Computing, pp. 1–24 (2018)
2. Koopmans, T.C., Beckmann, M.: Assignment problems and the location of economic activities. Econometrica **25**(1), 53–76 (1957)
3. Uluel, M., Ozturk, Z.K.: Solution approaches to multiobjective quadratic assignment problems. In: European Conference on Operational Research (2016)
4. Fritsche, G.: The cooperation of multi-objective evolutionary algorithms for many-objective optimization. Ph.D. dissertation, Universidade Federal do Paraná, Curitiba, PR, BR (2020)
5. Deb, K., Jain, H.: An evolutionary many-objective optimization algorithm using reference-point-based nondominated sorting approach, part i: Solving problems with box constraints. IEEE Trans. Evol. Comput. **18**(4), 577–601 (2014)
6. Burke, E., et al.: Hyper-heuristics: a survey of the state of the art. J. Oper. Res. Soc. **64**, 1695–1724 (2013)
7. Thompson, W.R.: On the likelihood that one unknown probability exceeds another in view of the evidence of two samples. Biometrika **25**(3–4), 285–294 (1933)
8. Li, K., Deb, K., Zhang, Q., Kwong, S.: An evolutionary many-objective optimization algorithm based on dominance and decomposition. IEEE Trans. Evol. Comput. **19**(5), 694–716 (2015)
9. Zhang, Q., Li, H.: MOEA/D: A multiobjective evolutionary algorithm based on decomposition. IEEE Trans. Evol. Comput. **11**(6), 712–731 (2007)
10. Zitzler, E., Laumanns, M., Thiele, L.: SPEA2: Improving the strength pareto evolutionary algorithm for multiobjective optimization, vol. 3242 (2001)
11. Deb, K., Pratap, A., Agarwal, S., Meyarivan, T.: A fast and elitist multiobjective genetic algorithm: NSGA-II. IEEE Trans. Evol. Comput. **6**(2), 182–197 (2002)
12. Zhang, Q., Liu, W., Li, H.: The performance of a new version of moea/d on cec09 unconstrained mop test instances. In: Congress on Evolutionary Computation, pp. 203–208. IEEE Press (2009)
13. Çela, E.: The Quadratic Assignment Problem: Theory and Algorithms, ser. Combinatorial Optimization. Springer, US (2013)
14. Sabar, N., Ayob, M., Kendall, G., Qu, R.: A dynamic multiarmed bandit-gene expression programming hyper-heuristic for combinatorial optimization problems. IEEE Trans. Cybern. **45**, 06 (2014)
15. Pillay, N., Qu, R.: Hyper-heuristics: theory and applications, 1st ed. Springer Nature (2018). https://doi.org/10.1007/978-3-319-96514-7
16. Drake, J.H., Kheiri, A., Özcan, E., Burke, E.K.: Recent advances in selection hyper-heuristics. Eur. J. Oper. Res. **285**(2), 405–428 (2020)
17. Talbi, E.-G.: Metaheuristics: From Design to Implementation. Wiley, Hoboken (2009)

18. Oliver, I., Smith, D., Holland, J.R.: Study of permutation crossover operators on the traveling salesman problem. In: Genetic Algorithms and Their Applications: Proceedings of the Second International Conference on Genetic Algorithms: July 28–31, at the Massachusetts Institute of Technology, Cambridge, MA. Hillsdale, NJ: L. Erlhaum Associates. (1987)

19. Larranaga, P., Kuijpers, C., Murga, R., Inza, I., Dizdarevic, S.: Genetic algorithms for the travelling salesman problem: a review of representations and operators. Artif. Intell. Rev. **13**, 129–170 (1999)

20. Russo, D.J., Roy, B.V., Kazerouni, A., Osband, I., Wen, Z.: A tutorial on thompson sampling. Found. Trends Mach. Learn. **11**, 1–96 (2018)

21. Zitzler, E., Laumanns, M., Bleuler, S.: A tutorial on evolutionary multiobjective optimization. In: Gandibleux, X., Sevaux, M., Sörensen, K., T'kindt, V. (Eds.) Metaheuristics for Multiobjective Optimisation, Springer, Berlin Heidelberg, pp. 3–37 (2004). https://doi.org/10.1007/978-3-642-17144-4_1

22. Dokeroglu, T., Cosar, A.: A novel multistart hyper-heuristic algorithm on the grid for the quadratic assignment problem. Eng. Appl. Artif. Intell. **52**, 10–25 (2016)

23. Dokeroglu, T., Cosar, A.: A novel multistart hyper-heuristic algorithm on the grid for the quadratic assignment problem. Eng. Appl. Artif. Intell. **52**, 10–25 (2016)

24. Senzaki, B.N.K., Venske, S.M., Almeida, C.P.: Multi-objective quadratic assignment problem: an approach using a hyper-heuristic based on the choice function. In: Cerri, R., Prati, R.C. (eds.) Intelligent Systems, pp. 136–150. Springer International Publishing, Cham (2020). https://doi.org/10.1007/978-3-030-61377-8_10

25. Drugan, M.M.: Stochastic pareto local search for many objective quadratic assignment problem instances. In: 2015 IEEE Congress on Evolutionary Computation (CEC), pp. 1754–1761, May 2015

26. Rahimi, I., Gandomi, A.: Evolutionary many-objective algorithms for combinatorial optimization problems: A comparative study. Arch. Comput. Methods Eng. **28**, 03 (2020)

27. de Almeida, C.P.: Transgenética computacional aplicada a problemas de otimização combinatória com múltiplos objetivos. Ph.D. dissertation, Federal University of Technology - Paraná, Brazil (2013)

28. Castro, O.R., Pozo, A.: A mopso based on hyper-heuristic to optimize many-objective problems. In: 2014 IEEE Symposium on Swarm Intelligence, pp. 1–8 (2014)

29. Walker, D.J., Keedwell, E.: Towards many-objective optimisation with hyper-heuristics: identifying good heuristics with indicators. In: Handl, J., Hart, E., Lewis, P.R., López-Ibáñez, M., Ochoa, G., Paechter, B. (eds.) PPSN 2016. LNCS, vol. 9921, pp. 493–502. Springer, Cham (2016). https://doi.org/10.1007/978-3-319-45823-6_46

30. Kuk, J., Gonçalves, R., Almeida, C., Venske, S., Pozo, A.: A new adaptive operator selection for nsga-iii applied to cec 2018 many-objective benchmark. In: 2018 7th Brazilian Conference on Intelligent Systems (BRACIS), pp. 7–12 (2018)

31. Fritsche, G., Pozo, A.: The analysis of a cooperative hyper-heuristic on a constrained real-world many-objective continuous problem. In: 2020 IEEE Congress on Evolutionary Computation (CEC), pp. 1–8 (2020)

32. Friesche, G., Pozo, A.: Cooperative based hyper-heuristic for many-objective optimization. In: Genetic and Evolutionary Computation Conference, ser. GECCO '19. New York, NY, USA: Association for Computing Machinery, pp. 550–558 (2019)

33. Knowles, J., Corne, D.: Instance generators and test suites for the multiobjective quadratic assignment problem. In: Fonseca, C.M., Fleming, P.J., Zitzler, E., Thiele, L., Deb, K. (eds.) EMO 2003. LNCS, vol. 2632, pp. 295–310. Springer, Heidelberg (2003). https://doi.org/10.1007/3-540-36970-8_21

34. Ishibuchi, H., Masuda, H., Tanigaki, Y., Nojima, Y.: Modified distance calculation in generational distance and inverted generational distance. In: Gaspar-Cunha, A., Henggeler Antunes, C., Coello, C.C. (eds.) EMO 2015. LNCS, vol. 9019, pp. 110–125. Springer, Cham (2015). https://doi.org/10.1007/978-3-319-15892-1_8

35. Nebro, A.J., Durillo, J.J., Vergne, M.: Redesigning the jmetal multi-objective optimization framework. In: Conference on Genetic and Evolutionary Computation, ser. GECCO Companion '15. New York, NY, USA: Association for Computing Machinery, pp. 1093–100 (2015)

36. Conover, W.: Practical nonparametric statistics, 3rd ed., ser. Wiley series in probability and statistics. New York, NY [u.a.]: Wiley (1999)

I2DE: Improved Interval Differential Evolution for Numerical Constrained Global Optimization

Mariane R. S. Cassenote[(⊠)][iD], Guilherme A. Derenievicz[iD],
and Fabiano Silva[iD]

Informatics Department, Federal University of Paraná, Curitiba, Brazil
{mrscassenote,guilherme,fabiano}@inf.ufpr.br

Abstract. Several hybrid approaches have been proposed to solve numerical constrained optimization problems. In this paper we present an Improved Interval Differential Evolution (I2DE) that uses structural information of the instance during the optimization process. We extend the math operations supported by a multi-interval core implementation that allows pruning infeasible solutions by using local consistency techniques and a backtrack-free local search. Furthermore, we propose a reformulation of interval evolutionary mutation strategies. A comprehensive experimental analysis is conducted over COCONUT and CEC2018 competition benchmarks and indicates that the hybridization between metaheuristics and constraint programming significantly improves the quality of the solutions. The experimental evaluation shows that our black-box version of I2DE outperformed several state-of-the-art solvers.

Keywords: Global optimization · Differential evolution · Interval methods

1 Introduction

In the last few decades, interval based solvers have been used to tackle *Numerical Constrained Global Optimization Problems* (NCOP) in a rigorously way [1]. In general, such methods are composed by a complete investigation of the search space, using the structure of the problem and consistency techniques from the constraint programming field to prune infeasible solutions. Despite the great progress of interval techniques, such methods remain impractical in instances with a large number of dimensions.

On the other hand, Differential Evolution (DE) has become one of the most used evolutionary algorithms to deal with large global optimization problems, due to its performance and simplicity [3,4,20]. As opposed to interval algorithms, DE based solvers are fast, but do not guarantee the global optimality. Also, DE uses the instance as a black-box model, where its structure is unknown.

A promising strategy to handle with NCOPs is the hybrid approach, where different methods are combined [26]. In this context, the solver *Interval Differential Evolution* (InDE) was proposed [5]. This method integrates the usual DE

© Springer Nature Switzerland AG 2021
A. Britto and K. Valdivia Delgado (Eds.): BRACIS 2021, LNAI 13073, pp. 186–201, 2021.
https://doi.org/10.1007/978-3-030-91702-9_13

approach with an interval solver called *Relaxed Global Optimization* (OGRe) [8]. This solver uses a structural decomposition of the NCOP instance to identify the amount of local consistency that guarantees a backtrack-free optimization. Although this is a strong theoretical result, achieving this amount of local consistency can be an intractable problem. Therefore, OGRe uses an *Interval Branch & Bound* (IB&B) method to tackle a relaxed form of the instance. Thus, the solution found by OGRe may be infeasible on the original NCOP instance. Moreover, the computational cost of the IB&B is prohibitive.

The InDE solver extended usual DE operators to the interval context, using OGRe's core to select a subset of variables on which the search process will occur, whilst the others are valuated by constraint propagation. In addition, local consistency techniques are applied to prune infeasible solutions. By combining theses techniques, InDE outperformed OGRe [5].

In this work, we present the *Improved Interval Differential Evolution* (I2DE) solver. We extend InDE and the operators supported by the OGRe's multi-interval core, which allows us to tackle a greater diversity of benchmark functions and real-world problems. Moreover, we present improved versions of three interval evolutionary mutation operations and incorporate several heuristics of state-of-the-art solvers. An extensive experimental analysis performed over the COCONUT [19] and CEC2018 Competition on Constrained Real-Parameter Optimization [27] benchmarks reveals that our I2DE significantly outperforms InDE, OGRe and a black-box version of I2DE, which suggests that hybridization between metaheuristics and structural decomposition is a promising research direction. Furthermore, our black-box solver outperformed several state-of-the-art metaheuristic solvers.

The remaining of this paper is organized as follows: Sect. 2 contains background definitions of interval methods and metaheuristics. Section 3 provides details of I2DE's features and the main improvements compared to InDE. The experimental analysis is presented in Sect. 4 and Sect. 5 concludes this work.

2   Background

A *Numerical Constrained Global Optimization Problem* (NCOP) consists of finding an assignment of values to a set of variables $V = \{x_1, x_2, \ldots, x_D\}$ that minimizes an objective function $f : \mathbb{R}^D \mapsto \mathbb{R}$ subject to a *constraint network* (CN) $\mathcal{N} = (V, \mathcal{D}, \mathcal{C})$, where $\mathcal{D}$ is the domain set of V and $\mathcal{C}$ is a set of constraints of the form $g_i(x_{i_1}, \ldots, x_{i_k}) \leq 0$. Domain sets can be represented by intervals or multi-intervals. Given a closed interval $X = [\underline{x}, \overline{x}]$, we call $\underline{x}, \overline{x} \in \mathbb{R}$ the *endpoints* of X; $\omega(X) = \overline{x} - \underline{x}$ denotes the *width* of X and $\mu(X) = (\underline{x} + \overline{x})/2$ denotes its *midpoint*. A closed interval can be defined by its width and midpoint as follows: $X = [\mu(X) - \omega(X)/2, \ \mu(X) + \omega(X)/2]$. A *multi-interval* $\mathcal{X} = \langle X_1, X_2, \ldots, X_k \rangle$ is an ordered set of disjointed intervals, where $i < j \implies \overline{x}_i < \underline{x}_j$. In this case, the domain set of a CN is a (multi-) interval *box* $(\mathcal{X}_1, \mathcal{X}_2, \ldots, \mathcal{X}_D)$.

In this paper, we tackle NCOPs which constraints can be decomposed into a set of ternary constraints[1] $x = y \circ z$ or $x = \diamond y$, where $\circ$ ($\diamond$) is a well-defined binary

---

[1] A constraint is said to be *ternary* if it involves at most three variables.

(unary) operation. Therefore, auxiliary variables are included in the network. For example, the constraint $x + 2 \cdot \sin(y)^2 \leq 0.5$ is encoded as $\{c_1 = x + a_1, \; a_1 = 2 \cdot a_2, \; a_2 = a_3{}^2, \; a_3 = \sin(y)\}$, where $A_i = (-\infty, +\infty)$ is the domain set of the auxiliary variable a_i and $C_1 = (-\infty, 0.5]$ is the interval constant that represents the original constraint's relation.

2.1   Interval Analysis Applied to Global Optimization

Interval Analysis is a method of numerical analysis introduced by Moore [16]. Given intervals X, Y and Z the *interval extension* of any binary (unary) operation $\circ$ ($\diamond$) well defined in $\mathbb{R}$ is defined by:

$$X \circ Y = \{x \circ y \mid x \in X, \, y \in Y \text{ and } x \circ y \text{ is defined in } \mathbb{R}\},$$
$$\diamond Z = \{\diamond z \mid z \in Z \text{ and } \diamond z \text{ is defined in } \mathbb{R}\}.$$

The sets $X \circ Y$ and $\diamond Z$ can be intervals, multi-intervals or empty sets. It is possible to compute $X \circ Y$ or $\diamond Z$ for all algebraic and the common transcendental functions only by analyzing the endpoints of X and Y or Z [10,16], e.g., $[\underline{x}, \overline{x}] + [\underline{y}, \overline{y}] = [\underline{x} + \underline{y}, \overline{x} + \overline{y}]$, $[\underline{x}, \overline{x}] \cdot [\underline{y}, \overline{y}] = [\min\{\underline{x}\underline{y}, \underline{x}\overline{y}, \overline{x}\underline{y}, \overline{x}\overline{y}\}, \max\{\underline{x}\underline{y}, \underline{x}\overline{y}, \overline{x}\underline{y}, \overline{x}\overline{y}\}]$, $2^{[\underline{z}, \overline{z}]} = [2^{\underline{z}}, 2^{\overline{z}}]$, etc. Given multi-intervals $\mathcal{X}$, $\mathcal{Y}$ and $\mathcal{Z}$, the operation $\mathcal{X} \circ \mathcal{Y}$ is the tightest multi-interval that contains $\{X \circ Y \mid X \in \mathcal{X}, Y \in \mathcal{Y}\}$, and $\diamond \mathcal{Z}$ is the tightest multi-interval that contains $\{\diamond Z \mid Z \in \mathcal{Z}\}$.

In the constraint programming field, a constraint is locally consistent if it satisfies some specific property within a given box. For instance, a ternary constraint $C : x = y \circ_1 z$ is *Generalized Arc-Consistent* (GAC) w.r.t. a box (X, Y, Z) iff $X \subseteq Y \circ_1 Z$, $Y \subseteq X \circ_2 Z$ and $Z \subseteq X \circ_3 Y$, where $\circ_2$ and $\circ_3$ are the inverse operations of $\circ_1$ that hold the condition $(x = y \circ_1 z) \iff (y = x \circ_2 z) \iff (z = x \circ_1 y)$. In other words, if C is GAC then given any value for one of its variables, one can extend such valuation for its remaining variables whilst satisfying C. This notion of local consistency was proposed by [15].

It is well known that acyclic[2] CN can be solved in a backtrack-free manner if GAC is achieved [6]. Such result can be extended to NCOP by encoding the objective function $f(x)$ as a new constraint $y = f(x)$; after enforcing GAC (by removing inconsistent values from the current box) we instantiate $y = \min Y$ and propagate this valuation over the entire network without encountering any conflicts [8]. However, GAC is not enough when the network is not acyclic.

Another notion of local consistency is *Relational Arc-Consistent* (RAC) [7]. A ternary constraint $C : x = y \circ_1 z$ is RAC w.r.t. a box (X, Y, Z) iff $X \supseteq Y \circ_1 Z$, $Y \supseteq X \circ_2 Z$ and $Z \supseteq X \circ_3 Y$, i.e., given any value for two variables of C, one can extend such valuation for its remaining variable whilst satisfying C.

In [8] it was proposed a decomposition of CNs that relates the amount of consistency necessary to ensure a backtrack-free solution. An *epiphytic decomposition* of a CN is a tuple $(\mathcal{A}, \Omega, t)$, where $\mathcal{A}$ is an ordered set of acyclic networks

---

[2] The structure of a CN can be represented by a hypergraph which vertices are the variables and for each constraint there is a hyperedge connecting its respective vertices. Therefore, a CN is acyclic if its hypergraph is Berge-acyclic [2].

obtained by removing from the CN a set of constraints Ω and $t : \Omega \mapsto V_\Omega$ is a function that associates each constraint $C \in \Omega$ with one of its variables $t(C)$ satisfying the following: if $t(C)$ belongs to the network N_i then (i) the remaining variables of C belongs to previous networks $N_{j<i}$; and (ii) there is no other constraint $C' \in \Omega$ such that $t(C')$ belongs to N_i. If the CN encodes a NCOP instance, the variable $y = f(c)$ that represents the objective function must be in the first acyclic network of $\mathcal{A}$.

It was shown that if a NCOP $\mathcal{P}$ encoded as a ternary CN is GAC and the constraints in the set Ω of its epiphytic decomposition are RAC, then $\mathcal{P}$ can be solved in a backtrack-free fashion. The proposed OGRe [8] and InDE [5] solvers are based on this relation and attempt to achieve the relational arc-consistency of the CN as a form of optimization. However, enforcing RAC is in general intractable. The approach proposed in OGRe approximates RAC by using an *Interval Branch and Pruning* scheme. If a constraint $(C : x = y \circ z) \in \Omega$ is not RAC under the current box, the domain of some variable in this constraint (excluding $t(C)$) is bisected (*branch*) and GAC is enforced on both sub-problems (*pruning*) before a new verification of the RAC property of Ω constraints. The algorithm continues in a recursive fashion. Therefore, although the ternary decomposition of the original NCOP instance increases the number of variables, not all of them are considered in the *branch* process, but only those of Ω constraints.

In each branch of the search tree, a backtrack-free local search occurs. First, the variable representing the objective function is instantiated with its minimum value in the current box. Next, the valuation is propagated over all the network, following the ordering $\mathcal{A}$ of the epiphytic decomposition. Note that acyclic networks of $\mathcal{A}$ are feasible if GAC was enforced, but the constraints in Ω may be infeasible. There are two main parameters that control OGRe's search procedure: the tolerance ε_Ω allowed in each Ω constraint, and the minimum granularity Δ which intervals can be bisected.

OGRe's approach is a variation of usual *Interval Branch and Bound* (IB&B) methods [1] that have been used in the last few decades to rigorously solve NCOP. Interval methods compute a set of atomic boxes that contains an optimum solution of the NCOP instance. Due to elevated computational cost, these algorithms remain inefficient in instances with many dimensions. On the other hand, a solution found by OGRe may be inexact (w.r.t. objective function cost or constraint violation), because RAC is approximated by a tolerance ε_Ω in Ω constraints. Besides that, find an acceptable value for ε_Ω is not trivial.

The multi-interval core proposed in OGRe is also used in InDE, including the local search procedure, GAC contractor, epiphytic decomposition of the CN, and multi-interval operations $+, -, *, /, \wedge$ and $\sqrt{}$.

2.2 Differential Evolution

Storn and Price [20] proposed Differential Evolution (DE) as an Evolutionary Algorithm that combines the coordinates of existing solutions with a particular probability to generate new candidate solutions. The classical DE consists of a loop of G generations over a population of NP individuals. An individual is an assignment of values to all variables of the instance, represented by a vector $\mathbf{x} = (a_1, a_2, \ldots, a_D)$, where $a_i \in X_i$ is a value from the domain of the variable x_i, $1 \le i \le D$. The fitness evaluation of an individual is responsible for determinate its quality throughout the evolutionary process.

The initial population is randomly generated according to a uniform distribution over $X_1 \times \cdots \times X_D$. During each generation, the operators of mutation, crossover and selection are performed on the population until a termination condition is satisfied, like a fixed maximum number of fitness evaluations ($MaxFEs$).

In the mutation phase, an operator is applied to generate a *mutant vector* $\mathbf{v}_i$ for each individual $\mathbf{x}_i$ (called *target vector*) of the population. The most popular mutation operator is DE/rand/1 and is defined by:

$$\mathbf{v}_i = \mathbf{r}_1 + F \cdot (\mathbf{r}_2 - \mathbf{r}_3), \tag{1}$$

where F is the scaling factor and $\mathbf{r}_1$, $\mathbf{r}_2$ and $\mathbf{r}_3$ are three mutually distinct individuals randomly selected from the population. Other popular mutation operators are used in this work, such as DE/current-to-rand/1 (Eq. 2) [12] and DE/current-to-$p$best/1 (Eq. 3) [29]:

$$\mathbf{v}_i = \mathbf{x}_i + s \cdot (\mathbf{r}_1 - \mathbf{x}_i) + F \cdot (\mathbf{r}_2 - \mathbf{r}_3), \tag{2}$$

$$\mathbf{v}_i = \mathbf{x}_i + F \cdot (\mathbf{r}_{\text{pbest}} - \mathbf{x}_i) + F \cdot (\mathbf{r}_1 - \mathbf{r}_2), \tag{3}$$

where s is a uniformly distributed random number between 0 and 1, p is a value in $[1, NP]$, and $\mathbf{r}_{\text{pbest}}$ is an individual randomly chosen from the p best individuals of the current population.

A crossover search operator is applied on the mutant individual $\mathbf{v}_i$ to produce a new solution $\mathbf{u}_i$, called *trial vector*, given a probability defined by the crossover rate CR. In the exponential crossover used in this work, we first choose an integer $d \in [1, D]$ to be the starting point of the target vector for crossover. We also determine another integer value l chosen from $[1, D]$ with probability $P(l \ge L) = CR^{L-1}$ for any $L > 0$ which denotes how many consecutive decision variables are selected from the mutant vector starting at d position. Then, the offspring is generated as follows:

$$\mathbf{u}_{ij} = \begin{cases} \mathbf{v}_{ij}, & \text{if } j \in \{m_D(d), m_D(d+1), \ldots, m_D(d+l-1)\}, \\ \mathbf{x}_{ij}, & \text{otherwise}, \end{cases}$$

where $m_D(n) = 1 + ((n-1) \mod D)$ allows to iterate cyclically through the vector. Finally, a selection operator is performed on $\mathbf{x}_i$ and $\mathbf{u}_i$, and the best one according to a fitness evaluation function is chosen for the next generation.

3   Improved Interval Differential Evolution

In this section we describe details of our approach, called I2DE, and emphasize the improvements and differences from the work presented in [5] which introduced the InDE solver, an Interval Differential Evolution approach that uses structural information for global optimization. One of the main contributions of our work is the extension of operations supported by OGRe's multi-interval core. Previously, it was only possible to tackle instances restricted to operations $+$, $-$, $*$, $/$, $\wedge$ and $\sqrt{}$. We extended the multi-interval implementation for handling the following additional operations: *log, exp, sin, cos, tan, abs, sign, max* and *min*. Considering the new operations, we implemented a new procedure to identify the tolerance ε_Ω needed to approximate RAC. For example, given a constraint $x = y \circ z$, this tolerance is the maximum distance between any value of the multi-interval $\mathcal{Y} \circ \mathcal{Z}$ to its closest value in the multi-interval $\mathcal{X}$. However, the operator $\circ$ may be not well-defined within the box $(\mathcal{Y}, \mathcal{Z})$. In this case, we compute the distance using the inverse operation of $\circ$ that results in the tightest multi-interval. For instance, we can not compute the distance for the constraint $x = y/z$ within the box $Z = [0,0]$, but we can do it for the inverse $y = x * z$.

In I2DE and InDE, an individual is an assignment of intervals to variables (a box). The population is a set of boxes that covers parts of the search space, instead of just points as in classical metaheuristics. In the optimization process, only the variables of constraints in the Ω set of an epiphytic decomposition are considered. This allows to apply local consistency techniques to prune infeasible solutions. OGRe's local search is used to compute the real parameter instantiation of the variables. However, unlike in [5], we use the original instance modeling to evaluate the individual's fitness value, instead of the ternary encoded CN. We propose improved formulations of three DE interval mutation operators. Interval adaptations of the main features of state-of-the-art solvers are implemented. Some details of I2DE's components are discussed below.

3.1 Interval Population

In the same way as in [5], the I2DE's initial population is generated by the top level branching tree of OGRe's IB&B. This strategy guarantees that the initial population covers all the search space. Although each branch of OGRe's IB&B is composed by a GAC multi-interval box, we iteratively split multi-intervals to obtain a set of interval boxes (individuals) that are added to the initial population until the number of individuals is *NP*. If numerical rounding errors make an individual to be considered inconsistent by OGRe's backtrack-free local search, this individual is replaced by a randomly generated individual from the initial domain of the instance.

The scheme in which the current population is subdivided between the subpopulations A and B of size $NP/2$ was mantained. The entire population is kept sorted, which allows A to contain the best individuals. In order to promote exploitation of the fitter solutions and accelerate the convergence process, we apply a pool of three strategies to each individual of this sub-population. The best one of the three trial individuals is compared to the parent individual in the selection operation. The other individuals are added to an archive, while in [5] they were discarded. In sub-population B, an adaptive scheme is used to choose the mutation strategy to be applied, as proposed in [24]. At the end of each generation, all individuals are sorted and divided between the sub-populations.

In InDE, when a multi-interval box is split into interval individuals, only the best fitting one is compared to the target vector, and the others are added to the archive. Similarly, individuals who are not selected for the next generation are also added to this archive with a maximum size of 100,000 individuals. When the archive is full, or every 50 generations, the current population and the archive are merged, the NP best individuals go to the new population, and the archive is emptied. Our tests revealed that this process is computationally expensive and contributes very little to the maintenance of diversity in the population. So I2DE only maintain individuals who are not selected for the next generation in a $NP \times 2.6$ archive, according to the strategy proposed in [29] and used in many state-of-art DE solvers [3,13,18].

Finally, we kept the linear population size reduction scheme proposed in [22] and applied in [5]. At each generation, the new population size is calculated and if it is different from the current size, the worst individuals are deleted from the population until it has the new size. When the population size reduces, the archive size is proportionally reduced.

3.2   Interval Operations

In IUDE [24] it was proposed the use of three mutation operators: DE/rand/1, DE/current-to-$p$best/1 and DE/current-to-rand/1. Just like in [5], we use this operators pool in the sub-population A in order to intensify the local search around the best individuals in the current population. The three generated trial vectors are compared among themselves and the mutation strategy with the best trial vector scores a win. At every generation, the success rate of each strategy is evaluated over the period of previous 25 generations. In the bottom sub-population B, the probability of employing a mutation strategy is equal to its recent success rate on the top sub-population A. Additionally, base and terminal vectors are selected from the top sub-population.

Since an interval is defined by its width (ω) and midpoint (μ), mutation operators can be applied over midpoints and extended to deal with interval widths. In [5], it was introduced interval versions of the three mutation operators mentioned above. However, our tests revealed that in practice numerical rounding errors may result in intervals with negative widths. Consequently, with the application of local consistency, the box is considered inconsistent and discarded.

In order to optimize the convergence process, we reformulated the mutation strategies proposed in InDE [5]. The interval version of DE/rand/1 combines r_1, r_2 and r_3 to generate the mutant vector v_i. The j-th element of v_i is defined by:

$$\mu(v_{ij}) = \mu(r_{1j}) + F_i \cdot (\mu(r_{2j}) - \mu(r_{3j})), \tag{4}$$

$$\omega(v_{ij}) = \omega(r_{1j}) \cdot \left(1 + F_i \cdot \left(\frac{\omega(r_{2j})}{\omega(r_{3j})} - 1\right)\right).$$

This formulation reduces numerical errors and avoid inconsistent intervals.

The other two mutation operators are defined in a similar way. The interval version of DE/current-to-rand/1 combines the target vector x_i with three randomly selected individuals r_1, r_2 and r_3. The mutant vector is defined by:

$$\mu(\mathbf{v}_{ij}) = \mu(\mathbf{x}_{ij}) + s_i \cdot (\mu(\mathbf{r}_{1j}) - \mu(\mathbf{x}_{ij})) + F_i \cdot (\mu(\mathbf{r}_{2j}) - \mu(\mathbf{r}_{3j})), \qquad (5)$$

$$\omega(\mathbf{v}_{ij}) = \omega(\mathbf{x}_{ij}) \cdot \left(1 + s_i \cdot \left(\frac{\omega(\mathbf{r}_{1j})}{\omega(\mathbf{r}_{ij})} - 1\right)\right) \cdot \left(1 + F_i \cdot \left(\frac{\omega(\mathbf{r}_{2j})}{\omega(\mathbf{r}_{3j})} - 1\right)\right),$$

where s_i is a random number between 0 and 1. This strategy does not use crossover operator, so the trial vector is a copy of the mutant vector $\mathbf{v}_i$.

Finally, the interval version of DE/current-to-$p$best/1 with archive uses the coordinates $\mathbf{x}_i$ and $\mathbf{r}_1$ in the same way as in Eq. 5, while $\mathbf{r}_2$ is randomly chosen from the union of the current population and the archive, and $\mathbf{r}_{pbest}$ is selected among the p best individuals in the population:

$$\mu(\mathbf{v}_{ij}) = \mu(\mathbf{x}_{ij}) + F_i \cdot (\mu(\mathbf{r}_{\text{pbest}j}) - \mu(\mathbf{x}_{ij})) + F_i \cdot (\mu(\mathbf{r}_{1j}) - \mu(\mathbf{r}_{2j})), \qquad (6)$$

$$\omega(\mathbf{v}_{ij}) = \omega(\mathbf{x}_{ij}) \cdot \left(1 + F_i \cdot \left(\frac{\omega(\mathbf{r}_{\text{pbest}j})}{\omega(\mathbf{r}_{ij})} - 1\right)\right) \cdot \left(1 + F_i \cdot \left(\frac{\omega(\mathbf{r}_{1j})}{\omega(\mathbf{r}_{2j})} - 1\right)\right).$$

Whereas in [5] the value of p remained fixed throughout the evolutionary process, we adopted the strategy proposed in [4]. After each generation g, the p value in the next generation $g + 1$ is computed as follows:

$$p = \left(\frac{p_{max} - p_{min}}{MaxFEs}\right) \cdot nfes + p_{min}, \qquad (7)$$

where p_{min} and p_{max} are, respectively, the minimum and maximum values of p, $nfes$ is the current number of fitness evaluations and $MaxFEs$ is the maximum number of fitness evaluations.

Additionally, in [5] if the width $\omega(\mathbf{u}_{ij})$ is greater than the width of the base individual $\omega(\mathbf{r}_{1j})$, it is updated by $\omega(\mathbf{u}_{ij}) := \omega(\mathbf{r}_{1j})$. Considering that this mechanism can decrease the population diversity by forcing the reduction of intervals, it is not used in I2DE.

It is known that the settings of values for the F and CR parameters is instance dependent and may change according to the region of the search space being visited. To make these parameters self-adaptive, InDE employed a scheme introduced in [22] that uses a pair of memory values $\langle M_{CR}, M_F \rangle$ for each mutation operator in order to store the settings that were successful in the last generations. All the memory pairs are stored in a vector of H positions. At each generation one of the positions is circularly updated based on the weighted Lehmer mean of the fitness differences between offspring and their parents.

In I2DE we incorporated some features proposed in [3, 4]. The first $H - 1$ positions of the vector are initialized with $\langle 0.8, 0.3 \rangle$. The last position is always set to $\langle 0.9, 0.9 \rangle$ and remains unchanged during the evolutionary process. At each generation one of the first $H - 1$ positions are circularly updated based on the Euclidean distance between the coordinates of offspring and their parents. Also, very low values of CR and very high values of F are not allowed in early stages of the search. It is important to note that only successful trial vectors in A are used in the adaptation of parameters, since only in this sub-population the three strategies are used for each individual.

3.3   Fitness Evaluations

To estimate the quality of the boxes that represent individuals in the population, we use OGRe's pruning and local search strategy. First, we enforce GAC on individuals from the initial population or resulting from the operators mentioned in Sect. 3.2. Then, we try the instantiation of the entire CN in a backtrack-free fashion starting by the initial valuation $y = \min Y$, where Y is the interval domain of the objective function $f(\mathbf{x})$. We allow the constraints in the Ω set of the epiphytic decomposition to be instantiated regardless of the tolerance ε_Ω required to satisfy them. In [5], the sum of all these tolerances is considered the *constraint violation* value $\phi(\mathbf{x})$ of the individual, while its *cost* is $f(\mathbf{x}) = \min Y$.

In this paper we assume that the instantiation of real parameters described above provides suitable reference points for evaluating interval individuals. However, to obtain the real values of $f(\mathbf{x})$ and $\phi(\mathbf{x})$, we only consider the instantiation of variables belonging to the original modeling of the instance, not the approximation given by the ternary CN employed in [5]. This allows us to compare results with recent black-box solvers that use the original instance modeling.

The consistent individuals obtained by the GAC contractor may contain multi-intervals. In [5], such multi-interval box was split into a set of interval individuals, adding to the population the one with the best fitness value (if it is better than the target vector) and saving all other generated interval individuals in the additional archive. As commented in the Sect. 3.1, our tests revealed that this mechanism is computationally expensive and does not contribute to maintaining the population diversity. So, instead of splitting the multi-intervals and adding them to the archive, we use the *interval hull* of the box (the smallest interval box that contains all the multi-intervals) and consider it as only one individual which is compared with its parent in the selection operation.

In order to compare two individuals, we apply the widely used ε constrained method [21]. The ε comparisons are defined as a lexicographic order in which $\phi(\mathbf{x})$ precedes $f(\mathbf{x})$. This precedence is adjusted by the parameter ε that is updated at each generation until the number of generations exceeds a predefined threshold. From this point the ε level is set to 0 to prefer solutions with minimum constraint violation.

4   Experimental Results

Our experimental evaluation considers four solvers: the improved approach proposed in this paper (I2DE); an implementation that uses the same DE interval operations and additional archive of the original InDE [5]; a black-box version of the I2DE (BBDE); and OGRe [8]. The aim of use the features of original InDE is to measure the impact of the modifications incorporated in our approach. The black-box version does not use interval representation neither structural decomposition as the other three white-box solvers. This comparison aims to evaluate the impact of using the instance structure in optimization process. In turn, the comparative analysis with OGRe is intended to provide a baseline with a search method that employs IB&B, local consistency and constraint propagation over the constraint network. It is important to note that local consistency and backtrack-free local search techniques are the same in I2DE, InDE and OGRe.

To ensure a fair comparison, the three DE solvers employ the same heuristics coming from state-of-the-art solvers for adapting and adjusting parameters, as well as the same number of fitness evaluations. Furthermore, the input parameters of the solvers were empirically defined after a series of preliminary experiments. It was set to $MaxFEs = 20000 \times D$. The maximal and minimal population size was defined as $NP_{max} = 15 \times D$ and $NP_{min} = 6$, respectively, where D is the number of variables on the original instance. Note that in [5], $MaxFEs$ and NP were defined in relation to $|V_\Omega|$, which resulted in a much larger budget and population. The DE/current-to-pbest/1 operator used $p_{max} = 0.25$ and $p_{min} = p_{max}/2$. The length of historical memory was $H = 5$. Parameters of ε level were $\theta = 0.7$, $cp = 4$ and $T = 0.85 \times G$, where G is the maximum number of generations. We used the exponential version of the classical crossover operator. GAC contractor was applied to a maximum of 1000 iterations. The timeout for OGRe was 10000s. Since OGRe parameters ε_Ω and Δ are not trivially configured, and to have a similar methodology of execution to DEs, we ran 25 different configurations for each instance using $\varepsilon_\Omega \in \{10^e \mid e = -5, \ldots, 1\}$ and $\Delta \in \{10^e \mid e = -6, \ldots, 0\}$, $\Delta \leq \varepsilon_\Omega$.

Although most experimental evaluations of metaheuristic-based solvers use CEC competition benchmarks, their instances are only available in source code to be used as black-box evaluation functions. Therefore, they do not provide the necessary formalization for use in solvers that exploit structural information of the NCOP instance, such as I2DE. For the main experimental evaluation we used the COCONUT Benchmark [19] due to its wide use in numerical optimization research and because it contains the AMPL (A Mathematical Programming Language) description necessary to explore the structure of the instances in a white-box approach.

However, to provide a baseline with some state-of-the-art solvers[3], we compared our black-box version of I2DE (BBDE) with solvers from CEC2018 Competition on Constrained Real-Parameter Optimization [27] in the 28 proposed benchmark functions with $D = \{10, 30, 50, 100\}$, totalizing 112 instances. The results of the CEC2018 solvers were obtained from the competition records[4]. BBDE used the same competition protocol, with 25 runs for instance.

The rank of Table 1 considers the CEC2018 [27] methodology based on mean and median score values on each instance, in which the best solver obtains the lowest total score. The results indicate that our black-box version is highly competitive with some of the most popular state-of-the-art solvers.

To investigate the performance of I2DE, experiments were conducted on 155 optimization problems from the COCONUT Benchmark [19] with different number of equality and inequality constraints and up to 75 dimensions. Note that

---

[3] LSHADE-IEpsilon [9], εMAgES [11], LSHADE44 [17], UDE [23], IUDE [24], LSHADE+IDE [25] and CAL-SHADE [28].

[4] https://www3.ntu.edu.sg/home/EPNSugan/index_files/CEC2018/CEC2018.htm, last accessed 18 Jun 2021.

Table 1. Comparison of BBDE with solvers from CEC2018.

Solver	Mean	Median	Total	Rank	Solver	Mean	Median	Total	Rank
BBDE	**381**	**380**	**761**	**1st**	LSHADE44	462	485	947	5th
IUDE	445	398	843	2nd	UDE	521	481	1002	6th
εMAgES	439	409	848	3rd	LSHADE+IDE	529	548	1077	7th
LSHADE-IEpsilon	445	449	894	4th	CAL-SHADE	647	575	1222	8th

it was only possible to tackle a larger number of functions than in [5] because our approach extends the implementation of OGRe's multi-interval core to other math operations, as commented in Sect. 3.

The first aspect to be analyzed is the impact of using structural decomposition. Transforming the original instance modeling into a ternary CN usually involves adding an auxiliary variable for each occurrence of math operators. From this, the variables of constraints in the Ω set, V_Ω, whose valuation is critical for the search process, are extracted. As noted in Sect. 3.2, I2DE and InDE only apply its operators on these variables, while the other valuations are assigned through OGRe's backtrack-free local search. Figure 1 illustrates the number of variables (and operators) from the original modeling, the ternary representation and the V_Ω size in the 155 instances. Although V_Ω presents a considerable reduction in number of variables compared to the ternary representation, it is evident that the search space is still larger than in the original instance modeling.

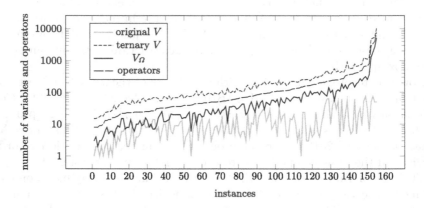

Fig. 1. Number of variables in original V, ternary V, V_Ω and operators per instance.

The solvers ran on a computer with Intel Xeon E5-4627v2 3.30 GHz processor, 256 GB of RAM and CentOS Linux 7. The average time of the 25 executions of the 155 instances was 7372.04 s for I2DE, 5031.19 s for InDE, 4.63 s for BBDE and 5619.52 s for OGRe. The processing times of I2DE, InDE and OGRe are significantly higher due to the local consistency process. Moreover, the execution time of InDE is smaller compared to I2DE because many candidate solutions resulting from InDE interval operations considered inconsistent are immediately discarded, without going through the local consistency process.

I2DE found feasible solutions on 119 instances, against 107 in InDE, 115 in BBDE and 64 in OGRe. Feasible solutions with value at most 10^{-8} greater than the optimal value are considered as optimal solutions. So, I2DE found 40 optimal solutions, while InDE, BBDE and OGRe found 35, 35 and 5, respectively.

In order to perform a comparative analysis among I2DE, InDE, BBDE and OGRe we used two methodologies. The first was the one applied in CEC2018 [27], the same used in Table 1. The second was proposed in CEC2020 Competition on Non-Convex Constrained Optimization Problems from the Real-World [14] and involves the objective function and constraints violation of the best, the mean, and the median solutions. The values are normalized and adjusted to calculate the performance measure. The total score is a weighted composition of best, mean and median with weights 50%, 30% and 20%, respectively.

To estimate the feasibility of the four solvers, we compared the average constraint violation values of the 25 runs for the 155 benchmark instances. As the pre-analysis rejected the null hypothesis that the distribution of violation values of each instance is normal, we chose a non-parametric test. The Friedman test with post-hoc Nemenyi test with significance level of 0.05 was applied to the violation results of each instance, comparing four sets of 25 values each time. The null hypothesis that the four solvers perform equally was rejected in all instances with p-value close to 10^{-6}. Based on the Nemenyi test rank on each instance, we counted the number of instances where each solver was better, equal or worse than the others. If the rank distance of two solvers was less than the critical distance of the test, we considered the solvers perform equally in the instance. Table 2 shows in its violation rows the number of instances where I2DE was better, equal or worse than the other three solvers. In all three cases, I2DE obtained more solutions with violation values smaller than the other solvers.

A second statistical test was conducted over the normalized and adjusted objective function value of best, mean and median solution of the CEC2020 methodology. The value of each solver in an instance is a composition of objective function value and the average constraint violations of the obtained solutions. As the pre-analysis rejected the null hypothesis that the distribution of each set of values is normal, we chose a non-parametric test. We applied the Wilcoxon signed rank test with significance level of 0.05 to compare the solvers on each set of values: best, mean and median. Table 2 shows the test results and the number of instances where I2DE was better, equal or worse than the other solver. Only in two cases the null hypothesis, that the two solvers perform equally, was not rejected: decision $\approx$. In the other cases, I2DE perform better: decision $+$.

According to both CEC2018 and CEC2020 ranking methodologies, the best solver will obtain the lowest total score value. Table 3 shows that the I2DE outperforms the other three approaches in both ranking methodologies. In comparison to InDE, the results point to a significant performance improvement due to the reformulation of interval mutation strategies and the maintenance of population diversity with the additional archive of solutions. Furthermore,

Table 2. Statistical tests results and comparison of I2DE with InDE, BBDE and OGRe based on best, mean, median and violation scores over 155 COCONUT instances.

Solvers	Criteria	Better	Equal	Worse	P-value	Decision
I2DE vs. InDE	Best	66	60	29	0.0002	+
	Mean	89	42	24	10^{-8}	+
	Median	85	46	24	10^{-6}	+
	Violation	39	110	6	$\sim 10^{-6}$	+
I2DE vs. BBDE	Best	60	36	59	0.9204	$\approx$
	Mean	64	33	58	0.0002	+
	Median	61	39	55	0.9987	$\approx$
	Violation	47	83	25	$\sim 10^{-6}$	+
I2DE vs. OGRe	Best	124	10	21	10^{-16}	+
	Mean	131	9	15	10^{-24}	+
	Median	127	7	21	10^{-20}	+
	Violation	108	40	7	$\sim 10^{-6}$	+

Table 3. Rank of I2DE, InDE, BBDE and OGRe according to CEC2018 and CEC2020 methodologies on COCONUT Benchmark.

Solver	CEC2018				CEC2020				
	Mean	Median	Total	Rank	Best	Mean	Median	Total	rank
I2DE	**249**	**229**	**478**	**1st**	**0.2153**	**0.0811**	**0.1348**	**0.1589**	**1st**
InDE	330	309	639	3rd	0.3068	0.1361	0.2345	0.2411	2nd
BBDE	317	320	637	2nd	0.2975	0.2340	0.3049	0.2799	3rd
OGRe	575	543	1118	4th	0.7306	0.8123	0.7314	0.7553	4th

the comparative analysis with BBDE and OGRe suggests that the hybridization between metaheuristics, constraint programming and structural decomposition is a promising research direction.

5   Conclusion

In this work we proposed the I2DE, an improved version of Interval Differential Evolution that uses structural information to solve global optimization problems. From exploration of the epiphytic decomposition, it becomes possible to concentrate the search process only on a subset of variables that have critical valuation, while all the others are instantiated by propagation through the constraint hypergraph. Additionally, our search is enhanced by a local consistency process that prunes values that certainly do not constitute the optimal solution.

One of the main contributions of our approach is the extension of the multi-interval core of OGRe. This allows us to tackle a greater diversity of benchmark functions and real-world problems that otherwise could not be adequately represented. We also proposed a reformulation of three DE interval mutation operations and incorporated heuristics from several state-of-the-art solvers that contributed to improve the performance of our approach.

The experimental analysis of the proposed I2DE on the 155 functions selected from the COCONUT Benchmark [19] showed that the reformulation of the interval mutation strategies and of the additional archive significantly improved the performance of the search method. Furthermore, the results obtained in comparison with OGRe and BBDE reveal that, although the exploration of the instance's structural information increases the size of the search space and the processing time, it considerably improves the quality of the solutions found. Considering that BBDE outperformed several state-of-the-art solvers and was overcome by I2DE, we shown that the use of structural instance information in the context of metaheuristics is a promising research direction.

Some future work includes the implementation of other contractors that help to efficiently prune the intervals without loss of solutions. In addition, we intend to develop a hybrid cooperative approach that joins our interval metaheuristics with exact methods that also use this representation for the solutions.

References

1. Araya, I., Reyes, V.: Interval branch-and-bound algorithms for optimization and constraint satisfaction: a survey and prospects. J. Glob. Optim. **65**(4), 837–866 (2016)
2. Berge, C.: Graphs and Hypergraphs. Elsevier Science Ltd., Oxford (1985)
3. Brest, J., Maučec, M.S., Bošković, B.: iL-SHADE: Improved L-SHADE algorithm for single objective real-parameter optimization. In: 2016 IEEE Congress on Evolutionary Computation (CEC), pp. 1188–1195. IEEE (2016)
4. Brest, J., Maučec, M.S., Bošković, B.: Single objective real-parameter optimization: Algorithm jSO. In: 2017 IEEE Congress on Evolutionary Computation (CEC), pp. 1311–1318. IEEE (2017)
5. Cassenote, M.R.S., Derenievicz, G.A., Silva, F.: Interval differential evolution using structural information of global optimization problems. In: Moura Oliveira, P., Novais, P., Reis, L.P. (eds.) EPIA 2019. LNCS (LNAI), vol. 11804, pp. 724–736. Springer, Cham (2019). https://doi.org/10.1007/978-3-030-30241-2_60
6. Cohen, D., Jeavons, P.: The power of propagation: when gac is enough. Constraints **22**, 3–23 (2016)
7. Dechter, R., van Beek, P.: Local and global relational consistency. In: Montanari, U., Rossi, F. (eds.) CP 1995. LNCS, vol. 976, pp. 240–257. Springer, Heidelberg (1995). https://doi.org/10.1007/3-540-60299-2_15
8. Derenievicz, G.A., Silva, F.: Epiphytic trees: relational consistency applied to global optimization problems. In: van Hoeve, W.-J. (ed.) CPAIOR 2018. LNCS, vol. 10848, pp. 153–169. Springer, Cham (2018). https://doi.org/10.1007/978-3-319-93031-2_11
9. Fan, Z., Fang, Y., Li, W., Yuan, Y., Wang, Z., Bian, X.: LSHADE44 with an improved ε constraint-handling method for solving constrained single-objective optimization problems. In: 2018 IEEE Congress on Evolutionary Computation (CEC), pp. 1–8. IEEE (2018)

10. Hansen, E., Walster, G.W.: Global optimization using interval analysis. Monographs and textbooks in pure and applied mathematics, New York (2004)
11. Hellwig, M., Beyer, H.G.: A matrix adaptation evolution strategy for constrained real-parameter optimization. In: 2018 IEEE Congress on Evolutionary Computation (CEC), pp. 1–8. IEEE (2018)
12. Iorio, A.W., Li, X.: Solving rotated multi-objective optimization problems using differential evolution. In: Webb, G.I., Yu, X. (eds.) AI 2004. LNCS (LNAI), vol. 3339, pp. 861–872. Springer, Heidelberg (2004). https://doi.org/10.1007/978-3-540-30549-1_74
13. Jou, Y.C., Wang, S.Y., Yeh, J.F., Chiang, T.C.: Multi-population modified L-SHADE for single objective bound constrained optimization. In: 2020 IEEE Congress on Evolutionary Computation (CEC), pp. 1–8. IEEE (2020)
14. Kumar, A., Wu, G., Ali, M.Z., Mallipeddi, R., Suganthan, P.N., Das, S.: A testsuite of non-convex constrained optimization problems from the real-world and some baseline results. Swar Evol. Comput. **56**, 100693 (2020)
15. Mackworth, A.K.: On reading sketch maps. In: Proceedings of the Fifth International Joint Conference on Artificial Intelligence, IJCAI 1977, pp. 598–606. MIT, Cambridge, MA (1977)
16. Moore, R.E.: Interval Analysis. Prentice-Hall Englewood Cliffs, N.J (1966)
17. Poláková, R.: L-SHADE with competing strategies applied to constrained optimization. In: 2017 IEEE Congress on Evolutionary Computation (CEC), pp. 1683–1689. IEEE (2017)
18. Sallam, K.M., Elsayed, S.M., Chakrabortty, R.K., Ryan, M.J.: Improved multi-operator differential evolution algorithm for solving unconstrained problems. In: 2020 IEEE Congress on Evolutionary Computation (CEC), pp. 1–8. IEEE (2020)
19. Shcherbina, O., Neumaier, A., Sam-Haroud, D., Vu, X.-H., Nguyen, T.-V.: Benchmarking global optimization and constraint satisfaction codes. In: Bliek, C., Jermann, C., Neumaier, A. (eds.) COCOS 2002. LNCS, vol. 2861, pp. 211–222. Springer, Heidelberg (2003). https://doi.org/10.1007/978-3-540-39901-8_16
20. Storn, R., Price, K.: Differential Evolution - a simple and efficient heuristic for global optimization over continuous spaces. J. Glob. Optim. **11**(4), 341–359 (1997)
21. Takahama, T., Sakai, S.: Constrained optimization by the ε constrained Differential Evolution with an archive and gradient-based mutation. In: 2010 IEEE Congress on Evolutionary computation, pp. 1–9. IEEE (2010)
22. Tanabe, R., Fukunaga, A.S.: Improving the search performance of SHADE using linear population size reduction. In: 2014 IEEE Congress on Evolutionary Computation (CEC), pp. 1658–1665. IEEE (2014)
23. Trivedi, A., Sanyal, K., Verma, P., Srinivasan, D.: A unified Differential Evolution algorithm for constrained optimization problems. In: 2017 IEEE Congress on Evolutionary Computation (CEC), pp. 1231–1238. IEEE (2017)
24. Trivedi, A., Srinivasan, D., Biswas, N.: An improved unified Differential Evolution algorithm for constrained optimization problems. IEEE (2018)
25. Tvrdík, J., Poláková, R.: A simple framework for constrained problems with application of L-SHADE44 and IDE. In: 2017 IEEE Congress on Evolutionary Computation (CEC), pp. 1436–1443. IEEE (2017)
26. Vanaret, C., Gotteland, J.B., Durand, N., Alliot, J.M.: Preventing premature convergence and proving the optimality in evolutionary algorithms. In: International Conference on Artificial Evolution, pp. 29–40. Springer (2013)
27. Wu, G., Mallipeddi, R., Suganthan, P.: Problem definitions and evaluation criteria for the CEC 2017 competition on constrained real-parameter optimization. Technical report (2017)

28. Zamuda, A.: Adaptive constraint handling and success history Differential Evolution for CEC 2017 constrained real-parameter optimization. In: 2017 IEEE Congress on Evolutionary Computation (CEC), pp. 2443–2450. IEEE (2017)
29. Zhang, J., Sanderson, A.C.: JADE: adaptive Differential Evolution with optional external archive. IEEE Trans. Evol. Comput. **13**(5), 945–958 (2009)

Improving a Genetic Clustering Approach with a CVI-Based Objective Function

Caio Flexa[✉], Walisson Gomes, Igor Moreira, Reginaldo Santos,
Claudomiro Sales, and Moisés Silva

Applied Electromagnetism Laboratory, Federal University of Pará,
01 Augusto Corrêa Street, Guamá, Belém, PA 66.075-110, Brazil
{caio.rodrigues,walisson.gomes,igor.moreira,moises.silva}@icen.ufpa.br,
{regicsf,cssj}@ufpa.br

Abstract. Genetic-based clustering meta-heuristics are important bioinspired algorithms. One such technique, termed Genetic Algorithm for Decision Boundary Analysis (GADBA), was proposed to support Structural Health Monitoring (SHM) processes in bridges. GADBA is an unsupervised, non-parametric approach that groups data into natural clusters by means of a specialized objective function. Albeit it allows a competent identification of damage indicators of SHM-related data, it achieves lackluster results on more general clustering scenarios. This study improves the objective function of GADBA based on a Cluster Validity Index (CVI) named Mutual Equidistant-scattering Criterion (MEC) to expand its applicability to any real-world problem.

Keywords: Genetic Algorithm · Automatic Clustering Algorithm · Decision Boundary Analysis · Mutual Equidistant-scattering Criterion

1 Introduction

The world is increasingly filled with fruitful data, most of which daily stored in electronic media. As such, there is a high potential of technique research and development for automated data retrieval, analysis, and classification [15]. Around 90% of the data produced up to 2017 were generated in 2015 and 2016, and the tendency is to biennially double this amount [17]. The exponential increase in size and complexity of Big Data are aspects worthy of attention.

Latent potentialities for decision-making based on insights learned from historical data are only actually exploited if pushed into practice. A comprehensive information extraction procedure is composed of two sub-processes: data management and data analysis [15]. Management supports data acquisition, handling, storage, and retrieval for analysis [18,29]; in contrast, data analysis refers to the evaluation and acquisition of intelligence from data.

It was pointed out by [17] that researchers of various subjects have been adopting methods from Machine Learning (ML) [5,10] and Data Mining (DM) [31]. Many of them did so in Big Data contexts such as stock data monitoring,

© Springer Nature Switzerland AG 2021
A. Britto and K. Valdivia Delgado (Eds.): BRACIS 2021, LNAI 13073, pp. 202–217, 2021.
https://doi.org/10.1007/978-3-030-91702-9_14

financial analysis, traffic monitoring, and Structural Health Monitoring (SHM) [14]. [6] demonstrated the relationships between the data management and analysis sub-processes in practice, with a specific focus on the application herein highlighted. On one hand, this work claims that Big Data is not restricted to a computerized manipulation of massive data streams; on the other hand, it emphasizes that SHM can learn *ipsis litteris* with the conscientious use of ML.

The problem of data grouping (i.e., data clustering) is one of the main tasks in ML [2,5] and DM [31], prevailing in any discipline involving multivariate data analysis [21]. It gained a prominent place in many applications lately, especially in speech recognition [11], web applications [35], image processing [3], outlier detection [23], bioinformatics [1], and SHM [9].

A wide variety of Genetic Algorithm (GA)-based clustering techniques have been proposed in recent times [25,28,42]. Their search ability is commonly exploited to find suitable prototypes in the feature space such that a per-chromosome measure of the clustering results is optimized in each generation. In [2], two conflicting functions were proposed and defined based on cluster cohesion and connectivity. The goal was to reach well-separated, connected, and compact clusters by means of two criteria in an efficient, multi-objective particle swarm optimization algorithm. More recently, [20] combined K-MEANS and a GA through a differentiate arrangement of genetic operators to conglomerate different solutions, with the intervention of fast hill-climbing cycles of K-MEANS.

An unsupervised, non-parametric, GA-based approach to support the SHM process in bridges termed Genetic Algorithm for Decision Boundary Analysis (GADBA), which was proposed by [36], aims to group data into natural clusters. The algorithm is also supported by a method based on spatial geometry to eliminate redundant clusters. Upon testing, GADBA was more efficient in the task of fitting the normal condition than its state-of-the-art counterparts in SHM contexts. However, due to the specialization of its objective function to SHM contexts, GADBA is lackluster on more general clustering scenarios.

This work aims to improve the objective function of GADBA to expand its application potential to a wider range real-world problems. In this sense, a version of GADBA based on the Mutual Equidistant-scattering Criterion (MEC) is proposed as a general-purpose clustering approach. Four clustering algorithms are compared against the new proposal: K-MEANS [26], Gaussian Mixture Models (GMM) [30], LINKAGE [22], and GADBA, of which the first three are well-known and explored in literature.

The remaining sections of this paper are divided as follows. Section 2 and Sect. 3 respectively define GADBA and MEC. Section 4 discusses the performance of the new proposal under some experimental evaluations. Finally, Sect. 5 summarizes and ends the paper.

2   Genetic Algorithm for Decision Boundary Analysis

Given a minimum (K_{min}) and maximum (K_{max}) number of clusters, clustering is done by the combination of a GA to dispose their centroids in the M-dimensional feature space, and a method called Concentric Hypersphere (CH) to agglutinate clusters, to choose an appropriate $K \in [K_{min}, K_{max}]$ [37].

The initial population $\mathbf{P}(t{=}0)$ is randomly created and each individual represents a set of K centroids, where K is randomly selected. The chromosome is then formed by concatenating K feature vectors (Fig. 1), whose values are initialized by randomly selecting K data points from the training set. A length ratio is defined as $\gamma_i = K_i/K_{max}$, where K_i is the number of active centroids in individual i. The role of γ is to define the number of active centroids for a given candidate solution, since a single individual might have enabled/disabled centroids during the recombination process.

The parent selection is based on tournament with reposition, where R individuals are randomly selected and the fittest one is chosen to recombine with other individual chosen in the same way. The recombination is conducted in three steps for each pair of parents P_i and P_j to generate a pair of descendants:

1. A random number $r \in [0,1]$ is compared with p_{rec} defined *a priori*. If $r \leq p_{rec}$, then two cut points π_1 and π_2 are selected such that $1 \leq \pi_1 < \pi_2 \leq min(K_i, K_j)$. The centroids in the range are then swapped. If $r > p_{rec}$, the parents remain untouched.
2. Likewise, two random numbers $r, T \in [0,1]$ are picked for each centroid position in the parents and, if $r \leq p_{pos}$, defined *a priori*, an arithmetic recombination is conducted as follows:

$$F_{\mathbf{x},t}^i = F_{\mathbf{x},t}^i + (F_{\mathbf{y},t}^j - F_{\mathbf{x},t}^i)T, \tag{1}$$

$$F_{\mathbf{x},t}^j = F_{\mathbf{x},t}^j + (F_{\mathbf{y},t}^i - F_{\mathbf{x},t}^j)T, \tag{2}$$

where $F_{\mathbf{x},t}^i$ and $F_{\mathbf{x},t}^j$ are the values in the tth position of $\mathbf{x}^{\text{th}}$ centroid from parents P_i and P_j, respectively. Similarly, $F_{\mathbf{y},t}^i$ and $F_{\mathbf{y},t}^j$ respectively correspond to the yth centroid of the i and j^{th} parents. A pair of parents can recombine even if they have a different number of genes.
3. Finally, the last step consists in arithmetically recombining the parents' length ratio to define the length ratios of the offspring individuals.

The mutation is the result of a personalized two-step process:

1. Let $T_{\mathbf{x}} = K_{max}^{-1}$ and T_r be a random number on interval $[0,1]$. The number of centroids to be enabled in an offspring individual is $K_{new} = \lceil T_r/T_{\mathbf{x}} \rceil$. When $K < K_{new} \leq K_{max}$, the missing positions are filled with the information of $K_{new} - K$ data points chosen at random.

$F(1,1)$	$F(1,...)$	$F(1,M)$	$F(2,1)$	$F(2,...)$	$F(2,M)$	$\cdots$	$F(K,1)$	$F(K,...)$	$F(K,M)$

Fig. 1. Chromosome organization in GADBA.

2. Each centroid position can be mutated with a probability p_{mut}, defined *a priori*, in which a Gaussian mutation is applied by using

$$F^j_{\mathbf{x},t} = F^j_{\mathbf{x},t} + \mathcal{N}(0,1), \tag{3}$$

where $\mathcal{N}(0,1)$ is a random number from a standard Gaussian distribution and $F^j_{\mathbf{x},t}$ is the value in the tth position of $\mathbf{x}^{\text{th}}$ centroid.

Survivor selection is based on elitism, where the parents $\mathbf{I}^{(t)}_p$ and offspring $\mathbf{I}^{(t)}_c$ are concatenated into $\mathbf{I}^{(t+1)}_p = \mathbf{I}^{(t)}_p \cup \mathbf{I}^{(t)}_c$, which is then sorted according to a fitness measure based on Pareto Front and Crowding Distance. The new population $\mathbf{P}(t+1)$ is composed by the $|\mathbf{P}|$ best individuals [8].

Parent selection, recombination, mutation, and survivor selection are repeated until a maximum number of iterations is reached and/or the difference of the current solution against the last one is smaller than a given threshold ϵ.

As mentioned, the CH algorithm is used to regularize the number of clusters encoded in the individuals. It is executed in each individual prior to their evaluation by determining the regions that limit each cluster in three steps:

1. For each cluster, its centroid is dislocated to the mean of its data points.
2. Each centroid is the center of a hypersphere whose radius will increase while the difference of density between two consecutive inflations is positive.
3. If more than one centroid is found inside a hypersphere, they are agglutinated into a centroid located at their mean point.

3   A New Objective Function

3.1   Basic Notations and Definitions

The objective of clustering is to find out the best way to split a given data set $\mathcal{X} \in \mathbb{R}^{N \times M}$, with N input vectors in an M-dimensional real-valued feature space $\{ x_i \mid x_i \in \mathbb{R}^M, 1 \leqslant i \leqslant N \}$, into K mutually disjoint subsets ($K \leq N$). Assume the vectors in $\mathcal{X}$ have hard labels marking them as members of one cluster. A set of prototypes Θ is described as a function of $\mathcal{X}$ and K as

$$\Theta \in \mathbb{R}^{K \times M} = \Theta(\mathcal{X}), \tag{4}$$

whereupon Θ contains K representative vectors $\{ \theta_\kappa \mid \theta_\kappa \in \mathbb{R}^M, 1 \leqslant \kappa \leqslant K \}$.

Let the hard label for cluster κ be

$$y_\kappa = (\overset{1}{0}, \overset{2}{0}, \dots, \overset{\kappa}{\overbrace{1}}, \dots, \overset{K}{0}).$$

The prototypes are computed whereby N label vectors are organized into a partition over the vectors in $\mathcal{X}$ such that

$$U \in \mathbb{Z}^{N \times M} = U(\mathcal{X}), \tag{5}$$

subject to

$$U = [\mu_{i\kappa}]_{N \times K} \in \{0, 1\}, \quad i = 1, \dots, N,$$

where $\mu_i = y_\kappa \Leftrightarrow x_i$ is in cluster κ.

Put another way, the membership of x_i to cluster κ, $\mu_{i\kappa}$, is either 1 if the ith object belongs to the κth cluster, or 0 otherwise. Accordingly, $\mu_{i\kappa} = 1$ for one value of κ only, such that [41]

$$\cup_{i=1}^{N} \mu_{i\kappa} x_i \neq \emptyset, \quad \kappa = 1, \dots, K;$$

$$\cup_{\kappa=1}^{K} \left(\cup_{i=1}^{N} \mu_{i\kappa} x_i \right) = \mathcal{X};$$

$$\left(\cup_{i=1}^{N} \mu_{i\kappa} x_i \right) \cap \left(\cup_{i=1}^{N} \mu_{il} x_i \right) = \emptyset, \quad \kappa, l = 1, \dots, K, \quad \text{and} \quad \kappa \neq l.$$

The resulting grouping in this structure is hard [24], since one object belonging to one cluster cannot simultaneously belong to another.

In general, data clustering involves finding $\{U, \Theta\}$ to partition $\mathcal{X}$ somehow. For a given initial Θ, the optimal set of prototypes can be represented by centroids, medians, medoids, and others, in which the optimized partition U is obtained by assigning each input vector to the cluster with the nearest prototype. Both U and Θ comprise a dual structure (if one of them is known, the other one will also be) named clustering solution.

3.2 Cluster Validation

Most researchers have some theoretical difficulty in describing what a cluster is without assuming an induction principle (i.e., a criterion) [21]. A classic definition for them is: "objects are grouped based on the principle of maximizing intra-class similarity and minimizing inter-class similarity". Another cluster definition involving density defines it as a connected, dense component such that high-density regions are separated by low-density ones [1, 19].

In clustering algorithms, K is usually assumed to be unknown. Since clustering is an unsupervised learning procedure (i.e., there is no prior knowledge on data distribution), the significance of the defined clusters must be validated for the data [33]. Therefore, one of the most challenging aspects of clustering is the quantitative examination of clustering results [31]. This procedure is performed by Cluster Validity Indices (CVIs), sometimes called criteria, which also targets hard problems such as cluster quality assessment and the degree wherewith a clustering scheme fits into a specific data set. The most common application of CVIs is to fine-tune K. Given $\mathcal{X}$, a specific clustering algorithm and a range of values of K, these steps are executed [10, 43]:

1. Successively repeat a clustering algorithm according to a number of clusters from a fixed range of values defined *a priori*: $K \in [K_{min}, K_{max}]$;
2. Obtain the clustering result $\{U, \Theta\}$ for each K in the range;

3. Calculate the validity index score for all solutions; and
4. Select K_{opt} for which data partitioning provides the best clustering result.

CVIs are considered to be independent of the clustering algorithms used [40] and usually fall into one of two categories: internal and external [27,32]. Internal validation does not require knowledge about the problem for it only uses information intrinsic to the data; hence, it has a practical appeal. Conversely, external validation is more accurate, but not always feasible [32]. Knowing this, it can be evaluated how well the achieved solution approaches a predefined structure based on previous and intuitive understanding regarding natural clusters.

3.3   Mutual Equidistant-Scattering Criterion

This work proposes the replacement of the objective function of GADBA with a CVI called MEC [12]. MEC is a non-parametric, internal validation index for crisp clustering. An immediate benefit of MEC is the absence of fine-tuning hyper-parameters, thus mitigating the user's effort in operational terms and enabling the use of GADBA to cluster real-world data whose structure is unknown.

MEC assumes that "objects belonging to the same data cluster will tend to be more equidistantly scattered among themselves compared to data points of distinct clusters" [12]. As such, the mean absolute difference $\mathcal{M}_\kappa$ is applied using multi-representative data in every clustering solution $\{U, \Theta\}$ obtained from a pre-determined K. MEC is weighted by a penalty of local restrictive nature to each cluster κ as well, while a global penalty is then applied *a posteriori*. Such penalties are a measure of intra-cluster homogeneity and inter-cluster separation.

Mathematical Formulations. The mean absolute difference is calculated between any possible pair of intra-cluster dissimilarities

$$D_\kappa^{ij} = \begin{cases} \text{dist}(\mathbf{x}_i, \mathbf{x}_j) & \text{if } \mu_{i\kappa} = 1, \ \mu_{j\kappa} = 1, \ i < j \\ 0 & \text{otherwise,} \end{cases} \quad \kappa = 1, \ldots, K, \quad (6)$$

where n_κ objects within the cluster are considered as representative data in formulation (thereby multi-representative). That is, D_κ is a strictly upper triangular matrix of order n_κ.

Nevertheless, only the pairwise distances matter in $\mathcal{M}_\kappa$. Thus, $\Upsilon(\cdot)$ reshapes all elements above the main diagonal of D_κ (Eq. 6) into a column vector

$$\begin{bmatrix} d_{1\kappa} \\ d_{2\kappa} \\ \vdots \\ d_{L_\kappa\kappa} \end{bmatrix} = \Upsilon(D_\kappa), \quad (7)$$

denoted by $L_\kappa = \frac{n_\kappa(n_\kappa-1)}{2}$ intra-cluster Euclidean distances. The key part of MEC is then defined in Eq. 8 as

$$\mathcal{M}_\kappa = \eta_\kappa^{-1} \sum_{i=1}^{L_\kappa-1} \sum_{j=i+1}^{L_\kappa} abs(d_{i\kappa} - d_{j\kappa}), \quad L_\kappa > 1, \tag{8}$$

where $L_\kappa > 1 \Leftrightarrow n_\kappa > 2$, $abs(\cdot)$ is the absolute value and $\eta_\kappa = \frac{L_\kappa(L_\kappa-1)}{2}$ stands for the total number of differences for a single cluster.

An exponential-like distance measure provides a robust property based on the analysis of the influence function [39]. [12] empirically observed that it works properly, particularly when we look for K_{opt} within a hierarchical data set. Therefore, a new homogeneity measure Σ_κ of non-negative exponential type was modelled as a penalty over $\mathcal{M}_\kappa$ as

$$\Sigma_\kappa = \begin{cases} \dfrac{1 - e^{-\sigma_\kappa^2}}{e^{-\sigma_\kappa^2}} & \text{if } \sigma_\kappa > 0 \\ 0 & \text{otherwise,} \end{cases} \tag{9}$$

where σ_κ^2 is the variance of d_κ.

One can observe that the homogeneity measure gets closer to zero with the approximation of the ideal model solution, where the criterion value is zero and, therefore, the loss of information is null. Thus, we have MEC defined as

$$\text{MEC}(K) = \lambda \sum_{\kappa=1}^{K} \Sigma_\kappa \times \mathcal{M}_\kappa, \tag{10}$$

where

$$\lambda = \begin{cases} K \dfrac{1}{\max\limits_{i \neq j} \{\text{dist}(\theta_i, \theta_j)\}} & \text{if } K > 1 \\ 1 & \text{otherwise.} \end{cases} \tag{11}$$

The measure of global separation and penalty λ, therefore, does not depend exclusively on the κth cluster, but on the greater distance between the pairs of representative points of each data cluster (e.g., centroids). In a few words, λ globally weights the result of the solution. The presence of K, in Eq. 11, is a simple way to avoid overfitting as a result of clustering solutions already sufficiently accommodated to the data. In addition to avoiding overfitting, an improvement over other indices is the possibility of evaluating the clustering tendency ($K=1$) without resorting to additional, external techniques [43].

To illustrate, Fig. 2 shows the MEC composition for three feasible cluster solutions, where each dotted line represents one measure of intra-cluster dissimilarity and each cluster is depicted by a quadratic centroid. The operating mechanism of MEC, which encompasses both homogeneity (Eq. 9) and separation (Eq. 11), is visualised for $K = 1, 2, 4$ (Figs. 2a, 2b, and 2c, respectively). The motivation is that the dissimilarity measures should be similar to each

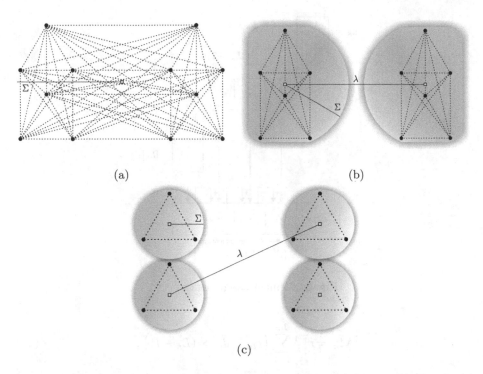

(a)   (b)

(c)

Fig. 2. MEC results for a small set of twelve data points: (a) $K = 1$; (b) $K = 2$; (c) $K = 4$.

other when looking at each cluster. In this case, Fig. 2a contains the least suitable solution among those shown graphically, as their dissimilarity measures are more divergent in magnitude than those in Fig. 2b and 2c. The four-cluster solution (Fig. 2c) is the best within the solution set, as the distances among objects are exactly the same in each cluster.

At last, it is worth noting that Eq. 10 should be minimized,

$$\hat{K} = \arg\min \mathrm{MEC}(K), \tag{12}$$

where $K \in [K_{min}, K_{max}]$ and $\hat{K}$ is inferred by the variation of K which determines the lowest MEC value, regardless of the clustering algorithm.

Improving the Time-Complexity of MEC. Equation 8 can be equivalently computed in terms of a log-linear time complexity as a function of L_κ, to improve the computational efficiency of MEC. To do so, Eq. 8 can be reformulated to generate an auxiliary vector, as well as in sorting d_κ with an algorithm of same complexity (e.g., HeapSort). In fact, the time complexity of MEC will be entirely dependent on the complexity of the chosen sorting algorithm. As such, we have a complexity of $\mathcal{O}(L_\kappa \log L_\kappa)$ with the Heap-Sort algorithm, or even $\mathcal{O}(N^2)$, by the reformulated

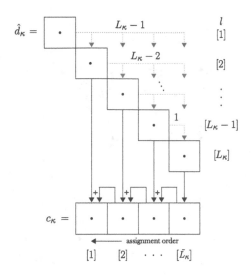

Fig. 3. Element assignment of c_κ.

$$\mathcal{M}_\kappa = \eta_\kappa^{-1} \sum_{l=1}^{\tilde{L}_\kappa} \left(c_{l\kappa} - \hat{d}_{l\kappa} \times (L_\kappa - l) \right), \qquad (13)$$

where $\hat{d}_\kappa$ is the increasing ordering of the values of d_κ and $\tilde{L}_\kappa = L_\kappa - 1 = |c_\kappa|$; c_κ is an auxiliary variable that consists of a cumulative and naturally ordered vector of $\hat{d}_\kappa$ defined as

$$
\begin{aligned}
c_{\tilde{L}_\kappa\kappa} &= \hat{d}_{L_\kappa\kappa} &\geq 0 \\
c_{\tilde{L}_\kappa-1,\kappa} &= c_{\tilde{L}_\kappa\kappa} + \hat{d}_{L_\kappa-1,\kappa} &\geq c_{\tilde{L}_\kappa\kappa} \\
&\vdots & \vdots & \quad \vdots \\
c_{1\kappa} &= c_{2\kappa} + \hat{d}_{2\kappa} &\geq c_{2\kappa}.
\end{aligned}
\qquad (14)
$$

Looking at Fig. 3, each square and value between square brackets depicts some vector position (l notation). In Eq. 13, the general form ($L_\kappa - l$) consists of the number of subtractions (Eq. 8) represented by arrows in the Figure, with $\hat{d}_{l\kappa}$ depending on its location. The ordered $\hat{d}_\kappa$ ensures that $\hat{d}_{l\kappa} \leq \hat{d}_{l+1,\kappa}$. By transitivity we have that, in Eq. 13,

$$\hat{d}_{l\kappa} \times (L_\kappa - l) \leq c_{l\kappa}, \quad l = 1, \ldots, \tilde{L}.$$

Hence, $\hat{d}_\kappa$ is sensibly less accessed, thus reducing the time complexity of MEC.

4   Results and Analyses

This section describes the results achieved by the five algorithms compared in this study: GADBA, its new version GADBA-MEC, K-MEANS, GMM, and

LINKAGE. Since none of the last three techniques automatically finds $\hat{K}$, the Calinski and Harabasz Cluster Validity Index (CVI) is used to optimize $\hat{K}$ through cluster validation (Sect. 3.2). Section 4.1 presents the methodology as how to, and by what means, the results were generated; Sect. 4.2 discusses the results highlighting the techniques that clustered the data; and the statistical significance of the results is analysed in Sect. 4.3.

4.1 Applied Methodology

The accuracy of the clustering algorithms is explained in a set of statistical indicators, such as absolute frequency, mean and standard deviation of $\hat{K}$, in twenty clustering validations for each data set (i.e., $N_r = 20$). The Mean Absolute Percentage Error (MAPE) was then estimated between the desired (K_{opt}) and optimized ($\hat{K}$) number of clusters in Sect. 4.3. It generally expresses accuracy as a percentage which is designated by

$$\text{MAPE} = \frac{100}{N_r} \sum_{t=1}^{N_r} \frac{abs\left(\hat{K}_t - K_{opt}\right)}{K_{opt}}. \tag{15}$$

Table 1 presents data sets from different benchmarks used for performance analysis when comparing clustering algorithms. To evaluate the algorithms, ten sets were selected as archetypes of real challenges faced in cluster validation (e.g., data hierarchy, clustering tendency, different densities/sizes).

Table 1. Properties of test data sets.

Data set	N	M	K_{opt}	Separation	Homogeneity	Distribution	Density	Overlap	Shape	Noise
	From [34]									
S&C$_1$	4500	2	9	Low	High	Uniform	High	–	Quadratic	–
S&C$_2$	3200	2	10	High	**High**	Uniform	**High**	–	Circular	–
	From [13]									
Dim-32	1024	32	16	**High**	High	Gaussian	High	Low	**Hyperspherical**	Low
G2-4-100	2048	4	2	Low	High	Gaussian	High	High	**Hyperspherical**	Low
	From [38]									
Hepta	212	3	7	High	High	–	High	–	**Hyperspherical**	–
GolfBall	4002	3	1	–	Low	–	Low	–	Hyperspherical	–
	From [12]									
One-G	1000	2	1	–	High	Gaussian	High	–	Circular	High
H$_1$	300	2	6(2)	Low	High	Uniform	High	–	Circular	–
H$_2$	180	2	6(5,3)	High	High	Gaussian	High	–	Circular	–
	From [4]									
Iris	150	4	3	**High**	High	–	High	**High**	**Hyperspherical**	–

The attributes in bold are effective for most clusters.

The GADBA-MEC algorithm works through some previously specified hyper-parameters. Considering an oscillation of the best fitness in the order of 1×10^{-4}, the number of generations needed to infer the convergence of the fitness value is 50. The crossover and mutation probabilities are 0.8 and 0.03, respectively. The ring size of the tournament method for individual selection is set to 3. The population size and the maximum number of clusters are 100 and 30, respectively.

All experiments presented herein were conducted on a computer with an Intel© Core™ i5 CPU @ 3.00 GHz with 8 GB of memory running MATLAB® 2017a. Most packages used in our tests are internal to MATLAB®.

4.2 Cluster Detection Results

One approach to evaluate the performance of the clustering algorithms is to analyse how frequently $\hat{K} = K_{opt}$. In this sense, Table 2 shows the frequency of K_{opt} with emphasis on the highest absolute frequency by algorithm in blue.

Only GADBA is inconsistent with K_{opt} overall due to its SHM-related objective function, as proven by the performance of GADBA-MEC. Moreover, GADBA is the most unstable algorithm, as shown by the standard deviation values. The only highlight of GADBA was reached in Iris, although this might be explained by its tendency of settling on lower K values. Thus, a new version is justified *ex post facto*, attesting to the generalization potential of GADBA-MEC.

Table 2. Cluster detection results taken from the data sets.

Data set	K_{opt}	GADBA-MEC	GADBA	K-MEANS	GMM	LINKAGE
S&C$_1$	9	20 9.00 ± 0.00	0 5.55 ± 4.99	13 9.45 ± 0.69	6 9.95 ± 0.76	20 9.00 ± 0.00
S&C$_2$	10	8 9.15 ± 1.04	0 3.00 ± 1.72	0 26.90 ± 2.34	1 12.10 ± 6.13	0 9.00 ± 0.00
Dim-32	16	5 17.05 ± 0.83	0 3.70 ± 5.55	20 16.00 ± 0.00	19 16.05 ± 0.22	20 16.00 ± 0.00
G2-4-100	2	20 2.00 ± 0.00	0 26.85 ± 2.87	20 2.00 ± 0.00	11 2.50 ± 0.61	20 2.00 ± 0.00
Hepta	7	17 7.15 ± 0.37	0 2.80 ± 0.95	7 7.75 ± 0.64	12 7.45 ± 0.60	20 7.00 ± 0.00
GolfBall	1	20 1.00 ± 0.00	0 26.50 ± 5.90	0 25.95 ± 5.11	0 28.45 ± 1.32	0 21.00 ± 0.00
One-G	1	20 1.00 ± 0.00	0 2.00 ± 0.00	0 6.30 ± 0.47	0 4.70 ± 1.49	0 10.00 ± 0.00
H$_1$	6(2)	20 6.00 ± 0.00	0 2.55 ± 0.94	0 2.00 ± 0.00	0 2.00 ± 0.00	0 2.00 ± 0.00
H$_2$	6(5,3)	20 6.00 ± 0.00	1 2.25 ± 0.91	12 6.40 ± 0.50	14 6.30 ± 0.47	20 6.00 ± 0.00
Iris	3	2 3.95 ± 0.39	4 2.90 ± 1.41	0 26.00 ± 3.84	1 9.10 ± 10.24	0 30.00 ± 0.00

An important highlight of the proposed version is the detection of low-separation hierarchical data in H_1. Virtually all other techniques tended to settle on expected sub-optimal K values. Contrastingly, in cases where GADBA-MEC did not reach the highest frequency (i.e., Dim-32, Hepta, and Iris), it at least approached the expected result in a stable manner, unlike GADBA.

For the rest of the algorithms, it should be noted that LINKAGE is deterministic, so its null standard deviation is expected. It was the second best in finding K_{opt}, although it failed to assess clustering tendency. In this regard, only GADBA-MEC determined that $\hat{K} = 1$ in GolfBall and One-G.

4.3 Statistical Significance Analyses

Friedman's test is a non-parametric statistical test analogue to the two-way ANOVA (analysis of variance) [1,16]. This statistical test is used to determine whether there are any statistically significant differences among algorithms from sample evidences. The samples to be considered are clustering algorithm performance results collected over the data sets, where the null hypothesis H_0, to be considered is that all algorithms obtained similar results. Friedman's test converts all results to ranks where all algorithms are classified for each problem according to its performance. As such, p-values can be computed for hypothesis testing. The p-value represents the probability of obtaining a result as extreme as the one observed, given H_0 [7]. Specifically, given the significance level $\alpha = 0.05$, the null hypothesis is rejected if $p < \alpha$.

Since we want to know which algorithms are significantly different from each other when H_0 is rejected, a *post-hoc* procedure is necessary to compare

Table 3. MAPE (%) taken from the data sets with emphasis on values above 100%.

Data set	K_{opt}	GADBA-MEC	GADBA	K-MEANS	GMM	LINKAGE
S&C$_1$	9	0.00	62.78	5.00	10.56	0.00
S&C$_2$	10	9.50	70.00	169.00	39.00	10.00
Dim-32	16	6.56	83.75	0.00	0.31	0.00
G2-4-100	2	0.00	1242.50	0.00	25.00	0.00
Hepta	7	2.14	60.00	10.71	6.43	0.00
GolfBall	1	0.00	2550.00	2495.00	2745.00	2000.00
One-G	1	0.00	100.00	530.00	370.00	900.00
H$_1$	6(2)	0.00	57.50	66.67	66.67	66.67
H$_2$	6(5,3)	0.00	62.50	6.67	5.00	0.00
Iris	3	31.67	36.67	766.67	243.33	900.00
Overall mean		4.99	432.57	404.97	351.13	387.67
Overall mean'		6.23	209.46	128.09	49.54	122.08

Table 4. Friedman's test on the data set results.

F	p-value
270.65	2.31×10^{-57}

Table 5. Friedman's *post-hoc* pairwise comparisons on the data set results, with emphasis on significant comparisons.

Algorithm	$\bar{R}_i$	GADBA	K-MEANS	GMM	LINKAGE
GADBA-MEC	28.02	41.23	29.12	23.65	18.38
GADBA	69.25		12.10	17.58	22.85
K-MEANS	57.15			5.47	10.75
GMM	51.67				5.27
LINKAGE	46.40				

Critical value of Friedman test ($\alpha = 0.05$): 9.49

all possible algorithm pairs. In this work, the procedure presented in [16] is employed, in which the means of critical values at α are compared to each absolute difference on mean ranks as $abs(\bar{R}_i - \bar{R}_j)$, $i \neq j$. The absolute difference must be greater than α to determine statistical significance.

In this section, we verify the significance of GADBA-MEC using the Friedman's test F. For this purpose, we calculate the MAPE of each data set. Once the algorithm with the smallest error is determined, the statistical significance test is applied to verify if the obtained difference is substantial. If this is the case, one can justify using one algorithm instead of another with more confidence.

Table 3 summarizes MAPE per data set emphasizing values above 100%. All algorithms had error rates above 100%, except GADBA-MEC with the lowest overall value (4.99%). Table 4 focuses on these errors, for which the Friedman's test rejects the null hypothesis for an obtained p-value $\ll \alpha = 0.05$. Accordingly, Friedman's *post-hoc* test shows that there are significant improvements of the proposed version in terms of MAPE (Table 5), as well as the fact that significant differences are shown in virtually all algorithm pairs.

5 Conclusions and Further Work

Genetic-based clustering approaches play an important role in natural computing. In this sense, GADBA was introduced as an efficient, bioinspired approach to cluster data in SHM. Despite its competitive performance identifying structural components, it produces poor results on more general clustering scenarios. For this reason, this study proposed the replacement of its objective function for MEC, a recently developed CVI based on mutual equidistant-scattering analysis.

GADBA-MEC outperforms conventional clustering algorithms when statistically evaluated across various data sets, attaining the expected number of clusters more often than others. The results showed that GADBA-MEC yielded better results in terms of cluster validation and MAPE errors, in particular when handling hierarchical data and data with low separation. Also, only GADBA-MEC is able to verify the clustering tendency in the data sets addressed.

As future work, we intend to expand GADBA-MEC to multi-objective optimization contexts. It is also relevant to apply GADBA-MEC in real-world problems to validate its efficiency in finding natural clusters. Finally, comparing other CVI's and bioinspired algorithms would be pertinent as well.

References

1. Alswaitti, M., Albughdadi, M., Isa, N.A.M.: Density-based particle swarm optimization algorithm for data clustering. ESWA **91**, 170–186 (2018)
2. Armano, G., Farmani, M.R.: Multiobjective clustering analysis using particle swarm optimization. Expert Syst. Appl. **55**, 184–193 (2016)
3. Bayá, A.E., Larese, M.G., Namías, R.: Clustering stability for automated color image segmentation. Expert Syst. Appl. **86**, 258–273 (2017)
4. Bezdek, J.C., Pal, N.R.: Some new indexes of cluster validity. TSMC-B (1998)
5. Campello, R.: Generalized external indexes for comparing data partitions with overlapping categories. Pattern Recogn. Lett. **31**(9), 966–975 (2010)
6. Cremona, C.: Big data and structural health monitoring. In: Challenges in Design and Construction of an Innovative and Sustainable Built Environment, 19th IABSE Congress Stockholm, pp. 1793–1801, September 2016
7. Daniel, W.W.: Applied nonparametric statistics. PWS-KENT, USA (1990)
8. Deb, K., Pratap, A., Agarwal, S., Meyarivan, T.: A fast and elitist multiobjective genetic algorithm: NSGA-ii. IEEE TEC **6**(2), 182–197 (2002)
9. Diez, A., Khoa, N.L.D., Makki Alamdari, M., Wang, Y., Chen, F., Runcie, P.: A clustering approach for structural health monitoring on bridges. JCSHM (2016)
10. Dziopa, T.: Clustering validity indices evaluation with regard to semantic homogeneity. In: FedCSIS 2016, Gdańsk, Poland, 11–14 September 2016, pp. 3–9 (2016)
11. Esfandian, N., Razzazi, F., Behrad, A.: A clustering based feature selection method in spectro-temporal domain for speech recognition. EAAI **25**(6), 1194–1202 (2012)
12. Flexa, C., Santos, R., Gomes, W., Sales, C., Costa, J.C.: Mutual equidistant-scattering criterion: a new index for crisp clustering. ESWA **128**, 225–245 (2019)
13. Fränti, P., Sieranoja, S.: K-means properties on six clustering benchmark datasets. Appl. Intell. **48**(12), 4743–4759 (2018)
14. Fumeo, E., Oneto, L., Anguita, D.: Condition based maintenance in railway transportation systems based on big data streaming analysis. PCS **53**, 437–446 (2015)
15. Gandomi, A., Haider, M.: Beyond the hype: big data concepts, methods, and analytics. Int. J. Inf. Manage. **35**(2), 137–144 (2015)
16. García, S., Fernández, A., Luengo, J., Herrera, F.: Advanced nonparametric tests for multiple comparisons in the design of experiments in computational intelligence and data mining: experimental analysis of power. IS **180**(10), 2044–2064 (2010)
17. Gardiner, A., Aasheim, C., Rutner, P., Williams, S.: Skill requirements in big data: a content analysis of job advertisements. JCIF **58**(4), 374–384 (2018)
18. Gil, D., Songi, I.Y.: Modeling and management of big data: challenges and opportunities. Futur. Gener. Comput. Syst. **63**, 96–99 (2016)

19. Güngör, E., Özmen, A.: Distance and density based clustering algorithm using gaussian Kernel. Expert Syst. Appl. **69**, 10–20 (2017)
20. Islam, M.Z., Estivill-Castro, V., Rahman, M.A., Bossomaier, T.: Combining k-means and a genetic algorithm through a novel arrangement of genetic operators for high quality clustering. Expert Syst. Appl. **91**, 402–417 (2018)
21. Jain, A.K.: Data clustering: 50 years beyond k-means. PRL **31**(8), 651–666 (2010)
22. Johnson, S.C.: Hierarchical clustering schemes. Psychometrika, pp. 241–254 (1967)
23. Langone, R., Reynders, E., Mehrkanoon, S., Suykens, J.A.: Automated structural health monitoring based on adaptive kernel spectral clustering. In: MSSP (2017)
24. Lingras, P., Chen, M., Miao, D.: Qualitative and quantitative combinations of crisp and rough clustering schemes using dominance relations. In: IJAR, pp. 238–258 (2014)
25. Lucasius, C., Dane, A., Kateman, G.: On k-medoid clustering of large data sets with the aid of a genetic algorithm: background, feasiblity and comparison. Anal. Chim. Acta **282**(3), 647–669 (1993)
26. MacQueen, J.B.: Some methods for classification and analysis of multivariate observations. In: BSMSP, vol. 1, pp. 281–297. University of California Press (1967)
27. Mary, S.A.L., Sivagami, A.N., Rani, M.U.: Cluster validity measures dynamic clustering algorithms. ARPN J. Eng. Appl. Sci. **10**(9), 4009–4012 (2015)
28. Maulik, U., Bandyopadhyay, S.: Genetic algorithm-based clustering technique. Pattern Recogn. **33**(9), 1455–1465 (2000)
29. McAfee, A., Brynjolfsson, E.: Big data: the management revolution (2012)
30. Mclachlan, G., Basford, K.: Mixture Models: Inference and Applications to Clustering, vol. 38, January 1988
31. Moulavi, D., Jaskowiak, P.A., Campello, R.J.G.B., Zimek, A., Sander, J.: Density-based clustering validation. In: 14th SIAM ICDM, Philadelphia, PA (2014)
32. Pagnuco, I.A., Pastore, J.I., Abras, G., Brun, M., Ballarin, V.L.: Analysis of genetic association using hierarchical clustering and cluster validation indices. Genomics **109**(5), 438–445 (2017)
33. Rubio, E., Castillo, O., Valdez, F., Melin, P., Gonzalez, C.I., Martinez, G.: An extension of the fuzzy possibilistic clustering algorithm using type-2 fuzzy logic techniques. Adv. Fuzzy Sys. **2017**, 7094046 (2017)
34. Salvador, S., Chan, P.: Determining the number of clusters/segments in hierarchical clustering/segmentation algorithms. In: 16th IEEE ICTAI, USA (2004)
35. Silva, J.A., Hruschka, E.R., Gama, J.: An evolutionary algorithm for clustering data streams with a variable number of clusters. ESWA **67**, 228–238 (2017)
36. Silva, M., Santos, A., Figueiredo, E., Santos, R., Sales, C., Costa, J.C.W.A.: A novel unsupervised approach based on a genetic algorithm for structural damage detection in bridges. Eng. Appl. Artif. Intell. **52**(C), 168–180 (2016)
37. Silva, M., Santos, A., Santos, R., Figueiredo, E., Sales, C., Costa, J.C.: Agglomerative concentric hypersphere clustering applied to structural damage detection. Mech. Syst. Signal Process. **92**, 196–212 (2017)
38. Ultsch, A.: Clustering with SOM: U*C. In: Proceedings of Workshop on Self-organizing Maps, pp. 75–82, January 2005
39. Wu, K.L., Yang, M.S.: Alternative c-means clustering algorithms. Pattern Recogn. **35**, 2267–2278 (2002)
40. Wu, K.L., Yang, M.S.: A cluster validity index for fuzzy clustering. PRL (2005)
41. Xu, R., Wunsch, D., II.: Survey of clustering algorithms. TNN **16**(3), 645–678 (2005)

42. Yang, C.L., Kuo, R., Chien, C.H., Quyen, N.T.P.: Non-dominated sorting genetic algorithm using fuzzy membership chromosome for categorical data clustering. Appl. Soft Comput. **30**, 113–122 (2015)

43. Zhao, Q.: Cluster Validity in Clustering Methods. Ph.D. thesis, UEF, June 2012

Improving Particle Swarm Optimization with Self-adaptive Parameters, Rotational Invariance, and Diversity Control

Matheus Vasconcelos[✉], Caio Flexa, Igor Moreira, Reginaldo Santos, and Claudomiro Sales

Federal University of Pará, 1 Augusto Corrêa St., Belém, PA 66.075-110, Brazil
matheusfv@ufpa.br

Abstract. Particle Swarm Optimization (PSO) algorithms are swarm intelligence methods that are effective in solving optimization problems. However, current techniques have some drawbacks: the particles of some PSO implementations are sensible to their input hyper-parameters, lack direction diversity in their movement, have rotational variance, and might prematurely converge due to rapid swarm diversity loss. This article addresses these issues by introducing Rotationally Invariant Attractive and Repulsive eXpanded PSO (RI-AR-XPSO) and Rotationally Invariant Semi-Autonomous eXpanded PSO (RI-SAXPSO) as improvements of Rotationally Invariant Semi-Autonomous PSO (RI-SAPSO) and eXpanded PSO (XPSO). Their swarm behavior was evaluated with classic functions in the literature and their accuracy was tested with the Congress on Evolutionary Computation (CEC) 2017 optimization problems, in whose results a statistical significance test was applied. The results obtained attest that strategies such as diversity control, automatic hyper-parameter adjustment, directional diversity, and rotational invariance improve performance without accuracy loss when adequately implemented.

Keywords: Particle swarm optimization · Global continuous optimization · Adaptive adjustment · Rotational invariance · Diversity control

1 Introduction

Heuristic algorithms are becoming increasingly more robust means to find suitable solutions to complex problems. Current approaches range from low-level heuristics—designed for specific issues—to complex, more general hyper-heuristics [4]. Particle Swarm Optimization (PSO) is an effective, swarm-intelligence technique based on the collective behavior of fish shoals and bird flocks used to solve optimization problems.

The parameter settings of non-deterministic algorithms such as meta-heuristics are known to be dependent on the problem. The same occurs with

© Springer Nature Switzerland AG 2021
A. Britto and K. Valdivia Delgado (Eds.): BRACIS 2021, LNAI 13073, pp. 218–233, 2021.
https://doi.org/10.1007/978-3-030-91702-9_15

PSO, for the adjustment of hyper-parameters affects the trajectory of the particles in the swarm. In some implementations, the cognitive and social coefficients, as well as the inertial weight, are dynamically selected [17]. In others, these hyper-parameters are manually set and remain fixed throughout the execution [15]. The authors of implementations with fixed hyper-parameters customarily disclose recommended values for them, defined after multiple analyzes on different problems, as in [14].

In PSO, the works of [7,10,18] demonstrate that particles tend to move in parallel to the coordinate axes. Using rotational invariance mitigates this dependency on the coordinate system wherein the objective function is defined, improving results across various problems. In terms of complex multimodal functions, experiments indicate that maintaining swarm diversity helps mitigate premature convergence and avoid sub-optimal solutions [11].

Motivated by these aspects, this paper aims to improve two PSO implementations termed Rotationally Invariant Semi-Autonomous PSO(RI-SAPSO) [14] and eXpanded PSO (XPSO) [21] by automatically updating hyper-parameters, promoting rotational invariance, and maintaining swarm diversity during execution.

The remaining sections are structured as follows. Section 2 describes the canonical PSO, as well as some properties of its variants. In Sect. 3, the framework of some PSO variants is exposed. Details on the improvements made to RI-SAPSO and XPSO are presented in Sect. 4. Section 5 presents the obtained experimental results and their statistical significance. Lastly, Sect. 6 concludes this paper with some remarks and future work.

2 Theoretical Background

The canonical PSO algorithm only performs simple mathematical operations. As such, it is computationally inexpensive in terms of time and space. PSO is a population-based meta-heuristic where each individual (i.e., particle) represents a candidate solution to an optimization problem in a D-dimensional space. In each iteration t of the algorithm, the ith particle of the swarm is associated to three vectors: the position vector $\vec{x}_i = [x_{i,1}, x_{i,2}, \ldots x_{i,D}]$, that represents the candidate solution; the velocity vector $\vec{v}_i = [v_{i,1}, v_{i,2}, \ldots v_{i,D}]$, that represents the direction and velocity of the particle in the search space; and local memory vector $\vec{p}_i = [p_{i,1}, p_{i,2}, \ldots p_{i,D}]$, which stores the best solution found. By its turn, the swarm is associated to the global memory vector $\vec{g} = [g_1, g_2, \ldots g_D]$, which stores the best solution found overall. Following the algorithm of [5], the position of the particles are updated as t increases according to

$$\vec{v}_i^{t+1} = \vec{v}_i^t + c_1\phi_1(\vec{p}_i^t - \vec{x}_i^t) + c_2\phi_2(\vec{g}^t - \vec{x}_i^t) \tag{1}$$

and

$$\vec{x}_i^{t+1} = \vec{x}_i^t + \vec{v}_i^{t+1}, \tag{2}$$

where c_1 and c_2 respectively represent the cognitive and social coefficients and ϕ_1 and ϕ_2 are random numbers sampled from a uniform distribution in $[0, 1]$.

Ulteriorly, [17] added the inertial weight w to (1), turning $\vec{v}_i^t$ into $w\vec{v}_i^t$. This significantly improved the canonical PSO and is widely used in its variants.

Making the algorithm rotationally invariant ensures that particles will move independently without influence of the coordinate system of the problem. Its presence (or lack thereof) might lie in how ϕ_1 and ϕ_2 are used in the velocity update equation. If they are vectors, the optimizer is variant in terms of scale, translation, and rotation. If they are scalar values, then the algorithm is rotationally invariant. To illustrate, a rotationally variant version of (1), which is rotationally invariant, can be defined as

$$\vec{v}_i^{t+1} = w^t \vec{v}_i^t + c_1 \overset{\scriptstyle\cdot t}{\phi_i} \odot (\vec{p}_i^t - \vec{x}_i^t) + c_2 \overset{\scriptstyle\cdot t}{\phi_i} \odot (\vec{g}^t - \vec{x}_i^t), \tag{3}$$

where $\overset{\scriptstyle\cdot t}{\phi_i}$ and $\overset{\scriptstyle\cdot t}{\phi_i}$ are two vectors containing random values sampled from a uniform distribution in $[0, 1]$ and $\odot$ represents the element-wise multiplication of vectors and matrices [8]. The mathematical tool presented in [20] can be used to prove whether a given PSO implementation is variant or invariant.

The work of [19] introduces the concept of directional diversity, i.e., the ability to carry out the stochastic search in various directions. Conversely, particles without directional diversity perform their search in fixed directions. This work proved that, unlike (3), (1) provokes loss in directional diversity. To demonstrate that the directional diversity and rotational invariance are not necessarily exclusive, [20] proposed a rotationally invariant PSO that maintains directional diversity named Diverse Rotationally Invariant PSO (DRI-PSO), whose velocity update equation is rotationally invariant in a stochastic manner but also diverse in direction. This was possible thanks to small, consistent perturbations in the direction of the local $(\vec{p} - \vec{x})$ and global swarm memories $(\vec{g} - \vec{x})$, multiplying these values with an independent, random rotation matrix W.

The ability to store gradient information allows the particles to individually perform asynchronous searches, mitigate the random walk effect and avoid wasting random efforts to reach local optima. When using the method described in [9] with problems of high D, the employment of gradient descent information might be difficult. Depending on how it is applied throughout the search, up to D objective function evaluations per particle might be performed.

The ability to control swarm diversity, firstly proposed to PSO in [13], is present in many implementations. This strategy, which entails monitoring the swarm diversity during the search, enables the algorithm to increase exploration when the diversity is low to avoid local optima. Equation (4) represents one form of defining swarm diversity, where $|L|$ is the diagonal size of the search space, $x_{i,j}^t$ is the j-th dimension of the i-th particle, and $\bar{x}_j^t$ is the j-th dimension of the mean position across all particles.

$$diversity(\vec{x^t}) = \frac{1}{n \times |L|} \times \sum_{i=1}^{n} \sqrt{\sum_{j=1}^{D} (x_{i,j}^t - \bar{x}_j^t)^2} \tag{4}$$

3   Related Work

Works related to this paper are associated with other meta-heuristics inspired on the canonical PSO. This section will focus on implementations that involve diversity control, gradient use, rotational invariance, directional diversity, rotation matrices, and hyper-parameter auto-adjustment.

The introduction of Gradient-Based PSO (GPSO) is described in [12], wherein the gradient directions are used as a deterministic approach for a precise local exploration around the best global position, thus strengthening the global exploration of the algorithm. In [6], where diversity control and gradient information were employed as strategies to switch between two PSO algorithms throughout the search, a new, hybrid approach termed Diversity-Guided PSO based on Gradient Search (DGPSOGS) was introduced.

The Semi-Autonomous PSO (SAPSO) [15] takes advantage of ideas present in Attraction and Repulsion PSO (ARPSO) [13], GPSO [12], and DGPSOGS [6] to provide a semi-autonomous particle swarm optimizer that uses gradient and diversity-control information to optimize unimodal functions. SAPSO attempts to reduce computational efforts related to local investigation with the aid of gradient information and provide a diversity control mechanism to avoid local optima. The performance of SAPSO and other PSO implementations are evaluated in a set of test functions based on optimization problems of De Jong's benchmark. Numerical results showed that the proposed method attained at least the same performance of other PSO implementations. Moreover, SAPSO achieved better results in terms of global minima found and fine-tuning of the final solution. However, problems in spaces of higher dimensionality caused longer execution times due to the employment of gradient-based information by each particle and the need to modify parameters depending on the problem at hand.

As an improvement to DRI-PSO, the Locally Convergent Rotationally Invariant PSO (LcRiPSO) was introduced in [2]. The authors pointed out several problematic situations: particles can get stuck in some areas of the search space, unable to change the value of one or more decision variables; poor performance is observed when the swarm is small or the dimensionality is high; convergence is not guaranteed even for local optima; and the algorithm is sensible to the rotation of the search space. Aiming to solve these issues, LcRiPSO contains a new general form of velocity update equation that contains a normal distribution function defined by the user around local and global memories.

The RI-SAPSO is proposed in the work of [14] as an improvement to SAPSO [15]. This PSO implementation inherits the rotational invariance of SAPSO and incorporates a rotation matrix generated by an exponential map to maintain directional diversity using an idea present in [20]. Besides mathematically proving that RI-SAPSO is rotationally invariant, benchmark tests with statistical significance demonstrated that the algorithms were capable of finding better solutions in most problems. Therefore, RI-SAPSO is the starting point of this work because despite its superior results against those of SAPSO, it still has long execution times when handling high-dimensional problems and employs fixed hyper-parameters during execution.

An expanded variant of PSO termed XPSO was proposed in [21]. It updates acceleration coefficients and the learning model based on fitness, given that updating these coefficients by only considering iterations might result in sub-par intelligence when solving complex problems. Moreover, XPSO expands the social part of each particle from one to two exemplars and applies the ability to forget some particles, which is based on a universal biological phenomenon wherein parts of the memory management system forget undesirable or useless information. Upon testing, this approach presented promising results with different types of objective function, despite its possible lack of rotational invariance as it employs the random components similarly to (3).

4   Improving RI-SAPSO and XPSO

This section details how XPSO and RI-SAPSO were changed into the Rotationally Invariant Attractive and Repulsive eXpanded PSO (RI-AR-XPSO) and the Rotationally Invariant Semi-Autonomous eXpanded PSO (RI-SAXPSO).

4.1   Rotationally Invariant Attractive and Repulsive eXpanded PSO

The implementation herein defined is termed Rotationally Invariant Attractive and Repulsive eXpanded PSO. Since XPSO has a vector of random numbers $\vec{\phi}^t$ for each component in its velocity update equation—similar to (3)—, its search strategy might have directional diversity, but still suffers from rotation variance, which can be proven by using the mathematical tool seen in [20]. As previously exposed, in addition to automatically adjusting the acceleration coefficients during the search, XPSO has an additional social coefficient. Here, the ith particle out of N particles will be related to three coefficients: $c_{1,i}$, which is the cognitive coefficient; and $c_{2,i}$ and $c_{3,i}$, which are social coefficients. The coefficients of each particle are defined using a Gaussian following the method described in [21].

A greater movement amplitude is preserved in each particle by using the stochastic rotation matrix W in the velocity equation, thus increasing directional diversity in the search [20]. Furthermore, it also promotes rotational invariance. As such, the velocity equation will be changed to use W similarly to [14].

The authors of XPSO justify that its ability to forget relates to maintaining swarm diversity throughout the search. In this regard, another improvement to be added was motivated by the attraction and repulsion strategy of [13], where the diversity monitoring will influence the behavior of the particles such that they will not forget during the attraction phase to accelerate convergence. Moreover, a lower swarm diversity bound will be dynamically defined in the new implementation to avoid that the repulsion phase precludes refining the solution at the end of the search. The d_{min}^t values will exponentially decrease at this point, with values defined by $f(x) = b^x$ (the best results were attained with x having values within $[1, 10]$ linearly increasing during the search). $b = 10^{-2}$ is the same value used by [14] as an upper swarm diversity bound.

The modifications made to XPSO change its velocity equation to

$$\vec{v}_i^{t+1} = w^t \vec{v}_i^t + dir \times \left[\vec{\Omega}_i^t + \vec{\Upsilon}_i^t + \vec{\Lambda}_i^t \right], \tag{5}$$

$$\vec{\Omega}_i^t = \begin{cases} c_{1,i} \dot{\phi}_i^t \left(\vec{p}_i^t - \vec{x}_i^t \right), & \text{if } dir = 1; \\ c_{1,i} \dot{\phi}_i^t \dot{W}_i^t \left(\vec{p}_i^t - \vec{x}_i^t \right), & \text{else} \end{cases} \tag{6}$$

$$\vec{\Upsilon}_i^t = \begin{cases} c_{2,i} \ddot{\phi}_i^t \left(\vec{l}_i^t - \vec{x}_i^t \right), & \text{if } dir = 1; \\ c_{2,i} \ddot{\phi}_i^t \dot{W}_i^t \left[\left(1 - \vec{f}_i \right) \vec{l}_i^t - \vec{x}_i^t \right], & \text{else} \end{cases} \tag{7}$$

$$\vec{\Lambda}_i^t = \begin{cases} c_{3,i} \dddot{\phi}_i^t \left(\vec{g}^t - \vec{x}_i^t \right), & \text{if } dir = 1; \\ c_{3,i} \dddot{\phi}_i^t \dddot{W}_i^t \left[\left(1 - \vec{f}_i \right) \vec{g}^t - \vec{x}_i^t \right], & \text{else} \end{cases} \tag{8}$$

where $\dot{\phi}_i^t$, $\ddot{\phi}_i^t$ and $\dddot{\phi}_i^t$ are three random vectors uniformly distributed within $[0, 1]$; $\vec{\mathcal{F}} = [f_{i,1}, f_{i,2}, \cdots, f_{i,D}]$ represents the amount of information that the i-th particle forgot of a given sample; $\vec{l}_i^t$ represents the best position visited between two neighbors (i.e., $\vec{l}_i^t$ will be equal to the most fit solution between $\vec{p}_{i-1}$ and $\vec{p}_{i+1}$); and dir controls the attraction and repulsion phases according to

$$dir = \begin{cases} -1, & \text{if } d < d_{min}^t \text{ and } StagG < StagG_{max}; \\ 1, & \text{otherwise,} \end{cases} \tag{9}$$

where the current swarm diversity value is represented by d, calculated in (4), and $StagG$ and $StagG_{max}$ respectively represent the number of iterations without updating the best global position and the maximum number of iterations without progress allowed. The elitist behavior employed by the hyper-parameter adjusting logic during execution also aids the update of dir. As exposed by [21], when $StagG \geq StagG_{max}$, it is assumed that the particles got stuck in a local minimum. As such, a hyper-parameter update takes place to change their search direction. Nevertheless, if the particles maintain low diversity, the implementation will trigger the repulsion phase to speed up this process. Conversely, the attraction phase is initiated when adequate diversity is maintained.

4.2   Rotationally Invariant Semi-autonomous eXpanded PSO

The implementation defined herein is termed Rotationally Invariant Semi-Autonomous eXpanded PSO. eXpanded refers to the automatic hyper-parameter adjustment and the expansion of the RI-SAPSO velocity equation inspired by XPSO. In [14], the variant employs parameters without dynamic update during execution and uses two main strategies: the diversity-guided attraction/repulsion logic and the replacement of the personal part of the velocity equation from $\vec{p}_i^t - \vec{x}_i^t$ to $\nabla f(\vec{x}_i^t)$ to use the gradient. As such, the improvements are: implement a hyper-parameter update strategy; employ adaptive diversity control; and employ three acceleration coefficients in the velocity update equation involving the best global position (c_1), the gradient (c_2), and the best personal position (c_3).

The implementation of automatic hyper-parameter update during execution is identical to what is performed in XPSO. The update scenarios of dir will be changed due to the difficulty it faces to converge when diversity is low, which hurts the results of RI-SAPSO in such situations. Therefore, (9) will be used herein as well. The upper swarm diversity bound was removed in this case because when the search triggers the repulsion phase in later iterations, this restriction will result in high swarm diversity, thus preventing the refinement of the solution. The lower swarm diversity bound is d_{min}^t, which will have the same values explained in Sect. 4.1. Unlike the original implementation, the switch from repulsion to attraction does not alter the individual decision of the particles. In other words, only dir changes from -1 to 1, mitigating gradient use.

Due to the smaller gradient descent use in the attraction phase, c_3 was added as a cognitive coefficient in (10) to accelerate convergence. In this case, if a particle is unable to decide with the gradient, it will determine with $\vec{g}$ and $\vec{p}$.

After detailing all modifications, the equation that governs velocity updates in RI-SAPSO had its social component

$$\vec{\Omega}_i^t = \begin{cases} I_i^t c_1 \dot{\phi}_i^t (\vec{g}^t - \vec{x}_i^t), & \text{if } dir = 1; \\ I_i^t c_1 \dot{\phi}_i^t \dot{W}_i^t (\vec{g}^t - \vec{x}_i^t), & \text{otherwise} \end{cases} \qquad (10)$$

changed in the new implementation to

$$\vec{\Omega}_i^t = \begin{cases} I_i^t c_{1,i} \dot{\phi}_i^t (\vec{g}^t - \vec{x}_i^t) + I_i^t c_{3,i} \dddot{\phi}_i^t (\vec{p}^t - \vec{x}_i^t), & \text{if } dir = 1; \\ I_i^t c_{1,i} \dot{\phi}_i^t \dot{W}_i^t (\vec{g}^t - \vec{x}_i^t), & \text{otherwise.} \end{cases} \qquad (11)$$

5   Experimental Simulations

This section presents tests performed with the two new algorithms side-by-side with their original counterparts in two scenarios. Firstly, the swarm diversity and the behavior of the convergence curve of the best solution are analyzed with unimodal and multimodal test functions found in the literature. Subsequently, tests are performed using a set of benchmark functions from Congress on Evolutionary Computation (CEC) 2017 [1], whose functions were randomly rotated or dislocated. The final results are evaluated considering time, Number of Function Evaluations (NFEs), and the minimal error rate—the difference between the solution found and the best one (i.e., $f(\vec{x}) - f(\vec{x}_{opt})$). Lastly, statistical tests are carried out to investigate how significant the differences are between algorithms. All results are the mean of 50 executions.

It is noteworthy that this work does not aim to solve the CEC 2017 functions, but rather to use them as a means of proving the concepts previously exposed.

5.1   Swarm Diversity and Solution Convergence

Five classical test functions found in the literature were selected to analyze behavioral swarm changes during the search: a unimodal one (Sphere), a non-convex unimodal one (Rosenbrock), and three multimodal functions (viz., Ackley, Griewank, and Rastrigin). In these tests, D was set to 30, N was set to 20 following [14], and T was set to 6000 following [21]. Existing algorithms were set with the hyper-parameters that were recommended in their respective studies: RI-AR-XPSO used the same settings as XPSO and RI-SAXPSO received a value range for each coefficient guided by [15]. Specific hyper-parameter settings are presented in Table 1. The results represent the average of 50 independent executions where the swarm diversity and the convergence curve of the best solution (in logarithmic scale) were evaluated. The only stopping criteria used was the maximum number of iterations T.

Table 1. Parameter settings utilized in the tested algorithms.

	c_1	c_2	c_3	w
RI-SAPSO	1.4962	10^{-2}	-	$w = 0.7298$
RI-SAXPSO	$min = 1$	$min = 10^{-5}$	$min = 1$	$w = 0.9 \rightarrow 0.4$
	$\mu = 1.4962$	$\mu = 10^{-2}$	$\mu = 1.4962$	
	$max = 4$	$max = 10^{-1}$	$max = 4$	
XPSO	$min = 0.5$	$min = 0.5$	$min = 0.5$	$w = 0.9 \rightarrow 0.4$
	$\mu = 1.35$	$\mu = 1.35$	$\mu = 1.35$	
	$max = 2.05$	$max = 2.05$	$max = 2.05$	
RI-AR-XPSO	$min = 0.5$	$min = 0.5$	$min = 0.5$	$w = 0.9 \rightarrow 0.4$
	$\mu = 1.35$	$\mu = 1.35$	$\mu = 1.35$	
	$max = 2.05$	$max = 2.05$	$max = 2.05$	

Figure 1 shows the performance of the modified and original algorithms. It can be seen that RI-SAXPSO quickly converges while maintaining high swarm diversity at the beginning of the search. It attains better results than RI-SAPSO, which only achieves similar results in the Ackley function (Fig. 1c) despite maintaining swarm diversity throughout the search. XPSO did not lose diversity quickly, but it was lower at the beginning of the search when compared to RI-AR-XPSO. In terms of solution convergence, XPSO best refines the final solution of Sphere (Fig. 1a), and loses diversity faster. XPSO solutions converge with inferior quality when compared to RI-AR-XPSO in the multimodal functions.

5.2   Congress on Evolutionary Computation 2017 Benchmark Functions

The CEC 2017 benchmark is composed of 30 functions, of which only Sum of different Power is not used in this work due to its unstable behavior in higher

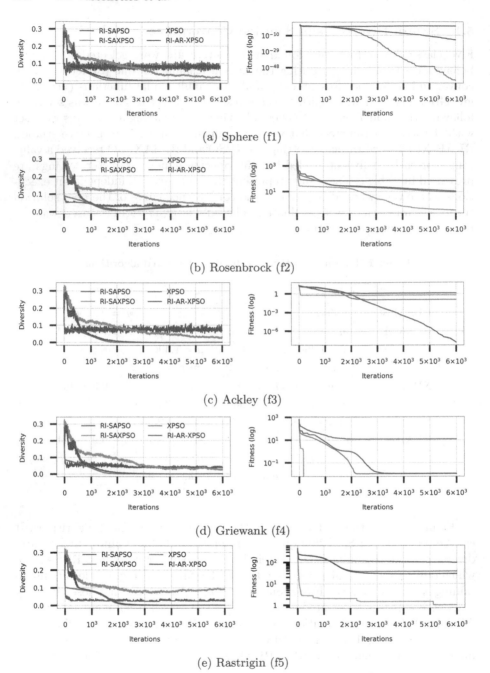

(a) Sphere (f1)

(b) Rosenbrock (f2)

(c) Ackley (f3)

(d) Griewank (f4)

(e) Rastrigin (f5)

Fig. 1. Swarm diversity and best solution convergence curve.

dimensionalities [1]. Hence, a total of 29 functions were used to test the algorithms: unimodal ones ($f1 - 2$), simple multimodal ones ($f3 - 9$), hybrid ones ($f10 - 19$), and composed ones ($f20 - 29$). The same settings found in Sect. 5.1 were used herein except for the stopping criteria: in addition to using $T = 6000$, the search is interrupted whenever the global minima is found or NFE $= 10^6$, which is the upper limit defined in [1]. The D of each function was set to 10, 30, and 50 to evaluate performance in different dimensionalities.

Table 2. Average error rates and algorithm ranks.

fi	D	RI-SAPSO	RI-SAXPSO	XPSO	RI-AR-XPSO
$f1$	10	6.141E+08 (4)	1.337E+03 (2)	1.924E+03 (3)	1.291E+03 (1)
	30	1.232E+09 (4)	5.076E+03 (3)	3.015E+03 (2)	2.398E+03 (1)
	50	1.193E+10 (4)	2.183E+04 (2)	4.137E+05 (3)	2.915E+03 (1)
$f2$	10	6.497E-12 (3)	1.701E+00 (4)	1.819E-14 (1)	6.696E-13 (2)
	30	8.479E+04 (4)	2.043E+02 (3)	5.830E+01 (2)	7.579E+00 (1)
	50	1.019E+06 (4)	2.485E+03 (1)	1.448E+04 (3)	1.140E+04 (2)
$f3$	10	1.113E+00 (4)	3.104E-13 (1)	9.308E-02 (2)	1.137E-01 (3)
	30	4.133E+03 (4)	1.097E+01 (1)	1.224E+02 (2)	1.273E+02 (3)
	50	1.218E+04 (4)	2.481E+01 (1)	2.641E+02 (2)	2.695E+02 (3)
$f4$	10	4.482E+01 (4)	2.514E+01 (3)	6.806E+00 (1)	8.218E+00 (2)
	30	2.500E+02 (4)	1.370E+02 (3)	5.056E+01 (2)	4.889E+01 (1)
	50	4.950E+02 (4)	2.784E+02 (3)	1.074E+02 (1)	1.092E+02 (2)
$f5$	10	3.124E+01 (4)	4.365E+00 (3)	1.221E-06 (2)	1.267E-07 (1)
	30	7.366E+01 (4)	3.088E+01 (3)	2.741E-01 (2)	8.438E-02 (1)
	50	7.958E+01 (4)	4.274E+01 (3)	2.533E+00 (2)	1.676E+00 (1)
$f6$	10	2.411E+01 (3)	2.579E+01 (4)	1.866E+01 (1)	1.889E+01 (2)
	30	2.159E+02 (4)	1.409E+02 (3)	9.389E+01 (2)	8.987E+01 (1)
	50	6.468E+02 (4)	3.122E+02 (3)	1.898E+02 (1)	1.939E+02 (2)
$f7$	10	1.110E+01 (3)	2.042E+01 (4)	7.661E+00 (2)	6.825E+00 (1)
	30	1.908E+02 (4)	1.190E+02 (3)	5.094E+01 (2)	5.092E+01 (1)
	50	4.751E+02 (4)	2.883E+02 (3)	1.073E+02 (2)	1.054E+02 (1)
$f8$	10	5.681E+01 (4)	8.847E+00 (3)	1.819E-14 (1)	6.594E-14 (2)
	30	4.720E+03 (4)	1.385E+03 (3)	2.224E+01 (2)	1.475E+01 (1)
	50	2.004E+04 (4)	7.121E+03 (3)	1.565E+02 (1)	1.857E+02 (2)
$f9$	10	1.049E+03 (4)	8.194E+02 (3)	3.247E+02 (2)	3.197E+02 (1)
	30	4.637E+03 (4)	4.496E+03 (3)	3.034E+03 (2)	2.813E+03 (1)
	50	8.065E+03 (4)	7.824E+03 (3)	5.724E+03 (1)	5.758E+03 (2)
$f10$	10	1.462E+02 (4)	5.785E+01 (3)	3.196E+00 (2)	3.010E+00 (1)
	30	1.655E+03 (4)	1.993E+02 (3)	1.028E+02 (2)	9.989E+01 (1)
	50	5.214E+02 (4)	3.147E+02 (3)	2.248E+02 (2)	2.104E+02 (1)

(*continued*)

Table 2. (*continued*)

fi	D	RI-SAPSO	RI-SAXPSO	XPSO	RI-AR-XPSO
$f11$	10	6.421E+07 (4)	5.948E+03 (1)	1.001E+04 (2)	1.239E+04 (3)
	30	2.867E+06 (4)	5.112E+04 (1)	1.666E+05 (2)	2.050E+05 (3)
	50	1.247E+10 (4)	3.754E+05 (1)	2.091E+06 (3)	1.041E+06 (2)
$f12$	10	2.302E+03 (2)	7.992E+02 (1)	6.438E+03 (4)	4.832E+03 (3)
	30	9.465E+08 (4)	9.169E+03 (1)	1.054E+04 (2)	1.202E+04 (3)
	50	6.819E+04 (4)	1.338E+04 (3)	6.867E+03 (2)	3.131E+03 (1)
$f13$	10	4.875E+05 (4)	8.270E+01 (3)	3.145E+01 (1)	3.201E+01 (2)
	30	1.033E+06 (4)	1.023E+03 (1)	8.734E+03 (3)	3.515E+03 (2)
	50	5.757E+05 (4)	4.518E+03 (1)	5.257E+04 (3)	4.178E+04 (2)
$f14$	10	9.803E+02 (4)	1.270E+02 (3)	4.034E+01 (2)	2.423E+01 (1)
	30	1.372E+08 (4)	1.826E+03 (1)	6.428E+03 (3)	3.949E+03 (2)
	50	2.583E+09 (4)	1.039E+04 (3)	4.574E+03 (2)	2.946E+03 (1)
$f15$	10	2.328E+02 (4)	6.265E+01 (2)	1.093E+02 (3)	5.961E+01 (1)
	30	1.401E+03 (4)	1.081E+03 (3)	7.314E+02 (2)	6.945E+02 (1)
	50	2.509E+03 (4)	2.188E+03 (3)	1.183E+03 (2)	1.181E+03 (1)
$f16$	10	1.359E+02 (4)	7.116E+01 (3)	3.114E+01 (2)	1.740E+01 (1)
	30	4.642E+03 (4)	6.600E+02 (3)	2.112E+02 (2)	1.857E+02 (1)
	50	2.012E+03 (4)	1.855E+03 (3)	9.730E+02 (1)	1.053E+03 (2)
$f17$	10	1.175E+07 (4)	4.453E+03 (1)	7.393E+03 (3)	6.324E+03 (2)
	30	7.251E+05 (4)	2.798E+04 (1)	1.193E+05 (3)	8.578E+04 (2)
	50	1.040E+08 (4)	5.880E+04 (1)	6.948E+05 (3)	6.522E+05 (2)
$f18$	10	4.299E+02 (4)	7.378E+01 (3)	5.379E+01 (2)	4.500E+01 (1)
	30	1.065E+08 (4)	5.381E+03 (1)	7.265E+03 (3)	5.617E+03 (2)
	50	3.887E+08 (4)	3.387E+03 (1)	1.563E+04 (3)	1.386E+04 (2)
$f19$	10	1.774E+02 (4)	9.491E+01 (3)	6.659E+01 (2)	1.481E+01 (1)
	30	8.454E+02 (4)	5.037E+02 (3)	2.481E+02 (2)	2.249E+02 (1)
	50	1.774E+03 (4)	1.219E+03 (3)	7.360E+02 (2)	5.733E+02 (1)
$f20$	10	1.939E+02 (2)	2.080E+02 (4)	1.981E+02 (3)	1.024E+02 (1)
	30	4.782E+02 (4)	3.415E+02 (3)	2.465E+02 (1)	2.468E+02 (2)
	50	8.361E+02 (4)	5.367E+02 (3)	3.062E+02 (2)	3.055E+02 (1)
$f21$	10	3.665E+02 (4)	1.235E+02 (3)	9.958E+01 (2)	9.451E+01 (1)
	30	4.785E+03 (4)	3.538E+03 (3)	3.584E+02 (2)	1.006E+02 (1)
	50	8.717E+03 (4)	8.240E+03 (3)	4.039E+03 (1)	4.628E+03 (2)
$f22$	10	3.518E+02 (4)	3.313E+02 (3)	3.087E+02 (2)	3.035E+02 (1)
	30	8.522E+02 (4)	6.245E+02 (3)	4.046E+02 (2)	3.925E+02 (1)
	50	1.422E+03 (4)	1.010E+03 (3)	5.581E+02 (1)	5.618E+02 (2)
$f23$	10	3.071E+02 (3)	3.540E+02 (4)	2.520E+02 (2)	2.139E+02 (1)
	30	7.860E+02 (4)	6.301E+02 (3)	4.731E+02 (1)	4.753E+02 (2)
	50	1.300E+03 (4)	9.482E+02 (3)	6.398E+02 (1)	6.518E+02 (2)
$f24$	10	4.780E+02 (4)	4.099E+02 (1)	4.227E+02 (2)	4.235E+02 (3)
	30	1.226E+03 (4)	3.935E+02 (1)	4.017E+02 (2)	4.074E+02 (3)
	50	1.610E+03 (4)	5.409E+02 (1)	6.098E+02 (2)	6.148E+02 (3)

(*continued*)

Table 2. (*continued*)

fi	D	RI-SAPSO	RI-SAXPSO	XPSO	RI-AR-XPSO
f25	10	4.944E+02 (3)	5.395E+02 (4)	2.739E+02 (1)	2.889E+02 (2)
	30	4.173E+03 (4)	3.133E+03 (3)	7.849E+02 (2)	7.631E+02 (1)
	50	9.589E+03 (4)	5.184E+03 (3)	1.263E+03 (2)	1.231E+03 (1)
f26	10	4.284E+02 (4)	4.012E+02 (3)	3.953E+02 (1)	3.966E+02 (2)
	30	7.909E+02 (4)	5.799E+02 (3)	5.343E+02 (1)	5.378E+02 (2)
	50	1.559E+03 (4)	1.004E+03 (3)	7.729E+02 (2)	7.503E+02 (1)
f27	10	4.731E+02 (2)	5.396E+02 (4)	5.013E+02 (3)	3.500E+02 (1)
	30	9.498E+02 (4)	3.717E+02 (1)	3.898E+02 (3)	3.840E+02 (2)
	50	2.501E+03 (4)	4.929E+02 (1)	5.702E+02 (2)	5.806E+02 (3)
f28	10	3.754E+02 (4)	3.151E+02 (3)	2.680E+02 (2)	2.568E+02 (1)
	30	1.700E+03 (4)	1.262E+03 (3)	6.379E+02 (2)	5.936E+02 (1)
	50	3.069E+05 (4)	2.954E+03 (3)	1.149E+03 (2)	1.044E+03 (1)
f29	10	5.634E+05 (4)	4.440E+05 (3)	1.941E+05 (2)	1.303E+05 (1)
	30	4.570E+05 (4)	8.392E+03 (1)	1.176E+04 (3)	8.725E+03 (2)
	50	5.318E+08 (4)	1.172E+06 (1)	2.532E+06 (3)	2.344E+06 (2)

Table 2 shows average error rates accompanied by rank values (from 1 to 4). Overall, RI-AR-XPSO attained the lowest error rates on most optimization problems. Despite the simplicity of this evaluation, Table 3 presents the rank-sum stratified by D. It shows that the proposed implementations surpassed the original ones overall, although RI-SAXPSO did not overcome XPSO.

Table 4 presents average computation times and the NFEs of each algorithm stratified by dimensionality. One might notice how RI-SAPSO was the worst, with higher execution time and NFEs. Despite the high number of NFEs, employment of rotation matrix, diversity control, gradient information, and automatic hyper-parameter adjustment during the execution, RI-SAXPSO performed best overall in terms of time. However, it took longer than XPSO and RI-AR-XPSO for 50 dimensions, indicating that using gradient information in high dimensionalities results in lower efficiency. The performance of XPSO attests that using two exemplars, the ability to forget, and the velocity update equation postpones the execution. In spite of using these techniques, RI-AR-XPSO forgets particles with less frequency thanks to the other properties involved, consequently attaining similar performance to XPSO despite the presence of diversity control.

Table 3. Total sum of the rankings, detailed by dimensionality.

D	RI-SAPSO	RI-SAXPSO	XPSO	RI-AR-XPSO
10	105	82	58	45
30	116	67	61	46
50	116	68	57	49
Overall	337	217	176	140

Table 4. Time (in seconds) and NFEs.

	D	RI-SAPSO	RI-SAXPSO	XPSO	RI-AR-XPSO
Time	10	15.58	10.29	15.71	15.85
	30	19.82	15.59	18.31	17.67
	50	29.56	24.49	20.82	20.31
NFEs	10	5.911E+05	2.362E+05	1.144E+05	1.197E+05
	30	5.713E+05	3.647E+05	1.200E+05	1.200E+05
	50	5.695E+05	4.267E+05	1.200E+05	1.200E+05

In terms of success rate, the evaluated algorithms never reached the global optima when optimizing the evaluated functions for 30 and 50 dimensions. When $D = 10$, XPSO obtained a global success rate of 8.07%, followed by RI-AR-XPSO with 1.86%, RI-SAXPSO with 0.07%, and RI-SAPSO with 0%. Therefore, it can be argued that XPSO finishes its execution faster than RI-AR-XPSO thanks to its higher success rate for 10 dimensions.

5.3   Statistical Hypothesis Testing

This section describes the Friedman's statistical significance test, as well as the *post-hoc* tests performed to demonstrate which differences are statistically significant in terms of the error rates obtained in Table 2.

Friedman's test with significance level $\alpha = 0.05$ was applied for multiple comparisons. In this work, each pair of algorithms is analyzed $(k \times k)$ considering the four PSO implementations $(k = 4)$ and a family of different hypotheses $(h = k(k - 1)/2 = 6)$. The null hypothesis of this experiment is that there is no significant difference in the results. The *post-hoc* procedure was applied to find which algorithm pairs attained significantly different results, returning a p-value for each comparison to determine the degree of rejection of each hypothesis. Furthermore, the p-values are adjusted due to the Family-Wise Error Rate (FWER), which is the probability of making one or more false discoveries between all hypotheses by performing multiple pairwise tests [16].

To calculate Friedman's statistics, the results seen in Table 2 are the rankings, and Table 5 highlights the mean ranks of each PSO implementation obtained by applying a ranking-based classification. The results obtained attest that the best algorithm among the four implementations is RI-AR-XPSO. Upon application of Friedman's test, a p-value of $1.085E-32$ and a Friedman's statistic of 151.79 were found. Since $p \ll \alpha$, there seems to be at least one significant difference.

Table 5. Average ranks

Algorithm	Average rank	Order
RI-AR-XPSO	1,61	1
XPSO	2,02	2
RI-SAXPSO	2,49	3
RI-SAPSO	3,87	4

Table 6. Friedman's p-values and adjusted p-values (Holm) for pairwise comparison tests between all algorithms.

Hypothesis			p-value	adjusted p-value
RI-SAPSO	vs	RI-SAXPSO*	1.83E-12	7.33E-12
RI-SAPSO	vs	XPSO*	3.25E-21	1.63E-20
RI-SAXPSO	vs	XPSO*	0.016058	0.032115
RI-SAPSO	vs	RI-AR-XPSO*	5.98E-31	3.58E-30
RI-SAXPSO	vs	RI-AR-XPSO*	6.14E-06	1.84E-05
XPSO	vs	RI-AR-XPSO*	0.034516	0.034516

The evidence of statistically significant differences prompted the execution of *post-hoc* tests to pinpoint which algorithm pairs obtained significantly different results. p-values of all six hypotheses were obtained by converting the mean ranks of each algorithm by a normal approximation. z-values were used to find the corresponding p-value of the standard normal distribution $\mathcal{N}(0,1)$. Friedman's z-value can be defined as

$$z_{ij} = \frac{R_i - R_j}{\sqrt{\frac{k(k+1)}{6n}}}. \tag{12}$$

The p-values found using (12) are in Table 6. As previously explained, these values are not suitable for direct comparison with the significance levels due to FWER. Thus, the adjusted p-values were used to directly compare the values with the significance levels. In this paper, an adjust procedure termed Holm [3] was employed. In Table 6, * indicates which algorithm performed better. In harmony with the results previously obtained, RI-AR-XPSO obtained the best results, and RI-SAXPSO only outperformed RI-SAPSO. All p-values are smaller than 0.05, meaning that all differences are statistically significant.

6   Final Considerations

This work presented two new PSO implementations with conveniences such as automatic hyper-parameter update during execution and employment of a rotation matrix to promote independence from the coordinate systems wherein the objective function is defined. Both RI-AR-XPSO and RI-SAXPSO boast an attraction/repulsion mechanism guided by a diversity control whose effect is exponentially reduced throughout the search, resulting in more exploration at the beginning and more exploitation at the end. Both algorithms were compared against their original counterparts in a varied set of problems and dimensionalities, and statistical significance tests indicated their superiority against the original implementations. RI-SAXPSO, which uses gradient-based information, consumed less time in problems with smaller dimensionalities. In these cases and those where the objective function is not very costly, this technique might be convenient.

In future work, new arrangements can be tested to use the gradient information more efficiently since the gradient-based techniques did not reach good performance. New approaches that consider swarm diversity information to automatically adjust hyper-parameters during execution can also be studied. The goal would be to adapt the hyper-parameter fine-tuning according to current swarm diversity status instead of just verifying whether the search is stagnated.

References

1. Awad, N., Ali, M., Liang, J.J., Qu, B., Suganthan, P.: Problem definitions and evaluation criteria for the CEC 2017 special session and competition on single objective real-parameter numerical optimization. Technical report (2016)
2. Bonyadi, M.R., Michalewicz, Z.: A locally convergent rotationally invariant particle swarm optimization algorithm. Swarm Intell. **8**(3), 159–198 (2014)
3. Derrac, J., García, S., Molina, D., Herrera, F.: A practical tutorial on the use of nonparametric statistical tests as a methodology for comparing evolutionary and swarm intelligence algorithms. SEC **1**(1), 3–18 (2011)
4. Drake, J.H., Kheiri, A., Özcan, E., Burke, E.K.: Recent advances in selection hyper-heuristics. Eur. J. Oper. Res. **285**(2), 405–428 (2020)
5. Eberhart, R., Kennedy, J.: Particle swarm optimization. In: Proceedings of the IEEE International Conference on Neural Networks, vol. 4, pp. 1942–1948 (1995)
6. Han, F., Liu, Q.: A diversity-guided hybrid particle swarm optimization based on gradient search. Neurocomputing **137**, 234–240 (2014)
7. Hansen, N., Ros, R., Mauny, N., Schoenauer, M., Auger, A.: Impacts of invariance in search: when CMA-ES and PSO face ill-conditioned and non-separable problems. Appl. Soft Comput. **11**(8), 5755–5769 (2011)
8. Horn, R.A., Johnson, C.R.: Matrix Analysis. Cambridge University Press, Cambridge (2012)
9. Horst, R., Tuy, H.: Global optimization: deterministic approaches. SBH (1996)
10. Janson, S., Middendorf, M.: On trajectories of particles in PSO. In: 2007 IEEE Swarm Intelligence Symposium, pp. 150–155. IEEE (2007)
11. Liu, X.F., Zhan, Z.H., Gao, Y., Zhang, J., Kwong, S., Zhang, J.: Coevolutionary particle swarm optimization with bottleneck objective learning strategy for many-objective optimization. IEEE TEC **23**(4), 587–602 (2019)
12. Noel, M.M.: A new gradient based particle swarm optimization algorithm for accurate computation of global minimum. ASC **12**(1), 353–359 (2012)
13. Riget, J., Vesterstrøm, J.S.: A diversity-guided particle swarm optimizer-the ARPSO. Department of Computer Science, Univ. of Aarhus, Aarhus, Denmark, Technical report 2 (2002)

14. Santos, R., Borges, G., Santos, A., Silva, M., Sales, C., Costa, J.C.A.: A rotationally invariant semi-autonomous particle swarm optimizer with directional diversity. Swarm Evol. Comput. **56**, 100700 (2020)
15. Santos, R., Borges, G., Santos, A., Silva, M., Sales, C., Costa, J.C.: A semi-autonomous particle swarm optimizer based on gradient information and diversity control for global optimization. Appl. Soft Comput. **69**, 330–343 (2018)
16. Santos, R., Borges, G., Santos, A., Silva, M., Sales, C., Costa, J.C.: Empirical study on rotation and information exchange in particle swarm optimization. In: SEC (2019)
17. Shi, Y., Eberhart, R.: A modified particle swarm optimizer. In: 1998 IEEE International Conference on Evolutionary Computation Proceedings (1998)
18. Spears, W., Green, D., Spears, D.: Biases in particle swarm optimization. IJSIR **1**, 34–57 (2010)
19. Wilke, D.N., Kok, S., Groenwold, A.A.: Comparison of linear and classical velocity update rules in particle swarm optimization: notes on diversity. In: IJNME (2007)
20. Wilke, D.N., Kok, S., Groenwold, A.A.: Comparison of linear and classical velocity update rules in particle swarm optimization: notes on scale and frame invariance. Int. J. Numer. Meth. Eng. **70**(8), 985–1008 (2007)
21. Xia, X., et al.: An expanded particle swarm optimization based on multi-exemplar and forgetting ability. Inf. Sci. **508**, 105–120 (2020)

Improving Rule Based and Equivalent Decision Simplifications for Bloat Control in Genetic Programming Using a Dynamic Operator

Gustavo F. V. de Oliveira[✉] and Marcus H. S. Mendes

ICET, Universidade Federal de Viçosa, Florestal, Brazil
{gustavo.viegas,marcus.mendes}@ufv.br

Abstract. Bloat is a common issue regarding Genetic Programming (GP), specially noted in Symbolic Regression (SR) problems. Due to this, GP tends to generate a huge amount of ineffective code that could be avoided or removed. Code editing is one of many approaches to avoid bloat. The objective in this strategy is to mutate or remove subtrees which do not contribute to the final solution. Two known methods of redundant code removal, the Rule Based Simplification (RBS) and Equivalent Decision Simplification (EDS) are extended in a new operator presented in this paper, called Dynamic Operator with RBS and EDS (DORE). This operator gives the algebraic simplification table used by RBS the potential to learn from reductions performed by EDS. An initial benchmark highlighted how the RBS table can grow as much as 86% with DORE, and reducing the time spent on simplification by 16.83%. Experiments with the other three SR problems were performed showing a considerable improvement on fitness of the generated programs, besides a slight reduction in the population of the average tree size.

Keywords: Genetic Programming · Bloat control · Code editing

1 Introduction

Symbolic Regression (SR) is one of the main applications and motivators to Genetic Programming (GP), a method for automatically generate computer programs from a high level definition of a problem [15]. Since early 90ś, many data-driven problems are modeled as SR problems [2], where no previous knowledge or pre-processing input is required. GP is well suited for resolving such problems since any algebraic function set can be effectively represented as trees and implemented as computer programs for the problem domain [9].

Bloat - the uncontrolled and excessive growth of individuals without a proportional gain of fitness - is a well-known issue and a field of study in GP. It is specially noted in SR. The large amount of inefficient code causes excessive consumption of computational resources, as well as many other practical issues [16], hiding the problems real complexity and domain.

© Springer Nature Switzerland AG 2021
A. Britto and K. Valdivia Delgado (Eds.): BRACIS 2021, LNAI 13073, pp. 234–248, 2021.
https://doi.org/10.1007/978-3-030-91702-9_16

There are many approaches for avoiding uncontrolled code growth in tree-based GP [5], some of which are presented as follows. The most simple and popular of them is implementing a Depth Limit [9], although is not a truly effective approach, as it can induce growth in some scenarios. Parsimony Pressure [9, 13] is another popular technique which adds a penalty term in fitness function to punish large trees. Pseudo-hill Climbing [6] attempts to guarantee the fitness to improve in population rejecting individuals least fit than its parents until one is finally accepted. Code Editing approach mutates or removes redundant sub-trees in individuals.

Regarding the code editing approach, Koza [9] proposed a simple method to simplify (to convert a tree into a smaller, equivalent, tree) using grammar rewrite rules, which further inspired the Rule Based Simplification (RBS) method [7, 11, 20]. RBS was extended with Equivalent Decision Simplification (EDS) [11], which recursively compares all sub-trees in an individual for equivalency with a small set of terminals.

This paper proposes the Dynamic Operator with RBS and EDS (DORE), which improves a code editing bloat control algorithm using both RBS and EDS to maximize its reduction potential without greater punishments in execution time. The main feature of DORE is to dynamically learn redundant expressions to be applied with RBS from EDS outputs. It also optimizes the access of RBS rules using the *any* keyword in a hash-table implementation, as well as introduces a warm-up stage to grow RBS rules before the evolutionary process begins.

The article is organized as follows: Sect. 2 provides a quick overview about code editing operators; In Sect. 3 an improved strategy using the previous operators is proposed; Sect. 4 shows how RBS and EDS were implemented and the technical resources utilized in benchmarks; Sect. 5 shows a performance comparison between the simplification operators, beside the empirical results and discussion over three SR benchmark problems; and finally in Sect. 6, the conclusion is presented and the possibilities for further research.

2   Background

Redundancy is one of the key contributors to inefficient code growth. EDS [11] was introduced to extend RBS as they complement each other. RBS removes redundant subtrees by replacing a tree for a smaller equivalent one by applying arithmetic rules such as $X/1 \rightarrow X$ and $0*X \rightarrow 0$. These rules must be known and provided before the evolutionary process starts, thus each rule must be explicitly specified. Another limitation by RBS is that it's rules must be specified exactly like it would appear in an individual. For example, the rule $X*0 \rightarrow 0$ would not be sufficient to simplify a tree $0*X$ unless the rule $0*X \rightarrow 0$ is specified.

EDS extends RBS in a manner that it simplifies trees without previous knowledge of algebraic rules. It can also remove redundancies that are only true in the training domain. The simplification by EDS is made by evaluating each subtree in an individual and comparing these subtrees with a set of small trees or terminals which is usually the result of simplifications, such as X, 1 and 0. In a SR problem, EDS is evaluated as follows [11]:

1. Determine a suitable set of simple trees S_{simple}.
2. Check all subtrees in the target for equivalence to a tree in S_{simple}.
3. If some subtree is equivalent to a tree in S_{simple}, and larger than it, replace that subtree with the simple tree.
4. Repeat this procedure recursively until it fails.

Finding a suitable set of trees S_{simple} for a problem is not an easy task. Usually the set of terminals is a natural fit for S_{simple}, but if it is already known that some subtrees must appear in the final output, such as in trigonometrical problems, they can be inserted in this set. However, the main issue of using EDS as a single operator of bloat control is the computational performance. With problems hard enough and individuals big enough, reducing a single generation can take as long as multiple generational evaluations. This scenario is not uncommon in simple SR problems. If an evaluated individual with EDS contains more subtrees than the population size, the evaluation function will be called upon as much as in the generational evaluation.

2.1 Simplification with RBS and EDS

The flow of simplification using both RBS and EDS is as follows [11]:

1. Let the genotype tree of individual i be t_i.
2. Apply RBS recursively to all nodes of t_i, until there is no node to which RBS can be applied, obtaining t'_i.
3. Apply EDS to all nodes of t'_i. If any node is translated, let the translated tree be t_i and go to (2). If there is no node to which EDS can be applied, finish and let t'_i be the final result.

Applying RBS before applying EDS is a good idea since it prevents EDS to be used for simplifying already known rules. Also, RBS execution time is lower than EDS, which will be explored further in this paper.

3 Proposed Improvements

The main idea of the improvements to the simplification with RBS and EDS flow is to increase RBS rules table R dynamically as soon as new rules, or more efficient ones, are discovered. This allows RBS to execute independently with no large impacts in execution time. A more robust RBS operator also helps in simplification itself, since any simplification made by EDS triggers new RBS calls, as well as new EDS calls. If simplifications made by EDS occurred in the first step of the simplification flow, then it would eliminate the need to reevaluate subtrees that would only be reduced by EDS.

The first improvement proposed is to allow the simplification rules table R, used by RBS, to have rules inserted in execution time. Each time a subtree is simplified by RBS using the keyword $\_ANY\_$, which denotes any subtree, a new rule with the original subtree as input is inserted in R. Afterwards, new calls to

RBS with the same subtree as input have $O(1)$ access guaranteed in a hash-table implementation.

Analogously, each simplification made with EDS creates a new rule in R table with the learned simplification. This way, there is no need to apply simplification with EDS to the same subtree in subsequent individuals. Algorithm 1 shows an example algorithm to simplify a generation using this improved simplification flow with RBS and EBS in a SR problem.

Another strategy adopted to grow even more the R table was the insertion of a warm-up step before the evolutionary process begins. This warm-up step consists of generating random trees in the training domain of a problem, with random sizes, and then simplifying them with the improved flow. Each successful reduction creates a new rule in the table, so the GP starts with a larger R table.

Algorithm 1: Example of a generation simplification using improved RBS and EDS flow

Data: P: generation population; R: RBS rules table
1 population$_{simplified}$ ← {}
2 **foreach** *individual I in P* **do**
3   **foreach** *subtree $I_{subtree}$ in I* **do**
4   subtree$_{simpl}$, simplification ← call RBS with $I_{subtree}$
5   **if** *simplification exists and not in R* **then** add simplification to R ;
6   subtree$_{simpl}$, simplification ← call EDS with subtree$_{simpl}$
7   **if** *simplification exists* **then**
8   **if** *simplification not in R* **then** add simplification to R ;
9   go back to the beginning of current loop
10   **end**
11       $I_{subtree}$ ← subtree$_{simpl}$
12   **end**
13   add I to population$_{simplified}$
14 **end**
15 **return** population$_{simplified}$

---

4 Methodology

All algorithms were implemented in Python 3.8.3 with the package DEAP [4] (Distributed Evolutionary Algorithm in Python) in version 1.3.1. Parallelism was used to evaluate fitness functions using the package Ray [12] in version 0.8.6. The benchmarks were conducted in a 2018 MacBook Pro with 2.2 GHz 6-Core Intel Core i7 and 16 GB 2400 MHz DDR4 memory.

4.1 Rule Based Simplification

The RBS simplification rules table R was implemented as a hash table. The keyword $\_ANY\_$ denotes any subtree and allows rules such $\_ANY\_ * 0 \rightarrow 0$ to be defined. In the original paper [11] this keyword was referenced as A and no further details were provided on how it was implemented. In this paper, every subtree evaluated by RBS consults the hash table directly and only if no rule with the subtree is described the _ANY_ rules are consulted. The access to the table R has complexity $O(1)$, except on the scenario where the _ANY_ keyword is included in the rule body, which makes the complexity $O(N)$ where N is the amount of rules in the table.

Reducing the amount of rules with _ANY_ is a slight improvement to this implementation as it maximizes the rate of $O(1)$ accesses. Another way to attain this goal is to define as many redundant rules as possible in R, for example prioritizing the definition of rules $1 * 0 \rightarrow 0$ and $X * 0 \rightarrow 0$ instead of _ANY_ * $0 \rightarrow 0$. This task can be hard when there is no previous knowledge of such simplifications in the domain of the problem.

4.2 Equivalent Decision Simplification

The subtrees set S_{simple} for comparison for equivalence in EDS was implemented as a simple list. Two subtrees is considered equivalent if all fitness values elapsed from this subtree in the domain has relative error $\epsilon < 0.005$. This threshold was chosen after some tests and set to all benchmark problems in both compared operators in this paper.

5 Experimental Results

This section is organized as follows: Sect. 5.1 shows a preliminary benchmark, validating the performance improvement on each simplification method; It is presented in Sect. 5.2 the results for three artificial SR problems using the proposed operator, comparing it with the original operator and baseline GP.

5.1 Simplification Methods Performance

A simple benchmark - using $cos(2\pi x)$ as objective function - was ran 50 times to compare the execution time of the five simplification strategies.

Experimental Setting. The experiment was the simplification of many independent individuals in each reduction strategy: traditional versions of RBS and EDS, RBS and EDS with dynamic rules allowed and RBS after the warm-up step. More details of each strategy were discussed in Sect. 3. The fitness function was defined as the RMSE (Root Mean Squared Error), of 20 uniform points in the interval $[-\pi, \pi]$. The function set used was $\{+. - .*, \div\}$, where $\div$ is the protected division satisfying $X/0 \rightarrow 1$. The terminal set is $\{X, 0, 1, \pi\}$. For each execution, 200 trees are generated and then cloned in each step to avoid any bias or propagation of simplifications. The reductions applied in one step are not carried on to the next. The procedure is described as:

1. Reduction with RBS is applied to the 200 cloned individuals.
2. Reduction with EDS is applied to the 200 cloned individuals.
3. With the dynamic insertions in the rules table R allowed, reduction with RBS is applied to the 200 cloned individuals.
4. With the dynamic insertions in the rules table R allowed, reduction with EDS is applied to the 200 cloned individuals.
5. With the dynamic insertions in the rules table R allowed, a warm-up step is done, and reduction with EDS is applied to the 200 cloned individuals.

The 200 individuals are generated in each of 50 executions with the method **Ramped Half and Half** and with depth limited to [5, 10]. The warm-up step applied RBS and EDS sequentially to 200 trees generated with **Grow** method and with depth limited to [1, 5]. The initial RBS rules table R is defined in Table 1.

Table 1. Initial simplification rules table used by RBS

$$\_ANY\_ + 0 \to \_ANY\_, \_ANY\_ + 0 \to \_ANY\_, \_ANY\_ * 1 \to \_ANY\_,$$
$$1 * \_ANY\_ \to \_ANY\_, \_ANY\_ * 0 \to 0, 0 * \_ANY\_ \to 0,$$
$$\_ANY\_ - \_ANY\_ \to 0, 1 - 1 \to 0, 0 - 0 \to 0, 0 + 0 \to 0,$$
$$\_ANY\_ - 0 \to \_ANY\_, 1 \div 0 \to 1, 0 \div 1 \to 0, 0 \div 0 \to 1, 1 \div 1 \to 1,$$
$$\_ANY\_ \div 0 \to 1, 0 \div \_ANY\_ \to 0, \_ANY\_ \div \_ANY\_ \to 1,$$
$$\_ANY\_ \div 1 \to \_ANY\_$$

Results and Discussion. Table 2 shows the arithmetic mean of the 50 executions in each strategy. Dynamic RBS is 4.84% faster than the default RBS. The RBS after the warm-up is shown to be 16.83% faster than the traditional RBS and provided a 11.52% improvement to dynamic RBS before the warm-up.

Table 2. Execution time comparison of the simplification operators

Simplification strategy	RBS	EDS	Dynamic RBS	Dynamic EDS	Dynamic RBS after warm-up
Avg. execution time (s)	5.49	11.48	5.23	9.61	4.69

This benchmark indicates that not only a bigger R table can improve the simplifications performance but also how much slower EDS is compared to RBS as well. With no improvements, EDS is 2.09 times slower than RBS. Dynamic EDS is 1.84 times slower than dynamic RBS before warm-up. Dynamic RBS after the warm-up can be twice as fast as the dynamic EDS. The warm-up increased the rules of table R, in average, from 19 entries to 135, to an around 86% improvement.

5.2   Symbolic Regression Problems

To validate the proposed improvements, 3 simple artificial SR problems were tested. These benchmarks were extracted in [5], selected from [8,10,17–19]. Since this paper presents a preliminary analysis of the operator, the benchmark problems are rather simple and harder problems will be explored in future works. The problems are presented in Table 3.

Experimental Setting. For each problem, 50 independent executions were performed, in 3 different approaches: **Base GP**: with no code editing bloat control operator, **Baseline Operator**: with the original simplification flow of RBS and EDS as in [11], and **DORE**: using the proposed dynamic operator.

Table 3. Artificial symbolic regression problems

Benchmark	Objective function	Function formula	Domain - Training
1	$f_1(x_1, x_2, x_3, x_4, x_5)$	$\frac{10}{5+\sum_{i=1}^{5}(x_i-3)^2}$	50 random points. $x_i \in [0,6]$
2	$f_2(x_1, x_2)$	$\frac{(x_1-3)^4+(x_2-3)^3+(x2-3)}{(x_2-2)^4+10}$	50 random points. $x_i \in [0,6]$
3	$f_3(x)$	$0, 3x sen(2\pi x)$	40 random points. $x \in [-2,2]$

The function set to all benchmarks is set to $\mathcal{F} = \{+, -, *, \div\}$ where $\div$ is the protected division satisfying $X/0 \to 1$. The terminal set to benchmarks 1 and 2 is $\tau = \{0, 1\}$ and the decision variables. For benchmark 3, the terminal set is $\tau = \{0, 1, \pi\}$ and the decision variable. To all benchmarks, the fitness function is the RMSE of the training domain points.

Besides the traditional GP parameters, two new parameters were introduced to fit the operator's context in more computational resource expansive problems. The first one is the percentage of population ρ which passes through the full simplification flow. The algorithm is developed in such way that the best $\rho\%$ individuals of the population are chosen each generation. The second additional parameter is the remaining best γ percentage of the population that was not previously chosen and passes through the reduction with RBS only. For example, if the population size is 100, and the values $\rho = 40\%$ and $\gamma = 20\%$, the 40 most fit individuals would pass through the complete reduction flow and the remaining best 20 trees would be reduced with RBS only.

These additional parameters are needed due the high computational cost of the reductions that would make unfeasible the operation of all individuals without a great impact in execution time. A large γ value does not have a high impact on the execution time, since RBS reduction is simpler. Nonetheless, the ρ value greatly impacts the total execution time. For the benchmarks these parameters were chosen empirically, testing values big enough that would not make the total GP execution time significantly slower than the Base GP.

A custom tournament selection, named as Partial Tournament, is used to reduce the number of repeated individuals in the crossover. Being P the population, the Partial Tournament does $|P|$ traditional K size tournaments, but limits the repeated number of each individual to two. Then, if necessary, the population is completed with individuals chosen at random up to $|P|$.

Table 4. GP settings used in symbolic regression benchmarks

Number of runs	50
Generations per run	50
Initialisation	Ramped half-and-half
Population size	200
Selection	Non-elitist Partial Tournament Selection, with $K = 5$
Crossover	*One Point Crossover* with 90% probability
Mutation	Uniform subtree mutation with 5% probability
Tree depth limit	Initial limit = 6; Subsequent limit = 17
Reduction threshold ρ	40%
Reduction threshold γ	60%

The GP settings used is presented in Table 4. The simplification rules used in RBS are as described in Table 1. The simple subtrees set S_{simple} used in EDS is the terminal set τ for each problem. The warm-up step used in DORE runs is defined as 200 full trees with depth limited in $[1, 5]$ range.

Results and Discussion. The median was preferred over the mean in this section since it is less sensitive to outliers and the data is not guaranteed to follow a normal distribution. In all of the data in any graph or table the median of the 50 executions is shown.

It was first analyzed how the fitness behaved over generations. Figure 1 shows how Base GP is worse than the others in all benchmarks. It also presents how Base GP fitness curve tends to flatten in each benchmark. DORE has slightly better results than the baseline operator. Figure 2 shows a fitness boxplot of the best individuals in each generation.

The size of individuals is a crucial concern to have bloat-controlled GP executions, beside the fitness stagnation. A tree depth limit was used, so it was not expected a completely uncontrolled growth on Base GP. As Fig. 3 shows, both operators had a much better size control on the population over generations than Base GP. The difference between DORE and baseline operator was tiny, with DORE having best results on the first two benchmarks.

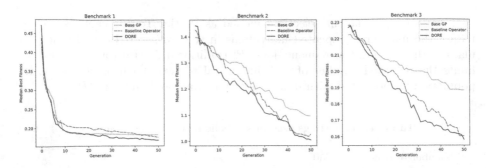

Fig. 1. Best fitness versus generations, for all benchmark problems

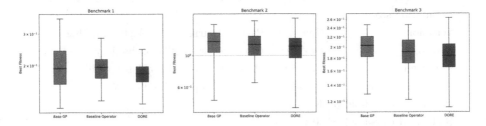

Fig. 2. Boxplots of the best of generation individuals, for all benchmark problems

Table 5 shows the validation and test domains used to analyze the best individuals from each generation for all benchmark problems. It is presented in Table 6 the median and median average deviation (MAD), as utilized in [1], of each set of points - training, validation, and test - using the RMSE as reference. Table 7 shows the same metrics for the tree size of these best individuals. As the data is not normally distributed, Table 6 and Table 7 also present the p-value of the Mann-Whitney U-test [3] considering DORE x Base GP and DORE x Baseline Operator datasets. The null hypothesis is that the distributions of both datasets are equal. DORE performed better regarding tree size in all benchmark problems, compared to the other two strategies. It also presented better results in training error in all benchmarks. The benchmark 3 test error was worse using an operator, which suggests that this kind of strategy is not well suited for more complex problems even though the generated trees were smaller.

Besides the best trees found for each generation, the best individuals of each run were also analyzed. Figure 4 shows the mean training fitness against the mean tree sizes. It is clear that DORE dominates the other strategies in all 3 benchmark problems regarding the fitness.

In benchmark 1, due to bloat, Base GP generates smaller but unfit individuals. These individuals are, in average, from earlier generations than the ones using code editing operators, see Fig. 5. The boxplot shows how Base GP converges earlier than the other two strategies. Thus, it is expected that trees generated in later generations have better fitness but also a larger size. DORE performed better than the other strategies in Benchmarks 2 and 3 and regarding both size

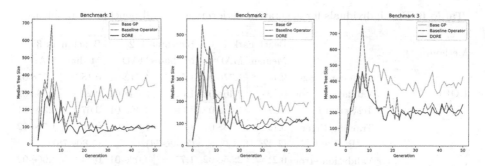

Fig. 3. Tree median size versus generation

Table 5. Validation and test domain for best of run individuals analysis

Benchmark	Validation domain	Test domain
1	100 random points $x_i \in [0,6]$	500 random points $x_i \in [0,6]$
2	100 random points $x_i \in [0,6]$	1156 points $x_1, x_2 \in (-0,25:0,2:6,35)$
3	50 random points $x \in [-2,2]$	2000 points $x \in (-2:0,001:2)$

and fitness, as shown in Table 6 and Table 7. In all benchmarks it is noted that the average size of the best individuals in DORE are better than the ones with the baseline operator, see Table 7.

Another point of interest analyzed was the reductions made between baseline operator and DORE. Each RBS and EDS simplification made was logged in each run in these strategies. The tables and graphs related to reductions used the arithmetic mean as an average value. Table 8 highlights how DORE increased the number of unique simplifications made by RBS. This result was expected since DORE tends to increase the reductions table used by RBS.

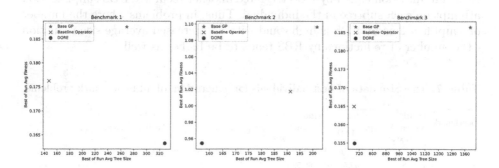

Fig. 4. Best of run average fitness against best of run average size

Table 6. Best individuals for generation performance, of all benchmark problems

Method		Benchmark 1		Benchmark 2		Benchmark 3	
		Median	MAD	Median	MAD	Median	MAD
DORE	Training Error	0.18	1.77e-02	1.16	1.71e-01	0.18	1.95e-02
	Validation error	0.22	2.46e-02	1.62	3.94e-01	0.26	4.58e-02
	Test error	0.21	1.48e-02	1.09	4.49e-01	0.67	4.39e-01
Base GP	Training Error (p-value)	0.19 (1.42e-16)	4.03e-02	1.26 (1.49e-34)	1.93e-01	0.20 (5.80e-78)	1.79e-02
	Validation error (p-value)	0.21 (1.28e-12)	2.55e-02	1.75 (1.73e-27)	4.02e-01	0.24 (5.23e-23)	1.96e-02
	Test error (p-value)	0.21 (4.12e-11)	1.12e-02	1.59 (7.96e-09)	8.89e-01	0.25 (3.78e-96)	9.80e-03
Baseline operator	Training error (p-value)	0.20 (2.09e-34)	2.38e-02	1.20 (1.19e-07)	1.79e-01	0.19 (1.10e-16)	2.10e-02
	Validation error (p-value)	0.22 (5.81e-05)	2.25e-02	1.62 (1.16e-01)	3.72e-01	0.25 (3.94e-14)	2.85e-02
	Test error (p-value)	0.21 (1.84e-04)	1.36e-02	1.24 (5.31e-02)	7.44e-01	0.33 (6.43e-32)	1.07e-01

The average value of total reductions made by RBS and EDS is shown in Table 9. These graphs present different scenarios. In benchmark 1, the total reductions made by both RBS and EDS in DORE was greater than in baseline operator. In benchmark 2, the reductions by RBS were greater in DORE but lesser by EDS. Finally, in benchmark 3, the total reductions made by both RBS and EDS was greater in the baseline operator. These different scenarios are directly related to the nature of each benchmark problem. Also, although the RBS reduction table in DORE had more entries than the baseline table, each simplification called by EDS triggered another recursive RBS simplification attempt on each subtree in the individual. Thus, in problems where the number of simplifications by EDS is high - and with an increased average size reduction - the number of reductions by RBS tends to be higher as well.

Table 7. Tree Size data of best individuals for generation, of all benchmark problems

Benchmark	DORE		Base GP			Baseline operator		
	Median of Avg. tree size	MAD	Median of Avg. tree size	MAD	p-value	Median of Avg. tree size	MAD	p-value
1	94.55	6.11e+01	282.19	2.01e+02	5.61e-128	103.26	7.04e+01	1.37e-03
2	101.88	5.55e+01	196.58	1.21e+02	2.09e-87	119.84	7.96e+01	4.53e-10
3	203.80	1.34e+02	367.09	2.51e+02	8.29e-62	229.19	1.61e+02	1.30e-04

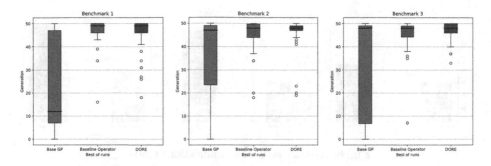

Fig. 5. Best of run generation boxplot

Table 8. Average value of unique RBS simplifications performed

Method	Benchmark 1	Benchmark 2	Benchmark 3
Baseline operator	866.34	756.1	1735.98
DORE	1667.04	1330.9	2066.62

Even though each benchmark presented a different reduction scenario, a pattern in the ratio between reductions by RBS and reductions by EDS could be noted, as is shown in Fig. 6. In all benchmark problems, except the proportion, the total amount of RBS reductions relative to the total amount of EDS reductions is bigger. It can also be noted that this ratio is greater in DORE than in the baseline operator, which is exactly the main goal of the proposed improvements.

The impact of the dynamically learned rules was analyzed and shown in Fig. 7. The graphs are histograms of the total amount of reductions made by RBS in each subtree size reduction percentage interval. The values highlighted in purple are from rules that could only be learned with DORE improvements. The average size reduction of these dynamically learned rules in each benchmark were 86.01%, 88.01% and 85.40%, respectively, compared to the average size reduction of 59.38%, 55.98% and 52.74% from the predefined rules. DORE not only provided more reductions to be applied but also better ones.

Table 9. Average total simplifications made by RBS and EDS

Method	Avg. Total RBS simplifications			Avg. Total EDS simplifications		
	Benchmark 1	Benchmark 2	Benchmark 3	Benchmark 1	Benchmark 2	Benchmark 3
Baseline Op	8060.32	12569.58	67622.52	104.8	165.92	680.82
DORE	16100.26	23830.96	39455.28	145.28	137.94	214.92

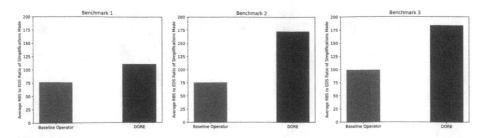

Fig. 6. Average ratio of simplifications made

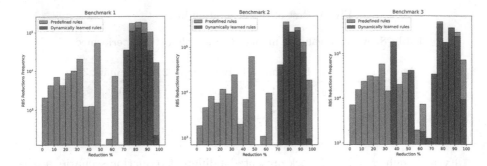

Fig. 7. RBS simplification frequency against size reduction percentage

Finally the execution time was analyzed, an important point of interest in this work as DORE execution time should be close to the baseline operator. Average execution time - using the arithmetic mean - for all benchmark problems is shown in Table 10. Since the values of parameters ρ and γ were defined empirically to make execution time between baseline operator and DORE close, the values presented are expected. However, the difference between Base GP and executions with operators is notable. Using RBS and EDS operators as a single bloat control method seems to be not adequate in hard or complex problems as the execution time tends to increase even more due to all the fitness function evaluation performed recursively in EDS step of the simplification flow.

Table 10. Average total execution time (in seconds) of each strategy, for all benchmark problems

Benchmark	Average execution time (s)		
	Base GP	Baseline operator	DORE
1	180.36	292.23	321.12
2	167.31	276.08	277.59
3	206.10	585.60	514.68

6   Conclusion and Future Work

This paper presented improvements to a simplification flow introduced in [11] to make it more reasonable to be applied in complex problems. Besides a meta-learning reduction, two parameters were introduced to optimize the number of trees to be reduced by RBS and EDS in harder problems than the one presented in the original work, without greater impact in execution time.

The resulting operator was able to improve the execution time performance of reduction by RBS, as well as the relative frequency of this reduction in the simplification flow, even with a limit of trees applied. A greater percentage of the population could pass through reduction by RBS with no significant penalty in execution time.

The results of the three artificial SR problems benchmark highlighted how DORE improved the fitness and slightly the tree size in the population during the evolutionary process. The best individuals in these experimental runs had significantly better results in both fitness and tree size using the dynamic operator.

It is noted that DORE is helpful to control bloat in GP but could be challenging to adopt in more complex problems due the computational burden. This issue should be addressed in future research. Probably tuning the parameters ρ and γ may be needed as well as limiting the size of the subtrees in S_{simple}.

Future work could measure the limits of the list S_{simple} and a robust strategy to handle random ephemeral constants, which was not explored in this paper. Parallelization of the reductions could improve the operator performance, making it more viable, although is not a trivial problem as the rule table is constantly updated by each EDS reduction performed. It would also be interesting to apply this operator with other bloat control methods, such as a equalization operator [14], using it as a tool to optimize a small set of individuals in a population.

References

1. Castelli, M., Manzoni, L., Mariot, L., Saletta, M.: Extending local search in geometric semantic genetic programming, pp. 775–787, August 2019. https://doi.org/10.1007/978-3-030-30241-2_64
2. Chen, C., Luo, C., Jiang, Z.: Block building programming for symbolic regression. Neurocomputing **275**, 1973–1980 (2018). https://doi.org/10.1016/j.neucom.2017.10.047
3. Fay, M.P., Proschan, M.A.: Wilcoxon-Mann-Whitney or t-test? On assumptions for hypothesis tests and multiple interpretations of decision rules. Stat. Surv. **4**, 1 (2010)
4. Fortin, F.A., De Rainville, F.M., Gardner, M.A., Parizeau, M., Gagné, C.: DEAP: evolutionary algorithms made easy. J. Mach. Learn. Res. **13**, 2171–2175 (2012)
5. Haeri, M.A., Ebadzadeh, M.M., Folino, G.: Statistical genetic programming for symbolic regression. Appl. Soft Comput. **60**, 447–469 (2017)
6. Hagiwara, M.: Pseudo-hill climbing genetic algorithm (PHGA) for function optimization. In: Proceedings of 1993 International Conference on Neural Networks (IJCNN-93-Nagoya, Japan), vol. 1, pp. 713–716 (1993). https://doi.org/10.1109/IJCNN.1993.714013

7. Hooper, D.C., Flann, N.S.: Improving the accuracy and robustness of genetic programming through expression simplification. In: Proceedings of the 1st Annual Conference on Genetic Programming, p. 428. MIT Press, Cambridge (1996)

8. Keijzer, M.: Improving symbolic regression with interval arithmetic and linear scaling. In: Ryan, C., Soule, T., Keijzer, M., Tsang, E., Poli, R., Costa, E. (eds.) EuroGP 2003. LNCS, vol. 2610, pp. 70–82. Springer, Heidelberg (2003). https://doi.org/10.1007/3-540-36599-0_7

9. Koza, J.R., Koza, J.R.: Genetic Programming: on the Programming of Computers by Means of Natural Selection, vol. 1. MIT press, Cambridge (1992)

10. McDermott, J., et al.: Genetic programming needs better benchmarks. In: Proceedings of the 14th Annual Conference on Genetic and Evolutionary Computation, pp. 791–798 (2012)

11. Naoki, M., McKay, B., Xuan, N., Daryl, E., Takeuchi, S.: A new method for simplifying algebraic expressions in genetic programming called equivalent decision simplification. In: Omatu, S., Rocha, M.P., Bravo, J., Fernández, F., Corchado, E., Bustillo, A., Corchado, J.M. (eds.) IWANN 2009. LNCS, vol. 5518, pp. 171–178. Springer, Heidelberg (2009). https://doi.org/10.1007/978-3-642-02481-8_24

12. Moritz, P., et al.: Ray: a distributed framework for emerging AI applications. CoRR (2017). http://arxiv.org/abs/1712.05889

13. Poli, R., McPhee, N.F.: Parsimony pressure made easy: solving the problem of bloat in GP. In: Borenstein, Y., Moraglio, A. (eds.) Theory and Principled Methods for the Design of Metaheuristics. NCS, pp. 181–204. Springer, Heidelberg (2014). https://doi.org/10.1007/978-3-642-33206-7_9

14. Silva, S., Dignum, S., Vanneschi, L.: Operator equalisation for bloat free genetic programming and a survey of bloat control methods. Genetic Program. Evol. Mach. 13, 197–238 (2012). https://doi.org/10.1007/s10710-011-9150-5

15. Sivanandam, S., Deepa, S.: Genetic Programming, pp. 131–163. Springer, Heidelberg (2008). https://doi.org/10.1007/978-3-540-73190-0_6

16. Trujillo, L., Muñoz, L., Galván-López, E., Silva, S.: Neat genetic programming: controlling bloat naturally. Inf. Sci. 333, 21–43 (2015). https://doi.org/10.1016/j.ins.2015.11.010

17. Uy, N.Q., Hien, N.T., Hoai, N.X., O'Neill, M.: Improving the generalisation ability of genetic programming with semantic similarity based crossover. In: Esparcia-Alcázar, A.I., Ekárt, A., Silva, S., Dignum, S., Uyar, A.Ş (eds.) EuroGP 2010. LNCS, vol. 6021, pp. 184–195. Springer, Heidelberg (2010). https://doi.org/10.1007/978-3-642-12148-7_16

18. Uy, N.Q., Hoai, N.X., O'Neill, M., McKay, R.I., Galván-López, E.: Semantically-based crossover in genetic programming: application to real-valued symbolic regression. Genet. Program Evolvable Mach. 12(2), 91–119 (2011)

19. Vladislavleva, E.J., Smits, G.F., Den Hertog, D.: Order of nonlinearity as a complexity measure for models generated by symbolic regression via pareto genetic programming. IEEE Trans. Evol. Comput. 13(2), 333–349 (2008)

20. Wong, P., Zhang, M.: Algebraic simplification of gp programs during evolution. In: Proceedings of the 8th Annual Conference on Genetic and Evolutionary Computation, pp. 927–934, GECCO 2006. Association for Computing Machinery, New York, NY, USA (2006). https://doi.org/10.1145/1143997.1144156

Lackadaisical Quantum Walk in the Hypercube to Search for Multiple Marked Vertices

Luciano S. de Souza[1]([✉])[ID], Jonathan H. A. de Carvalho[2][ID],
and Tiago A. E. Ferreira[1][ID]

[1] Departamento de Estatística e Informática, Universidade Federal
Rural de Pernambuco, Recife, Brazil
{luciano.serafim,tiago.espinola}@ufrpe.br
[2] Centro de Informática, Universidade Federal de Pernambuco, Recife, Brazil
jhac@cin.ufpe.br

Abstract. Adding self-loops at each vertex of a graph improves the performance of quantum walks algorithms over loopless algorithms. Many works approach quantum walks to search for a single marked vertex. In this article, we experimentally address several problems related to quantum walk in the hypercube with self-loops to search for multiple marked vertices. We first investigate the quantum walk in the loopless hypercube. We saw that neighbor vertices are also amplified and that approximately $1/2$ of the system energy is concentrated in them. We show that the optimal value of l for a single marked vertex is not optimal for multiple marked vertices. We define a new value of $l = (n/N) \cdot k$ to search multiple marked vertices. Next, we use this new value of l found to analyze the search for multiple marked vertices non-adjacent and show that the probability of success is close to 1. We also use the new value of l found to analyze the search for several marked vertices that are adjacent and show that the probability of success is directly proportional to the density of marked vertices in the neighborhood. We also show that, in the case where neighbors are marked, if there is at least one non-adjacent marked vertex, the probability of success increases to close to 1. The results found show that the self-loop value for the quantum walk in the hypercube to search for several marked vertices is $l = (n/N) \cdot k$.

Keywords: Quantum computing · Quantum walk · Quantum search algorithm

Acknowledgments to the Science and Technology Support Foundation of Pernambuco (FACEPE) Brazil, Brazilian National Council for Scientific and Technological Development (CNPq), and Coordenação de Aperfeiçoamento de Pessoal de Nível Superior - Brasil (CAPES) - Finance Code 001 by their financial support to the development of this research.

© Springer Nature Switzerland AG 2021
A. Britto and K. Valdivia Delgado (Eds.): BRACIS 2021, LNAI 13073, pp. 249–263, 2021.
https://doi.org/10.1007/978-3-030-91702-9_17

1   Introduction

According to Shenvi, Kempe, and Whaley [17], quantum walks provide one of the most promising features, an intuitive framework for building new quantum algorithms. They were pioneers in designing a quantum search algorithm on the hypercube based on quantum random walks [14]. Recent works have used the quantum walks to search weights and train artificial neural networks [19,20].

The topology of the structure where the walk is applied considerably affects the evolution of the walker [22]. Therefore, many works are developed to improve the performance of quantum walks, quantum search algorithms in different structures: one-dimensional, two-dimensional, and multidimensional grids, complete and bipartite graphs, among others [4,5,11,15].

Quantum walk modification proposals are also made to improve their performance. For example, Wong [24] added to each vertex of a two-dimensional grid a self-loop, so the walker has some probability of staying put, achieving an improvement over the algorithm without self-loop [3].

Rhodes [16] proposed an ideal weight for all vertex-transitive graphs with a single marked vertex such that the ideal self-loop weight is equal to the degree of the loopless graph divided by the total number of vertices. Potovcek [14] observed that the nearest neighbors are also presented with high probability and Nahimovs [10] that adjacent vertices can be hard to find by quantum walks.

In this way, we investigate whether the optimal value of $l = (d/N)$ for a single marked vertex is optimal for multiple marked vertices, where d is the degree of the loopless vertex and N is the number of vertices. We analyzed the quantum walk on hypercube without self-loop and with self-loop. We analyzed the quantum walk on the hypercube for multiple marked adjacent and non-adjacent vertices. Finally, we find an optimal value of l for a quantum walk in the hypercube with multiple marked vertices.

This paper is organized as follows. In Sect. 2, we present some concepts about quantum walks and specifically the quantum walk on the hypercube. In Sect. 3, we characterize the probability distribution along with the space, adjust the self-loop weight for multiple marked vertices, and search for adjacent marked vertices. Finally, in Sect. 4 is the conclusion.

2   Quantum Walk

The processing of quantum information is governed by quantum mechanics or quantum physics [18]. Quantum computing study the processing of this information [8,13,25]. Quantum walks are the quantum counterpart of classical random walks. Discrete and continuous-time quantum walks are the advanced tools used to build quantum algorithms [1,2]. The main feature that differentiates these two types of quantum walks is the timing used to applying the evolution operators. In the quantum walk in continuous time, the evolution operator is applied at any time, whereas the quantum walks in discrete time, the evolution operator is applied in discrete time steps [21]. The quantum walk evolution in the discrete-time process occurs by the successive applications of a unitary evolution operator U that acts on the Hilbert space

$$\mathcal{H} = \mathcal{H}^C \otimes \mathcal{H}^S.$$

The coin space $\mathcal{H}^C$ is the Hilbert space associated with a quantum coin, and the walker's space $\mathcal{H}^S$ is the Hilbert space associated with the position of the nodes in a graph, for example. The evolution operator U is defined in Eq. 1.

$$U = S(C \otimes I_N) \tag{1}$$

where, S is the shift operator, i.e., a permutation matrix that acts in the walker's space based on the state of the coin space. The unitary matrix C is the coin operator [17]. Therefore, the equation of evolution represented by a quantum walk at time t is given by

$$|\Psi(t)\rangle = U^t |\Psi(0)\rangle$$

.

2.1   Quantum Walk on the Hypercube

According to Venegas [21], the hypercube is defined as an undirected graph of degree n and $N = 2^n$ nodes. Each node is represented by an n-bit binary string. Two nodes $\vec{x}$ and $\vec{y}$ are connected by an edge if the Hamming distance between them is 1, i.e., $|\vec{x} - \vec{y}| = 1$. This means that $\vec{x}$ and $\vec{y}$ only differ in a single bit. The expression $|\vec{x}|$ is the Hamming weight of $\vec{x}$. The Hilbert space associated with the quantum walk on the hypercube is

$$\mathcal{H} = \mathcal{H}^n \otimes \mathcal{H}^{2^n},$$

where $\mathcal{H}^n$ is the Hilbert space associated with the quantum coin space, and $\mathcal{H}^{2^n}$ is the Hilbert space associated with nodes on the hypercube.

According to Shenvi [17], in a d-dimensional hypercube, the d directions specify the coin state. Kempe [7] defines that directions can be labeled by the n base-vectors $\{|0\rangle, |1\rangle, \ldots, |n-1\rangle\}$ on the hypercube which corresponding to the n vectors of Hamming weight 1. These n vectors are represented by the states $\{|e_0\rangle, |e_1\rangle, \ldots, |e_{n-1}\rangle\}$, where e_d has a 1 in the d-th bit. The shift operator S described in Eq. 2 acts mapping a state $|d, \vec{x}\rangle \to |d, \vec{x} \oplus \vec{e_d}\rangle$.

$$S = \sum_{d=0}^{n-1} \sum_{\vec{x}} |d, \vec{x} \oplus \vec{e_d}\rangle \langle d, \vec{x}| \tag{2}$$

The initial state of the quantum walk in the hypercube is defined according to Eq. 3 as an equal superposition over all N nodes and n directions.

$$|\Psi(0)\rangle = \frac{1}{\sqrt{n}} \sum_{d=0}^{n-1} |d\rangle \otimes \frac{1}{\sqrt{N}} \sum_{\vec{x}} |\vec{x}\rangle \tag{3}$$

According to Rhodes [16], the hypercube was the first graph in which quantum walks were researched. In their work, Shenvi [17] presented a quantum search algorithm based on the random walk quantum architecture. In this article, we are based on the approach used by Wong [24]. The pure quantum walk (without search) evolves by repeated applications from the evolution operator described in Eq. 1, where C is Grover's "diffusion" operator on the coin space and is given by

$$C = 2 \left| s^C \right\rangle \left\langle s^C \right| - I_n \tag{4}$$

where, I_n is the identity operator, n is the vertex degree loopless, and $\left| s^C \right\rangle$ is the equal superposition over all n directions [9,17], i.e.,

$$\left| s^C \right\rangle = \frac{1}{\sqrt{n}} \sum_{d=0}^{n-1} \left| d \right\rangle . \tag{5}$$

We include a query to the "Grover oracle", described in Eq. 6, at each step of the quantum walk.

$$U' = U \cdot (I_n \otimes Q) \tag{6}$$

where, $Q = I_N - 2 \left| \omega \right\rangle \left\langle \omega \right|$, and $\left| \omega \right\rangle$ means the marked vertex. The system is initiated according to the initial state presented in Eq. 3.

3   Analyzing the Quantum Walk on the Hypercube

In this section, we experimentally analyze the quantum walk on the hypercube searching for multiple marked vertices. The simulations and the obtained results are detailed in the following subsections.

3.1   Characterizing the Probability Distribution Along the Space

Previous works showed there is an amplification in the solution neighborhood, which interferes with the amplification of the solutions by the quantum walk on the hypercube [11,14,17]. Initially, it is necessary to understand how the probability amplitudes are distributed in the search space and how the quantum walk evolves in the hypercube over time considering the impacts caused by the solution neighborhood.

Figure 1 shows the probability of success after one hundred steps for the quantum walk in the hypercube with one, two, three, and four arbitrarily marked vertices. Although the search algorithm is able to amplify the probability amplitudes of the marked vertices, if a measurement is performed, the probability of finding one of the solutions is still unsatisfactory. Another interesting aspect that can be observed is that as the number of marked vertices increases, the speed of amplification the probability amplitudes also increases. However, it is necessary to increase the probability amplitudes of the marked vertices.

Figure 2 shows the probability distributions of the marked vertices only after the number of iterations necessary to reach the maximum value of the probability amplitude close to 1/2. As Potovcek et al. [14] noted in their work, we also note that the set of neighbors have a high probability. If we add the amplitudes of the neighbor's vertices, the values are compatible with the amplitude value of the marked vertex. We conclude that a considerable part of the energy, approximately 1/2, is retained in the neighbors of the marked vertices. Figure 2d, shows the probability distribution of four marked vertices. Note that the amplitudes of each vertex have their maximum and a neighborhood region. The x-axis distribution is the relative position of the position on the hypercube. It explains why even increasing the number of marked vertices, the success probabilities do not reach values above 1/2.

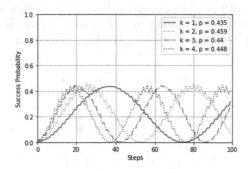

Fig. 1. Success probability after 100 steps in a hypercube with 1024 nodes. The solid blue curve is the success probability for one solution. The dotted orange curve is the success probability for two solutions. The dot-dashed green curve is the success probability for three solutions. The dotted red curve is the success probability for four solutions. (Color figure online)

Figure 3 shows the success probability for the quantum walk with one and four marked vertices after one hundred steps. Figure 3a shows the behavior of the success probability of one marked vertex, the solid blue curve, and its neighbors, which is the dotted orange curve. If a measurement is performed, the probability of getting a neighbor vertex is greater than getting a marked vertex. With probability above 90%, you get the solution or a vertex that is one step away from the solution. Figure 3b shows the behavior of the success probability of four marked vertices, the solid blue curve, and their neighbors, the dotted orange curve. Note that in a step when the probability of success of the marked vertices is high, the probability of success of the neighbors decreases, and in the next step, when the probability of success of the neighbors is high, the probability of success of the marked vertices decreases. Because of this behavior, if a measurement is performed, the probability of getting a neighbor is high. This happens in Fig. 3a but more smoothly.

Observing these results, we must consider the probability p of obtaining a marked vertex and the probability $p' = (1 - p)$ of obtaining an unmarked vertex which is the sum of the probabilities of the $(N - k)$ vertices, where k is the number of marked vertices. These results are shown in Table 1. Note the column of the value of p', which is composed of the value of the amplitudes of the neighbors and the amplitude of the vertices that are neither neighbors nor marked. The probability of the walker finding a region is high because the energy is concentrated in the neighboring region. It is concluded that the amplification of the neighborhood around the marked vertices interferes with the probability of success of finding a target vertex.

3.2   Adjusting the Self-loop Weight for Multiple Marked Vertices

Many works have been proposed with the purpose of improving the search capacity of quantum algorithms. According to Wong [23], adding a self-loop to each vertex boosts the success probability from 1/2 to 1. A modification to the initial state in the Eq. 3 and to Grover's coin in the Eq. 4 is needed so that the self-loop

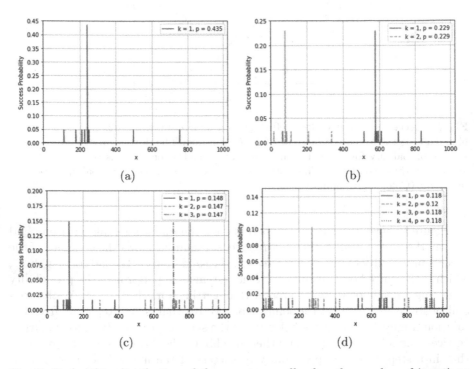

Fig. 2. Probability distribution of the quantum walk after the number of iterations necessary to reach the maximum value of the probability amplitude with $n = 10$ and $N = 1024$ vertices. The y-axis values are at different ranges to improve visualization. (a) solid blue bar show the probability distribution for one marked vertex. (b) solid blue bar and orange dashed bar show the probability distribution for two marked vertices. (c) solid blue bar, orange dashed bar and green dash-dot bar show the probability distribution for three marked vertices. (d) solid blue bar, orange dashed bar, green dash-dot bar and dotted red bar show the probability distribution for four marked vertices. (Color figure online)

Table 1. Probabilities of success of marked and unmarked vertices.

| Figure | p | Probabilities of success $p' = (1 - p)$ | |
		Neighbors	Neither
2a	43.5%	48.2%	8.3%
2b	45.8%	45.5%	8.7%
2c	44.2%	48.4%	7.4%
2d	47.4%	44.5%	8.1%
3a	43.5%	48.2%	8.3%
3b	40.5%	52.9%	6.6%

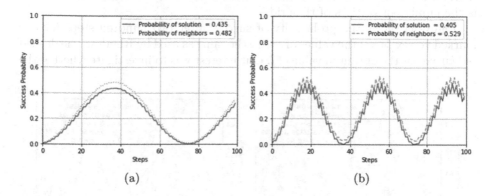

(a)   (b)

Fig. 3. Probability of success after 100 steps with $n = 10$ and $N = 1024$ vertices. (a) shows the probability of success for one marked vertex and its neighbors. (b) shows the probability of success for four marked vertices and their neighbors. (Color figure online)

can be added. The addition of the self-loop is described in Eq. 7. Thus, the coin space is now an $(n + 1)$-dimensional space [16].

$$|s^C\rangle = \frac{1}{\sqrt{n+l}} \left(\sqrt{l}\,|\circlearrowleft\rangle + \sum_{d=0}^{n-1} |d\rangle \right) \tag{7}$$

One of the concerns when adding a self-loop at each vertex is knowing the best self-loop value. More specifically, in the case of the quantum walk on the hypercube, Rhodes [16] proposed an optimal self-loop value

$$l = \frac{d}{N}, \tag{8}$$

where d is equal to the degree of the loopless graph and N is the number of vertices in the hypercube. Recently, two works showed that inserting the number of marked vertices in calculating the self-loop value optimizes quantum walks.

Carvalho [6] shows that the optimal value of the self-loop for quantum walks in D-dimensional grids with multiple marked vertices is

$$l = \frac{2Dm}{N},$$

where $2D$ is the number of movements the walker can do, not counting the self-loop, m the number of marked vertices, and N the number of vertices of the grid. Nahimovs [12] shows that for different types of two-dimensional grids - triangular, rectangular, and honeycomb the optimal self-loop value is also,

$$l = \frac{m \cdot d}{N}$$

where d is the degree of the vertex, m is the number of marked vertices, and N is the number of vertices of the grid.

Figure 4a shows the probability of success after two hundred steps for one marked vertex. Here, the values of l were the same as used by Rhodes. The dashed red curve has the optimum value of l. Our interest was to investigate whether the

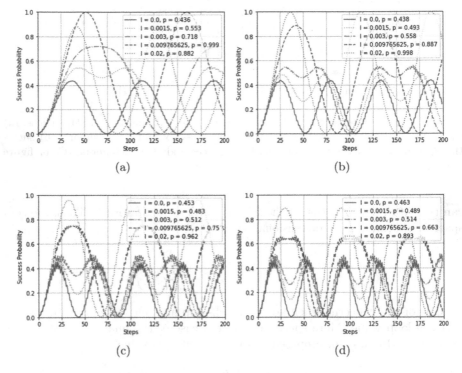

Fig. 4. Comparison between multiple self-loops values and $l = (n/N)$. (a) shows the success probability for one marked vertex. (b) shows the success probability for two marked vertices. (c) shows the success probability for three marked vertices. (d) shows the success probability for four marked vertices. (Color figure online)

value of l described in Eq. 8 also improved the walk results for a number $(k > 1)$ of marked vertices. For this, we performed three more experiments where we increased the number of marked vertices up to four. As we added the marked vertices the success probability of the dashed red curve decreased to 88.7% (4b) while the success probability of the dotted purple curve increased to 99.8% (4b) but then also decreased to 96.2% (4c) and 89.3% (4d). It indicates that a new value of l is required when the number of marked vertices increases. To find the optimal self-loop for multiple marked vertices, we defined a set of values in the form $l' = (\alpha \cdot l)$, where $\alpha \in \mathbb{N}$.

Figure 5 compares the probability of success for a set of marked vertices, $k = \{2, 3, 5, 14, 17\}$, these vertices were chosen randomly as well as their number. The self-loop values for these vertex numbers are $\alpha \cdot l$, where $l = (d/N)$ and $\alpha = \{1, 2, 3, ...\}$. Note that the curves have their maximum points exactly at the locations on the x-axis where the l' values are. We can conclude that the value of $(\alpha = k)$. Therefore, we can set the value of l for multiple marked vertices for the quantum walk in the hypercube,

$$l' = \frac{n}{N} \cdot k \tag{9}$$

where n is equal to the degree of the loopless vertex of the hypercube, N the number of vertices in the hypercube, and k the number of marked vertices. The self-loop value shown by Nahimovs [12] for the quantum search in various types of two-dimensional grids coincides with the optimal self-loop value for the search for a quantum walk in the hypercube.

Figure 5 shows that, as the values of l approach the optimal value, the probability of success of the curve also approaches its maximum value. We can observe this behavior in Table 2 which shows the probability of success for multiple values of l and multiple marked vertices. Consider the values of the main diagonal, which are the maximum success probabilities for each $l = (n/N) \cdot k$.

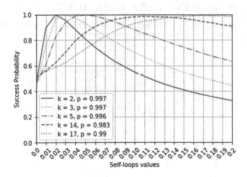

Fig. 5. Investigation to set the value of l for multiple marked vertices.

Table 2 shows the relationship between the self-loop value and the number of marked vertices. We observe the relationship between the self-loop value and the number of marked vertices. Note that when the values of l approach the optimal values for each number of marked vertices, there is an improvement in the probability amplitude. Figure 6 shows the probability of success after two hundred steps for multiple marked vertices. We can conclude that for cases where there is more than one marked vertex, the optimal value of $l = (n/N) \cdot k$.

3.3 Searching for Adjacent Marked Vertices

The results found in the previous sections refer to the search for non-adjacent marked vertices, i.e., $|\vec{\omega}_i - \vec{\omega}_j| \neq 1$ the Hamming distance from vertex $\vec{\omega}_i$ and all other marked vertices is different from 1. Nahimovs et al. [11] shows in their work that for quantum walks in the hypercube if the search space contains marked neighbors vertices, the search can be drastically affected. The authors considered two sets, one with two adjacent marked vertices and the other with two non-adjacent marked vertices. In the first case, the two adjacent marked vertices

Table 2. Probability of success for multiple values of l.

$l = (n/N) \cdot k$	Number of marked vertices									
	1	2	3	4	5	6	7	8	9	10
(n/N)*1	**0.999**	0.888	0.75	0.663	0.775	0.592	0.575	0.576	0.589	0.55
(n/N)*2	0.888	**0.998**	0.958	0.886	0.815	0.9	0.705	0.672	0.639	0.624
(n/N)*3	0.749	0.959	**0.998**	0.976	0.934	0.886	0.941	0.792	0.857	0.727
(n/N)*4	0.64	0.888	0.978	**0.998**	0.975	0.954	0.922	0.885	0.847	0.813
(n/N)*5	0.555	0.816	0.937	0.986	**0.996**	0.989	0.966	0.943	0.912	0.883
(n/N)*6	0.49	0.75	0.888	0.958	0.989	**0.996**	0.991	0.973	0.953	0.928
(n/N)*7	0.438	0.691	0.84	0.926	0.969	0.992	**0.996**	0.99	0.978	0.983
(n/N)*8	0.395	0.641	0.794	0.888	0.944	0.895	0.991	**0.993**	0.99	0.988
(n/N)*9	0.361	0.596	0.75	0.852	0.915	0.957	0.978	0.993	**0.994**	0.982
(n/N)*10	0.331	0.554	0.711	0.816	0.888	0.935	0.966	0.982	0.99	**0.996**

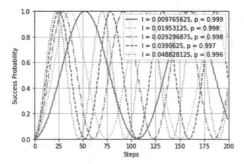

Fig. 6. Probability of success after 200 steps. Solid blue curve, $k = 1$. Dotted orange curve, $k = 2$. Green dash-dot curve, $k = 3$. Red dashed curve, $k = 4$. Dotted purple curve, $k = 5$ (Color figure online)

are $M = \{0, 1\}$. The absolute value of the overlap remained close to 1, and the probability remains close to the initial state probability. In the second case, the two non-adjacent marked vertices are $M = \{0, 3\}$. The behavior on this one is different, the same behavior as the solid blue curve in Fig. 3a.

As the addition of self-loop in the quantum walk in the hypercube improved the search for multiple non-adjacent marked vertices, we investigated the case where the marked vertices are adjacent. We consider ten sets of vertices, $M = [\{0, 1\}, \{0, 1, 2\}, \cdots, \{0, 1, 2, 4, 8, \cdots, 256, 512\}]$, i.e., all vertices adjacent to the vertex 0. We add one more vertex to the set of marked vertices on each new walk until the number of vertices in M is equal to the degree n of the vertex.

Figure 7 shows the probability of success after two hundred steps. Figure 7a shows the result for the value of $l = (n/N)$. The probability reaches its maximum when the number of vertices reaches $k = 4$ with a probability of success of 99.1%. Then the probability starts to decrease as k increases. Figure 7b shows the result for the value of $l = (n/N) \cdot k$. The probability reaches its maximum when the number of vertices reaches $k = 11$ with a success probability of 94.5%. Although the probability increases with a slower speed when $k = 5$, it already reaches 78.3%. This behavior is interesting for search spaces where the marked vertex density is high. Note the probability of the solid cyan curve. This behavior was found in work done by Nahimovs et al. [11] and was repeated here in our experiments. According to the authors, this is because the quantum walk has a stationary state.

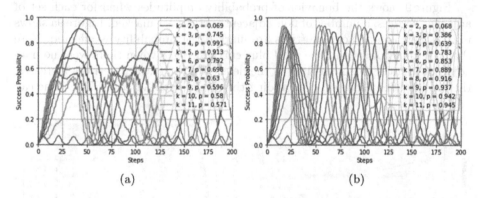

(a)   (b)

Fig. 7. Probability of success after 200 steps with $n = 10$ and $N = 1024$ vertices. Shows the probability of success for k adjacent marked vertices. (a) shows for $l = (n/N)$ and (b) for $l = (n/N) \cdot k$.

Figure 8 shows the comparison between what happens to the success probabilities in Fig. 7 when the number of k increases. Note the dotted orange curve, the probability of success grows to its maximum value when the value of $l = (n/N) \cdot k$. The same does not happen when $l = (n/N)$.

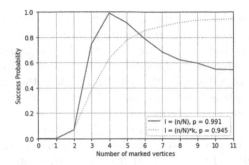

Fig. 8. Maximum probability reached for each number of marked vertices in the neighborhood after one hundred steps with $n = 10$ and $N = 1024$ vertices. Evaluating the interference of the number of adjacent marked vertices in the value of l. (Color figure online)

We considered before that the marked vertices were neighbors. Now, let us analyze the possibility that in addition to having marked vertices in the neighborhood, there are also marked vertices that are not neighbors. We run ten experiments, and each one starts with two adjacent marked vertices $M = \{0, 1\}$. In each experiment, a $i = \{1, 2, 3, \cdots\}$ non-adjacent vertex is randomly marked and the next marked neighbor, i.e., $M = \{0, 1, 2, ...\}$. Therefore, in the tenth experiment, there will be eleven adjacent and ten non-adjacent vertices.

Figure 9 shows the behavior of probability amplitudes when for each set of adjacent vertices, a number of non-adjacent vertices are marked. Figure 9a shows that as new non-adjacent vertices are marked the probability is affected. Note that the behavior seen in the solid blue curve in Fig. 8 when there were no non-adjacent vertices is similar, i.e., as the density of the marked vertices increases, the probabilities decrease, even adding the vertices non-adjacent. The same can

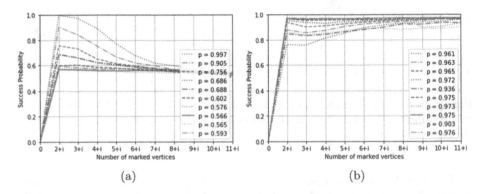

(a)   (b)

Fig. 9. Maximum probability reached for each number of marked vertices after one hundred steps with $n = 10$ and $N = 1024$ vertices. (a) shows the probability of success for k adjacent and non-adjacent marked vertices for $l = (n/N)$. (b) shows the probability of success for k adjacent and non-adjacent marked vertices for $l = (n/N) \cdot k$. (Color figure online)

be seen in the case of the dotted orange curves in Figs. 8 and 9b, i.e., when the density of the marked vertices increases, the probability also increases, this tells us that the value of $l = (n/N) \cdot k$ is optimal for high marked vertex densities.

Figure 10 shows the probability of success for the search of marked adjacent and non-adjacent vertices in the search space. We performed an experiment, where, at every hundred steps, an adjacent vertex and a non-adjacent vertex were marked, i.e., for each M set of adjacent vertices a vertex $i \notin M$ was marked randomly, then, $M' = \{0, 1, i_0\}, \{0, 1, i_0, 2, i_1\}, \cdots , \{0, 1, i_0, 2, i_1, 4, i_2, \cdots , 512, i_{10}\}$. Figure 10a shows the probability of success for $l = (n/N)$ and Fig. 10b shows the probability of success for $l = (n/N) \cdot k$. Note that the probability of success above 90% is achieved in a smaller number of steps.

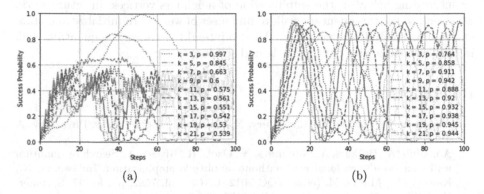

(a)   (b)

Fig. 10. Probability of success after 100 steps with $n = 10$ and $N = 1024$ vertices. (a) shows the probability of success for k adjacent and non-adjacent marked vertices for $l = (n/N)$. (b) shows the probability of success for k adjacent and non-adjacent marked vertices for $l = (n/N) \cdot k$. (Color figure online)

4   Conclusions

Many efforts are applied in order to improve the performance of quantum search algorithms. Quantum walks are the main tool for building these algorithms. We initially analyzed the quantum walk in the hypercube applying Grover's search and came to the conclusion that neighbor vertices affect the search performance, an observation that has been corroborated by other authors. We found that the walk could not improve its results even for a number of marked vertices equal to one. Many authors have developed works for adding self-loops in various types of graphs and grids of different dimensions. In this sense, we decided to investigate how to improve the quantum search in the hypercube using self-loops. Previous works defined the optimal self-loop value as $l = (d/N)$ for one marked vertex to the quantum walk on the hypercube. After performing experiments we saw that this value of l was not optimal for multiple marked vertices. We arrive at a value of $l = (n/N) \cdot k$ for an arbitrary number of vertices. This value is also used when searching in two-dimensional grids. Another aspect of the quantum walk in the hypercube is whether the marked vertex is adjacent or not, this interferes with the search performance. We then analyzed whether the value of $l = (n/N)$ and $l = (n/N) \cdot k$ had any positive effect when applied to the hypercube vertices. The results show that the value of $l = (n/N)$ is not optimal for the quantum walk in the hypercube with multiple marked vertices adjacent or not. It also shows that for a search space where there are marked adjacent vertices, just one non-adjacent marked vertex is sufficient for the value of $l = (n/N) \cdot k$ to be better. According to the results presented here, there is a greater than 90% probability that the measurement will collapse in one of the solutions. Recent works have used the quantum walks to search weights and train artificial neural networks [19,20]. The quantum walk in the hypercube has an interesting behavior, the amplification of neighbors vertices. In future work, we intend to use this quantum walk to find a set of weights to initialize and train classical artificial neural networks. We also intend to analyze the quantum walk in the hypercube with multiple weighted self-loops.

References

1. Aharonov, Y., Davidovich, L., Zagury, N.: Quantum random walks. Phys. Rev. A **48**(2), 1687 (1993)
2. Ambainis, A., Bačkurs, A., Nahimovs, N., Ozols, R., Rivosh, A.: Search by quantum walks on two-dimensional grid without amplitude amplification. In: Iwama, K., Kawano, Y., Murao, M. (eds.) TQC 2012. LNCS, vol. 7582, pp. 87–97. Springer, Heidelberg (2013). https://doi.org/10.1007/978-3-642-35656-8_7
3. Ambainis, A., Kempe, J., Rivosh, A.: Coins make quantum walks faster. arXiv preprint quant-ph/0402107 (2004)
4. Bezerra, G., Lugão, P., Portugal, R.: Quantum walk-based search algorithms with multiple marked vertices. Phys. Rev. A **103**(6), 062202 (2021)
5. de Carvalho, J.H.A., de Souza, L.S., de Paula Neto, F.M., Ferreira, T.A.E.: Impacts of multiple solutions on the Lackadaisical Quantum Walk search algorithm. In: Cerri, R., Prati, R.C. (eds.) BRACIS 2020. LNCS (LNAI), vol. 12319, pp. 122–135. Springer, Cham (2020). https://doi.org/10.1007/978-3-030-61377-8_9
6. Carvalho, J.H.A., Souza, L.S., Paula Neto, F.M., Ferreira, T.A.E.: On applying the lackadaisical quantum walk algorithm to search for multiple solutions on grids. arXiv preprint quant-ph/2106.06274 (2021)
7. Kempe, J.: Quantum random walks hit exponentially faster. arXiv preprint quant-ph/0205083 (2002)
8. McMahon, D.: Quantum Computing Explained. Wiley, New York (2007)
9. Moore, C., Russell, A.: Quantum walks on the hypercube. In: Rolim, J.D.P., Vadhan, S. (eds.) RANDOM 2002. LNCS, vol. 2483, pp. 164–178. Springer, Heidelberg (2002). https://doi.org/10.1007/3-540-45726-7_14
10. Nahimovs, N.: Lackadaisical Quantum Walks with multiple marked vertices. In: Catania, B., Královič, R., Nawrocki, J., Pighizzini, G. (eds.) SOFSEM 2019. LNCS, vol. 11376, pp. 368–378. Springer, Cham (2019). https://doi.org/10.1007/978-3-030-10801-4_29
11. Nahimovs, N., Santos, R.A.M., Khadiev, K.R.: Adjacent vertices can be hard to find by quantum walks. Mosc. Univ. Comput. Math. Cybern. **43**(1), 32–39 (2019)

12. Nahimovs, N., Santos, R.A.: Lackadaisical quantum walks on 2D grids with multiple marked vertices. arXiv preprint arXiv:2104.09955 (2021)
13. Nielsen, M.A., Chuang, I.: Quantum Computation and Quantum Information. AAPT, Cambridge (2002)
14. Potoček, V., Gábris, A., Kiss, T., Jex, I.: Optimized quantum random-walk search algorithms on the hypercube. Phys. Rev. A **79**(1), 012325 (2009)
15. Rhodes, M.L., Wong, T.G.: Quantum walk search on the complete bipartite graph. Phys. Rev. A **99**(3), 032301 (2019)
16. Rhodes, M.L., Wong, T.G.: Search on vertex-transitive graphs by lackadaisical quantum walk. Quantum Inf. Process. **19**(9), 1–16 (2020)
17. Shenvi, N., Kempe, J., Whaley, K.B.: Quantum random-walk search algorithm. Phys. Rev. A **67**(5), 052307 (2003)
18. Singh, J., Singh, M.: Evolution in quantum computing. In: 2016 International Conference System Modeling & Advancement in Research Trends (SMART), pp. 267–270. IEEE (2016)
19. Souza, L.S., Carvalho, J.H.A., Ferreira, T.A.E.: Quantum walk to train a classical artificial neural network. In: 2019 8th Brazilian Conference on Intelligent Systems (BRACIS), pp. 836–841. IEEE (2019)
20. Souza, L.S., Carvalho, J.H.A., Ferreira, T.A.E.: Classical artificial neural network training using quantum walks as a search procedure. IEEE Trans. Comput. (2021)
21. Venegas-Andraca, S.E.: Quantum walks: a comprehensive review. Quantum Inf. Process. **11**(5), 1015–1106 (2012)
22. Wang, H., Zhou, J., Wu, J., Yi, X.: Adjustable self-loop on discrete-time quantum walk and its application in spatial search. arXiv preprint arXiv:1707.00601 (2017)
23. Wong, T.G.: Grover search with lackadaisical quantum walks. J. Phys. A: Math. Theor. **48**(43), 435304 (2015)
24. Wong, T.G.: Faster search by lackadaisical quantum walk. Quantum Inf. Process. **17**(3), 1–9 (2018)
25. Yanofsky, N.S., Mannucci, M.A.: Quantum Computing for Computer Scientists. Cambridge University Press, Cambridge (2008)

On the Analysis of CGP Mutation Operators When Inferring Gene Regulatory Networks Using ScRNA-Seq Time Series Data

José Eduardo H. da Silva[1(✉)], Heder S. Bernardino[1], Itamar L. de Oliveira[1], Alex B. Vieira[1], and Helio J.C. Barbosa[1,2]

[1] Universidade Federal de Juiz de Fora, Juiz de Fora, MG, Brazil
{jehenriques,heder,itamar.leite,alex.borges}@ice.ufjf.br
[2] Laboratório Nacional de Computação Científica, Petrópolis, RJ, Brazil
hcbm@lncc.br

Abstract. Gene Regulatory Networks (GRNs) inference from gene expression data is a hard task and a widely addressed challenge. GRNs can be represented as Boolean models similarly to digital circuits. Cartesian Genetic Programming (CGP), often used for designing circuits, can thus be adopted in the inference of GRNs. The main CGP operator for generating candidate designs is mutation, making its choice important for obtaining good results. Although there are many mutation operators for CGP, to the best of our knowledge, there is no analysis of them in the GRN inference problem. An evaluation of the Single Active Mutation (SAM) and the Semantically-Oriented Mutation Operator (SOMO) is performed here for GRNs inference. Also, a combination of both operators is proposed. We use a benchmark single-cell RNA-Sequencing time series data and its evaluation pipeline to measure the performance of the approaches. The experiments indicate that (i) combining SOMO and SAM provides the best results, and (ii) the results obtained by the proposal are competitive with those from state-of-the-art methods.

1 Introduction

Systems Biology is an interdisciplinary research area that focuses on the computational and mathematical analysis of interactions among the components of a biological system [28]. All cellular activities are controlled by their genes through a network that forms proteins from DNA. The gene expression depends on the relationships in this network, known as Gene Regulatory Network (GRN) [13].

Several methods have been developed for the inference of GRNs, such as PIDC [3], that uses partial information decomposition (PID) to identify regulatory relationships between genes, GENIE3 [12], winner of the DREAM5 network challenge, that decomposes the prediction of a regulatory network among p genes into p different regression problems, and GRNBoost2 [19], that is based on GENIE3, and uses a Gradient Boosting Machine (GBM) regression [7]. GENIE3

© Springer Nature Switzerland AG 2021
A. Britto and K. Valdivia Delgado (Eds.): BRACIS 2021, LNAI 13073, pp. 264–279, 2021.
https://doi.org/10.1007/978-3-030-91702-9_18

and GRNBoost2 are provided in the same Python package Arboreto [19]. Also, evolutionary computation techniques have been applied to infer gene regulatory networks. For instance, one can find the application of Genetic Programming and Evolutionary Strategies in this context [15, 26].

GRNs can be modeled as continuous models in the form of differential equations or as discrete models, such as Bayesian and Boolean networks. For discrete models, a common scheme for representing the genes' interactions is an alphabet of two symbols $\Gamma = \{0,1\}$, where 0 means inhibition and 1 means activation. A model for this type of network is similar to digital circuits, which can be evolved using Cartesian Genetic Programming (CGP) [18], as presented in [24,25].

Although it is possible to use recombination operators, CGP mostly uses mutation operators to create genetic variation in its offspring. There is no general-purpose crossover operator with good performance for CGP [17]. Point-mutation, in which a gene is randomly chosen and changed to another value, and Single Active Mutation (SAM) [9] that guarantees a phenotypic difference after the mutation, are two common mutation approaches used on CGP.

Recently, the Semantically-Oriented Mutation Operator (SOMO) was developed for designing evolutionary circuits [11]. SOMO can obtain feasible circuits using less computational effort than other approaches from the literature, and this advantage helps to deal with the scalability limitation of CGP. However, SOMO uses only a specific population initialization in order to optimize the circuit in terms of reducing the number of logic elements. There is not a proper optimization step. Also, point-mutation has a parameter, μ_r, which defines the percentage of the genotype to be mutated. SOMO, on the other hand, has more sensitive parameters that control the genetic variation of the offspring.

Therefore, in this paper, we propose the evaluation and analysis of CGP when applied to the context of the gene regulation networks inference by using benchmark curated models with single-cell RNA-Sequencing (scRNA-Seq) time-series gene expression data technology. Here, we provide a critical assessment of the advantages and disadvantages of using SOMO, with and without an appropriate optimization scheme with SAM. Furthermore, we perform a parameter sensitivity analysis of SOMO. Also, the results obtained by the proposal in the computational experiments are better than those found by other GP approaches and similar to those reached by the state-of-the-art algorithm GENIE3, one of the techniques more widely used in the literature for modeling GRNs. Our results show that it is important to have an optimization step to obtain correct GRNs and that SOMO is very sensitive to its parameters.

2 Problem Definition

The usual research aim is to infer the network topology from given gene expression data [1]. The gene expression data can be given as a steady-state or time-series matrix. For the former, each measure represents the expression level of a given gene in a certain condition and, for the latter, the gene expression measurements is carried out in each time point. Considering N the total number of genes and S the total number of time-points measured, a gene expression dataset

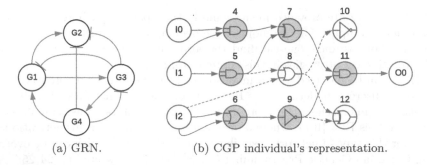

(a) GRN.   (b) CGP individual's representation.

Fig. 1. Illustrations of a GRN with four genes (G1-G4) (a) and an individual of CGP (b). The representation is composed by three primary inputs (I0, I1, and I2) and one output (O0). Continuous lines defines the phenotype. In this representation, $n_c = n_r$ = 3 and the function set Γ = {AND, OR, NOT, XOR}.

is represented by a S x N matrix, where each row vector s (s = 1, ..., S) represents a N-dimensional transcriptome, and each column vector y (y = 1, ..., N) corresponds to a S-dimensional gene profile in the total cell population [4]. The goal of the network inference method is to use the data matrix to predict a set of regulatory interactions between any two genes from the total of N genes. The final output is in the form of a graph with N nodes and a set of edges [4].

Figure 1a shows an example of a GRN with four genes (nodes) and their regulatory relationships (edges). Blue pointed arrows represent an activation and orange lines, inhibition. For example, G1 activates G2, and G4 inhibits G3.

Discrete and continuous models of GRNs are often used to understand the process. When considering continuous models, it is common to use differential equations to model the regulatory relationships between genes, such as presented in [8,24]. This type of model uses the concentrations of the macromolecules such as RNAs and proteins (both are gene products) and other biological species concentration, modeling them through the time-rate-of-change of their concentration variables. Regulatory interactions take the form of functional and differential relationships between concentration variables [22]. More specifically, gene regulation is modeled by rate equations that express the rate of production of a system component as a function of the concentrations of other components.

Also, it is possible to model GRNs through a discrete model, such as Boolean Networks. Boolean Networks use Boolean Algebra to discover the relationship between genes. Moreover, Boolean networks assume each gene g at a time point t to be in one of two states, active or inactive, according to its gene expression data at time t [2], therefore, one needs to binarize the data, leading to information loss. Given a threshold, the data is discretized in 1 (activation) or 0 (inhibition). Boolean-based models simplify the structure and dynamics of gene regulation. Inferred networks provide a quantitative measure of gene regulatory mechanisms [22]. This model, despite its simplicity, can represent, through its dynamics, several biologically significant phenomena. Furthermore, it is possible to obtain several practical uses, such as the identification of drugs for cancer treatment through the inference of the relationships between genes from experimental data such as gene expression profiles [16].

There are several techniques for measuring the expression of a gene. For instance, single-cell RNA-Sequencing (scRNA-seq) is attractive for GRN inference due to its production of thousands of independent measurements [14] and there are methods that sort cells along "trajectories" describing the development or progress of the cell [10,21,23,27], called pseudotime, which is a measure of how far a cell has moved through biological progress. Also, when using scRNA-Seq data technologies, it is common to observe a dropout, which occurs when a gene is observed at a low or moderate expression level in one cell but it is not detected in another cell of the same type. Dropouts make it difficult to get correct GRNs.

3   Cartesian Genetic Programming

Cartesian Genetic Programming is a Genetic Programming technique in which programs are Directed Acyclic Graphs (DAGs) encoded by a matrix of processing nodes, with n_c columns and n_r rows. [17]. The genes are integer values and, for each gene, there are inputs and operation/function that the node performs.

Given a node in the matrix, the nodes at its left side can be used as inputs. There is a user-defined parameter (levels-back) that limits the number of columns at the left side where inputs can be selected to constrain the connectivity of the graph. The number of inputs that a node can receive is called arity. Also, the genotype contains nodes that contribute directly to the output, called active nodes, and those that do not, the inactive nodes. The phenotype is composed only of the active nodes and the genotype-phenotype mapping is done by recursively determining the nodes that contribute to each output, starting on the output and ending on the primary inputs. The function set is user-defined and problem-dependent. For example, logic functions or gates are used when designing digital circuits. Figure 1b presents a CGP individual with three primary inputs and one output. The functions are logical ones. Grey nodes are active and white nodes, inactive. The most common search technique used in CGP is the $(1 + \lambda)$ Evolutionary Strategy (ES) [17], where λ is the number of new solutions generated at each iteration. In this case, the best individual, the one with more matches concerning its truth-table, between the parent and the λ new generated solutions is selected for the next generation.

3.1   Mutation Operators

CGP normally uses only mutation to generate new individuals and crossover operators have received little attention in the CGP [17]. The two most adopted mutation approaches are point mutation [17] and SAM [9].

Point mutation is a simple approach that randomly selects a node and a gene and changes it to another valid value. However, the changes may occur in inactive nodes leading to a lack of modifications in the phenotype.

SAM, proposed for reducing the number of wasted objective function evaluations, operates by (i) randomly selecting one node and one of its elements (inputs or function) and (ii) changing its value to another valid value. Steps (i) and (ii) are repeated until an active node is modified. As a result, SAM ensures that one active node is changed.

Algorithm 1. Semantically-oriented mutation operator [11].

Input: A CGP individual p consisting of $|C|$ nodes
Output: A mutated individual p'

1: $(C, E, C_{PI}, C_{PO}, \Psi) \leftarrow$ decode(p); ▷ decode p as a DAG (C, E) with C_{PI} leaves
   and C_{PO} roots (outputs); $\Psi \colon C \to \Gamma$
2: $N \leftarrow$ activeNodes; $c \leftarrow$ selectNodeRandomly$(N \ / \ C_{PI})$
3: **if** $(\mathrm{rand}(0,1) < p_f \wedge (c \notin C_{PO})$ **then** ▷ mutate node function
4:     $\Psi(c) \leftarrow \mathrm{rand}(0, \Gamma\text{-}1)$
5: **else** ▷ mutate node connection
6:     change connection and function of p_q inactive nodes
7:     $e \leftarrow$ selectInputEdgeRandomly$(\{(x, c) \in E | x \in N\})$
8:     $n \leftarrow$ identifyBestNode$(c, e, (C,E), \Psi)$; $E \leftarrow (E \ / \ \{e\}) \cup \{(n, c)\}$
9: **end if**
10: **return** $p' \leftarrow$ encode(C,E,C_{PI},C_{PO},Ψ)

SOMO [11] was developed for the design of digital circuits using CGP, where the purely stochastic mutation operator is replaced and operates in the phenotype space. The SOMO steps are presented in Algorithm 1, which can be summarized as: (i) all inactive nodes have their inputs changed randomly to another valid value; functions are changed with a probability p_f; (ii) a random node (c) is chosen, in order to be mutated using SOMO; (iii) c has its function modified randomly with a probability p_f; (iv) a random input (e) is chosen from c. Step (i) ensures that new genetic material is generated before performing the actual mutation. With the random input (e) chosen, SOMO identifies the best input considering all previous nodes in the genotype and then performs the mutation.

The identification of the most suitable node is based on semantics. Initially, it calculates the score of every node of the genotype that may be connected to the mutated node c. If more nodes receive the same score, the node closest to the program inputs is preferred. The score reflects the Hamming distance (HD). In the sequence, all left-sided nodes from the mutated node c are connected at e and simulated in three different ways: (a) using the function of the node being connected, (b) forcing e to logic zero $(val_{e=0}^{[0]})$, and (c) forcing e to a logic one $(val_{e=1}^{[0]})$. The desired input value is denoted as req and it can be equal to '0', '1' or 'X', where 'X' means that it does not matter what Boolean value the input e takes. The value of req is determined by using the ternary operator Θ and the reduction operator $\odot$, defined in Equations 1 and 2. The term $[o]$ in superscript points to a Boolean value associated with a program output node o.

$$\Theta(t, v_0, v_1) = \begin{cases} 'X', & v_0 = v_1 \\ '0', & v_0 = t \\ '1', & v_1 = t \end{cases} \quad (1) \qquad \odot(a, b) = \begin{cases} a, & a \neq 'X' \\ b, & \text{otherwise} \end{cases} \quad (2)$$

Algorithm 2. Semantically-oriented mutation operator with SAM.

1: (C, E, C_{PI}, C_{PO}, (Ψ)) ← decode(p); ▷ decode p as a DAG (C, E) with C_{PI} leaves
   and C_{PO} roots (outputs); Ψ: C → Γ
2: N ← activeNodes
3: c ← selectNodeRandomly(N / C_{PI})
4: **if** feasible == False **then**
5:     perform SOMO   ▷ Described in Algorithm 1
6: **else**   ▷ perform SAM
7:     **while** an active node is not selected and mutated **do**
8:         c ← selectNodeRandomly(N / C_{PI})
9:         g ← rand(0, arity-1)
10:        **if** g == arity-1 **then**   ▷ mutate node function
11:            $\Psi(c)$ ← rand(0, Γ-1)
12:        **else**
13:            Mutate node connection of c
14:        **end if**
15:        **if** c ∈ N **then**
16:            An active node was selected and mutated
17:        **end if**
18:    **end while**
19: **end if**
20: **return** p' ← encode(C,E,C_{PI},C_{PO},Ψ)

In SOMO, the mutated input e of c is connected to the best node in the genotype of the current generation. Differently from the standard CGP, SOMO uses $\lambda = 1$ and a different population initialization. In SOMO, the population is started with a candidate solution having no active gate in order to maximize the efficiency and minimize the number of active gates of the evolved solutions [11].

4 Proposal and Methods

Here, we evaluate different approaches using SOMO aiming at discovering if using an appropriate optimization step helps it to obtain better GRNs. We propose the use of SOMO with SAM as (i) SOMO can obtain a feasible solution quickly but with many logic elements, and (ii) SAM is appropriate for reducing the number of logic elements. The proposal is presented in Algorithm 2.

In [11], the percentage of inactive nodes to be mutated (p_q) is equal to 100. Here, we analyze the performance of SOMO when using $p_q = 50\%$ with SAM as optimization step. In addition, as highlighted in [11], SOMO usually gets stuck in local optima. To avoid this issue, we use a restart strategy for exploiting different locations of the search space. The restart strategy consists of initializing a new population and restart the search for every k objective function evaluations.

The criteria for evaluating and comparing methods are varied, but [20] presented a set of benchmark problems when considering scRNA-Seq data for inferring GRNs. This set is composed of toy, curated, and real datasets. For toy and curated models, we have the ground-truth network, which facilitates the evaluation of new methods. Also, there is an extensive comparison between several algorithms from the literature using the pipeline provided in [20].

In this paper, we use curated scRNA-seq data in the form of pseudotime-series. These datasets were made considering 2,000 cells with three configurations: 0%, 50%, and 70% dropout rates. For each configuration, 10 datasets are given. Furthermore, each problem has a different number of pseudotimes.

As each pseudotime gives information about one possible cell trajectory, for each pseudotime one GRN is inferred. Then, the final GRN is given by merging the partial GRNs discovered for each pseudotime. This final GRN is evaluated considering the pipeline presented in [20] which considers the area under the precision-recall curve (AUPRC) and the area under the receiver operating characteristic curve (AUROC) values for comparison. AUPRC is calculated as the area under the precision-recall (PR) curve and shows the trade-off between precision and recall across different decision thresholds. The x-axis of a PR curve is the recall and the x-axis is the False-Positive Ratio (FPR). Considering TP as True Positive, FP as False Positive and FN as False Negative, we can define Precision $= \frac{TP}{TP+FP}$ and Recall $= \frac{TP}{TP+FN}$. For AUROC, the ROC curve is plotted considering the False-Positive Ratio (FPR) on the x-axis and the Recall, on the y-axis. Recall was defined previously and FPR $= \frac{FP}{TN+FP}$. An excellent model has AUC close to 1, which means it has good separability. For many real-world datasets, particularly medical datasets, the fraction of positives is often less than 0.5, meaning that AUPRC has a lower baseline value than AUROC [6].

5   Computational Experiments

Computational experiments were conducted to analyze the performance of CGP mutation operators when applied to the inference of GRNs using the benchmark scRNA-Seq time-series data from [20]. Here, we highlight whether obtaining feasible solutions in a faster way has a positive impact on the quality of the solutions. Also, we analyze the importance of reducing the number of logic elements. The computational experiments are composed of two parts: (i) the performance of CGP variants are compared and, (ii) the results obtained by the two best CGP approaches are compared to those found by GENIE3.

The problems considered in our analysis are presented in Table 1, with information of the number of genes (#Genes) and the number of pseudotimes (#Pseudotimes). These problems are curated models with 2,000 cells and are presented in three different configurations considering 0%, 50%, and 70% dropout.

In order to remove or soften the effects of technical and biological variations as well as obtain a single curve that represents gene expression over time, we use data approximation via cubic smoothing splines implemented in Python and distributed through the library *csaps*[1]. Then, the data is binarized with Bikmeans through Gene Expression Data Pre-Processing Tool (GEDPROTOOLS)[2]. All methods were implemented in C++ and the source codes are available [3].

---

[1] https://csaps.readthedocs.io/en/latest/index.html.
[2] http://lidecc.cs.uns.edu.ar/files/gedprotools.zip.
[3] https://github.com/ciml.

Table 1. Problems used in the experiments.

Problem	#Genes	#Pseudotimes
HSC	11	4
mCAD	5	2
VSC	8	5

The experiments were run in a Ubuntu Server 20.04 LTS (HVM) with 16 vCPUs Intel(R) Xeon(R) CPU E5-2666 v3 @ 2.90GHz and 30GB RAM and the results are evaluated using the pipeline presented in [20] which considers the AUPRC and AUROC values for comparison.

5.1   Comparative Analysis of the CGP Techniques

We compare the standard CGP with SAM and the CGP with SOMO in different approaches: (i) CGP: Standard CGP using SAM for obtaining the first feasible solution and for optimizing; (ii) SOMO: CGP using SOMO without optimization step; (iii) SOMO-SAM: CGP using SOMO with optimization step; (iv) SOMO-SAM-R: CGP using SOMO with optimization step and evolutionary search restart; (v) SOMO-SAM-PQ50: CGP using SOMO with optimization step and $p_q = 50\%$. The suffix R means that the evolutionary process is restarted with a different initial population every 1,000 evaluations.

Here, we aim to evaluate the performance of SOMO with an appropriate optimization step. SOMO is able to obtain feasible solutions in a faster way than other approaches but this solution usually has many logic elements.

For the standard CGP, we use $\lambda = 4$ and random population initialization. When considering SOMO, we use $\lambda = 2$, $p_f = 0$ and the population initialization suggested in [11]. Also, in [11] the value of n_c is variable, considering a multiple of the number of standard gates required to implement each circuit. However, the definition of the number of gates to implement a circuit cannot be easily defined a priori and, hence, we fixed n_c as usually considered when using the standard CGP and used only $n_c = 100$. The other parameters are $n_r = 1$ and $lb = n_c$. For each problem, 5 independent runs were performed with a maximum of 100,000 objective function evaluations. For SOMO-SAM-R, the number of evaluations to restart is 1,000 and was defined during preliminary experiments, analyzing the average number of evaluations needed to find a feasible solution. Also, the accumulated number of evaluations is used in the stop criteria.

Tables 2 and 3 present the results of the CGP methods, respectively, for AUPRC and AUROC. The best (maximum), first quantile (Q1), Median, Mean, third quantile (Q3), and standard deviation values of the results are shown, and the best values are highlighted in boldface. Also, the methods with an asterisk have statistical difference, when considering Dunn's test for 95% of confidence.

Based on these results, one can see that the results of SOMO-SAM and SOMO-SAM-R are equal in some cases. This is expected as the first feasible

Table 2. AUPRC results for all problems. The suffix after the problem's name is the dropout rate.

Algorithm	Best	Q1	Median	Mean	Q3	Worst	Std.
			HSC-0				
CGP	0.4032	0.2535	0.2658	0.2901	0.3198	0.2241	5.63E-02
SOMO	**0.4637**	0.2691	0.2881	0.2986	0.3048	0.2253	6.01E-02
SOMO-SAM	0.4177	0.2679	0.2760	0.3023	0.3261	0.2396	**5.61E-02**
SOMO-SAM-R	0.4265	0.2634	0.2763	0.3044	0.3329	**0.2438**	5.74E-02
SOMO-SAM-PQ50	0.4626	**0.2960**	**0.3082**	**0.3277**	**0.3532**	0.2434	6.08E-02
			HSC-50				
CGP	0.4167	0.2998	0.3527	0.3440	0.3876	0.2509	5.28E-02
SOMO*	0.3738	0.3014	0.3093	0.3093	0.3282	0.2450	**3.46E-02**
SOMO-SAM	0.4290	0.3050	0.3771	0.3565	0.3872	**0.2718**	5.42E-02
SOMO-SAM-R	0.4325	0.2812	0.3569	0.3480	**0.4114**	0.2523	6.58E-02
SOMO-SAM-PQ50*	**0.4759**	**0.3275**	**0.3831**	**0.3660**	0.3952	0.2608	5.88E-02
			HSC-70				
CGP	0.3502	0.2714	0.2857	0.2915	0.2979	0.2462	**3.25E-02**
SOMO	**0.4490**	0.2682	0.2891	0.3074	**0.3419**	0.2284	6.41E-02
SOMO-SAM	0.3806	0.2608	0.2785	0.2962	0.3082	0.2562	4.39E-02
SOMO-SAM-R	0.4166	0.2751	0.2859	0.2964	0.3050	0.2417	4.42E-02
SOMO-SAM-PQ50	0.4001	**0.2891**	**0.3055**	**0.3151**	0.3361	**0.2597**	4.00E-02
			mCAD-0				
CGP	0.7508	0.5719	0.6452	0.6540	0.7508	0.5291	8.65E-02
SOMO	**0.8361**	0.5675	0.6049	0.6770	**0.8361**	0.5371	1.32E-01
SOMO-SAM	0.7844	**0.6238**	**0.6522**	**0.6871**	0.7844	**0.5843**	**8.22E-02**
SOMO-SAM-R	0.7844	**0.6238**	**0.6522**	**0.6871**	0.7844	**0.5843**	**8.22E-02**
SOMO-SAM-PF50	0.7631	0.5917	0.6369	0.6642	0.7631	0.5506	8.50E-02
			mCAD-50				
CGP*	0.6561	0.6081	0.6403	0.6281	0.6561	0.5515	**3.31E-02**
SOMO*	0.6020	0.5374	0.5737	0.5689	0.6020	0.5212	3.36E-02
SOMO-SAM*	0.6614	**0.5874**	**0.6536**	**0.6282**	**0.6614**	**0.5669**	3.85E-02
SOMO-SAM-R*	0.6614	**0.5874**	**0.6536**	**0.6282**	**0.6614**	**0.5669**	3.82E-02
SOMO-SAM-PF50	**0.6645**	0.5861	0.6431	0.6203	0.6431	0.5651	3.33E-02
			mCAD-70				
CGP	0.7624	0.5766	0.6452	0.6466	0.6926	0.5596	**7.13E-02**
SOMO	**0.8361**	0.5799	0.6073	0.6577	0.7407	0.5402	1.06E-01
SOMO-SAM	0.7960	**0.6274**	**0.6522**	**0.6793**	**0.7462**	**0.5843**	7.53E-02
SOMO-SAM-R	0.7960	**0.6274**	**0.6522**	**0.6793**	**0.7462**	**0.5843**	7.53E-02
SOMO-SAM-PF50	0.7747	0.6109	0.6369	0.6618	0.7284	0.5648	7.32E-02
			VSC-0				
CGP	0.4683	**0.2789**	0.3138	**0.3217**	0.3287	0.2338	6.67E-02
SOMO	0.3930	0.2660	0.2892	0.3011	0.3282	0.2205	5.23E-02
SOMO-SAM	0.3860	0.2643	0.3131	0.3134	**0.3634**	**0.2368**	**5.16E-02**
SOMO-SAM-R	0.4222	0.2456	0.2640	0.3021	0.3546	0.2293	6.91E-02
SOMO-SAM-PF50	**0.4951**	0.2749	**0.3286**	0.3290	0.3624	0.2287	7.41E-02
			VSC-50				
CGP	0.3730	0.2275	0.2590	0.2709	0.3071	0.1938	5.32E-02
SOMO	0.4035	**0.2665**	**0.2952**	**0.3024**	**0.3376**	**0.2218**	**5.10E-02**
SOMO-SAM	0.4645	0.2325	0.2541	0.2765	0.2696	0.2018	7.47E-02
SOMO-SAM-R	0.4317	0.2383	0.2634	0.2828	0.3144	0.2209	6.15E-02
SOMO-SAM-PF50	**0.4784**	0.2226	0.2675	0.2860	0.3147	0.2117	7.81E-02
			VSC-70				
CGP	**0.4671**	**0.3069**	0.3431	**0.3607**	**0.4395**	**0.2457**	7.50E-02
SOMO	0.4208	0.2479	0.2978	0.3007	0.3442	0.2235	6.69E-02
SOMO-SAM	0.4357	0.3014	**0.3674**	0.3496	0.3858	0.2409	6.10E-02
SOMO-SAM-R	0.4361	0.2596	0.3172	0.3213	0.3735	0.2191	6.94E-02
SOMO-SAM-PF50	0.4278	0.3013	0.3292	0.3289	0.3621	0.2329	**5.73E-02**

Table 3. AUROC results for all problems. The suffix after the problem's name is the dropout rate. Algorithms marked with an asterisk have statistical difference.

Algorithm	Best	Q1	Median	Mean	Q3	Worst	Std.
			HSC-0				
CGP	0.6376	0.5160	0.5413	0.5517	0.5951	0.4531	5.54E-02
SOMO	0.6186	0.5171	0.5598	0.5538	0.5739	0.5066	**3.78E-02**
SOMO-SAM	**0.7060**	0.5224	0.5654	0.5726	0.5967	0.4975	6.34E-02
SOMO-SAM-R	0.6969	0.5484	0.5568	0.5858	0.6108	**0.5103**	5.98E-02
SOMO-SAM-PQ50	0.6962	**0.5870**	**0.5960**	**0.6060**	**0.6467**	0.5089	5.71E-02
			HSC-50				
CGP	0.6827	0.5684	0.6197	0.6097	0.6465	0.5039	5.47E-02
SOMO	0.6625	0.5525	0.5834	0.5822	0.6207	0.4906	5.46E-02
SOMO-SAM	0.7109	0.5861	0.6268	0.6255	0.6542	**0.5412**	**5.12E-02**
SOMO-SAM-R	0.7141	0.5524	0.6006	0.6085	**0.6758**	0.5055	7.14E-02
SOMO-SAM-PQ50	**0.7596**	**0.6071**	**0.6493**	**0.6419**	0.6658	0.5391	5.64E-02
			HSC-70				
CGP	0.6172	0.5171	0.5488	0.5552	0.5912	0.4966	4.14E-02
SOMO	0.6564	0.5218	**0.5970**	0.5828	**0.6324**	0.5092	5.59E-02
SOMO-SAM	0.6516	0.5192	0.5616	0.5628	0.6161	0.4652	6.07E-02
SOMO-SAM-R	0.6777	0.5243	0.5604	0.5593	0.5705	0.5089	4.58E-02
SOMO-SAM-PQ50	**0.6861**	**0.5734**	0.5815	**0.5982**	0.6053	**0.5588**	**3.84E-02**
			mCAD-0				
CGP	0.6264	0.3750	0.5165	0.4978	0.6264	0.3242	1.22E-01
SOMO	0.6703	0.4148	0.4615	0.5176	0.6703	0.3681	1.28E-01
SOMO-SAM	**0.6923**	**0.4959**	**0.5330**	**0.5742**	**0.6923**	**0.4231**	1.01E-01
SOMO-SAM-R	**0.6923**	**0.4959**	**0.5330**	**0.5742**	**0.6923**	**0.4231**	1.01E-01
SOMO-SAM-PQ50	0.6484	0.4217	0.4973	0.5203	0.6484	0.3736	1.12E-02
			mCAD-50				
CGP	**0.5440**	0.4286	0.4863	0.4736	**0.5440**	0.3352	7.49E-02
SOMO*	0.4560	0.3626	0.4093	0.4055	0.4560	0.3352	5.11E-02
SOMO-SAM*	**0.5440**	**0.4341**	**0.4973**	0.4852	**0.5440**	**0.3791**	6.13E-02
SOMO-SAM-R*	**0.5440**	**0.4341**	**0.4973**	0.4852	**0.5440**	**0.3791**	6.13E-02
SOMO-SAM-PQ50	0.5055	0.4286	0.4918	0.4659	0.5055	0.3681	**4.66E-02**
			mCAD-70				
CGP	0.6319	0.4135	0.5055	0.4934	0.5412	0.3626	8.99E-02
SOMO	0.6703	0.4313	0.4753	0.5027	0.5632	0.3681	1.01E-01
SOMO-SAM	**0.6978**	**0.4973**	**0.5522**	**0.5604**	**0.6195**	**0.4231**	8.79E-02
SOMO-SAM-R	**0.6978**	**0.4973**	**0.5522**	**0.5604**	**0.6195**	**0.4231**	8.79E-02
SOMO-SAM-PQ50	0.6538	0.4602	0.500	0.5176	0.5797	0.3736	**8.75E-02**
			VSC-0				
CGP	0.6805	0.5014	0.5541	0.5608	0.6280	0.4211	7.78E-02
SOMO	0.6382	0.5006	**0.5581**	0.5440	0.5839	0.4236	**6.30E-02**
SOMO-SAM	0.6878	0.5089	0.5496	**0.5659**	0.6222	0.4707	7.06E-02
SOMO-SAM-R	0.6512	0.4848	0.5154	0.5462	**0.6394**	**0.4301**	8.36E-02
SOMO-SAM-PQ50	**0.7220**	**0.5167**	0.5508	0.5603	0.5982	0.4276	8.25E-02
			VSC-50				
CGP	0.6854	0.4280	0.5272	0.5113	0.5742	0.3415	1.01E-01
SOMO	0.6415	**0.4754**	**0.5488**	**0.5372**	**0.6085**	0.3699	8.32E-02
SOMO-SAM	0.7154	0.4467	0.4732	0.5049	0.5327	0.3659	1.01E-01
SOMO-SAM-R	0.6585	0.4591	0.5106	0.5217	0.5817	0.3894	**7.97E-02**
SOMO-SAM-PQ50	**0.7309**	0.4213	0.4699	0.5091	0.5602	**0.4114**	1.03E-02
			VSC-70				
CGP	**0.7707**	**0.5583**	0.6130	**0.6199**	**0.6929**	**0.4593**	9.13E-02
SOMO	0.7057	0.5071	0.5325	0.5480	0.5872	0.4317	**8.53E-02**
SOMO-SAM	0.7122	0.5630	**0.6276**	0.6023	0.6648	0.400	8.88E-02
SOMO-SAM-R	0.7415	0.5018	0.5780	0.5745	0.6433	0.4065	9.74E-02
SOMO-SAM-PQ50	0.7154	0.5545	0.5992	0.5841	0.6457	0.4057	9.95E-02

Table 4. Algorithm counting considering the best values of median. Values between parenthesis is the algorithm counting considering the statistical equality.

Algorithm	AUPRC Count	AUROC Count
CGP	0(8)	0(8)
SOMO	1(8)	3(9)
SOMO-SAM	**4**(8)	**4**(9)
SOMO-SAM-R	3(8)	3(9)
SOMO-SAM-PQ50	**4**(8)	2(8)

solution was obtained before reaching the first restart point (1,000 evaluations). In general, most CGP approaches have no statistical difference when compared to each other, both in AUPRC and AUROC. When considering the median, we count the times that each algorithm reached the best result. Values in parentheses consider the statistical tests and when there is no statistical difference, all methods score. This counting is shown in Table 4.

Also, the approaches using an optimization step performed better than the other ones. The restart scheme helps to escape from local optima but it does not provide good results. Furthermore, the results when using $p_q = 50\%$ are better than the standard values presented in [11]. The SOMO population initialization is essential for a good performance of the algorithm as well as using $\lambda = 1$.

Finally, based on Table 4, one can conclude that the best methods are SOMO-SAM and SOMO-SAM-PQ50 when AUPRC is considered. For AUROC, the best algorithms are SOMO-SAM and SOMO. However, here we consider one approach using the standard SOMO with SAM as optimization step (SOMO-SAM) and another approach varying the parameter p_q, namely SOMO-SAM-PQ50.

5.2   Comparative Analysis with GENIE3

As observed in Sect. 5.1, the two best CGP variants are SOMO-SAM and SOMO-SAMPQ50, and we compare them here with GENIE3. According to [20], GENIE3 is the best algorithm for inferring GRNs. Performance profiles (PPs) [5] were used in order to analyze the relative performance of the algorithms. Considering p as a particular problem (model) and s as a particular solver, $\rho(p, s)$ is defined as the performance ratio within a factor of τ of the best possible ratio.

From PPs, it is possible to extract: (i) the approach that obtained the best results for most problems (largest $\rho(1)$), (ii) the most reliable approach (smaller τ such that $\rho(\tau) = 1$), and (iii) the best overall performance (largest area under the performance profiles curves). Moreover, boxplots of the results are presented in Fig. 2. Also, Kruskal Wallis statistical test and Dunn's *post hoc* test were carried out and the results show that there is statistical difference only when comparing CGP approaches with GENIE3.

Based on the PPs presented in Fig. 3 is possible to conclude that: (i) GENIE3 has the best performance in most of the problems (largest $\rho(1)$), followed by

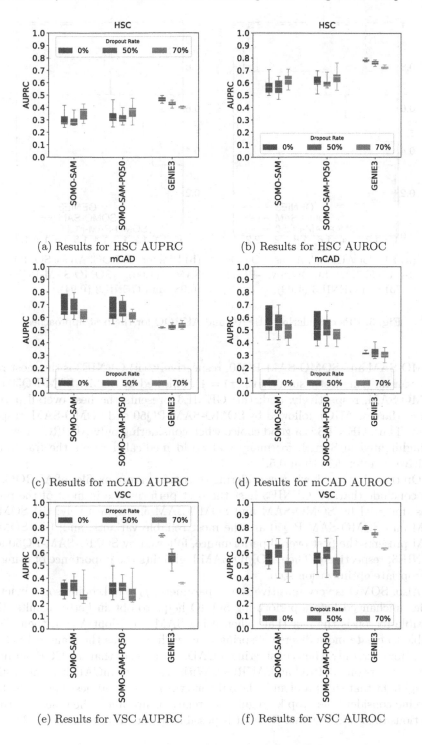

(a) Results for HSC AUPRC

(b) Results for HSC AUROC

(c) Results for mCAD AUPRC

(d) Results for mCAD AUROC

(e) Results for VSC AUPRC

(f) Results for VSC AUROC

Fig. 2. Results for the problems considering all scenarios.

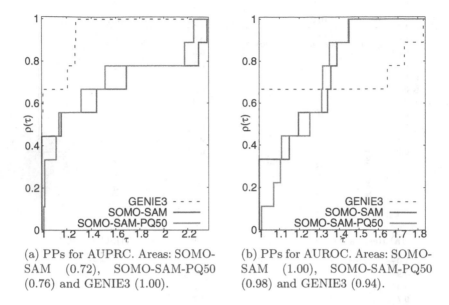

(a) PPs for AUPRC. Areas: SOMO-SAM (0.72), SOMO-SAM-PQ50 (0.76) and GENIE3 (1.00).

(b) PPs for AUROC. Areas: SOMO-SAM (1.00), SOMO-SAM-PQ50 (0.98) and GENIE3 (0.94).

Fig. 3. PPs considering AUPRC and AUROC for the best approaches.

SOMO-SAM and SOMO-SAM-PQ50, respectively. (ii) GENIE3 is the most reliable variant (smallest τ such that $\rho(\tau) = 1$), followed by SOMO-SAM-PQ50 and SOMO-SAM, respectively, and; (iii) GENIE3 presents the best overall performance (largest AUC), followed by SOMO-SAM-PQ50 and SOMO-SAM, respectively. Thus, GENIE3 is a good choice when considering only AUPRC. However, as highlighted in Sect. 4, for many real-world medical datasets the fraction of positives is often less than 0.5.

On the other hand, when considering the performance profiles of AUROC, we can conclude that: (i) GENIE3 hast the best performance in most of the problems, followed by SOMO-SAM and SOMO-SAM-50, respectively; (ii) SOMO-SAM and SOMO-SAM-PQ50 are the most reliable variants, and; (iii) SOMO-SAM presents the best overall performance, followed by SOMO-SAM-PQ50 and GENIE3, respectively. Then, SOMO-SAM highlights the importance of using an appropriate optimization step.

Also, SOMO is very sensitive to the parameter p_q, i.e. keeping some inactive nodes unchanged before performing SOMO helps to obtain better results. One possible reason is the fact that, when using SAM, we adopt $\lambda = 4$, then, CGP is able to create more diverse offspring than with $\lambda = 1$ as the standard SOMO uses. However, only when considering mCAD, approaches that use CGP obtained better results on AUPRC and AUROC. With respect to mCAD it is interesting to highlight that this problem is the only one with 2 pseudotimes. The evaluation pipeline considers the "top k" regulatory relationships and, when merging the 2 solutions to obtain the final GRN, it is possible to construct a smaller GRN (less number of regulatory relationships) that consider the most important regulatory relationships leading to a better evaluation.

The statistical tests show that there is no statistical difference between the CGP approaches. The statistical difference is observed only when comparing the CGP variants with GENIE3, reinforcing the superiority of GENIE3 as shown in the AUPRC boxplots.

6   Conclusions and Future Work

Here, we analyze the performance of CGP mutation operators applied when inferring GRN using benchmark curated scRNA-seq time series data. These approaches include mainly two aspects: (i) a proposal using SAM to optimize the solution obtained by SOMO, and (ii) a parameter sensitivity analysis for SOMO.

We compared the CGP approaches, and the best-obtained CGP methods with GENIE3. The results show that modifying p_q to 50% helped CGP to obtain better results, as a more diverse offspring is generated. GENIE3 performed better in most problems, except in mCAD. The mCAD problem has only two pseudo-times and the CGP solutions' merge can obtain a smaller GRN that is better evaluated in the pipeline. Also, there is no statistical difference between the CGP approaches, but there is when comparing the CGP variants with GENIE3. According to PPs, GENIE3 obtained the best results in general but one can highlight that both SOMO-SAM and SOMO-SAM-PQ50 are more reliable when AUROC is considered. Furthermore, the use of SOMO for obtaining faster first feasible solutions and SAM as an appropriate optimization step helps CGP to obtain better results when inferring gene regulatory networks when compared to the standard SOMO and to the standard CGP. Thus, one can conclude that using SAM with SOMO improves the inferred GRNs.

As future work, we intend to explore even more the SOMO parameters, considering other values of p_q and p_f in the context of the inference of GRNs and the impact of n_c when adopting values of $\lambda > 1$. Also, we intend to investigate the impact of the number of pseudotimes in the merging step of CGP.

Acknowledgements. We thank the support provided by FAPERJ, FAPESP, FAPEMIG, CAPES, CNPq, UFJF, and Amazon AWS.

References

1. Aalto, A., Viitasaari, L., Ilmonen, P., Mombaerts, L., Gonçalves, J.: Gene regulatory network inference from sparsely sampled noisy data. Nat. Commun. 11(1), 1–9 (2020)
2. Banf, M., Rhee, S.Y.: Computational inference of gene regulatory networks: approaches, limitations and opportunities. Biochimica et Biophysica Acta (BBA) Gene Regul. Mech. 1860(1), 41–52 (2017)
3. Chan, T.E., Stumpf, M.P., Babtie, A.C.: Gene regulatory network inference from single-cell data using multivariate information measures. Cell Syst. 5(3), 251–267 (2017)

4. Chen, S., Mar, J.C.: Evaluating methods of inferring gene regulatory networks highlights their lack of performance for single cell gene expression data. BMC Bioinf. **19**(1), 1–21 (2018)
5. Dolan, E.D., Moré, J.J.: Benchmarking optimization software with performance profiles. Math. Programm. **91**(2), 201–213 (2002)
6. Draelos, R.: Measuring performance: Auprc and average precision (2019). glassboxmedicine.com/2019/03/02/measuring-performance-auprc/
7. Friedman, J.H.: Greedy function approximation: a gradient boosting machine. Ann. Statist., pp. 1189–1232 (2001)
8. Gebert, J., Radde, N., Weber, G.W.: Modeling gene regulatory networks with piecewise linear differential equations. Eur. J. Oper. Res. **181**(3), 1148–1165 (2007)
9. Goldman, B.W., Punch, W.F.: Reducing wasted evaluations in cartesian genetic programming. In: Krawiec, K., et al. (eds.) EuroGP 2013. LNCS, vol. 7831, pp. 61–72. Springer, Heidelberg (2013). https://doi.org/10.1007/978-3-642-37207-0_6
10. Haghverdi, L., Büttner, M., Wolf, F.A., Buettner, F., Theis, F.J.: Diffusion pseudotime robustly reconstructs lineage branching. Nat. Methods **13**(10), 845 (2016)
11. Hodan, D., Mrazek, V., Vasicek, Z.: Semantically-oriented mutation operator in cartesian genetic programming for evolutionary circuit design. In: Proceedings of the 2020 Genetic and Evolutionary Computation Conference, pp. 940–948 (2020)
12. Irrthum, A., Wehenkel, L., Geurts, P., et al.: Inferring regulatory networks from expression data using tree-based methods. PloS One **5**(9), e12776 (2010)
13. Jackson, C.A., Castro, D.M., Saldi, G.A., Bonneau, R., Gresham, D.: Gene regulatory network reconstruction using single-cell RNA sequencing of barcoded genotypes in diverse environments. Elife **9**, e51254 (2020)
14. Liu, S., Trapnell, C.: Single-cell transcriptome sequencing: recent advances and remaining challenges. F1000Research **5** (2016)
15. Ma, B., Jiao, X., Meng, F., Xu, F., Geng, Y., Gao, R., Wang, W., Sun, Y.: Identification of gene regulatory networks by integrating genetic programming with particle filtering. IEEE Access **7**, 113760–113770 (2019)
16. McCall, M.N.: Estimation of gene regulatory networks. Postdoc J. Postdoc. Res. Postdoc. Affairs, **1**(1), 60 (2013)
17. Miller, J.F.: Cartesian genetic programming. CGP, pp. 17–34 (2011)
18. Miller, J.F., Thomson, P., Fogarty, T.: Designing electronic circuits using evolutionary algorithms. arithmetic circuits: A case study (1997)
19. Moerman, T., et al.: GRNBoost2 and Arboreto: efficient and scalable inference of gene regulatory networks. Bioinformatics **35**(12), 2159–2161 (2019)
20. Pratapa, A., Jalihal, A.P., Law, J.N., Bharadwaj, A., Murali, T.: Benchmarking algorithms for gene regulatory network inference from single-cell transcriptomic data. Nat. methods **17**(2), 147–154 (2020)
21. Qiu, X., Mao, Q., Tang, Y., Wang, L., Chawla, R., Pliner, H.A., Trapnell, C.: Reversed graph embedding resolves complex single-cell trajectories. Nat. Methods **14**(10), 979 (2017)
22. Huynh-Thu, V.A., Sanguinetti, G.: Gene regulatory network inference: an introductory survey. In: Sanguinetti, G., Huynh-Thu, V.A. (eds.) Gene Regulatory Networks. MMB, vol. 1883, pp. 1–23. Springer, New York (2019). https://doi.org/10.1007/978-1-4939-8882-2_1
23. Setty, M., et al.: Wishbone identifies bifurcating developmental trajectories from single-cell data. Nat. Biotech. **34**(6), 637–645 (2016)
24. da Silva, J.E.H., et al.: Inferring gene regulatory network models from time-series data using metaheuristics. In: IEEE Congress on Evolutionary Computer (CEC), pp. 1–8. IEEE (2020)

25. da Silva, J.E.H., Bernardino, H.S., de Oliveira, I.L.: Inference of gene regulatory networks from single-cell RNA-sequencing data using cartesian genetic programming (under review). In: Bioinformatics, pp. 1–8. Oxford (2021)
26. Streichert, F., et al.: Comparing genetic programming and evolution strategies on inferring gene regulatory networks. In: Deb, K. (ed.) GECCO 2004. LNCS, vol. 3102, pp. 471–480. Springer, Heidelberg (2004). https://doi.org/10.1007/978-3-540-24854-5_47
27. Trapnell, C., et al.: The dynamics and regulators of cell fate decisions are revealed by pseudotemporal ordering of single cells. Nat. Biotechnol. **32**(4), 381 (2014)
28. Wang, R.S., Saadatpour, A., Albert, R.: Boolean modeling in systems biology: an overview of methodology and applications. Phys. Biol. **9**(5), 055001 (2012)

Online Selection of Heuristic Operators with Deep Q-Network: A Study on the HyFlex Framework

Augusto Dantas[(✉)] and Aurora Pozo

Department of Computer Science, Federal University of Paraná (UFPR),
Curitiba, USA
{aldantas,aurora}@inf.ufpr.br

Abstract. General and adaptive strategies have been a highly pursued goal of the optimization community, due to the domain-dependent set of configurations (operators and parameters) that is usually required for achieving high quality solutions. This work investigates a Deep Q-Network (DQN) selection strategy under an online selection Hyper-Heuristic algorithm and compares it with two state-of-the-art Multi-Armed Bandit (MAB) approaches. We conducted the experiments on all six problem domains from the HyFlex Framework. With our definition of state representation and reward scheme, the DQN was able to quickly identify the good and bad operators, which resulted on better performance than the MAB strategies on the problem instances that a more exploitative behavior deemed advantageous.

Keywords: Hyper-Heuristic · Reinforcement Learning · Combinatorial Optimization

1 Introduction

For many complex optimization problems, the use of heuristic approaches is often required to achieve feasible solutions in reasonable computational time [2]. One drawback of heuristics is that their performance heavily rely on the configuration setting, which must be tuned for the problem-domain at hand [2].

Because of that, the optimization community has investigated several adaptive search methodologies [2], initially only for parameter tunning, but then it expanded for automatically controlling the heuristic operators to be used. These strategies are normally termed in the literature as Hyper-Heuristics (HH) [3], in which the algorithm explore the search space of low-level heuristics. It can also be found as Adaptive Operator Selection (AOS) [7], usually when the selection occurs at a certain step within a meta-heuristic.

Moreover, with the advance and success of Machine Learning (ML) techniques, there has been an increasing interest in using novel ML for guiding the optimization search in several ways, including the selection of heuristic operators [9]. Due to the stochastic and iterative nature of optimization heuristics, Reinforcement Learning [17] techniques have been widely investigated for HH and

© Springer Nature Switzerland AG 2021
A. Britto and K. Valdivia Delgado (Eds.): BRACIS 2021, LNAI 13073, pp. 280–294, 2021.
https://doi.org/10.1007/978-3-030-91702-9_19

AOS applications. However, most of them are traditionally simple additive reinforcement strategies, such as Probability Matching (PM) and Adaptive Pursuit (AP) [7], that use the received feedback to update a probability vector. Others are based on selection rules, that takes into account the feedback and the frequency of appliances in order to deal with the exploration versus exploitation dilemma (e.g., Choice Function and Multi Armed Bandit based strategies [7]).

Although those approaches presented good overall results, they lack a state representation according to the traditional RL definition [17], in which an agent learns a policy (directly or not) by interacting with an environment based on the observed state and the received feedback (reward or penalty). This work investigates a selection Hyper-Heuristic that uses a Deep Q-Network to choose the heuristics. The selection agent is updated while solving an instance using the Q-learning algorithm [19] with an Artificial Neural Network as function approximator [17]. In this way, we model the task of selecting the heuristics as a Markov Decision Process (MDP) [14], which implicates that the state representation must contain enough information for the agent to take an action.

Using a MDP-based strategy for this selection task has been shown to be advantageous over stateless strategies [18]. In fact, there are a few works that have successfully applied Q-Learning for HH and AOS. Handoko et al. [8] defined a discrete state space that relates to fitness improvement and diversity level. Then, the Q-learning updates the state-action values which are used to select the crossover operator of an evolutionary algorithm applied on the Quadratic Assignment Problem. The experimental results demonstrated that the approach is competitive with classical credit assignment mechanisms (AP, PM and MAB), while being less sensitive to the number of operators.

Similarly, Buzdalova et al. [4] applied Q-Learning to select crossover and mutation operators for the Traveling Salesman Problem. The state is defined by a 2-tuple containing the current generation and the fitness improvement of the current best individual over the initial best individual, both values discretized into 4 intervals. Their approach outperformed a random selection, indicating that the agent was able to learn a working policy while solving the instances. Mosadegh et al. [11] proposed a Simulated Annealing (SA) based HH that uses Q-Learning to select the moving operators. Each action consists on three operators, and the state is the number of times that the previous actions succeed (i.e., the operator generated an accepted solution under the SA conditions). The approach was significantly superior to other versions of SA and two software packages, with respect to both the quality of the solution and the computation time.

One limitation of these works is the use of a discrete state space, which may limit the representation of the search stage [18]. However, when defining a continuous state representation, the classical Q-Learning becomes infeasible due to the high dimensional Q-Table. Therefore, a function approximation model is necessary to estimate the state-action values [17]. The work from Teng et al. [18] defined a continuous state space that includes landscape measures about the current population and some parent-oriented features. Then, a Self-Organizing Neural Network is trained offline to select the crossover operator. The performance of this approach was competitive with other selection mechanisms (including a tabular Q-Learning) and even better on some instances, thus highlighting the advantages of using a continuous MDP-based selection strategy.

In Sharma et al. [15], the authors used a Double Deep Q-Network to select mutation operators of a Differential Evolution algorithm applied on several CEC2005 benchmark functions. The target network, which is trained offline during the training phase, receives as input 99 continuous features, where 19 are related to the current population and the reaming characterize the performance of each operator so far. This approach outperformed other non-adaptive algorithms and was competitive with state-of-the-art adaptive approaches.

In contrast with these works, we modeled a continuous state representation that consider only the past performance of the operators (within a certain memory), and the DQN is only trained online to learn a selection policy on each instance. We compare this approach with two state-of-the-art MAB based selection rules, namely the Dynamic MAB (DMAB) [5] and the Fitness-Rate-Rank MAB (FRRMAB) [10]. The MAB problem can be seen as a special case of Reinforcement Learning with only a single state [17].

We have present a preliminary work in a workshop paper [6], in which we report the results on two problem domains (the Vehicle Routing and the Traveling Salesman problems). Later, we found a flaw in our FRRMAB implementation. Here, we expand the results to all six problems from the Hyper-Heuristics Flexible framework (HyFlex) [12] and with all implementations revised. Moreover, we also present some analysis on the behavioral aspect of the selection mechanisms.

The remainder of this paper is organized as follows: Sect. 2 explains the concepts of HHs and describes the selection strategies that we compared: DMAB, FRRMAB and DQN. The experimental setup and results are given in Sect. 3. Finally, we draw some conclusions and indicate future works in Sect. 4.

2   Selection Hyper-Heuristic

According to Burke et al. [3], Hyper-Heuristics can be divided in two groups: selection HH, where it selects from a set of predefined low-level heuristics (llh); generative HHs, where the algorithm uses parts of llhs to construct new ones. Moreover, they can also be classified by its source of learning feedback: online, offline, and no-learning. This work is about an online learning selection HH.

As a search methodology, selection HHs explore the search space of low-level heuristics (e.g., evolutionary operators) [3]. To avoid getting stuck into local optima solutions, good HHs must know which is the appropriate low-level heuristic to explore a different area of the search space at the time [3]. We used in this work a standard selection Hyper-Heuristic algorithm, as shown in Algorithm 1. Iteratively, it selects and applies a low-level heuristic on the current solution and computes the reward. Then, the acceptance criteria decides if the new solution is accepted and, at last, the HH calls the update method of the corresponding selection model.

Algorithm 1: Selection Hyper-Heuristic

Input: A initial solution ϕ with size n
Output: The best found solution
repeat
    heuristic ← SelectHeuristic()
    ϕ' ← ApplyHeuristic(ϕ, heuristic)
    reward ← GetReward(f(ϕ), f(ϕ'))
    if AcceptSolution(ϕ') **then**
       | $\phi \leftarrow \phi'$
    end
    UpdateSelectionModel(reward)
until *stopping criteria is not met*

The reward is defined as the FIR value (Eq. 2) and was kept the same for all selection strategies. Since our goal is to investigate the learning ability of the selection strategies, the acceptance criteria accepts all solutions. In this way, the actions of the agent always reflects a change in the environment.

The selection is made according to the employed selection strategy. In this work, we compared three strategies: Dynamic MAB, Fitness-Rate-Rank MAB, and Deep Q-Network.

2.1 Dynamic Multi-Armed Bandit

A MAB framework is composed of N arms (e.g., operators) and a selection rule for selecting an arm at each step. The goal is to maximize the cumulative reward gathered over time [16]. Among several algorithms to solve the MAB, the Upper Confidence Bound (UCB) [1] is one of the most known in the literature, as it provides asymptotic optimality guarantees. The UCB chooses the arm that maximizes the following rule

$$p_{i,t} + C\sqrt{\frac{2log(\sum_{j=1}^{N} n_{j,t})}{n_{i,t}}} \quad (1)$$

where $n_{i,t}$ is the number of times the ith arm has been chosen, and $p_{i,t}$ the average reward it has received up to time t. The scaling factor C gives a balance between selecting the best arm so far ($p_{i,t}$, i.e., exploitation) and those that have not been selected for a while (second term in the Eq. 1, i.e., exploration).

However, the UCB algorithm was designed to work in static environments. This is not the case in the Hyper-Heuristic context, where the quality of the low-level heuristics can vary along the HH iterations [7]. Hence, the Dynamic MAB, proposed by [5], incorporates the *Page-Hikley* (PH) statistical test to deal with this issue. This mechanism resembles a context-drifting detection, but is related to the performance of the operators throughout the execution of the algorithm. Once a change in the reward distribution is detected, according to the PH test, the DMAB resets the empirical value estimates and the confidence intervals (p and n in Eq. 1, respectively) of the UCB [5].

2.2   Fitness-Rate-Rank Multi-Armed Bandit

The Fitness-Rate-Rank MAB [10] proposes the use of Fitness Improvement Rate (FIR) to measure the impact of the application of an operator i at time t, which is defined as

$$FIR_{i,t} = \max \left(0, \frac{pf_{i,t} - cf_{i,t}}{pf_{i,t}} \right) \tag{2}$$

where $pf_{i,t}$ is the fitness value of the original solution, and $cf_{i,t}$ is the fitness value of the offspring.

Moreover, the FFRMAB uses a sliding window of size W to store the indexes of past operators, and their respective FIRs. This sliding window is organized as a First-in First-out (FIFO) structure and reflects the state of the search process. Then, the empirical reward $Reward_i$ is computed as the sum of all FIR values for each operator i in the sliding window.

In order to give an appropriate credit value for an operator, the FRRMAB ranks all the computed $Reward_i$ in descending order. Then, it assigns a decay value to them based on their rank value $Rank_i$ and on a decaying factor $D \in [0, 1]$

$$Decay_i = D^{Rank_i} \times Reward_i \tag{3}$$

The D factor controls the influence for the best operator (the smaller the value, the larger influence). Finally, the Fitness-Rate-Rank (FRR) of an operator i is given by

$$FRR_{i,t} = \frac{Decay_i}{\sum_{j=1}^{N} Decay_j} \tag{4}$$

These $FRR_{i,t}$ values are set as the value estimate $p_{i,t}$ in the UCB Eq. (1). Also, the $n_{i,t}$ values considers only the amount of time that the operator appears in the current sliding window. This differs from the traditional MAB and other variants such as the DMAB, where the value estimate $p_{i,t}$ is computed as the average of all rewards received so far.

2.3   Deep Q-Network

The classic Q-learning algorithm keeps a table that stores the Q-values (the estimate value of performing an action at current state) of all state-action pairs [19]. This table is then updated accordingly to the feedback the agent receives upon interacting with the environment. However, in a continuous state space, keeping the Q-table is not feasible due to the high dimensionality of the problem [17]. Instead, we can use a function approximation model (called the Q-model) that gives the estimate Q-values. In DQN, the Q-model is defined by an Artificial Neural Network (ANN), in which the inputs are the current observed state representation, and the output layer yields the predicted Q-values for the current state-action pairs. For this task, we used the MultiLayer Perceptron Regressor from the Scikit-learn library [13].

With these estimated Q-values, the agent selects the next action (low-level heuristic) according to its exploration policy. We used the ϵ-greedy policy, that selects a random action with probability ϵ, and selects the action with the highest Q-value with probability $1 - \epsilon$. Thus, ϵ is a parameter that controls the degree of exploration of the agent and is usually set to a small value [17].

After performing the action, receiving the reward and observing the next state, the Q-model is updated by running one iteration of gradient descent on the Artificial Neural Network, with the following target value

$$\text{target} = \text{reward} + \gamma \max_{a'} Q\left(s', a'\right) \tag{5}$$

where s' is the next state after performing the action, and $\max_{a'} Q\left(s', a'\right)$ is the highest Q-value of all possible actions from state s'. The discount factor γ ($[0,1]$) controls the influence of the future estimate rewards.

We defined the state representation as the normalized average rewards of each operator. For this, we used the same sliding window structure from FRRMAB. Hence, if we have 10 available low-level heuristics, for example, the state is represented as a vector of 10 values ranging $[0,1]$, where each value is the average of the past W rewards (window size) of an operator. The idea is to investigate if the past observed rewards can be representative enough to allow the DQN to learn a proper selection policy.

3   Experiments

We conducted the experiments on all 6 problems from the HyFlex Framework [12]: One Dimensional Bin Packing (BP), Flow Shop (FS), Personal Scheduling (PS), Boolean Satisfiability (MAX-SAT), Traveling Salesman Problem (TSP), and Vehicle Routing Problem (VRP). The HyFlex provides 10 instances of each domain and 4 types of low-level heuristics: mutational, ruin-and-recreate, local search, and crossover. We included all operators but the crossover group into the selection pool. We refer to the documentation for more details [12].

We executed each selection strategy 31 times on every instance with different random seeds. We set the stopping criteria as 300 s of CPU running time on a Intel(R) Core(TM) i7-5930K CPU @ 3.50 GHz. The number of runs and stopping criteria were set following the Cross-Domain Heuristic Search Challenge[1] competition rules, for which the HyFlex was originally developed. Table 1 displays the parameters setting that we used throughout the experiments.

Next we present the performance comparison of the selection strategies for each problem domain. For this, we compared the mean performance obtained by each approach on all instances using the Friedman hypothesis test and a pairwise post-hoc test with the Bergmann correction. The results of these tests are shown in the pipe graphs (e.g. Fig. 1), where the approaches are displayed according to their rank (the smaller the better), and the connected bold lines indicate the approaches that are statistically equivalent ($p < 0.05$).

---

[1] http://www.asap.cs.nott.ac.uk/external/chesc2011/.

Table 1. Parameters setting

	Parameter	Value
UCB	C	8
FRRMAB	W	100
	D	1
DQN	γ	0.9
	W	100
	ϵ	0.05
	ANN hidden layers	(30, 20)
	Learning rate	0.001
	Solver	Adam

Additionally, we also compared them on each instance individually with the Kruskal-Wallis hypothesis test, followed by a pairwise Dunn's test, with the lowest ranked approach set as the control variable. The tables, such as Table 2, report the average and standard deviation of the best solution found by each selection strategy in the 31 runs. Bold values indicate that the corresponding approach achieved a better performance with statistical difference ($p < 0.05$), and gray background highlights all approaches that were statistically equivalent to the approach with the best rank on that instance.

Moreover, we also contrasted the approaches in terms of the selection behavior during the search. For this, we divided the search into 10 phases with equal number of iterations, and computed the average frequency that each operator was selected on that phase, resulting on the line graphs such as Fig. 2.

3.1 Bin Packing

On Bin Packing, both DQN and FRRMAB were statistically equivalent when considering all instances performance, as displayed in Fig. 1 where the DQN ranked better.

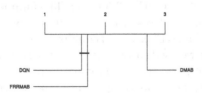

Fig. 1. Friedman ranking with post-hoc tests on BP instances

In fact, if we observe the individual performance shown in Table 2, we can notice that their performance was dependent on the type of the instance, meaning that neither of them could generalize for all instances.

Table 2. Performance comparison on Bin Packing

ID	Instance	DMAB	DQN	FRRMAB
0	falkenauer-falk1000-1	0.0757 (0.0016)	0.0556 (0.0041)	**0.0503 (0.0029)**
1	falkenauer-falk1000-2	0.0744 (0.002)	0.0538 (0.0049)	**0.0479 (0.0027)**
2	schoenfield-schoenfieldhard1	0.0283 (0.0006)	**0.0275 (0.0005)**	0.0284 (0.0005)
3	schoenfield-schoenfieldhard2	0.0459 (0.0083)	0.0312 (0.0011)	0.0315 (0.0005)
4	2000-10-30-instance1	0.0275 (0.0017)	0.0228 (0.0063)	0.0245 (0.0022)
5	2000-10-30-instance2	0.0289 (0.002)	0.0262 (0.0081)	**0.0247 (0.0007)**
6	trip1002-instance1	0.181 (0.0114)	**0.162 (0.0285)**	0.1838 (0.0086)
7	trip2004-instance1	0.1848 (0.0074)	**0.1626 (0.0234)**	0.1842 (0.0069)
8	testdual4-binpack0	0.1207 (0.0019)	0.1105 (0.0044)	**0.1044 (0.0036)**
9	testdual7-binpack0	0.0421 (0.0009)	0.0422 (0.0013)	**0.0377 (0.0014)**

Figure 2 displays the average appliances of each operator for instance 0, that the DQN was worse than FRRMAB. We can observe that the DQN presents a exploitative behavior, giving high emphasis on the early prominent operators. The FRRMAB, on the other hand, was able to detect a change in operator performance.

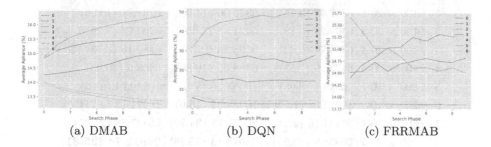

|(a) DMAB | (b) DQN | (c) FRRMAB |

Fig. 2. Average selection of operators on BP instance 0

However, this exploitative behavior was advantageous on some instances, such as instance 7 as shown in Fig. 3. The DQN detected at the initial phases the set of prominent operators, while the FRRMAB needed more time to do so.

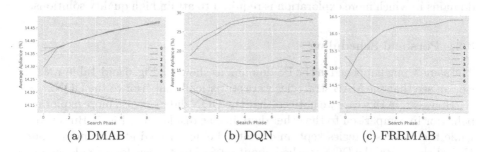

|(a) DMAB | (b) DQN | (c) FRRMAB |

Fig. 3. Average selection of operators on BP instance 7

3.2   Flow Shop

The DQN presented very poor performance on Flow Shop, where the FRRMAB and DMAB were both statistically better, as displayed in Fig. 4. Even when comparing the instances individually (Table 3), the DQN was not competitive in any instance.

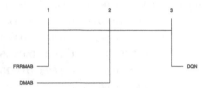

Fig. 4. Friedman ranking with post-hoc tests on FS instances

Table 3. Performance comparison on Flow Shop

ID	Instance	DMAB	DQN	FRRMAB
0	100 × 20-1	6389.13 (10.0)	6399.13 (8.68)	6385.39 (5.15)
1	100 × 20-2	6337.42 (8.75)	6346.06 (14.0)	6331.71 (10.96)
2	100 × 20-3	6409.29 (8.18)	6424.68 (7.85)	6406.74 (7.26)
3	100 × 20-4	6393.77 (6.64)	6405.26 (9.13)	6389.16 (7.73)
4	100 × 20-5	6473.29 (7.83)	6488.65 (10.86)	6468.9 (10.93)
5	200 × 10-1	10544.1 (8.63)	10553.26 (8.16)	10540.94 (9.53)
6	200 × 10-2	10974.58 (10.18)	10988.35 (15.6)	10971.13 (9.71)
7	500 × 20-1	26444.16 (29.13)	26477.71 (34.79)	26437.58 (21.05)
8	500 × 20-2	26938.9 (33.22)	26978.13 (35.76)	**26912.74 (25.31)**
9	500 × 20-4	26753.84 (27.49)	26787.61 (27.47)	26742.58 (26.18)

Figure 5 shows that, although the DQN kept selecting a few operators at around 10% of times, its exploitative behavior of giving high emphasis on the top two operators resulted on the poor performance on Flow Shop. Therefore, the reward scheme and/or the state representation that we defined were not satisfiable for domains in which more exploration is required to attain high quality solutions.

3.3   Personal Scheduling

The same thing happened in the PS domain, as shown in Fig. 6 and Table 4, where we can see that the MAB strategies were statistically superior to the DQN.

However, in this case we noticed that the DQN did present a more explorative behavior in comparison to the other domains, as displayed in Fig. 7. But even so, while the MAB strategies kept an operator being selected at maximum around 15% of the times, the DQN reached about 30% of preference for a single operator on some phases.

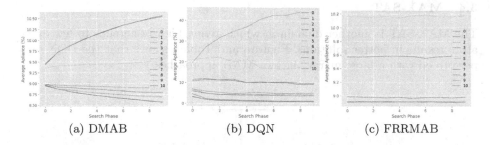

Fig. 5. Average selection of operators on FS instance 0

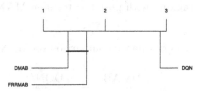

Fig. 6. Friedman ranking with post-hoc tests on PS instances

Table 4. Performance comparison on Personal Scheduling

ID	Instance	DMAB	DQN	FRRMAB
0	BCV-A.12.2	2369.19 (106.61)	2641.16 (430.57)	2433.52 (122.82)
1	BCV-3.46.1	3343.71 (11.9)	3345.77 (26.87)	3348.48 (13.67)
2	ORTEC02	385.35 (20.47)	714.55 (624.1)	390.35 (19.42)
3	Ikegami-3Shift-DATA1	21.32 (2.96)	44.45 (70.7)	21.19 (3.14)
4	Ikegami-3Shift-DATA1.1	23.48 (2.8)	54.61 (63.06)	24.03 (2.55)
5	Ikegami-3Shift-DATA1.2	24.65 (3.01)	35.32 (7.65)	24.58 (2.96)
6	CHILD-A2	1141.35 (38.43)	1676.26 (1295.02)	1133.19 (33.15)
7	ERRVH-A	2278.26 (36.05)	2646.29 (413.8)	2282.48 (45.74)
8	ERRVH-B	3273.9 (49.7)	3545.71 (363.42)	3307.42 (74.79)
9	MER-A	9960.58 (176.96)	13146.9 (14931.08)	9921.45 (115.97)

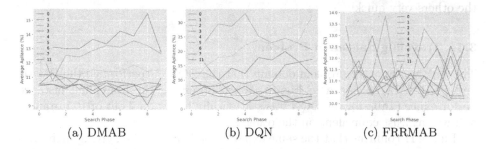

Fig. 7. Average selection of operators on PS instance 0

3.4 MAX-SAT

The MAX-SAT is another domain in which giving high preference to few operators results in better solutions. Figure 8 and Table 5 shows that the DQN outperformed with a large marge the two MAB approaches.

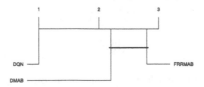

Fig. 8. Friedman ranking with post-hoc tests on MAX-SAT instances

Table 5. Performance comparison on MAX-SAT

ID	Instance	DMAB	DQN	FRRMAB
0	sat05-457	269.97 (6.05)	**16.0 (6.75)**	272.06 (5.05)
1	sat05-488	277.81 (5.52)	**38.58 (4.81)**	277.74 (6.23)
2	sat05-486	273.87 (5.22)	**30.97 (4.2)**	273.94 (6.41)
3	instance_n3_i3_pp	156.13 (4.41)	**10.77 (2.54)**	159.68 (4.42)
4	instance_n3_i3_pp_ci_ce	165.16 (3.46)	**7.1 (1.87)**	168.65 (4.76)
5	instance_n3_i4_pp_ci_ce	257.1 (5.2)	**24.9 (8.69)**	258.16 (6.1)
6	HG-3SAT-V250-C1000-1	44.58 (1.83)	**5.9 (0.59)**	48.81 (1.86)
7	HG-3SAT-V250-C1000-2	43.1 (2.4)	**5.84 (0.68)**	47.61 (2.1)
8	HG-3SAT-V300-C1200-2	63.97 (2.48)	**7.74 (1.37)**	67.55 (2.71)
9	t7pm3-9999	368.81 (5.83)	**218.32 (2.94)**	367.94 (4.37)

Interestingly, on some instances such as the one displayed in Fig. 9, the three approaches identified the same set of operators as the best ones, meaning that the simple fitness-based reward scheme was able to give the selection strategy useful information in order to select the proper operators at each time. But again, the DQN selects with higher frequency the better heuristics and discard the others very quickly.

3.5 Traveling Salesman Problem

The TSP was another problem domain that the DQN outperformed the DMAB and FRRMAB, as shown in Fig. 10. When observing the individual instance performance (Table 6), the DQN was statistically better on 7 out of 10 instances and was at least equivalent on the others.

Figure 11 confirms that the same exploitative behavior, i.e., giving high preference for one to three heuristics, allows the DQN to attain high quality solution on some domains.

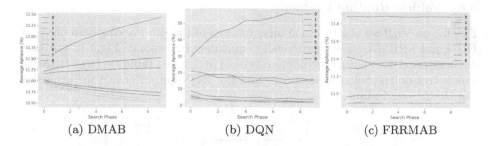

(a) DMAB   (b) DQN   (c) FRRMAB

Fig. 9. Average selection of operators on MAX-SAT instance 0

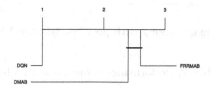

Fig. 10. Friedman ranking with post-hoc tests on TSP instances

Table 6. Performance comparison on TSP

ID	Instance	DMAB	DQN	FRRMAB
0	pr299	49747.28 (249.17)	**49260.04 (250.86)**	50154.84 (411.53)
1	pr439	135745.64 (3762.28)	**111659.98 (840.66)**	134431.84 (3313.73)
2	rat575	7011.83 (16.16)	7003.9 (12.78)	7038.14 (12.67)
3	u724	43535.11 (94.64)	**43325.02 (86.48)**	43748.77 (131.53)
4	rat783	9161.79 (20.45)	9153.62 (16.45)	9194.59 (19.78)
5	pcb1173	60699.53 (221.59)	**59922.78 (277.29)**	61313.72 (321.36)
6	d1291	61735.75 (1209.04)	61399.89 (1465.7)	61700.42 (1100.66)
7	u2152	78938.98 (954.45)	**71886.59 (1490.88)**	79205.61 (1067.75)
8	usa13509	25101651.94 (184874.36)	**23545009.35 (1517146.7)**	25041853.19 (169712.13)
9	d18512	793982.73 (3448.01)	**720240.45 (46824.84)**	792602.46 (3492.93)

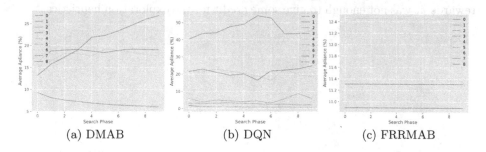

(a) DMAB   (b) DQN   (c) FRRMAB

Fig. 11. Average selection of operators on TSP instance 0

3.6   Vehicle Routing Problem

Finally, on VRP we observed a similar pattern of the Bin Packing domain: both DQN and FRRMAB were statistically equivalent on general (Fig. 12), but they outperformed one another on different type of instances (Table 7). The FRRMAB was better on the Solomon instances and the DQN outperformed the others on the Homberger instances.

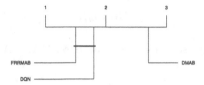

Fig. 12. Friedman ranking with post-hoc tests on VRP instances

Table 7. Performance comparison on VRP

ID	Instance	DMAB	DQN	FRRMAB
0	Solomon-RC207	5689.22 (46.49)	5738.31 (114.05)	**5579.54 (35.3)**
1	Solomon-R101	26120.17 (534.92)	25421.07 (504.9)	**25065.58 (445.35)**
2	Solomon-RC103	16686.23 (427.47)	16742.45 (414.33)	**15859.84 (60.77)**
3	Solomon-R201	6725.91 (267.25)	6766.63 (154.91)	**6097.2 (329.91)**
4	Solomon-R106	18288.04 (532.87)	17901.7 (338.36)	**17604.26 (403.66)**
5	Homberger-C1_10_1	361216.81 (4113.6)	**338425.55 (17118.67)**	358482.72 (5404.71)
6	Homberger-RC2_10_1	112072.12 (2187.45)	109000.24 (11591.1)	111727.57 (1787.41)
7	Homberger-R1_10_1	245814.38 (3857.74)	239111.47 (13276.12)	244623.53 (2885.65)
8	Homberger-C1_10_8	316065.31 (4010.8)	**296789.73 (24428.51)**	314762.88 (5444.14)
9	Homberger-RC1_10_5	223601.08 (2705.0)	216119.92 (13217.05)	221983.66 (3076.55)

Figure 13 shows the frequencies of operator selection on one Homberger instance. Such as in other VRP instances, we observed a pattern in which the DQN gave really high emphasis (more than 80%) for a single operator, which is a local search heuristic. This happens because only the fitness improvement is rewarded and, although the Q-Learning update rule (5) already implicitly deals with delayed rewards, it was not enough to give some credit for the exploration heuristics.

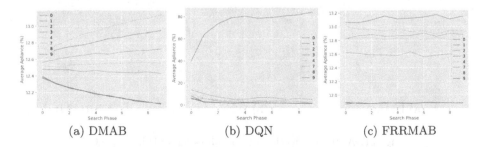

(a) DMAB   (b) DQN   (c) FRRMAB

Fig. 13. Average selection of operators on VRP instance 5

4 Conclusion

This work investigated the use of a heuristic selection strategy using Deep Q-Network. In comparison with the MAB-based strategies, the DQN selects its actions based on a state representation, which can give more insights about the current search stage.

We performed the experiments on six problem domains from the HyFlex framework and compared our approach with DMAB and FRRMAB strategies. We could observe that our approach can detect earlier the good and bad operators. However, it is slower to detect a change in performance of the heuristics. Therefore, for the domains and instances that there exists dominant operators, it outperformed the other selection approaches. On the other hand, it performed worse than both MABs when a wider diversity of operators was necessary.

Hence, further studies should be pursued to improve the exploration of DQN, so it can become more adaptive to different domains. This can be done either by improving the state representation, so it can give different information to the agent (such as stagnation, concept drift, etc.), or by investigating different reward schemes, so it can also rewards operators that do not necessarily improve the fitness function.

Acknowledgements. This work was financially supported by Conselho Nacional de Desenvolvimento Científico e Tecnológico (CNPq) and by Coordenação de Aperfeiçoamento de Pessoal de Nível Superior (CAPES).

References

1. Auer, P., Cesa-Bianchi, N., Fischer, P.: Finite-time analysis of the multiarmed bandit problem. Mach. Learn. **47**(2), 235–256 (2002)
2. Blum, C., Puchinger, J., Raidl, G.R., Roli, A.: Hybrid metaheuristics in combinatorial optimization: a survey. Appl. Soft Comput. **11**(6), 4135–4151 (2011)
3. Burke, E.K., Hyde, M., Kendall, G., Ochoa, G., Özcan, E., Woodward, J.R.: A Classification of Hyper-heuristic Approaches, pp. 449–468. Springer, US, Boston, MA (2010). https://doi.org/10.1007/978-1-4419-1665-5_15
4. Buzdalova, A., Kononov, V., Buzdalov, M.: Selecting evolutionary operators using reinforcement learning: initial explorations. In: Proceedings of the Companion Publication of the 2014 Annual Conference on Genetic and Evolutionary Computation, pp. 1033–1036 (2014)
5. DaCosta, L., Fialho, A., Schoenauer, M., Sebag, M.: Adaptive operator selection with dynamic multi-armed bandits. In: Proceedings of the 10th Annual Conference on Genetic and Evolutionary Computation, pp. 913–920. GECCO '08, Association for Computing Machinery, New York, NY, USA (2008)
6. Dantas, A., Rego, A.F.D., Pozo, A.: Using deep q-network for selection hyper-heuristics. In: Proceedings of the Genetic and Evolutionary Computation Conference Companion, pp. 1488–1492. GECCO '21, Association for Computing Machinery, New York, NY, USA (2021)
7. Fialho, Á.: Adaptive Operator Selection for Optimization. Université Paris Sud - Paris XI (Dec, Theses (2010)

8. Handoko, S.D., Nguyen, D.T., Yuan, Z., Lau, H.C.: Reinforcement learning for adaptive operator selection in memetic search applied to quadratic assignment problem. In: Proceedings of the Companion Publication of the 2014 Annual Conference on Genetic and Evolutionary Computation, pp. 193–194. GECCO Comp '14, Association for Computing Machinery, New York, NY, USA (2014)
9. Karimi-Mamaghan, M., Mohammadi, M., Meyer, P., Karimi-Mamaghan, A.M., Talbi, E.G.: Machine learning at the service of meta-heuristics for solving combinatorial optimization problems: A state-of-the-art. European Journal of Operational Research (2021)
10. Li, K., Fialho, Á., Kwong, S., Zhang, Q.: Adaptive operator selection with bandits for a multiobjective evolutionary algorithm based on decomposition. IEEE Trans. Evol. Comput. **18**(1), 114–130 (2014)
11. Mosadegh, H., Ghomi, S.F., Süer, G.A.: Stochastic mixed-model assembly line sequencing problem: Mathematical modeling and q-learning based simulated annealing hyper-heuristics. Eur. J. Oper. Res. **282**(2), 530–544 (2020)
12. Ochoa, G., et al.: HyFlex: A Benchmark Framework for Cross-domain Heuristic Search, vol. 7245, pp. 136–147 (2012)
13. Pedregosa, F., et al.: Scikit-learn: Machine learning in Python. J. Mach. Learn. Res. **12**, 2825–2830 (2011)
14. Puterman, M.L.: Chapter 8 markov decision processes. In: Handbooks in Operations Research and Management Science, Stochastic Models, vol. 2, pp. 331–434. Elsevier (1990)
15. Sharma, M., Komninos, A., López-Ibáñez, M., Kazakov, D.: Deep reinforcement learning based parameter control in differential evolution. In: Proceedings of the Genetic and Evolutionary Computation Conference, pp. 709–717 (2019)
16. Soria-Alcaraz, J.A., Ochoa, G., Sotelo-Figeroa, M.A., Burke, E.K.: A methodology for determining an effective subset of heuristics in selection hyper-heuristics. Eur. J. Oper. Res. **260**(3), 972–983 (2017)
17. Sutton, R.S., Barto, A.G.: Reinforcement Learning, Second Edition: An Introduction. MIT Press (2018)
18. Teng, T.-H., Handoko, S.D., Lau, H.C.: Self-organizing neural network for adaptive operator selection in evolutionary search. In: Festa, P., Sellmann, M., Vanschoren, J. (eds.) LION 2016. LNCS, vol. 10079, pp. 187–202. Springer, Cham (2016). https://doi.org/10.1007/978-3-319-50349-3_13
19. Watkins, C.J.C.H., Dayan, P.: Q-learning. Mach. Learn. **8**(3), 279–292 (1992)

Knowledge Representation, Logic and Fuzzy Systems

A Systematic Approach to Define Semantics for Prioritised Logic Programs

Renan Cordeiro[1], Guilherme Fernandes[1], João Alcântara[1(✉)],
and Henrique Viana[2]

[1] Department of Computer Science, Federal University of Ceará, Fortaleza, Brazil
{renandcsc,guisalesfer}@alu.ufc.br, jnando@lia.ufc.br
[2] Instituto Federal de Educação, Ciência e Tecnologia do Ceará (IFCE),
CE-040, Km 137,1 s/n, Aracati, Rodovia, CE, Brazil
henrique.viana@ifce.edu.br

Abstract. We propose in this paper a framework of prioritised logic programs (*PLP*) to represent priority information explicitly in a program. Differently of others approaches, we do not restrain the preference relation only to literals, but to sets of literals. As consequence, we can express in *PLP*s sophisticated forms of preferences without changing the programs or introducing new atoms to obtain artificially the intended preferences. Besides, inspired on various developments in the literature on preference, we present a comprehensive and systematic treatment to deal with preferences in logic programming. In fact, we introduced 32 different criteria (semantics) to establish preference between partial stable models as well as those semantics whose definition depends on partial stable models. We show some properties of our framework; in particular, we guarantee these semantics for *PLP* generalise their counterparts for logic programs without preferences.

Keywords: Logic programming · Preferences · Semantics

1 Introduction

In many applications of Artificial Intelligence and Commonsense Reasoning, we have to deal with uncertain, vague, inaccurate, doubtful and even contradictory information as well as with certain, precise and reliable information. In this scenario, conflicts can naturally arise and, it urges to develop mechanisms to resolve them. Adopting some criteria to establish preferences between these conflicting information is an effective way of reasoning with them. Somehow we have to prioritise the information to separate the wheat from the chaff. For representing and reasoning with preferences, several prioritised systems have been proposed (see [1] for a survey).

When moving our eyes to preferences in Logic Programming, we see a plethora of proposals (see [2–12] for a non exhaustive list). As witnessed by the multitude of formalisms with preferences, a key issue here is the lack of

© Springer Nature Switzerland AG 2021
A. Britto and K. Valdivia Delgado (Eds.): BRACIS 2021, LNAI 13073, pp. 297–312, 2021.
https://doi.org/10.1007/978-3-030-91702-9_20

consensus on the role played by preferences. Roughly speaking, we find in the literature various works with diverging intuitions, and as a consequence, conceptually different formal techniques to deal with preference. While there are many proposals on preferences in the reasoning process and knowledge representation, there is still much space for improvement.

Indeed, as noticed in [13], one of the main problems a user faces when expressing preferences is any preference representation language based on the direct assessment of user preferences over the complete set of options is simply infeasible. We have to resort to compact preference representation languages which represent partial descriptions of preferences and rank-order the possible options.

With such a motivation in view, in this paper, we generalise the notion of Prioritised Logic Programs (*PLPs*) proposed in [4] to encompass not only the relation of priority between literals (atoms or its negation), but also sets of literals. The resulting framework offers a preference representation language expressive enough to represent complex forms of preference in a compact manner. While in [4], in order to represent a conjunction of preferences, they have to include new rules in the program and preferences between new atoms, we will surmount this problem by representing conjunctions of preferences directly.

A second fundamental challenge is to select the most preferable models given the priorities between the sets of literals. Which criteria should be regarded? In order to decide a model I_1 is more preferable than a model I_2, someone would require any set of literals associated with I_1 should have higher priority than any set of literals associated with I_2. For others, it suffices to guarantee the worst-ranked sets of literals associated with I_1 should have higher priority than those worst-ranked sets of literals associated with I_2. Depending on the criteria employed to select the most preferable models, we would obtain different semantics for *PLPs*. Overall, there is no consensual way to define these semantics and, we will find many disconnected approaches. In this work, we will not cling to a unique path, but present comprehensive and systematic criteria to define semantics for them. As result, we will have a clearer notion of what we mean when selecting the most preferable models.

The rest of the paper develops as follows. In Sect. 2, we will introduce the basic concepts related to Normal Logic Programs (*NLP*) and their semantics. The main contributions of the paper can be found in Sect. 3, where we will present the Prioritised Logic Programs (*PLPs*) and comprehensive and systematic criteria to define semantics for them. Next, we show how complex forms of priorities can be expressed straightforwardly in *PLPs*. In Sect. 5, we prove some properties of our proposal; in particular, we guarantee these semantics for *PLP* generalise their counterparts for logic programs without preferences. Section 6 provides an overview of related works. Finally, Sect. 7 summarises our contributions and pointed out future works.

2   Preliminaries

In the sequel, we will consider propositional normal logic programs, which we will call logic programs or simply programs from now on.

Definition 1. *A Normal Logic Program (NLP), P, is a set of rules of the form* $a \leftarrow a_1, \ldots, a_m, \text{not } b_1, \ldots, \text{not } b_n$ *(m, n $\in \mathbb{N}$), where a, a_i (1 $\leq i \leq m$) and b_j (1 $\leq j \leq n$) are atoms;* **not** *represents default negation, and* **not** b_j *is a default literal. A literal is an atom or a default literal. We say a is the head of the rule, and $a_1, \ldots, a_m, \text{not } b_1, \ldots, \text{not } b_n$ is its body. If P has no occurrence of default literals, it is called a Definite Logic Program. The Herbrand Base of P is the set HB_P of all atoms occurring in P. The set of literals of P is $Lit_P = HB_P \cup \{\text{not } a \mid a \in HB_P\}$.*

A wide range of logic programming semantics can be defined based on the 3-valued interpretations (for short, interpretations) of programs:

Definition 2. *[14] A set $I \subseteq Lit_P$ is a 3-valued interpretation of an NLP P if $\not\exists a \in HB_P$ with $\{a, \text{not } a\} \subseteq I$. We say I is a model of P iff for each rule $a \leftarrow a_1, \ldots, a_m, \text{not } b_1, \ldots, \text{not } b_n \in P$, if $\{a_1, \ldots, a_m, \text{not } b_1, \ldots, \text{not } b_n\} \subseteq I$, then $a \in I$. We refer to $T_I = \{a \in HB_P \mid a \in I\}$ as the set of true atoms w.r.t. I and to $F_I = \{a \in HB_P \mid \text{not } a \in I\}$ as the set of false atoms w.r.t. I. If $a \notin T_I \cup F_I$, then a is undefined w.r.t. I.*

Let I be a 3-valued interpretation of a program P; in order to define the semantics for normal logic programs, we will resort to the reduct of P with respect to I (written as P/I), which is the definite logic program built by the execution of the following steps:

1. Remove any $a \leftarrow a_1, \ldots, a_m, \text{not } b_1, \ldots, \text{not } b_n \in P$ with $\{b_1, \ldots, b_n\} \cap I \neq \emptyset$;
2. Afterwards, remove any occurrence of **not** b_i from P such that **not** $b_i \in I$.
3. Then, replace any occurrence of **not** b_i left by a special atom **u** (**u** $\notin HB_P$).

We say I is the least model of a definite logic program P if among the models of P, T_I is minimal (w.r.t. set inclusion) and F_I is maximal (w.r.t. set inclusion). Note that P/I is a definite logic program. As consequence, P/I has a unique least model [14], denoted by $\Omega_P(I)$, such that for any $a \in HB_P$

- $a \in \Omega_P(I)$ iff $a \leftarrow a_1, \ldots, a_m \in P/I$ and $\{a_1, \ldots, a_m\} \subseteq \Omega_P(I)$;
- **not** $a \in \Omega_P(I)$ iff for every $a \leftarrow a_1, \ldots, a_m \in P/I$, it holds $\{\text{not } a_1, \ldots, \text{not } a_m\} \cap \Omega_P(I) \neq \emptyset$.

We now specify the *NLP* semantics to be examined in this paper.

Definition 3. *Let P be an NLP and I be an interpretation:*

- *I is a partial stable (PS) model of P iff $I = \Omega_P(I)$ [14].*
- *I is a well-founded model of P iff I is a PS model of P with minimal T_I [14].*
- *I is a regular model of P iff I is a PS model of P with maximal T_I [15].*
- *I is a stable model of P iff I is a PS model of P where for each $a \in HB_P$, $a \in T_I \cup F_I$ [14].*
- *I is an L-stable model of P iff I is a PS model of P with minimal (w.r.t. set inclusion) $\{a \in HB_P \mid a \notin T_I \cup F_I\}$ [15].*

Example 1. Consider the *NLP P*:

$$a \leftarrow \text{not } b \ b \leftarrow \text{not } a, \text{not } c \ c \leftarrow \text{not } b, \text{not } d \ c \leftarrow \text{not } c \ d \leftarrow \text{not } c$$

With respect to P, we have a) Partial stable models: $I_1 = \emptyset$, $I_2 = \{a, \text{not } b\}$, $I_3 = \{d\}$, $I_4 = \{b, d, \text{not } a\}$, $I_5 = \{a, c, \text{not } b, \text{not } d\}$ and $I_6 = \{a, d, \text{not } b\}$; b) Well-founded model: I_0; c) Regular models: I_4, I_5 and I_6; d) Stable model and L-Stable model: I_5.

In the next section, we will offer a comprehensive and systematic treatment to deal with preferences in logic programming.

3   Semantics for Prioritised Logic Programs

The semantics of Normal Logic Programs seen in Definition 3 have multiple models in general. Preference is then introduced to select among them the intended models of a program, but at the semantic level. In [4], they conceived the notion of Prioritised Logic Programs (*PLPs*) by establishing a prioritisation mechanism between literals (atoms or its negation) to represent and reason with preferences in logic programming at the syntactic level.

Here we generalise this notion of *PLP* to encompass not only the relation of priority between literals, but also sets of literals. The idea is to employ these priorities to select the most preferable partial stable models. We will focus on partial stable models, because it is the base to define the remaining semantics in Definition 3. In this work, the priorities will be established between sets of literals, which we will be called as options:

Definition 4 (Options). *Given an NLP P, an option o from P is a subset of Lit_P. By $\mathcal{O}_P$ we mean the set of all options from P. We say o is inconsistent, if there is $a \in HB_P$ such that $\{a, \text{not } a\} \subseteq o$. Otherwise, o is consistent.*

Thus in a program representing a menu based on fish (f), meat m, white wine (w), red wine (r), cake (c) and ice cream (i), $\{f, \text{not } m, r\}$, $\{i, \text{not } i\}$ and $\emptyset$ are possible options. Not rarely, some options have a higher priority than others:

Definition 5 (Priorities). *Given an NLP P, a reflexive and transitive relation is defined on $\mathcal{O}_P$. For any o_1 and o_2 from $\mathcal{O}_P$, $o_1 \preceq o_2$ is called a priority, and we say o_2 has a higher or equal priority than o_1. We write $o_1 \prec o_2$ if $o_1 \preceq o_2$ and $o_2 \npreceq o_1$, and say o_2 has a strictly higher priority than o_1. Furthermore, o_1 is indifferent to o_2, denoted by $o_1 \approx o_2$, when both $o_1 \preceq o_2$ and $o_2 \preceq o_1$ hold.*

From an *NLP* and a set of priorities, we build a prioritised logic program:

Definition 6 (Prioritised Logic Programs). *A prioritised logic program (PLP) is a pair (P, Φ), where P is an NLP and Φ is a set of priorities on $\mathcal{O}_P$.*

In order to enforce the priority relation in Φ is reflexive and transitive, while maintaining a compact representation of priorities, we resort to the closure of Φ:

Definition 7. (Closure of Φ). *Let (P, Φ) be a PLP. For each o_1, o_2 in $\mathcal{O}_P$, the closure Φ^* of Φ is defined as follows:*

- $o_1 \preceq o_1 \in \Phi^*$;
- *if $o_1 \preceq o_2 \in \Phi$, then $o_1 \preceq o_2 \in \Phi^*$;*
- *if $o_1 \preceq o_2 \in \Phi^*$ and $o_2 \preceq o_3 \in \Phi^*$, then $o_1 \preceq o_3 \in \Phi^*$.*
- $o_1 \preceq o_2 \in \Phi^*$ *and $o_1 \neq \emptyset$ and $o_2 \neq \emptyset$ and $e \notin o_1 \cup o_2$, then $o_1 \cup \{e\} \preceq o_2 \cup \{e\} \in \Phi^*$.*

The closure Φ^* of Φ is the set of priorities derived reflexively and transitively using priorities in Φ. Besides, we introduced in Φ^* an inertia principle to keep the preference representation in Φ compact. In the remainder of this paper, when we say $o_2 \not\prec o_1$ according to Φ^*, we mean $o_2 \preceq o_1 \notin \Phi^*$ or $o_1 \preceq o_2 \in \Phi^*$. Among the many possible options, we are particularly interested in those best/worst-ranked:

Definition 8 (best/worst-ranked option). *Let (P, Φ) be a PLP and $I \subseteq Lit_P$. We say o is a) a best-ranked option in I w.r.t. Φ iff $o \subseteq I$ and $\not\exists o' \subseteq I$ such that $o \prec o' \in \Phi^*$. b) a worst-ranked option in I w.r.t. Φ iff $o \subseteq I$ and $\not\exists o' \subseteq I$ such that $o' \prec o \in \Phi^*$.*

Given two set of options, we have to determine which one is preferable; we introduce below four criteria with this purpose:

Definition 9. *Let (P, Φ) be a PLP and $\mathcal{O}_1$ and $\mathcal{O}_2$ subsets of $\mathcal{O}_P$. We say*

- $\mathcal{O}_1 \subseteq^{\forall\forall} \mathcal{O}_2$ *in Φ iff $\forall o_1 \in \mathcal{O}_1$ and $\forall o_2 \in \mathcal{O}_2$, it holds $o_2 \not\prec o_1$ in Φ^*.*
- $\mathcal{O}_1 \subseteq^{\forall\exists} \mathcal{O}_2$ *in Φ iff $\forall o_1 \in \mathcal{O}_1$, it holds $\exists o_2 \in \mathcal{O}_2$ such that $o_2 \not\prec o_1$ in Φ^*.*
- $\mathcal{O}_1 \subseteq^{\exists\forall} \mathcal{O}_2$ *in Φ iff $\exists o_1 \in \mathcal{O}_1$ such that $\forall o_2 \in \mathcal{O}_2$, it holds $o_2 \not\prec o_1$ in Φ^*.*
- $\mathcal{O}_1 \subseteq^{\exists\exists} \mathcal{O}_2$ *in Φ iff $\exists o_1 \in \mathcal{O}_1$ and $\exists o_2 \in \mathcal{O}_2$ such that $o_2 \not\prec o_1$ in Φ^*.*

Besides, different preference semantics have been proposed in the literature. We recall some of the most well-known of them:

Definition 10 (Preference relation). *Let (P, Φ) be a PLP, I_1 and I_2 be PS models of P, $\mathcal{B}_1$ and $\mathcal{B}_2$ are respectively the sets of all best-ranked options in I_1 and I_2 w.r.t. Φ, and $\mathcal{W}_1$ and $\mathcal{W}_2$ are respectively the sets of all worst-ranked options in I_1 and I_2 w.r.t. Φ. Consider also $x \in \{st, opt, pes, opp\}$ and $y \in \{\forall\forall, \forall\exists, \exists\forall, \exists\exists\}$. By $I_1 \subseteq_x^y I_2$ w.r.t. Φ, we mean the preference relation $\subseteq_x^y$ over the set of PS models of P is defined as follows:*

1. $I_1 \subseteq_x^y I_1$ *w.r.t. Φ;*
2. $I_1 \subseteq_x^y I_2$ *w.r.t. Φ*
   a) *(**Strong Semantics**): if $x = st$ and $\mathcal{B}_1 \subseteq^y \mathcal{W}_2$ in Φ.*
   b) *(**Optimistic Semantics**): if $x = opt$ and $\mathcal{B}_1 \subseteq^y \mathcal{B}_2$ in Φ.*
   c) *(**Pessimistic Semantics**): if $x = pes$ and $\mathcal{W}_1 \subseteq^y \mathcal{W}_2$ in Φ.*
   d) *(**Opportunistic Semantics**): if $x = opp$ and $\mathcal{W}_1 \subseteq^y \mathcal{B}_2$ in Φ.*
3. *If $I_1 \subseteq_x^y I_2$ w.r.t. Φ and $I_2 \subseteq_x^y I_3$ w.r.t. Φ, then $I_1 \subseteq_x^y I_3$ w.r.t. Φ.*

When it is clear from the context, we will drop the reference to Φ and say $\mathcal{O}_1 \subseteq^{\forall\forall} \mathcal{O}_2$ and $I_1 \subseteq^y_x I_2$. Inspired by [4], we can define a preference relation $I_1 \sqsubseteq^y_x I_2$ which consider just their exclusive portions given by $I_1 - I_2$ and $I_2 - I_1$.

1. *$I_1 \sqsubseteq^y_x I_2$ iff $I_1 - I_2 \subseteq^y_x I_2 - I_1$.*
2. *If $I_1 \sqsubseteq^y_x I_2$ and $I_2 \sqsubseteq^y_x I_3$, then $I_1 \sqsubseteq^y_x I_3$.*

Note we have four criteria to choose the best/worst-ranked options (Definition 9), four criteria to choose the most preferable semantics (Definition 10). We also can decide if a *PS* model I_1 is preferred to a *PS* model I_2 by regarding the whole models or just their exclusive portions given by $I_1 - I_2$ and $I_2 - I_1$. Hence, we have at disposal 32 manners of selecting the preferred partial stable models:

Definition 11 (Preferred Partial Stable Model). *Let (P, Φ) be a PLP. A PS model I of P is a Preferred Partial Stable Model (or PPS model, for short) of (P, Φ) w.r.t. $\subseteq^y_x$ (resp. $\sqsubseteq^y_x$) if $I \subseteq^y_x I'$ (resp. $I \sqsubseteq^y_x I'$) implies $I' \subseteq^y_x I$ (resp. $I' \sqsubseteq^y_x I$) (w.r.t Φ) for any PS model I' of P. The set of all PPS of (P, Φ) w.r.t. $\subseteq^y_x$ (resp. $\sqsubseteq^y_x$) is written as $\mathcal{PPS}(P, \Phi)_{\subseteq^y_x}$ (resp. $\mathcal{PPS}(P, \Phi)_{\sqsubseteq^y_x}$).*

As the remaining semantics for *NLP* seen in this paper are based on *PS* models (Definition 3), we are equipped with 32 different manners of selecting the most preferred models according to them.

Definition 12. *Let (P, Φ) be an PLP and I be an interpretation:*

- *I is a preferential well-founded model of (P, Φ) w.r.t. $\subseteq^y_x$ (resp. $\sqsubseteq^y_x$) iff I is a PPS model of (P, Φ) w.r.t. $\subseteq^y_x$ (resp. $\sqsubseteq^y_x$) with minimal T_I.*
- *I is a preferential regular model of (P, Φ) w.r.t. $\subseteq^y_x$ (resp. $\sqsubseteq^y_x$) iff I is a PPS model of (P, Φ) w.r.t. $\subseteq^y_x$ (resp. $\sqsubseteq^y_x$) with maximal T_I.*
- *I is a preferential stable model of (P, Φ) w.r.t. $\subseteq^y_x$ (resp. $\sqsubseteq^y_x$) iff I is a PPS model of (P, Φ) w.r.t. $\subseteq^y_x$ (resp. $\sqsubseteq^y_x$) where for each $a \in HB_P$, $a \in T_I \cup F_I$.*
- *I is a preferential L-stable model of (P, Φ) w.r.t. $\subseteq^y_x$ (resp. $\sqsubseteq^y_x$) iff I is a a PPS model of (P, Φ) w.r.t. $\subseteq^y_x$ (resp. $\sqsubseteq^y_x$) with minimal (w.r.t. set inclusion) $\{a \in HB_P \mid a \notin T_I \cup F_I\}$ $\{a \in HB_P \mid M(a) = \mathbf{u}\}$.*

Let us consider the following example:

Example 2. Let P be the *NLP* in Example 1 and $\Phi = \{\emptyset \preceq \{a\}, \{a\} \preceq \{d\}, \{d\} \preceq \{b\}, \{b, d\} \preceq \{b\}, \{b, \text{not } a\} \preceq \{\text{not } a\}, \{a, d\} \preceq \{d\}\}$. Then the best-ranked options for I_1, I_2, I_3, I_4, I_5 and I_6 are respectively $\mathcal{B}_1 = \emptyset, \mathcal{B}_2 = \{\{a\}, \{\text{not } b\}, \{a, \text{not } b\}\}, \mathcal{B}_3 = \{\{d\}\}, \mathcal{B}_4 = \{\{b\}, \{\text{not } a\}\}, \mathcal{B}_5 = 2^{I_5}$ and $\mathcal{B}_6 = \{\{d\}, \{\text{not } b\}, \{d, \text{not } b\}\}$. Regarding only the optimistic approach for illustrative purposes, we have $\mathcal{PPS}(P, \Phi)_{\subseteq^{\forall\forall}_{opt}} = \{I_4\}$; $\mathcal{PPS}(P, \Phi)_{\subseteq^{\forall\exists}_{opt}} = \{I_2, I_4, I_5, I_6\}$; $\mathcal{PPS}(P, \Phi)_{\subseteq^{\exists\forall}_{opt}} = \{I_3, I_4, I_6\}$ and $\mathcal{PPS}(P, \Phi)_{\subseteq^{\exists\exists}_{opt}} = \{I_2, I_3, I_4, I_5, I_6\}$. In this example, notice at least one *PS* model of P is discarded in $\mathcal{PPS}(P, \Phi)_{\subseteq^y_{opt}}$ for any $y \in \{\forall\forall, \forall\exists, \exists\forall, \exists\exists\}$.

Many attempts have been made to empower logic programming with preferences. The distinguishing aspect of our work is that we offer a comprehensive and systematic way of dealing with preferences in logic programming inspired by well-known criteria found in the literature on preference. In the next section, we will examine how expressive our proposal is.

4 Expressing Priorities in *PLP*s

As it is not feasible in general when expressing preferences to compare all possible pairs of options or evaluate them individually (the number of options increases exponentially with the number of variables), we have to resort to compact preference representation languages to represent these partial descriptions preferences. Now we will show our preference representation language although compact is robust enough to express well-known complex forms of preferences. In Subsects. 4.1, 4.2, 4.3, 4.4 and 4.5, we will tackle respectively how to represent priorities between conjunctive knowledge, priorities between disjunctive knowledge, conditional priorities, priorities between rules and bipolar priorities.

4.1 Priorities Between Conjunctive Knowledge

Users may also express their partial preferences in term of comparative statements (as in "*I prefer white wine to red wine*"); this preference can be altered when interacting with other options (as in "*I prefer meat accompanied by a red wine to meat accompanied by a white wine*"); the preference for white wine can even be regained with the addition of more options (as in "*I prefer meat and cheesecake accompanied by a white wine to meat and cheesecake accompanied by a red wine*"). Thus, the conjunction of options can interfere with the users' preference and change the overall outcome.

In [4], conjunctive preferences $(e_1, \ldots, e_m) \preceq (e_1', \ldots, e_n')$ cannot in general be expressed directly as the preference relation applied there only involves a pair of individual literals. In order to represent these conjunctive preferences in a *PLP* (P, Φ), the authors have to add the rules $e_0 \leftarrow e_1, \ldots e_m$ and $e_0' \leftarrow e_1', \ldots e_n'$ to P with the newly introduced atoms e_0 and e_0', and the priority $(e_0 \preceq e_0')$ to Φ. In short, conjunctive preferences are represented in [4] by changing the program P with two new atoms and two new rules before introducing $(e_0 \preceq e_0')$, which expresses indirectly the desired conjunctive preference.

In our approach, we neither need to change the program P nor to add new artificial atoms to express conjunctive preferences. Preferences as $(e_1, \ldots, e_m) \preceq (e_1', \ldots, e_n')$ can just be represented as $\{e_1, \ldots, e_m\} \preceq \{e_1', \ldots, e_n'\}$ and included straightforwardly in Φ.

4.2 Priorities Between Disjunctive Knowledge

Disjunctive preferences as $(e_1 \vee \cdots \vee e_m) \preceq (e_1' \vee \cdots \vee e_n')$ can be represented in a *PLP* (P, Φ) by ensuing $\Phi' \subseteq \Phi$ and $\Phi' = \{e_i \preceq e_j' \mid 1 \leq i \leq m \text{ and } 1 \leq j \leq n\}$.

4.3 Conditional Priorities

It is not always the case a preference is absolute. In the so-called conditional preference, users may also wish to express specific preferences in particular contexts (e.g., *If red wine is served, I prefer meat to fish*). It is denoted by $\gamma : \phi \preceq \psi$ and interpreted as "given a context γ, prefer ψ to ϕ". Assuming $\gamma = c_1 \wedge \cdots \wedge c_o$, $\phi = e_1 \wedge \cdots \wedge e_m$ and $\psi = e'_1 \wedge \cdots \wedge e'_n$, we can represent the conditional preference $\gamma : \phi \preceq \psi$ as the conjunctive preference $\{c_1, \ldots, c_o, e_1, \ldots, e_m\} \preceq \{c_1, \ldots, c_o, e'_1, \ldots, e'_n\}$ in Φ.

Conditional preferences can occur disguised as in "if $\gamma = c_1 \wedge \cdots \wedge c_o$ is true, then prefer $\psi = e'_1 \wedge \cdots \wedge e'_n$"; it interpreted as if γ is true, then prefer ψ to not e'_j with $1 \leq j \leq n$. This is a particular case of our representation when $\phi = \text{not } e'_j$, i.e., such a preference can be represented in (P, Φ) by including in Φ the preferences $\{c_1, \ldots, c_o, \text{not } e'_j\} \preceq \{c_1, \ldots, c_o, e'_1, \ldots, e'_n\}$ with $1 \leq j \leq n$.

4.4 Priorities Between Rules

Preferences can also be established between (conflicting) rules as in $e \leftarrow e_1, \ldots, e_m \preceq e' \leftarrow e'_1, \ldots, e'_n$. Its intended meaning is the conclusion of e' via $e' \leftarrow e'_1, \ldots, e'_n$ will block the conclusion of e via $e \leftarrow e_1, \ldots, e_m$. Such a preference can be represented in (P, Φ) by including $\{e_1, \ldots, e_m\} \preceq \{e'_1, \ldots, e'_n\}$ in Φ.

4.5 Bipolar Preferences

Preferences can express not only what is satisfactory (positive preferences), but also what can be considered tolerable or unacceptable (negative preferences). These two forms of preferences (bipolar preferences) have been conjointly and compactly expressed in various works [16–21]. Roughly speaking, the intended meaning of introducing positive preferences is to express wishes which should be satisfied as best as possible, whilst negative preferences are intended to enhance the idea that what is not rejected or excluded is tolerated. Thus, when interpreting $\preceq$ as a bipolar priority by $\{e_1, \ldots, e_m\} \preceq \{e'_1, \ldots, e'_n\}$, we mean we are giving priority not only to outcomes satisfying $\{e_1, \ldots, e_m\} \preceq \{e'_1, \ldots, e'_n\}$, but to those not falsifying $\{e_1, \ldots, e_m\} \preceq \{e'_1, \ldots, e'_n\}$, i.e., it either should contain $\{e'_1, \ldots, e'_n\}$, but not $\{e_1, \ldots, e_m\}$ (positive preference), or should contain $\{\text{not } e_i\}$, but not $\{\text{not } e'_j\}$ with $1 \leq i \leq m$ and $1 \leq j \leq n$ (negative preference).

Bipolar preferences as $\{e_1, \ldots, e_m\} \preceq \{e'_1, \ldots, e'_n\}$ can be represented in (P, Φ) by including in Φ the conjunctive preference $\{e_1, \ldots, e_m\} \preceq \{e'_1, \ldots, e'_n\}$ and the preferences $\{\text{not } e'_j\} \preceq \{\text{not } e_i\}$ with $1 \leq i \leq m$ and $1 \leq j \leq n$.

Before moving to the next section, we will emphasise our decision of employing a representation language to express priorities between sets of literals instead of a language to express priorities only between literals as in [4]. Although the approach of [4] suffices to express in a *PLP* (P, Φ) the complex forms of preferences showed in this section, there is a price to pay: the program P should be changed with the introduction of new atoms as well as the set Φ of priorities. In our approach, as the priority relation is given in term of sets of literals, we neither need to change the program P nor to add new artificial atoms to express priorities between conjunctive knowledge, priorities between disjunctive knowledge, conditional priorities, priorities between rules and bipolar priorities.

5   Results

Now we will prove some properties found in our proposal. The main results are both the preference relations $\subseteq_x^y$ and $\sqsubseteq_x^y$ are consistent (Proposition 1) and have neither a monotonic nor an antitonic behaviour w.r.t. Φ (Theorem 3); Proposition 2 and Theorem 1 establish conditions under which the relations of preference are simplified; when no priority is available, Prioritised Logic Programs and Normal Logic Programs have the same semantics (Theorem 2) and inconsistent options are useless (Theorem 4). Next we define the notion of cyclic relation:

Definition 13 (Cyclic Relation). *A relation $\sqsubseteq$ is cyclic if and only if its induced strict preference relation is cyclic, i.e., there exists a chain $\mathcal{O} \ldots \mathcal{O}'$ such that $\mathcal{O} \sqsubset \mathcal{O}' \sqsubset \cdots \sqsubset \mathcal{O}$. Otherwise $\sqsubseteq$ is acyclic.*

If either $\subseteq_x^y$ or $\sqsubseteq_x^y$ were cyclic, the resulting preference relation would be inconsistent. In particular, we would find *PLP*s without preferred partial stable models at all. For instance, suppose an *NLP* whose partial stable models are M_1, M_2 and M_3 such that $M_1 \sqsubset_x^y M_2$, $M_2 \sqsubset_x^y M_3$ and $M_3 \sqsubset_x^y M_1$. Clearly the resulting *PLP* would not have any preferred partial stable model. Fortunately, both $\subseteq_x^y$ and $\sqsubseteq_x^y$ are acyclic:

Proposition 1. *For any $x \in \{st, opt, pes, opp\}$ and any $y \in \{\forall\forall, \forall\exists, \exists\forall, \exists\exists\}$, both the preference relations $\subseteq_x^y$ and $\sqsubseteq_x^y$ are acyclic.*

Proof. By absurd, suppose $\subseteq_x^y$ is cyclic. This means there exists a chain $\mathcal{O} \ldots \mathcal{O}'$ such that $\mathcal{O} \subset_x^y \mathcal{O}' \subset_x^y \cdots \subset_x^y \mathcal{O}$. As $\subseteq_x^y$ is reflexive (Definition 10), it holds $\mathcal{O} \not\subset_x^y \mathcal{O}$. Besides, by transitivity, we obtain $\mathcal{O}' \subseteq_x^y \mathcal{O}$. It is an absurd as $\mathcal{O} \subset_x^y \mathcal{O}'$. The proof for $\sqsubseteq_x^y$ is similar.

Notice in this last result the importance of imposing reflexivity and transitivity in Definition 10 to ensure every *PLP* will have at least one preferable *PS* model. Now we will show when the best-ranked options of the *PS* models have the same priority and the worst-ranked options of the *PS* models have the same priority, the relations of preference are simplified:

Proposition 2. *Let (P, Φ) be a PLP, I_1 and I_2 be subsets of Lit_P, $\mathcal{B}_1$ and $\mathcal{B}_2$ be respectively the set of best-ranked options in I_1 and I_2 w.r.t. Φ, and $\mathcal{W}_1$ and $\mathcal{W}_2$ be respectively the set of worst-ranked options in I_1 and I_2 w.r.t. Φ. Assume for each o_1 and o_2 in $\mathcal{B}_i$ and for each o_1' and o_2' in $\mathcal{W}_i$ with $i \in \{1, 2\}$, it holds $o_1 \approx o_2$ and $o_1' \approx o_2'$. Consider $\mathcal{X} = \mathcal{B}_1$ if $x \in \{st, opt\}$ and $\mathcal{X} = \mathcal{W}_1$ if $x \in \{pes, opp\}$, and $\mathcal{Y} = \mathcal{B}_2$ if $x \in \{opt, opp\}$ and $\mathcal{Y} = \mathcal{W}_2$ if $x \in \{st, pes\}$. Then for any*

$x \in \{st, opt, pes, opp\}$, $\forall o \in \mathcal{X}$ and $\forall o' \in \mathcal{Y}$, it holds $o' \not\prec o \in \Phi$ iff $\forall o \in \mathcal{X}$, it holds $\exists o' \in \mathcal{Y}$ such that $o' \not\prec o \in \Phi$ iff $o_2 \not\prec o_1 \in \Phi$ iff $\exists o \in \mathcal{X}$ such that $\forall o' \in \mathcal{Y}$, it holds $o' \not\prec o \in \Phi$ iff $o_2 \not\prec o_1 \in \Phi$ iff $\exists o \in \mathcal{X}$ and $\exists o' \in \mathcal{Y}$ such that $o' \not\prec o \in \Phi$.

Proof. For any $x \in \{st, opt, pes, opp\}$, let o_1 be an element of $\mathcal{X}$ and o_2 be an element of $\mathcal{Y}$. Given the elements of $\mathcal{X}$ are indifferent to each other and the elements of $\mathcal{Y}$ are also indifferent to each other, we obtain

- $o_2 \not\prec o_1 \in \Phi$ iff $\forall o \in \mathcal{X}$ and $\forall o' \in \mathcal{Y}$, it holds $o' \not\prec o \in \Phi$;
- $o_2 \not\prec o_1 \in \Phi$ iff $\forall o \in \mathcal{X}$, it holds $\exists o' \in \mathcal{Y}$ such that $o' \not\prec o \in \Phi$;
- $o_2 \not\prec o_1 \in \Phi$ iff $\exists o \in \mathcal{X}$ such that $\forall o' \in \mathcal{Y}$, it holds $o' \not\prec o \in \Phi$;
- $o_2 \not\prec o_1 \in \Phi$ iff $\exists o \in \mathcal{X}$ and $\exists o' \in \mathcal{Y}$ such that $o' \not\prec o \in \Phi$.

i.e., $\forall o \in \mathcal{X}$ and $\forall o' \in \mathcal{Y}$, it holds $o' \not\prec o \in \Phi$ iff $\forall o \in \mathcal{X}$, it holds $\exists o' \in \mathcal{Y}$ such that $o' \not\prec o \in \Phi$ iff $o_2 \not\prec o_1 \in \Phi$ iff $\exists o \in \mathcal{X}$ such that $\forall o' \in \mathcal{Y}$, it holds $o' \not\prec o \in \Phi$ iff $o_2 \not\prec o_1 \in \Phi$ iff $\exists o \in \mathcal{X}$ and $\exists o' \in \mathcal{Y}$ such that $o' \not\prec o \in \Phi$. $\qquad\square$

A binary relation $\sqsubseteq$ on a set $\mathcal{O}$ is a *total preorder* if it is reflexive, transitive and for each o_1 and o_2 in $\mathcal{O}$, it holds $o_1 \sqsubseteq o_2$ or $o_2 \sqsubseteq o_1$. For all $x \in \{st, opt, pes, opp\}$, if the relation $\preceq$ is a total preorder, $\subseteq_x^{\forall\forall}$, $\subseteq_x^{\forall\exists}$, $\subseteq_x^{\exists\forall}$ and $\subseteq_x^{\exists\exists}$ ($\sqsubseteq_x^{\forall\forall}$, $\sqsubseteq_x^{\forall\exists}$, $\sqsubseteq_x^{\exists\forall}$ and $\sqsubseteq_x^{\exists\exists}$) will collapse into each other.

Theorem 1. *Let (P, Φ) be a PLP such that the preference relation $\preceq$ in Φ is a total preorder on $\mathcal{O}_P$. Then for any $x \in \{st, opt, pes, opp\}$, I is a PPS model of (P, Φ) w.r.t. $\subseteq_x^{\forall\forall}$ (resp. $\sqsubseteq_x^{\forall\forall}$) iff I is a PPS model of (P, Φ) w.r.t. $\subseteq_x^{\forall\exists}$ (resp. $\sqsubseteq_x^{\forall\exists}$) iff I is a PPS model of (P, Φ) w.r.t. $\subseteq_x^{\exists\forall}$ (resp. $\sqsubseteq_x^{\exists\forall}$) iff I is a PPS model of (P, Φ) w.r.t. $\subseteq_x^{\exists\exists}$ (resp. $\sqsubseteq_x^{\exists\exists}$).*

Proof. Let I_1 and I_2 be *PS* models of P, $\mathcal{B}_1$ and $\mathcal{B}_2$ be respectively the set of best-ranked options in I_1 and I_2 w.r.t. Φ, and $\mathcal{W}_1$ and $\mathcal{W}_2$ be respectively the set of worst-ranked options in I_1 and I_2 w.r.t. Φ. Consider $\mathcal{X} = \mathcal{B}_1$ if $x \in \{st, opt\}$ and $\mathcal{X} = \mathcal{W}_1$ if $x \in \{pes, opp\}$, and $\mathcal{Y} = \mathcal{B}_2$ if $x \in \{opt, opp\}$ and $\mathcal{Y} = \mathcal{W}_2$ if $x \in \{st, pes\}$.

It suffices to show for any $x \in \{st, opt, pes, opp\}$, $I_1 \subseteq_x^{\forall\forall} I_2$ iff $I_1 \subseteq_x^{\forall\exists} I_2$ iff $I_1 \subseteq_x^{\exists\forall} I_2$ iff $I_1 \subseteq_x^{\exists\exists} I_2$ and $I_1 \sqsubseteq_x^{\forall\forall} I_2$ iff $I_1 \sqsubseteq_x^{\forall\exists} I_2$ iff $I_1 \sqsubseteq_x^{\exists\forall} I_2$ iff $I_1 \sqsubseteq_x^{\exists\exists} I_2$:

As $\preceq$ in Φ is a total preorder on $\mathcal{O}_P$, for each o_1 and o_2 in $\mathcal{B}_i$ and for each o'_1 and o'_2 in $\mathcal{W}_i$ with $i \in \{1, 2\}$, it holds $o_1 \approx o_2$ and $o'_1 \approx o'_2$. Let $I \subseteq_x^{\forall\forall} I'$; there are three possibilities:

1. $I_2 = I_1$. Then $I_1 \subseteq_x^{\forall\exists} I_2$, $I_1 \subseteq_x^{\exists\forall} I_2$ and $I_1 \subseteq_x^{\exists\exists} I_2$.
2. $\forall o \in \mathcal{X}$ and $\forall o' \in \mathcal{Y}$, it holds $o' \not\prec o \in \Phi$. Then by Proposition 2, a) $\forall o \in \mathcal{X}$, it holds $\exists o' \in \mathcal{Y}$ such that $o' \not\prec o \in \Phi$ and b) $\exists o \in \mathcal{X}$ such that $\forall o' \in \mathcal{Y}$, it holds $o' \not\prec o \in \Phi$ and c) $\exists o \in \mathcal{X}$ and $\exists o' \in \mathcal{Y}$ such that $o' \not\prec o \in \Phi$. Hence, $I_1 \subseteq_x^{\forall\exists} I_2$ and $I_1 \subseteq_x^{\exists\forall} I_2$ and $I_1 \subseteq_x^{\exists\exists} I_2$.
3. $I_1 \subseteq_x^{\forall\forall} I$ and $I \subseteq_x^{\forall\forall} I_2$. This means there exists a sequence

$$I_1 = J_1 \subseteq_x^{\forall\forall} J_2 \subseteq_x^{\forall\forall} \cdots \subseteq_x^{\forall\forall} J_m = I \subseteq_x^{\forall\forall} J_{m+1} \subseteq_x^{\forall\forall} \cdots \subseteq_x^{\forall\forall} J_n = I_2$$

such that $\forall i \in \{1, \ldots, n-1\}$, we have $J_1 \subseteq_x^{\forall\forall} J_2$ iff the condition 2 of Definition 10 holds. Again by Proposition 2, condition 2 of Definition 10 also holds for $J_1 \subseteq_x^{\forall\exists} J_2$, $J_1 \subseteq_x^{\exists\forall} J_2$ and $J_1 \subseteq_x^{\exists\exists} J_2$ for each $i \in \{1, \ldots, n-1\}$. Hence, by condition 3 of Definition 10, we obtain $I_1 \subseteq_x^{\forall\exists} I_2$ and $I_1 \subseteq_x^{\exists\forall} I_2$ and $I_1 \subseteq_x^{\exists\exists} I_2$.

Similarly for $I_1 \subseteq_x^{\forall\exists} I_2$, $I_1 \subseteq_x^{\exists\forall} I_2$ and $I_1 \subseteq_x^{\exists\exists} I_2$.
The proof of $I_1 \sqsubseteq_x^{\forall\forall} I_2$ iff $I_1 \sqsubseteq_x^{\forall\exists} I_2$ iff $I_1 \sqsubseteq_x^{\exists\forall} I_2$ iff $I_1 \sqsubseteq_x^{\exists\exists} I_2$ is similar. $\square$

Corollary 1. *Let (P, Φ) be a PLP such that the preference relation $\preceq$ in Φ is a total preorder on $\mathcal{O}_P$. Then $\forall x \in \{st, opt, pes, opp\}$, and $\forall y, \forall z \in \{\forall\forall, \forall\exists, \exists\forall, \exists\exists\}$*

- *I is a preferential well-founded model of (P, Φ) w.r.t. $\subseteq_x^y$ (resp. $\sqsubseteq_x^y$) iff I is a preferential well-founded model of (P, Φ) w.r.t. $\subseteq_x^z$ (resp. $\sqsubseteq_x^z$);*
- *I is a preferential regular model of (P, Φ) w.r.t. $\subseteq_x^y$ (resp. $\sqsubseteq_x^y$) iff I is a preferential regular model of (P, Φ) w.r.t. $\subseteq_x^z$ (resp. $\sqsubseteq_x^z$);*
- *I is a preferential stable model of (P, Φ) w.r.t. $\subseteq_x^y$ (resp. $\sqsubseteq_x^y$) iff I is a preferential stable model of (P, Φ) w.r.t. $\subseteq_x^z$ (resp. $\sqsubseteq_x^z$);*
- *I is a preferential L-stable model of (P, Φ) w.r.t. $\subseteq_x^y$ (resp. $\sqsubseteq_x^y$) iff I is a preferential L-stable model of (P, Φ) w.r.t. $\subseteq_x^z$ (resp. $\sqsubseteq_x^z$).*

Proof. It follows from Theorem 1 and Definition 12. $\square$

When no priority is available ($\Phi = \emptyset$), the semantics for (P, Φ) collapse into the semantics for P (and vice versa):

Theorem 2. *Let (P, Φ) be a PLP such that $\Phi = \emptyset$, i.e., (P, Φ) corresponds to the NLP P. Then $\forall x \in \{st, opt, pes, opp\}$ and $\forall y \in \{\forall\forall, \forall\exists, \exists\forall, \exists\exists\}$, it holds*

- *I is a PS model of P iff I is a PPS model of (P, Φ) w.r.t. $\subseteq_x^y$ iff I is a PPS model of (P, Φ) w.r.t. $\sqsubseteq_x^y$.*
- *I is a well-founded model of P iff I is a preferential well-founded model of (P, Φ) w.r.t. $\subseteq_x^y$ iff I is a preferential well-founded model of (P, Φ) w.r.t. $\sqsubseteq_x^y$.*
- *I is a regular model of P iff I is a preferential regular model of (P, Φ) w.r.t. $\subseteq_x^y$ iff I is a preferential regular model of (P, Φ) w.r.t. $\sqsubseteq_x^y$.*
- *I is a stable model of P iff I is a preferential stable model of (P, Φ) w.r.t. $\subseteq_x^y$ iff I is a preferential stable model of (P, Φ) w.r.t. $\sqsubseteq_x^y$.*
- *I is an L-stable model of P iff I is a preferential L-stable model of (P, Φ) w.r.t. $\subseteq_x^y$ iff I is a preferential L-stable model of (P, Φ) w.r.t. $\sqsubseteq_x^y$.*

Proof. Since $\Phi = \emptyset$, $\Phi^* = \{o \preceq o \mid o \in \mathcal{O}_P\}$ according to Φ^*. This means for every $o, o' \in \mathcal{O}_P$, it holds $o \not\prec o'$. Thus for any $I \subseteq Lit_P$, we obtain $\mathcal{B}_I = \mathcal{W}_I = 2^I$, in which $\mathcal{B}_I$ is the set of all best-ranked options in I and $\mathcal{W}_I$ is the set of all worst-ranked options in I w.r.t. Φ. Then for any PS models I_1 and I_2 of P, for any $o_1 \in \mathcal{B}_{I_1} = \mathcal{W}_{I_1} = 2^{I_1}$ and for any $o_2 \in \mathcal{B}_{I_2} = \mathcal{W}_{I_2} = 2^{I_2}$, it holds $o_2 \not\prec o_1$ and $o_1 \not\prec o_2$ according to Φ^*. Hence, for any $y \in \{\forall\forall, \forall\exists, \exists\forall, \exists\exists\}$, we obtain

$$\mathcal{B}_{I_1} = \mathcal{W}_{I_1} = 2^{I_1} \subseteq^y \mathcal{B}_{I_2} = \mathcal{W}_{I_2} = 2^{I_2} \ and$$
$$\mathcal{B}_{I_2} = \mathcal{W}_{I_2} = 2^{I_2} \subseteq^y \mathcal{B}_{I_1} = \mathcal{W}_{I_1} = 2^{I_1}$$

(1)

From Eq. (1), we infer for each $x \in \{st, opt, pes, opp\}$ and for all $y \in \{\forall\forall, \forall\exists, \exists\forall, \exists\exists\}$, it holds $I_1 \subseteq_x^y I_2$ and $I_2 \subseteq_x^y I_1$.

Consequently, for each $x \in \{st, opt, pes, opp\}$ and for all $y \in \{\forall\forall, \forall\exists, \exists\forall, \exists\exists\}$, it holds I is a *PS* model of P iff I is a *PPS* model of (P, Φ) w.r.t. $\subseteq_x^y$.

After replacing above I_1 by $I_1 - I_2$ and I_2 by $I_2 - I_I$, we obtain the same result for $\sqsubseteq_x^y$.   □

The following results are immediate:

Corollary 2. *Let (P, Φ) be a PLP such that $\Phi = \emptyset$, i.e., (P, Φ) corresponds to the NLP P. Then $\forall x \in \{st, opt, pes, opp\}$ and $\forall y \in \{\forall\forall, \forall\exists, \exists\forall, \exists\exists\}$, it holds*

- *I is a well-founded model of P iff I is a preferential well-founded model of (P, Φ) w.r.t. $\subseteq_x^y$ iff I is a preferential well-founded model of (P, Φ) w.r.t. $\sqsubseteq_x^y$.*
- *I is a regular model of P iff I is a preferential regular model of (P, Φ) w.r.t. $\subseteq_x^y$ iff I is a preferential regular model of (P, Φ) w.r.t. $\sqsubseteq_x^y$.*
- *I is a stable model of P iff I is a preferential stable model of (P, Φ) w.r.t. $\subseteq_x^y$ iff I is a preferential stable model of (P, Φ) w.r.t. $\sqsubseteq_x^y$.*
- *I is an L-stable model of P iff I is a preferential L-stable model of (P, Φ) w.r.t. $\subseteq_x^y$ iff I is a preferential L-stable model of (P, Φ) w.r.t. $\sqsubseteq_x^y$.*

Proof. These results follow from Definitions 3 and 12 and Theorem 2.

Both the relations $\subseteq_x^y$ and $\sqsubseteq_x^y$ have neither a monotonic nor an antitonic behaviour w.r.t. Φ:

Theorem 3. *Let (P, Φ_1) and (P, Φ_2) be PLPs. If $\Phi_1 \subseteq \Phi_2$, then $\forall x \in \{st, opt, pes, opp\}$ and $\forall y \in \{\forall\forall, \forall\exists, \exists\forall, \exists\exists\}$, it does not hold in general*

1. *$\mathcal{PPS}(P, \Phi_1)_{\subseteq_x^y} \subseteq \mathcal{PPS}(P, \Phi_2)_{\subseteq_x^y}$ (resp. $\mathcal{PPS}(P, \Phi_1)_{\sqsubseteq_x^y} \subseteq \mathcal{PPS}(P, \Phi_2)_{\sqsubseteq_x^y}$) (monotonic).*
2. *$\mathcal{PPS}(P, \Phi_2)_{\subseteq_x^y} \subseteq \mathcal{PPS}(P, \Phi_1)_{\subseteq_x^y}$ (resp. $\mathcal{PPS}(P, \Phi_2)_{\sqsubseteq_x^y} \subseteq \mathcal{PPS}(P, \Phi_1)_{\sqsubseteq_x^y}$) (antitonic).*

Proof. Let P be the *NLP* $\{a \leftarrow \text{not } b \quad b \leftarrow \text{not } a\}$, whose *PS* models are $\emptyset$, $\{a\}$ and $\{b\}$. Firstly, we will show a counterexample to the statement $\forall x \in \{st, opt, pes, opp\}$ and $\forall y \in \{\forall\forall, \forall\exists, \exists\forall, \exists\exists\}$, it holds $\mathcal{PPS}(P, \Phi_1)_{\subseteq_x^y} \subseteq \mathcal{PPS}(P, \Phi_2)_{\subseteq_x^y}$ (resp. $\mathcal{PPS}(P, \Phi_1)_{\sqsubseteq_x^y} \subseteq \mathcal{PPS}(P, \Phi_2)_{\sqsubseteq_x^y}$) :

Let $\Phi_1 = \emptyset$ and $\Phi_2 = \{\emptyset \preceq \{b\}, \{b\} \preceq \{a\}\}$. We have $\forall x \in \{st, opt, pes, opp\}$ and $\forall y \in \{\forall\forall, \forall\exists, \exists\forall, \exists\exists\}$, it holds $\mathcal{PPS}(P, \Phi_1)_{\subseteq_x^y} = \mathcal{PPS}(P, \Phi_1)_{\sqsubseteq_x^y} = \{\emptyset, \{a\}, \{b\}\}$ and $\mathcal{PPS}(P, \Phi_2)_{\subseteq_x^y} = \mathcal{PPS}(P, \Phi_2)_{\sqsubseteq_x^y} = \{\{a\}\}$. Consequently,

$$\mathcal{PPS}(P, \Phi_1)_{\subseteq_x^y} = \mathcal{PPS}(P, \Phi_1)_{\sqsubseteq_x^y} \nsubseteq \mathcal{PPS}(P, \Phi_2)_{\subseteq_x^y} = \mathcal{PPS}(P, \Phi_2)_{\sqsubseteq_x^y}$$

Now we will show a counterexample to $\forall x \in \{st, opt, pes, opp\}$ and $\forall y \in \{\forall\forall, \forall\exists, \exists\forall, \exists\exists\}$, it holds $\mathcal{PPS}(P, \Phi_2)_{\subseteq_x^y} \subseteq \mathcal{PPS}(P, \Phi_1)_{\subseteq_x^y}$ (resp. $\mathcal{PPS}(P, \Phi_2)_{\sqsubseteq_x^y} \subseteq \mathcal{PPS}(P, \Phi_1)_{\sqsubseteq_x^y}$):

Let $\Phi_1 = \{\emptyset \preceq \{b\}, \{b\} \preceq \{a\}\}$ and $\Phi_2 = \{\emptyset \preceq \{b\}, \{b\} \preceq \{a\}, \{a\} \preceq \emptyset\}$. We have $\forall x \in \{st, opt, pes, opp\}$ and $\forall y \in \{\forall\forall, \forall\exists, \exists\forall, \exists\exists\}$, it holds $\mathcal{PPS}(P, \Phi_1)_{\subseteq_x^y} = \mathcal{PPS}(P, \Phi_1)_{\sqsubseteq_x^y} = \{\{a\}\}$ and $\mathcal{PPS}(P, \Phi_2)_{\subseteq_x^y} = \mathcal{PPS}(P, \Phi_2)_{\sqsubseteq_x^y} = \{\emptyset, \{a\}, \{b\}\}$. Consequently,

$$\mathcal{PPS}(P, \Phi_2)_{\subseteq_x^y} = \mathcal{PPS}(P, \Phi_2)_{\sqsubseteq_x^y} \not\subseteq \mathcal{PPS}(P, \Phi_1)_{\subseteq_x^y} = \mathcal{PPS}(P, \Phi_1)_{\sqsubseteq_x^y}$$

We end up this section by showing inconsistency options do not interfere in the process of selecting the *PPS* models:

Theorem 4 (Invariant to Inconsistency). *Let* (P, Φ_1) *and* (P, Φ_2) *be PLPs such that* $\Phi_2 = \{o_1 \preceq o_2 \mid o_1 \preceq o_2 \in \Phi_1$ *and both* o_1 *and* o_2 *are consistent*$\}$. *Then* $\forall x \in \{st, opt, pes, opp\}$ *and* $\forall y \in \{\forall\forall, \forall\exists, \exists\forall, \exists\exists\}$, *it holds* $\mathcal{PPS}(P, \Phi_1)_{\subseteq_x^y} = \mathcal{PPS}(P, \Phi_2)_{\subseteq_x^y}$ *(resp.* $\mathcal{PPS}(P, \Phi_1)_{\sqsubseteq_x^y} = \mathcal{PPS}(P, \Phi_2)_{\sqsubseteq_x^y}$*)*.

Proof. Let I_1 and I_2 be a *PS* model of P. We will show $I_1 \in \mathcal{PPS}(P, \Phi_1)_{\subseteq_x^y}$ iff $I_1 \in \mathcal{PPS}(P, \Phi_2)_{\subseteq_x^y}$ (resp. $I_1 \in \mathcal{PPS}(P, \Phi_1)_{\sqsubseteq_x^y}$ iff $I_1 \in \mathcal{PPS}(P, \Phi_2)_{\sqsubseteq_x^y}$).

As no *PS* model is inconsistent, the inconsistent options are neither in the set $\mathcal{B}_{I_j}$ of all best-ranked options of I_j nor in the set $\mathcal{W}_{I_j}$ of all worst-ranked options of I_j with $j \in \{1, 2\}$. This means for each $o_1 \in \mathcal{B}_{I_1} \cup \mathcal{W}_{I_1}$ and for each $o_2 \in \mathcal{B}_{I_2} \cup \mathcal{W}_{I_2}$, it holds $o_2 \not\prec o_1$ according to Φ_1^* iff $o_2 \not\prec o_1$ according to Φ_2^*. Then $\forall y \in \{\forall\forall, \forall\exists, \exists\forall, \exists\exists\}$, it holds $\mathcal{B}_{I_1} \subseteq^y \mathcal{B}_{I_2}$ in Φ_1 iff $\mathcal{B}_{I_1} \subseteq^y \mathcal{B}_{I_2}$ in Φ_2, $\mathcal{B}_{I_1} \subseteq^y \mathcal{W}_{I_2}$ in Φ_1 iff $\mathcal{B}_{I_1} \subseteq^y \mathcal{W}_{I_2}$ in Φ_2, $\mathcal{W}_{I_1} \subseteq^y \mathcal{B}_{I_2}$ in Φ_1 iff $\mathcal{W}_{I_1} \subseteq^y \mathcal{B}_{I_2}$ in Φ_2 and $\mathcal{W}_{I_1} \subseteq^y \mathcal{W}_{I_2}$ in Φ_1 iff $\mathcal{W}_{I_1} \subseteq^y \mathcal{W}_{I_2}$ in Φ_2, i.e., $\forall x \in \{st, opt, pes, opp\}$ and $\forall y \in \{\forall\forall, \forall\exists, \exists\forall, \exists\exists\}$, it holds $I_1 \subseteq_x^y I_2$ in Φ_1 iff $I_1 \subseteq_x^y I_2$ in Φ_2. Hence, $I_1 \in \mathcal{PPS}(P, \Phi_1)_{\subseteq_x^y}$ iff $I_1 \in \mathcal{PPS}(P, \Phi_2)_{\subseteq_x^y}$.

Similarly we can prove $I_1 \in \mathcal{PPS}(P, \Phi_1)_{\sqsubseteq_x^y}$ iff $I_1 \in \mathcal{PPS}(P, \Phi_2)_{\sqsubseteq_x^y}$. $\square$

6 Related Work

The first studies on preference information in non-monotonic reasoning involved the idea of specificity [22]: given two conflicting conclusions obtained from the same initial information but resorting to different inference rules, one should prefer the more specific. Following a related point of view, in [23], they introduced preference between predicates in the context of *Circumscription*. Preference (usually based on simplicity measures) is also employed in [24] to select the most preferable explanations in abductive systems.

With respect to preferences in Logic Programming, we find a plethora of proposals (see [2–12] for a non exhaustive list). In [2] and [3], they extend respectively the well-founded semantics and the answer sets semantics [25] for logic programs to represent preference between rules in the language and resort to such information to obtain new conclusions. Unlike the semantics in [3], which allows multiple preferred answer sets even for some fully prioritised programs, the semantics in [10] selects at most one preferred answer set for all fully prioritised programs. Furthermore, for a large class of programs guaranteed to have an answer set, the existence of a preferred answer set is also guaranteed.

In ordered logic programming [5], CR-prolog [6], logic programming with ordered disjunction [7], answer set optimization [8], possibilistic answer set programming [9], preference is handled to evaluate the preferred answer sets by specifying the precedence over the rules or the literals in rules heads.

A logic programming language *PrefLog* based on an infinite-valued logic has been conceived in [11] to support operators for expressing preferences. Despite the infinite-valued truth domain, it can be defined a terminating bottom-up proof procedure for implementing a significant fragment of the language. For continuous operators, a least Herbrand model is guaranteed to exist.

In [12], it is proposed a logic programming paradigm to combine non-monotonic reasoning with epistemic preferential reasoning. In this proposal, the relation of preferences between atoms can occur in the body of the rules and can be employed to possibly generate new models.

An approach closer in spirit to ours is [4], where the authors introduced a framework of prioritised logic programming to represent priority information explicitly in a program. Differently of us, however, they restrain the preference relation to literals only, whereas we have defined the preference relation over sets of literals. As consequence, in order to express in a $PLP\,(P, \Phi)$ the complex forms of preferences expounded in Sect. 4, they have to change the program P with the introduction of new atoms as well as the set Φ of priorities. Furthermore, they defined a unique semantics to determine the preferred answer sets. In contradistinction, we introduced 32 different criteria (semantics) to establish preference between partial stable models (which allow defining 32 different ways of selecting the preferred answer sets). In this sense, our work offers a more comprehensive and systematic treatment to deal with preferences in logic programming.

7   Conclusion and Future Works

Prioritised Logic Programs (*PLPs*) [4] are obtained from Normal Logic Programs (*NLP*) by including priorities between literals (an atom or its negation). In this paper, we have presented a more general notion of *PLP* when compared with that version in [4]. Our proposal has two main distinguishing features: a) it encompasses not only priorities between literals (atoms or its negation), but also between sets of literals. As consequence, we can represent compactly complex forms of preference. b) Concerning its semantics, we have provided a comprehensive and systematic mechanism to select the most Preferable Partial Stable Models (*PPS* Models). Indeed, by combining well-known criteria found in the literature, we have defined 32 different manners of selecting the *PPS* models. As the most important semantics in *NLP* are based on partial stable models, we are also equipped with 32 different manners of selecting the most preferred models according to them. Besides, we have proved some properties of these *PLPs* and guaranteed these semantics for *PLP* generalise their counterparts for *NLP*.

Future directions include investigating computational complexity issues on *PLPs* and exploiting the consequence of introducing preference on paraconsistent semantics found in logic programming. We also intend to study how to work with dynamic preferences along with the resulting semantics.

References

1. Pigozzi, G., Tsoukias, A., Viappiani, P.: Preferences in artificial intelligence. Ann. Math. Artif. Intell. **77**(3), 361–401 (2016)
2. Brewka, G.: Well-founded semantics for extended logic programs with dynamic preferences. J. Artif. Intell. Res. **4**, 19–36 (1996)
3. Brewka, G., Eiter, T.: Preferred answer sets for extended logic programs. Artif. Intell. **109**(1–2), 297–356 (1999)
4. Sakama, C., Inoue, K.: Prioritized logic programming and its application to commonsense reasoning. Artif. Intell. **123**(1–2), 185–222 (2000)
5. Schaub, T., Wang, K.: A semantic framework for preference handling in answer set programming. arXiv preprint cs/0301023 (2003)
6. Balduccini, M., Gelfond, M.: Logic programs with consistency-restoring rules. In: International Symposium on Logical Formalization of Commonsense Reasoning, AAAI 2003 Spring Symposium Series, vol. 102 (2003)
7. Brewka, G., Niemelä, I., Syrjänen, T.: Logic programs with ordered disjunction. Comput. Intell. **20**(2), 335–357 (2004)
8. Brewka, G.: Answer sets and qualitative optimization. Logic J. IGPL **14**(3), 413–433 (2006)
9. Nicolas, P., Garcia, L., Stéphan, I., Lefèvre, C.: Possibilistic uncertainty handling for answer set programming. Ann. Math. Artif. Intell. **47**(1), 139–181 (2006)
10. Gabaldon, A.: A selective semantics for logic programs with preferences. In: del Cerro, L.F., Herzig, A., Mengin, J. (eds.) JELIA 2012. LNCS (LNAI), vol. 7519, pp. 215–227. Springer, Heidelberg (2012). https://doi.org/10.1007/978-3-642-33353-8_17
11. Rondogiannis, P., Troumpoukis, A.: Expressing preferences in logic programming using an infinite-valued logic. In: Proceedings of the 17th International Symposium on Principles and Practice of Declarative Programming, pp. 208–219 (2015)
12. Zhang, Z.: Introspecting preferences in answer set programming. In: 34th International Conference on Logic Programming (ICLP 2018) (2018)
13. Kaci, S.: Working with preferences: Less is more. Springer Science and Business Media, 2011. https://doi.org/10.1007/978-3-642-17280-9_5
14. Przymusinski, T.: The well-founded semantics coincides with the three-valued stable semantics. Fundam. Inf. **13**(4), 445–463 (1990)
15. Eiter, T., Leone, N., Sacca, D.: On the partial semantics for disjunctive deductive databases. Ann. Math. Artif. Intell. **19**(1), 59–96 (1997)
16. Van Der T., Leendert, W.E.: Parameters for utilitarian desires in a qualitative decision theory. Appl. Intelli. **14**(3), 285–301 (2001)
17. Benferhat, S., Dubois, D., Kaci, S., Prade, H.: Bipolar representation and fusion of preferences on the possibilistic logic framework. KR, **2**, 421–432 (2002)
18. Benferhat, S., Dubois, D., Kaci, S., Prade, H.: Bipolar possibilistic representations. In: 18th International Conference on Uncertainty in Artificial Intelligence, pp. 45–52 (2002)
19. Lang, J., Van der Torre, L., Weydert, E.: Utilitarian desires. Auton. Agents Multiagent Syst. **5**(3), 329–363 (2002)
20. Benferhat, S., Dubois, D., Kaci, S., Prade, H.: Bipolar possibility theory in preference modeling: representation, fusion and optimal solutions. Inf. Fusion **7**(1), 135–150 (2006)
21. Bistarelli, S., Pini, M., Rossi, F., Venable, K.: From soft constraints to bipolar preferences: modelling framework and solving issues. J. Exp. Theoret. Artif. Intell. **22**(2), 135–158 (2010)

22. Poole, D.: On the comparison of theories: preferring the most specific explanation. In: 9th International Joint Conference on Artificial Intelligence, vol. 85, pp. 144–147 (1985)
23. Lifschitz, V.: Computing circumscription. In: International Joint Conference on Artificial Intelligence (IJCAI), vol. 85, pp. 121–127 (1985)
24. Eiter, T., Gottlob, G.: The complexity of logic-based abduction. J. ACM (JACM) **42**(1), 3–42 (1995)
25. Gelfond, M., Lifschitz, V.: Classical negation in logic programs and disjunctive databases. New Gener. Comput. **9**(3-4), 365–385 (1991)

Active Learning and Case-Based Reasoning for the Deceptive Play in the Card Game of Truco

Daniel P. Vargas[(✉)], Gustavo B. Paulus, and Luis A. L. Silva

Graduate Program in Computer Science, Federal University of Santa Maria – UFSM, Santa Maria, RS, Brazil

Abstract. Deception is an essential behavior in many card games. Despite this fact, it is not trivial to capture the intent of a human strategist when making deceptive decisions. That is even harder when dealing with deception in card games, where components of uncertainty, hidden information, luck and randomness introduce the need of case-based decision making. Approaching this problem along with the investigation of the game of Truco, a quite popular game in Southern regions of South America, this work presents an approach that combines active learning and Case-Based Reasoning (CBR) in which agents request a human specialist to review a reused game action retrieved from a case base containing played Truco hands. That happens when the agents are confronted with game situations that are identified as opportunities for deception. The goal is to actively capture problem-solving experiences in which deception can be used, and later employ such case knowledge in the enhancement of the deceptive capabilities of the Truco agents. Experimental results show that the use of the learned cases enabled different kinds of Truco agents to play more aggressively, being more deceptive and performing a larger number of successful bluffs.

Keywords: Deception · Case-based reasoning · Active learning

1 Introduction

Deception involves a deliberate attempt to introduce in another person a false belief or belief in which the deceiver considers false [1, 2]. Such deceptive behavior can be modeled as a) concealment, aiming to hide/omit the truth and b) simulation, whose purpose is to show the untruth [3]. To conceal the deceiver acts by withholding information and omitting the truth. To simulate, in addition to the retention of genuine information, unreal information is presented as being legitimate. Among the various forms of deception is the bluff, where deception and bluffing is interchangeably used in this work. In the context of a card game, a bluff is an action where players, to deceit their opponents, seek to make an illusory impression of strength when they hold weak hands. Alternatively, players may try to show that their strong hands have little value in the game.

To card games with hidden, stochastic and imperfect information, acting deceptively is an essential strategy for players to succeed. The use of deceptive moves can also

© Springer Nature Switzerland AG 2021
A. Britto and K. Valdivia Delgado (Eds.): BRACIS 2021, LNAI 13073, pp. 313–327, 2021.
https://doi.org/10.1007/978-3-030-91702-9_21

be closely related to the nature of some popular games, and the entertainment that it introduces in the game disputes. To do so, players should be able to identify what the best opportunities for the use of deception are, considering the strength of their hand and their betting history in order to make themselves as unpredictable as possible [4]. In general, real-world situations present complex characteristics for the modeling of deceptive agents, such as the need for learning and decision-making with a small number of training examples, for instance.

For the development of agents capable of acting deceptively in card games, this work explores Case-based Reasoning (CBR) [5]. With relevant explanatory capabilities, CBR combines learning and problem-solving with the use of specific knowledge captured in the form of cases. In particular, this technique has supported the development of agents which competitively play Poker [6, 7]. In this line of research, this paper extends past work [8–10] in the CBR modeling of a popular game in the Southern regions of South America, a game that is under-investigation in Artificial Intelligence (AI): the game of Truco [11].

CBR allows continuous learning by retaining concrete problem-solving experiences in a reusable case base. Despite this fact, it is not simple to capture and label the intention of human players when making deceptive moves in card game. To approach this problem, active learning [12] is investigated in the analysis of Truco opportunities for being deceptive, and the consequent collection of such problem-solving experiences in a case base. Then, the acquired case knowledge is used to equip different kinds of CBR agents to make deceptive actions. In the proposed approach, the case learning is focused on the review of decisions and retention of cases in the case base. As a result of the implemented solution reuse policy, whenever a game action is reused by the agent, if a certain pre-established learning criterion is met, the agent requests for a human expert to review the reused game action and the current game state. If the solution presented by the reuse policy is not considered to be the most effective according to the judgment of the domain specialist, the expert suggests a game action to be played (deceptive or not). With attention to the capture and reuse of deceptive game actions from human players, the contributions of this paper are: i) the exploration of active learning to support the retention of case problem situations in which deceptive moves can be used, ii) the performance evaluation of deceptive Truco agents configured according to alternative solution reuse policies, and iii) the analysis of the resulting game playing behavior of the implemented agents when using case bases storing the collected problem-solving experiences.

2 Background to This Work

CBR [5] combines learning and problem-solving with the use of knowledge obtained from concrete problem-solving experiences. Learning in CBR aims to acquire, modify or improve different knowledge repositories [13], where the enhancement of the case base is often sought in different applications. In doing so, it is possible to explore the automatic case elicitation (ACE) technique [14]. This technique focuses on the system's ability to explore its domain in real time and automatically collect new cases. Another technique is the learning by observation, also referred as demonstration learning or imitation learning. In such learning modality, the system learns to perform a certain behavior by observing an expert to act [15]. The first learning stage is the acquisition of cases from the expert demonstrations. The second stage is the resolution of a problem using the case base collected from the observations [16]. An alternative to learning by observation is the active learning, where the goal is to obtain greater quality in the learning process considering the smallest possible number of labeled instances. Active learning tries to overcome the labeling and data distribution bottlenecks by allowing the learner to intelligently choose the most representative instances. This model allows requesting that a human expert present a solution to the problem. Later it allows adding the resolved instances to the training set.

2.1 CBR, Active Learning and Games

Active learning and CBR have been explored in a number of digital game applications. In the SMILe - Stochastic Mixing Iterative Learning [17] game, SMILe controls the agent while observing the specialist behavior. When the game iteration ends, SMILe uses the collected observations to train a new policy that can be used in subsequent game iterations. The DAgger – Dataset Aggregation [18] algorithm enhances SMILe by preventing the agent from selecting actions using outdated policies. In doing so, the agent updates a single policy learned each iteration. In both SMILe and DAgger, the control to determine whether the player is the agent or the specialist is defined probabilistically.

The SALT algorithm – Selective Active Learning from Traces [19, 20] allows the learner agent to perform a task, and when it is determined that the agent has left the space for which it has training data, the control is assigned to an expert. As in SMILe and DAgger, the focus is on the collection of training data for the set of states that are expected to be found during testing. Unlike SMILe and DAgger, control in SALT is assigned to the specialist only when the agent leaves the state-space of the training set. The training data is generated only when the specialist is in control, reducing the specialist' cognitive load.

With regard to expert consultation strategies, in [21], the retrieval of most similar cases is used to determine a game action to be taken according to a vote. Considering the average similarity value of the cases retrieved from the last five decisions made in the game and a coefficient obtained from a linear regression, which determines whether the similarities are increasing or decreasing during the last performed movements, the CBR agent gives the game control to the human expert. This happens whenever the mean similarity is increasingly moving away from the space of known situations. The expert plays until the states of the game are familiar again. To avoid continuous changes between the CBR agent and the human specialist, each one has to perform certain minimum play before giving control to the other.

In [22] and [15], a similarity threshold value is used to determine when the human specialist is consulted. Then the specialist automatically gives control to the CBR agent after performing a move in the game. Moreover, the retention of cases in the case base only happens when the human specialist is in control. Unlike passively acquired cases, which can result in the retention of redundant cases in the case base, the use of active learning in these games allowed the learning of certain situations that would not be observed in a purely passive manner. To achieve a reasonable expert imitation, active learning required a considerably lower number of cases than when a fully passive approach was used.

In contrast to these past works, this work actively learns only in the resolution of the required problems, which are identified as deception game opportunities. In addition to use a similarity threshold, the condition to query the human specialist is combined with a strategy that employs hand strength and probability to determine whether a situation is opportune for the use of deceptive moves. Instead of using active learning to collect any kind of expert experience of game playing, this work direct such learning to the improvement of the deceptive capabilities of card playing agents.

The AI research has also investigated the effectiveness of CBR in the modeling of card games, mainly with respect to the game of Poker [7, 23]. Considering deception-related Poker strategies, however, only [7] explicitly addresses this issue. There, the developed agent, whose case base starts empty, performs the random play strategy to populate the case base. With respect to the game of Truco, [10] addresses the case retention problem, especially considering the lack of large numbers of cases. It investigates alternative learning techniques such as ACE, Imitation Learning, and Active Learning to enable an agent to learn how to act in situations in which past case knowledge is limited. Through the assistance of a human player, the purpose of the active learning technique is to guide the agent in its use of any kind of game action whenever the agent had not encountered similar game situations stored in the case base. Despite this research, Truco matches disputed amongst the agents implemented according to the analyzed learning techniques showed that, unlike the automatic retention and the retention of new cases strategies, which demonstrated an improvement in the agents' performance, the active learning technique did not show an improvement in the agents' performance. Unlike [10], which performed a broad collection of case situations in Truco, this paper investigates the use of active learning in the analysis of deceptive game opportunities and, for those in which the expert decided that it was worth acting deceptively, the collection of new problem-solving experiences.

[8, 9] address the indexing of the Truco case base through the organization of cases into different clusters. Using such clusters, the goal was to identify game actions along with game states in which such actions are performed. In addition, it is proposed a two-step solution reuse model, which is further explored in our work. The model involves a step that retrieves the most similar cases for a given query, where a reuse criterion is used in the choice of the group of cases that is more similar to the current query situation (extra cluster reuse criterion). After selecting this group of cases, a filtering is performed in order to select only the retrieved cases that belong to the chosen group. Based on these filtered cases, a second reuse step can use another reuse criterion to choose the game action that is used to solve the current problem (intra cluster reuse criterion). The reuse policies that showed to be the most effective according to their experiments are described in Table 1.

Considering the cases retrieved from a given query, the number of points solution criterion (NPS) involves the reuse of game actions, where the game action choice is supported on the amount of earned points due to the use of that action in the game. The probability victory (PV) criterion involves the choice of either clusters (PVC) or game actions (PVS) to be reused (or both in the PVCS), where the reuse is based on the calculation of the chances of victory for each of the different game actions recorded in

Table 1. Reuse policies used by the implemented Truco agents.

Reuse policy	Criterion for the choice of clusters	Criterion for the choice of game actions	Reuse model
Number Points Solution (NPS)	–	Number Points	Standard reuse model
Probability Victory with Cluster and Number Points Solution (PVCNPS)	Probability Victory	Number Points	Two-step reuse model
Probability Victory Solution (PVS)	–	Probability Victory	Standard reuse model
Probability Victory with Cluster and Solution (PVCS)	Probability Victory	Probability Victory	Two-step reuse model

the retrieved cases. These policies were thoughtfully explored in the development of the Truco playing agents investigated in this paper.

2.2 The Card Game of Truco

Truco is a widely practiced card game in Southern regions of South America [11]. The AI techniques covered in this work were investigated with the use of matches disputed between two opposing players. Such blind Truco version (Truco "Cego") uses 40 of the 48 cards in the Spanish deck, as the four eights and four nines are removed. The deck is divided into "Black" cards, which are the cards with figures (King – 12, Horse – 11 and Sota – 10), and "White" cards that are from ace to seven.

In Truco, the dispute takes place through successive hands that are initially worth a point. Each player receives three cards to play one hand. A hand can be divided into two phases of dispute: ENVIDO and TRUCO. In each stage, players have different ways to increase the number of points that are played. Each hand can be played in a best of three rounds, in which the player who plays the highest card in each round wins. Finally, the match comes to an end when a player reaches twenty-four points.

ENVIDO is a dispute that takes place during the first round of a hand. Such a dispute is based on the sum of the value of each one of the player's cards. For ENVIDO, each card is worth the value presented in it, with the exception of "black" cards that are not computed in the sum of points. ENVIDO has the following betting modalities: ENVIDO, REAL_ENVIDO, and FALTA_ENVIDO, which the player can bet before playing the first card on the table. If a player advances any one of these bets, the opponent can accept or deny the ENVIDO dispute. There is a special case of ENVIDO, which is called FLOR. The FLOR occurs when a player has three cards with the same suit. The FLOR bet cancels any ENVIDO modality previously advanced since it increases the value of the ENVIDO dispute. As in ENVIDO, FLOR allows one to fight back (e.g. CONTRA_FLOR) if the opponent also has three cards of the same suit.

When the ENVIDO dispute ends, the TRUCO phase begins. At this stage, one to four points are disputed during the three rounds of the hand (one for each card in the hand). In one round, each player drops a card at the table starting with the hand player's or the winner from the previous round. These cards are confronted according to a Truco ranking involving each card. The player who wins two of the three rounds wins the hand. Unlike bet actions for ENVIDO, which can only be placed during the first round of each hand, TRUCO bets can be placed at any time during a hand dispute. In addition, if a player decides to go to the deck, the opponent receives a score of points equivalent to the points in dispute in that TRUCO stage.

Similar to other card games, such as in the different variations of Poker, for example, the game of Truco involves different degrees of deception/bluffing. These strategies allow players to win hands and even matches in situations where they do not own strong cards for the ENVIDO and TRUCO disputes. Most importantly, human players in real-life Truco matches employ deceptive actions with certain frequency. Among other reasons, this behavior makes the game more fun, even if such bluffs don't necessarily result in better results in the game.

3   Active Learning and CBR in the Card Game of Truco

Agents can employ CBR to learn game strategies for playing Truco. In our work, whenever such agents take the game turn, they evaluate the current state of the game. To do it, a query containing the game state information is formed. Then, the K-NN algorithm is executed along with a similarity function that averages case attribute similarities to perform the retrieval of past cases from the case base. After retrieval, the selected cases are used to generate a game move which is played in the current game situation. The reuse is supported by a reuse policy which defines, among other criteria, the number of similar cases considered in the solution choice and the minimum similarity value (threshold, set to 98% in this work) so that the solutions represented in the retrieved cases are reused in the resolution of the current problem. At the end of such problem-solving procedure, the system can decide whether the derived problem-solving experience is worth retaining as a new case in the case base.

3.1   The Case Base Formation

A web-based system was developed to permit the collection of Truco cases, where these cases were the result of Truco matches played between two human opponents who had various levels of Truco experience. At the end of each disputed Truco hand, a new case (i.e. a hand of Truco) was stored in the case base. In our project, 147 matches were played among different players using this system. In total, 3,195 cases were collected and stored into a case base called BASELINE. To represent the cases, a set of attributes captured the main information and game actions employed in the Truco disputes. Table 2 summarizes these attributes.

The played cards were recorded according to a numerical codification. The encoding uses a nonlinear numerical scale ranging from 1 to 52. Code 1 is assigned the cards with the lowest value (all 4's). Code 52 is assigned the highest value card, which is the ace

Table 2. Attributes for representing a Truco case.

Attributes	Description
Case identification	Identify the hand and match to which the hand belongs
Player who starts actions on each hand	Player 1 or 2. The player that starts the hand dispute is known as "hand player"
Received and played cards	Cards received and played during the 3 hand rounds by each player organized as: "high card, medium card and low card"
Scoreboard	Points from each player when starting and finishing the hand
Points regarding the ENVIDO dispute	Point score for the ENVIDO dispute
Played bets/actions	Bets and game actions that each player performed during the hand dispute
Winners for ENVIDO, TRUCO and the hand rounds	Player 1 or 2 who won every hand dispute
Amount of points earned/lost in each dispute	Amounts of points awarded to the winners of each dispute that occurred in the hand
...	...

of spades. Then it was explored both in the representation of cases and in the similarity evaluations. In effect, the codification is based on both the categories identified in [24] and the Truco knowledge from our research group participants. Each value in this encoding represents the relative strength of the Truco cards.

To collect deceptive game information to support the development of the active learning task (only used during the active case learning) through the course of each played Truco match, other set of attributes were added into the case representation model. These attributes are described in Table 3.

With respect to the deceptive actions performed by Truco players, case attributes to represent the deception information were used to measure the similarity of the current game situation in relation to the cases stored in a LEARNING case base. The purpose was to determine whether such case base had enough records of problem opportunities for using deceptive actions in order to solve the game problems encountered in the matches in which the active learning tasks were executed.

3.2 Game Actions and Deception

Truco has various kinds of game actions. To support the analysis of deceptive Truco behaviors, we classified as *aggressive* the Truco playing actions involving betting or raising an opponent's bet. Similarly, *passive* are the actions in which the player should decide either accepting or denying an opponent's bet. In addition, *aggressive* game actions can be labeled as either *honest* or *deceptive*. In *aggressive* moves, the player can

Table 3. Attributes regarding the deceptive actions made in the Truco match.

Attributes	Deception type	Description
Deceptive actions performed with success	Bet/Raise Slow playing	Number of deceptive moves that achieved the goal
Deceptive actions performed without success	Bet/Raise with complete game information	Number of deceptive moves that didn't have the expected effect
Revealed deceptive actions		Due to the cards played by the agent, number of deceptive actions that could have been detected by the opponent
Opponent's deceptive actions		Due to the cards played by the opponent, number of deceptive actions that were discovered by the agent

most effectively employ deception. In *passive* moves, the player has the opportunity to detect the opponent's deception since the opponent is either betting or increasing a bet. In Table 4, we analyze such deceptive game actions in Truco.

Table 4. Possible types of deceptive game actions in Truco.

Game action	Condition	Goal	Deception characteristic
Bet/Raise on ENVIDO-type bets	Low probability of having a better hand than the opponent	To give the impression that the deceiver has a strong hand. Consequently, to induce the opponent not to accept the bet or raise the bet	Concealment/simulation; omit the reality and show the untruth
Bet/Raise on TRUCO-type bets			
Bet/Raise with perfect information on TRUCO-type bets	Deceivers are sure they cannot defeat the opponent's card in the last hand round		
Slow playing on ENVIDO-type bets	High probability of having a better hand than the opponent	To induce the opponent to bet so that the deceiver can raise the bet in order to maximizing the number of points been disputed	Concealment; hides/omits the truth
Slow playing on TRUCO-type bets			
Bet/Raise with perfect information on ENVIDO-type bets	Low probability of having more points than the opponent	To induces the opponent not to accept the bet	Concealment/simulation; omit the reality and show the untruth

3.3 Hand Strength

Truco is played with 40 of the 48 cards of the Spanish deck, where there are 9,880 possible hands. With this, it is possible to sort and classify each hand according to their strength for the ENVIDO and TRUCO disputes. The ENVIDO hand strength is directly based on the ENVIDO points. To calculate the strength of a TRUCO hand, the relative strength and importance of each card that forms the hand have to be considered. A method to calculate such hand strength can be derived from the analysis of two components: a) the strength of the two highest value hand cards and b) the strength of the two lower value hand cards. This method considers the Truco rules since a hand dispute is played in a best of three rounds.

The two highest hand cards a player possesses (high and medium cards, see Table 2) are more important in the estimation of final hand strength. To have two high cards in a hand tends to increase the player's chances of winning in the best-of-3 competition. A low card among these higher hand cards has a high negative impact on the final hand strength. On the other hand, a low card between the two lowest hand cards (medium and low cards) should also have a negative impact on the final hand strength. However, this impact is not as severe in the calculation of the hand strength as it is the impact of owning a low card between the two highest ones.

The method explores the calculation of means between the hand card numerical encodings (i.e. the non-linear encoding from 1 to 52). The first calculates a harmonic mean (1) between the two highest hand cards (high and medium cards). When one value much lower than another is used in this type of harmonic mean calculation, the final result of the computed mean tends to be reduced toward the lowest value.

$$M_1 = \frac{2}{^1/_{HighCard} + ^1/_{MediumCard}} \tag{1}$$

The second uses the calculation of a weighted arithmetic mean (2) between the two lower hand cards (medium and low cards). In this case, the weight attributed for the highest card between these two lowest hand cards was set to double the weight of the lowest card. The use of a weighted arithmetic mean also allows expressing the impact of a low card on the hand strength. However, the weight of having a high card between the two lower ones should be greater than the weight of having a lower card among these two lower cards.

$$M_2 = \frac{(2 * MediumCard) + LowCard}{2 + 1} \tag{2}$$

To reach the final value of the hand strength, a weighted arithmetic mean (3) is calculated between the results obtained from the two mean values computed with (1) and (2).

$$M_3 = \frac{(2 * M_1) + M_2}{2 + 1} \tag{3}$$

According to numerical tests, it was not possible to identify hand situations in which our calculations of hand strength presented unsatisfactory results. Qualitatively, either higher or lower values than those obtained by the use of our method could be argued as relevant in some situations. In these situations, even without the use of our method, the strength of the considered hands is subject to debate, especially when we considered the Truco rules and the different ways of deceptively playing in this game.

3.4 Triggering the Expert Consultation

As part of the proposed active learning approach, two strategies to trigger a human specialist consultation are proposed in the work.

First, the coverage of a case base is used as a trigger to consult the expert. To do it, the similarity between the current game situation and the cases learned through active learning is computed when a query is emitted. The query is performed on the case base containing the newly retained cases: the LEARNING case base. Such query considers the case attributes that are relevant for each type of decision in the game. In addition, the case attributes referring to the previously taken deception decisions in the match (Table 3) are considered in the similarity computations. The 98% similarity threshold was used to determine whether the LEARNING case base had sufficient coverage to resolve the current query situation.

Second, the trigger for the expert consultation is also directed to the identification of a problem opportunity to play deceptively. To define whether a particular decision-making scenario is characterized as an opportunity for such bluffing, the number of possibilities of certain Truco events is computed. With this, for example, it is possible to determine the probability of an opponent having more ENVIDO points than the points of the agents ENVIDO, using the card that had already been played by the opponent and the position of the agent on the table. Moreover, in each moment of the hand competition and in each type of game move, whenever there is a probability of success lower than 50%, the game situation can be classified as an "opportunity for deception". Similarly, when the probability of success is higher than 85%, the agent may also adopt a slow playing deceptive move. Such estimate of the winning odds is computed in each new decision state of the game. It is updated according to the information revealed throughout the hand dispute.

The following example shows how this probability calculation is performed. Given the following agent' cards: (3♠, 12♦, 6♦), which are removed from the deck, it is possible to calculate that the opponent can have $C_{37,3} = 7,770$ possible hands. Then the strength of each possible opponent hand is compared with the strength of the agent's hand in each of these 7,770 card combinations. To do so, our method for calculating the hand strength is used. The result is that the agent has a better hand than the opponent in 5,106 hands. By computing the probability, there is a 66% chance of having a better hand than the opponent's hand.

4 Experiments and Results

The developed experiments aimed to evaluate the effectiveness of the proposed active learning and CBR approach in the collection and exploration of deception cases in the stochastic and imperfect information game of Truco. These experiments were organized as follows: a) case learning, covering the acquisition of cases through active learning, b) agent performance, referring to the analysis of agent victories with and without the use of the collected cases, and c) agent behavior, concerning the evaluation of the set of decisions taken by the agents with and without the use of the collected cases. The tested Truco playing agents were implemented according to the four different solution reuse policies listed in Table 1. Different case bases were used by them: a) the initially collected case base (BASELINE, storing 3,195 cases), collected from matches played amongst human players, b) the resulting case base later built in this work (ACTIVE, storing 5,013 cases), which increased the BASELINE case base with the new cases collected through active learning. To analyze the different game playing strategies adopted by the agents, according to the reuse policies along with their respective case bases, a number of evaluation attributes was used (Table 5).

Table 5. Game attributes observed in the analysis of the implemented agents.

Game attributes	Description
Matches	Total number of disputed matches
Wins	Total number of victorious matches by each agent
Game actions	Total number of ENVIDO/TRUCO game actions (passive + aggressive) by each agent
Aggressive game actions	Total number of ENVIDO/TRUCO aggressive game actions by each agent
Bluffs	Total number of played bluffs by each agent*
Successful bluffs	Total number of successful played bluffs by each agent*

*Bluffs: ENVIDO/TRUCO game actions in which the hand has either less than 50% or more than 85% (slow playing) winning chances.

Using the attributes described in Table 5, the analyzed strategies were the following: i) *Honest-deceptive*: indicating the rate of deceptiveness, it expresses the relationship between the total number of deceptive game moves and the total number of game moves. The higher the value is, the more deceptive the agent behavior is; ii) *Successful bluff*: indicating the rate of bluff effectiveness, it corresponds to the relationship between the number of successful bluffs and the total number of bluffs; and iii) *Passive-aggressive*: indicating the rate of aggressiveness, it captures the relationship between the number of aggressive moves of the ENVIDO/TRUCO-type and the total number of ENVIDO/TRUCO-type game moves, including when the agent does not bet. The higher the value is, the more aggressive the agent behavior is.

4.1 The Active Learning Experiment

To build the case base via active learning, 148 Truco matches were played between the agents implemented according to the tested reuse policies. The reuse policies used by each agent were randomly chosen at the beginning of each match. Only one of the players in each match had the ability to consult the expert human player, where the expert player was the first author of this paper.

During the collection of such cases via active learning, the agents computed their decisions by using the BASELINE case base. In these learning matches, whenever the learning criterion related to the detection of deception problem opportunities was satisfied, the learning algorithm presented information about the game to the specialist. Then, the expert reviewed whether the reused game action provided an effective solution for the current problem. With that, the expert player decided whether to maintain the decision recommended by the automated reuse policy or to perform another game action, deceptive or not. The expert decision was stored as a new case in a separate case base containing situations and decisions (LEARNING case base). New cases were stored in this case base only when the specialist performed an intervention by changing the game action suggested by the reuse policy. In total, 1,818 new cases were stored in the LEARNING case base. When the reviewed game actions were used by the agents, they won 79% of the disputes played during this learning experiment.

4.2 The Evaluation Experiments

Due to luck and randomness, the quality of the Truco cards received by each player is likely to have a large variation. To reduce this imbalance in the evaluation experiments, the dispute model described by the Annual Computer Poker Competition (ACPC) [25] was adopted. This model employs duplicate matches, in which the same set of hands is distributed in two sets of matches. In doing so, players reverse their positions at the table when playing the second match. Because all players receive the same set of cards, this dispute model allows a fair assessment of agents' ability.

In the first set of tests, a competition between the four implemented agents was developed, where all of them competed against the others in a total of 300 Truco matches. The agents only used the BASELINE case base in all these matches. The results in Fig. 1 (A) indicate that the PVCNPS and NPS agents achieved the best performance. Regarding the analysis of their deceptive characteristics, even the BASELINE case base, which did not yet retained the cases collected via active learning, allowed these agents to deceptively play. In fact, that BASELINE case base collected from human players already stored deceptive problem-solving experiences which were reused by the agents throughout these matches.

In the second set of tests, the test setup was similar to the previous one. However, the tested agents only used the ACTIVE case base. So the aim was to analyze whether the cases collected through the proposed active learning approach permitted to improve the agents' deceptive capabilities, and how such behavior change was expressed in the different kinds of tested agents. The results in Fig. 1 (B) indicate that the use of new cases collected via active learning enabled the PVS and PVCS agents to have the most significant performance improvement. While only the PVS and PVCS agents improved their aggressiveness rates, all tested agents increased their deceptiveness and successful bluff rates.

In the third set of tests, each one of the four implemented agents was now configured to use different case bases: BASELINE and ACTIVE. In a total of 200 played matches, each kind of agent implemented with the use of one of these case bases played against

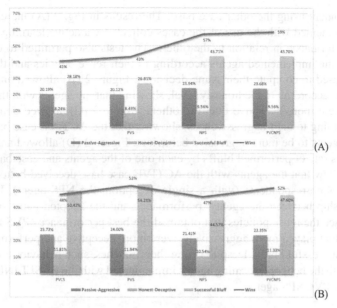

(A)

(B)

Fig. 1. Competition results between (A) agents implemented with the BASELINE case base only and (B) agents implemented with the ACTIVE case base only.

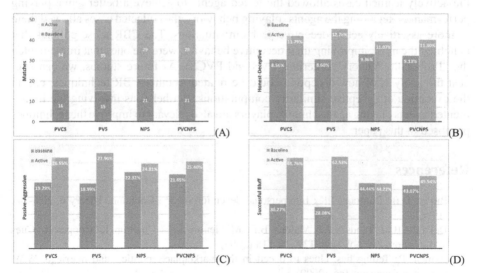

(A)

(B)

(C)

(D)

Fig. 2. Competition results between agents implemented with the BASELINE case base versus agents implemented with the ACTIVE case base.

its correspondent using the other case base. The results in Fig. 2 (A) indicate that the agents implemented with the use of the cases collected via active learning achieved a superior performance in relation to the others. The tests also permitted to observe the behavior of the implemented agents according to their reuse policies and the different case bases used to compute their game decisions. Figure 2 (B) allowed analyzing the tested agents according to the honesty level, showing that the agents with the ACTIVE case base were more deceptive than the others. Figure 2 (C) allowed comparing the agents according to their aggressiveness, showing that the ACTIVE case base enabled the tested agents to be more aggressive behaviors. Figure 2 (D) allowed analyzing the assertiveness rate of performed bluffs by each one of the agents and case bases. Again, the results show that the agents with the ACTIVE case base deceived better than their BASELINE's correspondents (with a single exception: the NPS agent). In addition, the relationship between the agents' performance and the adopted game behaviors is apparent since the reuse policies that obtained the best performance (PVS and PVCS) were those that played more aggressively, were more deceptive and performed a larger number of successful bluffs. Despite losing their matches, such behavior could also be observed with the better performing agents implemented with the BASELINE case base: the NPS and PVCNPS agents.

5 Final Remarks

This work investigates the integration of active learning and CBR, two different but complementary AI techniques, aiming to permit card playing agents to make better decisions when faced with problem opportunities to deceive. The experiments show that the actively learned cases allowed the tested agents to achieve a better game playing performance. Regarding the agents' playing behaviors, the collected cases allowed them to more assertively act in deceptive problem situations. The CBR reuse policies that benefited the most, improving their deceptive behavior, were the ones that implemented the "Probability Victory" criterion (PVS and PVCS). As future studies, we can suggest the analysis of how deception could be related to other CBR techniques, e.g. in the execution of deceptive similarity computations. Further tests involving the implemented agents playing against human players are also relevant to improve the techniques proposed in this paper.

References

1. Buller, D.B., Burgoon, J.K.: Interpersonal deception theory. Commun. Theory **6**, 203–242 (1996)
2. DePaulo, B.M., Lindsay, J.J., Malone, B.E., Muhlenbruck, L., Charlton, K., Cooper, H.: Cues to deception. Psychol. Bull. **129**, 74–118 (2003)
3. Ekman, P.: Telling lies: clues to deceit in the marketplace, politics, and marriage. W.W. Norton & Company, Inc. (2009)
4. Billings, D., Davidson, A., Schaeffer, J., Szafron, D.: The challenge of poker. Artif. Intell. J. **134**, 201–240 (2002)
5. Mantaras, De., et al.: Retrieval, reuse, revision and retention in case-based reasoning. Knowl. Eng. Rev. **20**, 215–240 (2005)

6. Rubin, J., Watson, I.: Computer poker: a review. Artif. Intell. **175**, 958–987 (2011)
7. Sandven, A., Tessem, B.: A case-based learner for Poker. In: The Ninth Scandinavian Conference on Artificial Intelligence (SCAI 2006), Helsinki, Finland (2006)
8. Paulus, G.B., Assunção, J.V.C., Silva, L.A.L.: Cases and clusters in reuse policies for decision-making in card games. In: IEEE 31st International Conference on Tools with Artificial Intelligence (ICTAI 2019), Portland, OR, pp. 1361–1365 (2019)
9. Paulus, G.B.: Cases and clusters for the development of decision-making reuse policies in card games (written in Portuguese). In: Programa de Pós-Graduação em Ciência da Computação, vol. Master in Computer Science, p. 132. Universidade Federal de Santa Maria (2020)
10. Moral, R.C.B., Paulus, G.B., Assunção, J.V.C., Silva, L.A.L.: Investigating case learning techniques for agents to play the card game of Truco. In: XIX Brazilian Symposium on Computer Games and Digital Entertainment (SBGames 2020), Recife, Brazil, pp. 107–116 (2020)
11. Winne, L.L.: Truco. Ediciones Godot, Ciudad Autónoma de Buenos Aires (2017)
12. Settles, B.: Active Learning Literature Survey. Department of Computer Sciences, University of Wisconsin–Madison (2009)
13. Richter, M.M.: Knowledge containers. In: Watson, I. (ed.) Readings in Case-Based Reasoning. Morgan Kaufmann Publishers, San Francisco (2003)
14. Neto, H.C., Julia, R.M.S.: ACE-RL-Checkers: decision-making adaptability through integration of automatic case elicitation, reinforcement learning, and sequential pattern mining. Knowl. Inf. Syst. **57**(3), 603–634 (2018). https://doi.org/10.1007/s10115-018-1175-0
15. Floyd, M.W., Esfandiari, B.: Supplemental observation acquisition for learning by observation agents. Appl. Intell. **48**(11), 4338–4354 (2018). https://doi.org/10.1007/s10489-018-1191-5
16. Ontanon, S., Floyd, M.: A comparison of case acquisition strategies for learning from observations of state-based experts. In: The 26th International Florida Artificial Intelligence Research Society Conf. (FLAIRS 2013), Florida, USA (2013)
17. Ross, S., Bagnell, D.: Efficient reductions for imitation learning. In: Yee Whye, T., Mike, T. (eds.) The Thirteenth International Conference on Artificial Intelligence and Statistics, vol. 9, pp. 661–668. PMLR (2010)
18. Ross, S., Gordon, G., Bagnell, J.A.: A reduction of imitation learning and structured prediction to no-regret online learning. In: The 14th International Conference on Artificial Intelligence and Statistics (AISTATS), Ft. Lauderdale, FL, pp. 627–635 (2011)
19. Packard, B., Ontanon, S.: Policies for active learning from demonstration. In: 2017 AAAI Spring Symposium Series. Stanford University (2017)
20. Packard, B., Ontanon, S.: Learning behavior from limited demonstrations in the context of games. In: The 31st Int. Florida Artificial Intelligence Research Society Conf. (FLAIRS 2018), Florida, USA (2018)
21. Miranda, M., Sánchez-Ruiz, A.A., Peinado, F.: Towards human-like bots using online interactive case-based reasoning. In: Bach, K., Marling, C. (eds.) ICCBR 2019. LNCS (LNAI), vol. 11680, pp. 314–328. Springer, Cham (2019). https://doi.org/10.1007/978-3-030-29249-2_21
22. Floyd, M.W., Esfandiari, B.: An active approach to automatic case generation. In: McGinty, L., Wilson, D.C. (eds.) ICCBR 2009. LNCS (LNAI), vol. 5650, pp. 150–164. Springer, Heidelberg (2009). https://doi.org/10.1007/978-3-642-02998-1_12
23. Rubin, J., Watson, I.: Case-based strategies in computer Poker. AI Commun. **25**, 19–48 (2012)
24. Sobrinho, M.G.: Manual do jogo do Truco Cego (Flor de Abóbora). Martins Livreiro Editora Ltda., Porto Alegre (2004)
25. ACPC: Annual Computer Poker Competition. http://www.computerpokercompetition.org/ (2018)

$ASPIC^?$ and the Postulates
of Non-interference and Crash-Resistance

Rafael Silva$^{(\boxtimes)}$ and João Alcântara

Department of Computer Science, Federal University of Ceará, Fortaleza, Brazil
{rafaels,jnando}@lia.ufc.br

Abstract. We introduce an interrogation mark ? in $ASPIC^+$ languages as a plausibility operator to enhance any defeasible conclusion does not have the same status as an irrefutable one. The resulting framework, dubbed $ASPIC^?$, is tailored to make a distinction between strong inconsistencies and weak inconsistencies. The aim is to avoid the former and to tolerate the latter. This means the extensions obtained from the $ASPIC^?$ framework are free of strong conflicts, but tolerant to weak conflicts. Then, in the current study, we show $ASPIC^?$ satisfy reasonable properties. In particular, we focus on the property that a conflict between two arguments should not interfere with the acceptability of other unrelated arguments. With this purpose in mind, we prove under which conditions the important principles of Non-interference and Crash-Resistance hold in $ASPIC^?$.

Keywords: Argumentation · Paraconsistency · Conflict-tolerance

1 Introduction

As noticed in [1], contradictions can be considered under the mantle of many points of views: as a consequence of the only correct description of a contradictory world, as a temporary state of our knowledge, as the outcome of a particular language which we have chosen to describe the world, as the result of conflicting observational criteria, as the superposition of world-views, or as the result from the best theories available at a given moment. Indeed, in [2], it is argued that inconsistency is a natural companion to defeasible methods of reasoning and that paraconsistency (the property of a logic admitting non-trivial inconsistent theories) should play a role in the formalisation of these methods. In fact, they introduced an interrogation mark ? as a plausibility operator to enhance any defeasible conclusion do not have the same status as an irrefutable one, obtained from deduction.

Inspired by these ideas, we will present the $ASPIC^?$ framework by extending the $ASPIC^+$ framework [3], one of the most important formalisms to represent and reason with structured argumentation. In $ASPIC^?$ (as well as in $ASPIC^+$), we identify two types of rules: strict (irrefutable) and defeasable. Unlike $ASPIC^+$, the distinguishing characteristic of $ASPIC^?$ is the conclusion of any defeasible rule will be a plausible (?-suffixed) formula ϕ?. The intended

© Springer Nature Switzerland AG 2021
A. Britto and K. Valdivia Delgado (Eds.): BRACIS 2021, LNAI 13073, pp. 328–343, 2021.
https://doi.org/10.1007/978-3-030-91702-9_22

meaning is the conclusion of ϕ? will not necessarily prevent the conclusion of $\neg\phi$?; it is required an argument with conclusion $\neg\phi$, which can only be obtained from a strict rule, to attack the conclusion $\neg\phi$? of an argument. Thus, to produce a strong conflict between the conclusions of the arguments, at least in one of them, the conclusion should be a ?-free formula obtained via a strict rule.

In *ASPIC*?, strong inconsistencies as in $\{\phi, \neg\phi\}$ (or $\{\phi?, \neg\phi\}$) are distinguished from those weak inconsistencies as in $\{\phi?, \neg\phi?\}$: the first should be avoided; the second can be tolerated. This means the (weak) conflict between ϕ? and $\neg\phi$? can be accommodated in the same extension. Hence, the extensions in *ASPIC*? will be free of strong conflicts, but tolerant to weak conflicts.

Given that much current work on structured argumentation [3–5] combines strict and defeasible inference rules, unexpected results can arise when two arguments based on defeasible rules have contradictory conclusions. This is particularly critical (see [6]) if the strict inference rules include the *Ex Falso* principle (that an inconsistent set implies anything), because for any formula ϕ, an argument concluding $\neg\phi$ can be constructed from these two arguments. As consequence, any other argument is potentially under threat!

In order to solve this problem for *ASPIC*$^+$, Wu [7,8] requires that in each argument, the set of conclusions of all its sub-arguments are classically consistent. Another approach was taken in [6], in which they replace classical logic as the source for strict rules by the (weaker) paraconsistent logic presented in [9] to invalidate the *Ex Falso* principle as a valid strict inference rule.

Here we will also exploit how to avoid the application of the *Ex Falso* principle in *ASPIC*? by combining these two solutions: 1) as in [6], we resort to paraconsistent reasoning to tolerate conflicts; our differential is we tolerate only weak conflicts. 2) as in [7,8], we require for each argument, the set of conclusions of all its sub-arguments are consistent; our differential is that we eliminate only those arguments whose sets of conclusions lead to a strong conflict.

Then, we show *ASPIC*? satisfies reasonable properties. In particular, we focus on the property that a conflict between two arguments should not interfere with the acceptability of other unrelated arguments. With this purpose in mind, we prove under which conditions the important principles of Non-interference and Crash-Resistance [10] hold in *ASPIC*?.

The rest of the paper is organised as follows: in Sect. 2, *ASPIC*? framework is presented. Then, we introduce the corresponding argumentation framework with two kinds of defeats (strong and weak) and its semantics. Section 3 is focused on proving the satisfaction of the principles of Non-interference and Crash-Resistance. Finally, we summarise our contributions and future developments.

2   The *ASPIC*? Framework

An *Abstract Argumentation Framework AF* [11] is a pair $(\mathcal{A}, \mathcal{D})$ in which $\mathcal{A}$ is a set of arguments and $\mathcal{D} \subseteq \mathcal{A} \times \mathcal{A}$ is a relation of *defeat*. An argument A defeats $\mathcal{B}$ if $(A, B) \in \mathcal{D}$. The *ASPIC*$^+$ framework [3,12] gives structure to the arguments and defeat relation in an AF. In this section, we introduce in $ASPIC^+$ languages an interrogation mark ? as a plausibility operator to enhance defeasible conclusions do not have the same status as those irrefutable. The resulting framework, $ASPIC^?$, is tailored to distinguish strong inconsistencies from weak inconsistencies. The aim is to avoid the former and to tolerate the latter. We start by defining the argumentation systems specified by $ASPIC^?$:

Definition 1 (Argumentation System). *An argumentation system is a tuple $AS = (\mathcal{L}, ^-, \mathcal{R}, n)$, in which*

- *$\mathcal{L} = \mathcal{L}^* \cup \mathcal{L}^?$ is a logical language with a unary negation symbol $\neg$ and a unary plausibility symbol ? such that*
  - *$\mathcal{L}^*$ is a ?-free logical language with a unary negation symbol.*
  - *$\mathcal{L}^? = \{\phi? \mid \phi \in \mathcal{L}^*\}$.*
- *$^-$ is a function from $\mathcal{L}$ to $2^{\mathcal{L}}$, such that*
  - *φ is a contrary of ψ if $\varphi \in \overline{\psi}$, $\psi \notin \overline{\varphi}$;*
  - *φ is a contradictory of ψ (denoted by $\varphi = -\psi$), if $\varphi \in \overline{\psi}$, $\psi \in \overline{\varphi}$;*
- *$\mathcal{R} = \mathcal{R}_s \cup \mathcal{R}_d$ is a set of strict ($\mathcal{R}_s$) and defeasible ($\mathcal{R}_d$) inference rules of the form $\phi_1, \ldots, \phi_n \rightarrow \phi$ and $\phi_1, \ldots, \phi_n \Rightarrow \psi?$ respectively (in which $\phi_1, \ldots, \phi_n, \phi$ are meta-variables ranging over wff in $\mathcal{L}$ and ψ is a meta-variable ranging over wff in $\mathcal{L}^*$), and $\mathcal{R}_s \cap \mathcal{R}_d = \emptyset$.*
- *n is a partial function such that $n : \mathcal{R}_d \longrightarrow \mathcal{L}$.*

For any formula $\phi \in \mathcal{L}^*$, we say $\psi \in -\phi$ if $\psi = \neg\phi$ or $\psi = \neg\phi?$ or $\phi = \neg\psi$ or ($\phi = \neg\gamma$ and $\psi = \gamma?$); we say $\psi \in -\phi?$ if $\psi = \neg\phi$ or $\phi = \neg\psi$.

Intuitively, contraries can be used to model well-known constructs like negation as failure in logic programming. Note for any $\phi \in \mathcal{L}^*$, ϕ and $\phi?$ are contradictories of $\neg\phi$; whilst, only ϕ is a contradictory of $\neg\phi?$. This means $\phi?$ is not a contradictory of $\neg\phi?$. A set as $\{\phi, -\phi\}$ (or $\{\phi?, -\phi?\}$) is intended to represent a *strong inconsistency*, and $\{\phi?, \neg\phi?\}$ is intended to represent a *weak inconsistency*. We will refer to these two kinds of inconsistencies (strong and weak) as epistemic inconsistencies or simply inconsistencies.

It is also required a knowledge base to provide premises for the arguments.

Definition 2 (Knowledge Base). *A knowledge base in an argumentation system $AS = (\mathcal{L}, ^-, \mathcal{R}, n)$ is a set $\mathcal{K} \subseteq \mathcal{L}$ consisting of two disjoint subsets $\mathcal{K}_n$ (the axioms) and $\mathcal{K}_p$ (the ordinary premises).*

Axioms are certain knowledge and cannot be attacked, whilst, ordinary premises are uncertain and can be attacked. Now we can define an argumentation theory:

Definition 3. *An argumentation theory $(AS, \mathcal{K})$ is a pair in which AS is an argumentation system and $\mathcal{K}$ is a knowledge base in AS.*

In $ASPIC^?$, arguments are constructed recursively from an argumentation theory by the successive application of construction rules:

Definition 4 (Argument). *An argument A on the basis of an argumentation theory $(AS, \mathcal{K})$ and an argumentation system $(\mathcal{L}, ^-, \mathcal{R}, n)$ is*

1. ϕ *if* $\phi \in \mathcal{K}$ *with* $\mathtt{Prem}(A) = \{\phi\}$, $\mathtt{Conc}(A) = \phi$, $\mathtt{Sub}(A) = \{\phi\}$, $\mathtt{DefR}(A) = \emptyset$, $\mathtt{Rules}(A) = \emptyset$, $\mathtt{TopRule}(A) =$ *undefined*.
2. $A_1, \ldots, A_n \rightarrow \psi$ *if* $A_1, \ldots, A_n$ *are arguments s.t. there is a strict rule* $\mathtt{Conc}(A_1), \ldots, \mathtt{Conc}(A_n) \rightarrow \psi \in \mathcal{R}_s$; $\mathtt{Prem}(A) = \mathtt{Prem}(A_1) \cup \cdots \cup \mathtt{Prem}(A_n)$; $\mathtt{Conc}(A) = \psi$; $\mathtt{Sub}(A) = \mathtt{Sub}(A_1) \cup \cdots \cup \mathtt{Sub}(A_n) \cup \{A\}$; $\mathtt{Rules}(A) = \mathtt{Rules}(A_1) \cup \cdots \cup \mathtt{Rules}(A_n) \cup \{\mathtt{Conc}(A_1), \ldots, \mathtt{Conc}(A_n) \rightarrow \psi\}$; $\mathtt{TopRule}(A) = \mathtt{Conc}(A_1), \ldots, \mathtt{Conc}(A_n) \rightarrow \psi$.
3. $A_1, \ldots, A_n \Rightarrow \psi$? *if* $A_1, \ldots, A_n$ *are arguments such that there exists a defeasible rule* $\mathtt{Conc}(A_1), \ldots, \mathtt{Conc}(A_n) \Rightarrow \psi$? $\in \mathcal{R}_d$; $\mathtt{Prem}(A) = \mathtt{Prem}(A_1) \cup \cdots \cup \mathtt{Prem}(A_n)$; $\mathtt{Conc}(A) = \psi$?; $\mathtt{Sub}(A) = \mathtt{Sub}(A_1) \cup \cdots \cup \mathtt{Sub}(A_n) \cup \{A\}$; $\mathtt{Rules}(A) = \mathtt{Rules}(A_1) \cup \cdots \cup \mathtt{Rules}(A_n) \cup \{\mathtt{Conc}(A_1), \ldots, \mathtt{Conc}(A_n) \Rightarrow \psi?\}$; $\mathtt{TopRule}(A) = \mathtt{Conc}(A_1), \ldots, \mathtt{Conc}(A_n) \Rightarrow \psi$?.

For any argument A we define $\mathtt{Prem}_n(A) = \mathtt{Prem}(A) \cap \mathcal{K}_n$; $\mathtt{Prem}_p(A) = \mathtt{Prem}(A) \cap \mathcal{K}_p$; $\mathtt{DefR}(A) = \{r \in \mathcal{R}_d \mid r \in \mathtt{Rules}(A)\}$ *and* $\mathtt{StR}(A) = \{r \in \mathcal{R}_s \mid r \in \mathtt{Rules}(A)\}$.

Example 1. Consider the argumentation system $AS = (\mathcal{L}, ^-, \mathcal{R}, n)$, in which

- $\mathcal{L} = \mathcal{L}^* \cup \mathcal{L}^?$ with $\mathcal{L}^* = \{a, b, f, w, \neg a, \neg b, \neg f, \neg w, \sim a, \sim b, \sim f, \sim w, \sim \neg a, \sim \neg b, \sim \neg f, \sim \neg w\}$. The symbols $\neg$ and $\sim$ respectively denote strong and weak negation.
- For any $\phi \in \mathcal{L}^*$ and any $\psi \in \mathcal{L}$, (1) $\phi \in \overline{\psi}$ iff (a) $\psi = \neg\phi$ or $\psi = \neg\phi$? or $\phi = \neg\psi$ or ($\phi = \neg\gamma$ and $\psi = \gamma$?); or (b) $\psi = \sim \phi$ or ($\psi = \sim \phi$?). (2) ϕ? $\in \overline{\psi}$ iff (a) $\psi = \neg\phi$ or $\phi = \neg\psi$; or (b) $\psi = \sim \phi$.
- $\mathcal{R}_s = \{\neg f \rightarrow \neg w; b \rightarrow a\}$ and $\mathcal{R}_d = \{a \Rightarrow \neg f?; b, \sim \neg w \Rightarrow w?; \neg f? \Rightarrow \neg w?\}$.

Let $\mathcal{K}$ be the knowledge base such that $\mathcal{K}_n = \emptyset$ and $\mathcal{K}_p = \{b, \sim \neg w\}$. The arguments defined on the basis of $\mathcal{K}$ and AS are $A_1 = [b]$, $A_2 = [\sim \neg w]$, $A_3 = [A_1 \rightarrow a]$, $A_4 = [A_3 \Rightarrow \neg f?]$, $A_5 = [A_1, A_2 \Rightarrow w?]$ and $A_6 = [A_4 \Rightarrow \neg w?]$.

An argument A is *for* ϕ if $\mathtt{Conc}(A) = \phi$; it is *strict* if $\mathtt{DefR}(A) = \emptyset$; *defeasible* if $\mathtt{DefR}(A) \neq \emptyset$; *firm* if $\mathtt{Prem}(A) \subseteq \mathcal{K}_n$; *plausible* if $\mathtt{Prem}(A) \cap \mathcal{K}_p \neq \emptyset$. An argument is *fallible* if it is defeasible or plausible and *infallible* otherwise. We write $S \vdash \phi$ if there is a strict argument for ϕ with all premises taken from S, and $S \mathrel{\vdash\!\sim} \phi$ if there is a defeasible argument for ϕ with all premises taken from S. The next definition will be repeatedly employed in Sect. 3:

Definition 5. *Let $AT = (AS, \mathcal{K})$ be an argumentation theory with argumentation system $AS = (\mathcal{L}, ^-, \mathcal{R}, n)$. For a formula $\phi \in \mathcal{L}$, we define $\mathtt{Atoms}(\phi) = \{a \mid a$ is an atom occurring in $\phi\}$. For a set $\mathcal{F} \subseteq \mathcal{L}$ of formulas in $\mathcal{L}$, we define $\mathtt{Atoms}(\mathcal{F}) = \bigcup_{\phi \in \mathcal{F}} \mathtt{Atoms}(\phi)$; furthermore, for a set of atoms $\mathfrak{A}$, $\mathcal{F}_{|\mathfrak{A}} = \{\phi \in \mathcal{F} \mid \phi$ contains only atoms in $\mathfrak{A}\}$. For a strict rule $s = \phi_1, \ldots, \phi_n \rightarrow \Psi$, we define $\mathtt{Atoms}(s) = \mathtt{Atoms}(\{\phi_1, \ldots, \phi_n, \psi\})$. For a defeasible rule $d = \phi_1, \ldots, \phi_n \Rightarrow \Psi$, we define $\mathtt{Atoms}(d) = \mathtt{Atoms}(\{\phi_1, \ldots, \phi_n, \psi\})$. For a set $\mathcal{S} = \{s_1, \ldots, s_n\}$ of* *strict rules, we define* $\texttt{Atoms}(\mathcal{S}) = \texttt{Atoms}(s_1) \cup \cdots \cup \texttt{Atoms}(s_n)$. *For a set* $\mathcal{D} = \{d_1, \ldots, d_n\}$ *of defeasible rules, we define* $\texttt{Atoms}(\mathcal{D}) = \texttt{Atoms}(d_1) \cup \cdots \cup \texttt{Atoms}(d_n)$. *For an argumentation system* $AS = (\mathcal{L}, ^-, \mathcal{R}, n)$, *we define* $\texttt{Atoms}(AS) = \texttt{Atoms}(\mathcal{R}_d) \cup \texttt{Atoms}(\{n(r) \mid r \in \mathcal{R}_d \text{ and } n(r) \text{ is defined}\})$, *in which* $\mathcal{R}_d \subseteq \mathcal{R}$ *is the set of defeasible rules in* $\mathcal{R}$. *For an argumentation theory* $AT = (AS, \mathcal{K})$, *we define* $\texttt{Atoms}(AT) = \texttt{Atoms}(AS) \cup \texttt{Atoms}(\mathcal{K})$. *For an argument* A, *we define* $\texttt{Atoms}(A) = \texttt{Atoms}(\texttt{StR}(A)) \cup \texttt{Atoms}(\texttt{DefR}(A))$. *Finally, for a set* $\mathcal{A} = \{A_1, \ldots, A_n\}$ *of arguments, we define* $\texttt{Atoms}(\mathcal{A}) = \texttt{Atoms}(A_1) \cup \cdots \cup \texttt{Atoms}(A_n)$.

Let $AT = (AS, \mathcal{K})$ be an argumentation theory with $AS = (\mathcal{L}, ^-, \mathcal{R}, n)$, and $\mathcal{A}$ the set of all arguments constructed from $\mathcal{K}$ in AS. Assume $\mathcal{K} = (\mathcal{K}_n, \mathcal{K}_p)$, such that $\mathcal{K}_n = \{a, b, c\}$ and $\mathcal{K}_p = \emptyset$. Consider $\mathcal{R} = \{a \to d, b \Rightarrow e\}$. The resulting arguments are $A = [a]$, $B = [b]$, $C = [c]$, $D = [A \to d]$, and $E = [B \Rightarrow e]$. We have $\texttt{Atoms}(D) = \{a, d\}$, $\texttt{Atoms}(AT) = \{a, b, c, e\}$. Note those atoms occurring only in the strict rules (as d) are not considered as atoms in $\texttt{Atoms}(AT)$.

2.1   Attacks and Defeats

In $ASPIC^?$ arguments are related to each other by attacks (as in $ASPIC^+$) and by weak attacks:

Definition 6 (Attacks). *Consider the arguments A and B. We say A attacks B iff A undercuts, undermines and rebuts B, in which*

- *A undercuts B (on B') iff* $\texttt{Conc}(A) \in \overline{n(r)}$ *for some $B' \in \texttt{Sub}(B)$ such that B''s top rule r is defeasible.*
- *A undermines B (on ϕ) iff* $\texttt{Conc}(A) \in \overline{\phi}$ *and $\phi \in \texttt{Prem}_p(B)$. In such a case, A contrary-undermines B iff* $\texttt{Conc}(A)$ *is a contrary of ϕ.*
- *A rebuts B (on B') iff* $\texttt{Conc}(A) \in \overline{\phi?}$ *for some $B' \in \texttt{Sub}(B)$ of the form $B_1'', \ldots, B_n'' \Rightarrow \phi?$. In such a case, A contrary-rebuts B iff* $\texttt{Conc}(A)$ *is a contrary of $\phi?$.*

We say A weakly attacks B iff A weakly undermines or weakly rebuts B, in which

- *A weakly undermines B (on $\phi?$ (resp. $\neg\phi?$)) iff* $\texttt{Conc}(A) = \neg\phi?$ *(resp.* $\texttt{Conc}(A) = \phi?$*) for an ordinary premise $\phi?$ (resp. $\neg\phi?$) of B.*
- *A weakly rebuts B (on B') iff* $\texttt{Conc}(A) = \neg\phi?$ *(resp.* $\texttt{Conc}(A) = \phi?$*) for some $B' \in \texttt{Sub}(B)$ of the form $B_1'', \ldots, B_n'' \Rightarrow \phi?$ (resp. $B_1'', \ldots, B_n'' \Rightarrow \neg\phi?$).*

Example 2. Recalling Example 1, we have A_5 weakly rebuts A_6 and A_6 weakly rebuts A_5. Besides, A_6 contrary-undermines A_2 and A_5 on $\sim \neg w$. If in addition, one had the argument $A_7 = [A_4 \to \neg w?]$, then A_7 (like A_6) would weakly rebut A_5 on A_5; however, A_7 (unlike A_6) would not be weakly rebutted by A_5.

Definition 7 *(SAF).* *A structured argumentation framework SAF defined by an argumentation theory $AT = (AS, \mathcal{K})$ is a tuple $\langle \mathcal{A}, \mathcal{C}, \mathcal{C}', \preceq \rangle$, in which*

- *$\mathcal{A}$ is the set of all arguments A constructed from $\mathcal{K}$ in AS such that it satisfies Definition 4 and* $\texttt{Atoms}(A) \subseteq \texttt{Atoms}(AT)$;

- $(X, Y) \in C$ iff X attacks Y and $(X, Y) \in C'$ iff X weakly attacks Y;
- $\preceq$ is a preference ordering on $\mathcal{A}$.

The restriction $\mathtt{Atoms}(A) \subseteq \mathtt{Atoms}(AT)$ is to avoid including in the SAF arguments built from strict rules without relation to $\mathtt{Atoms}(AT)$. Next, we define the corresponding defeat relation:

Definition 8 (Defeat). *[3] Let $A, B \in \mathcal{A}$ and A attacks B. If A undercut, contrary-rebut or contrary-undermine attacks B on B' then A is said to preference-independent attack B on B'; otherwise A is said to preference-dependent attack B on B'. A defeats B iff for some B' either A preference-independent attacks B on B' or A preference-dependent attacks B on B' and $A \not\prec B'$.*

As observed in the previous definition, a preference-dependent attack from one argument to another only succeeds (as a defeat) if the attacked argument is not stronger than the attacking argument. Thus, if an argument A preference-dependent attacks B and B is preferred over A, then the attack of A to B does not succeed, and B is not defeated by A.

2.2 Abstract Argumentation Frameworks with Two Kinds of Defeats

As SAFs have two kinds of attacks, the associated abstract argumentation frameworks have to couple with two kinds of defeats:

Definition 9 (Argumentation frameworks with two kinds of defeats). *An abstract argumentation framework with two kinds of defeats (AF_2) corresponding to a $SAF = \langle \mathcal{A}, C, C', \preceq \rangle$ is a tuple $(\mathcal{A}, \mathcal{D}, \mathcal{D}')$ such that $\mathcal{D} = \{(X, Y) \in C \mid X \text{ defeats } Y\}$ and $\mathcal{D}' = \{(X, Y) \in C' \mid X \not\prec Y\}$. For $A \in \mathcal{A}$, we define*

$$A^+ = \{B \in \mathcal{A} \mid (A, B) \in \mathcal{D} \cup \mathcal{D}'\} \text{ and } A^- = \{B \in \mathcal{A} \mid (B, A) \in \mathcal{D} \cup \mathcal{D}'\}.$$

Given a SAF SA defined by an argumentation theory AT and an AF_2 AF corresponding to SA, we will refer to AF as the resulting AF_2 from AT.

Example 3 (Example 2 continued). Let $\preceq= \{(A_6, A_2)\}$, i.e., $A_6 \prec A_2$ be a preference ordering on $\mathcal{A} = \{A_1, A_2, A_3, A_4, A_5, A_6\}$. In the $SAF(\mathcal{A}, C, C', \preceq)$ defined by AT, we have $C = \{(A_6, A_2)\}$ and $C' = \{(A_5, A_6), (A_6, A_5)\}$. As (A_6, A_2) is a preference independent attack, we obtain $\mathcal{D} = C$ and $\mathcal{D}' = C'$.

Traditional approaches to argumentation semantics ensure arguments attacking each other are not tolerated in the same set, which is said to be conflict free. In $ASPIC^?$, we distinguish a strong conflict from a weak conflict. The aim is to avoid strong conflicts, which are carried over by the defeat relation $\mathcal{D}$, and to tolerate weak conflicts, which are carried over by the relation $\mathcal{D}'$. We say the resulting set is compatible:

Definition 10 (Compatible sets). *Let $AF = (\mathcal{A}, \mathcal{D}, \mathcal{D}')$ be an AF_2 and $S \subseteq \mathcal{A}$. A set S is compatible (in AF) if $\forall A \in S$, $\not\exists B \in S$ such that $(B, A) \in \mathcal{D}$.*

Compatible sets do not contain the arguments A and B if A attacks B (or vice versa); however they can accommodate weak attacks between their members. This means in Example 3, the set $\{A_5, A_6\}$ is compatible, but $\{A_2, A_6\}$ is not. Now we are entitled to define semantics to deal with compatible sets:

Definition 11 (Semantics). *Let $AF = (\mathcal{A}, \mathcal{D}, \mathcal{D}')$ be an AF_2 and $S \subseteq \mathcal{A}$ be a compatible set of arguments. Then $X \in \mathcal{A}$ is* acceptable *with respect to S iff*

- *$\forall Y \in \mathcal{A}$ such that $(Y, X) \in \mathcal{D} : \exists Z \in S$ such that $(Z, Y) \in \mathcal{D}$ and*
- *$\forall Y \in \mathcal{A}$ such that $(Y, X) \in \mathcal{D}' : \exists Z \in S$ such that $(Z, Y) \in \mathcal{D} \cup \mathcal{D}'$.*

We define $f_{AF}(S) = \{A \in \mathcal{A} \mid A$ is acceptable w.r.t. $S\}$. For a compatible set S in AF, we say 1) S is an admissible set *of AF iff $S \subseteq F_{AF}(S)$; 2) S is a* complete extension *of AF iff $f_{AF}(S) = S$; 3) S is a* preferred extension *of AF iff it is a set inclusion maximal complete extension of AF; 4) S is the* grounded extension *iff it is the set inclusion minimal complete extension of AF; 5) S is a* stable extension *iff S is complete extension of AF and $\forall Y \notin S$, $\exists X \in S$ s.t. $(X, Y) \in \mathcal{D} \cup \mathcal{D}'$. 6) S is a* semi-stable extension *iff it is a complete extension of AF such that there is no complete extension S_1 of AF in which $S \cup S^+ \subset S_1 \cup S_1^+$.*

Notice for an argument X to be acceptable w.r.t. S, if $(Y, X) \in \mathcal{D}$, there should exist an argument $Z \in S$ such that $(Z, Y) \in \mathcal{D}$, i.e., a weak defeat as $(Z, Y) \in \mathcal{D}'$ is not robust enough to defend a defeat as $(Y, X) \in \mathcal{D}$. Otherwise, X defends a weak defeat $(Y, X) \in \mathcal{D}'$ if $\exists Z \in S$ such that Z (weak) defeats Y.

Example 4 (Example 3 continued).
Regarding the AF_2 constructed in Example 3, we obtain

- Complete Extensions: $\{A_1, A_3, A_4\}$, $\{A_1, A_3, A_4, A_5\}$, $\{A_1, A_3, A_4, A_6\}$, $\{A_1, A_3, A_4, A_5, A_6\}$;
- Grounded Extension: $\{A_1, A_3, A_4\}$;
- Preferred Extension: $\{A_1, A_3, A_4, A_5, A_6\}$
- Stable/Semi-stable Extensions: $\{A_1, A_3, A_4, A_6\}$, $\{A_1, A_3, A_4, A_5, A_6\}$

3   The Postulates of Non-interference and Crash-Resistance

In this section, we show under which conditions, $ASPIC^?$ satisfies the property that a conflict between two arguments should not interfere with the acceptability of other unrelated arguments. Let us illustrate it via the following example:

Example 5. [6] Let $\mathcal{R}_d = \{p \Rightarrow q; r \Rightarrow \neg q; t \Rightarrow s\}$ be a set of defeasible rules in $ASPIC^+$ (their conclusions are not ?-suffixed formulas), $\mathcal{K}_p = \emptyset$ and $\mathcal{K}_n = \{p, r, t\}$, while $\mathcal{R}_s$ consists of all propositionally valid inferences. The corresponding AF includes the arguments $A_1 = [p]$, $A_2 = [A_1 \Rightarrow q]$, $B_1 = [r]$, $B_2 = [B_1 \Rightarrow \neg q]$, $C = [A_2, B_2 \rightarrow \neg s]$, $D_1 = [t]$ and $D_2 = [D_1 \Rightarrow s]$. We have C defeats D_2 if $C \not\prec D_2$. This is problematic as s can be any formula. Hence, any defeasible argument unrelated to A_2 or B_2 can, depending on $\preceq$, be defeated by C owing to the explosiveness of classical logic as the source for $\mathcal{R}_s$.

This property is guaranteed by proving the postulates (originally conceived in [10]) of Non-interference (Definition 22) and Crash-Resistance (Definition 25) hold in $ASPIC^?$. Unlike [7,8], however, we will not eliminate all arguments whose set of conclusions of all its sub-arguments is contradictory, but only those whose set of conclusions contains strong contradictions as $\{\phi, \neg\phi\}$, $\{\phi, \neg\phi?\}$ or $\{\neg\phi, \phi?\}$. Weak contradictions as $\{\phi?, \neg\phi?\}$ will not lead to the elimination of the argument (see Definition 14). We proceed by introducing several definitions and lemmas before proving the satisfaction of these postulates:

Definition 12 (Consistency). *Let $\mathcal{A} = \{A_1, \ldots, A_n\}$ be a set of arguments on the basis of an argumentation theory $(AS, \mathcal{K})$ and an argumentation system $AS = (\mathcal{L}, ^-, \mathcal{R}, n)$. An argument $A \in \mathcal{A}$ is inconsistent iff $\forall\phi \in \mathcal{L}$, it holds $\{\texttt{Conc}(A') \mid A' \in \texttt{Sub}(A)\} \vdash \phi$. Otherwise, A is consistent. The set $\mathcal{A}$ is inconsistent if $\forall\phi \in \mathcal{L}$, it holds $\texttt{Concs}(\texttt{Sub}(A_1)) \cup \ldots \cup \texttt{Concs}(\texttt{Sub}(A_n)) \vdash \phi$. Otherwise $\mathcal{A}$ is consistent.*

A strict rule as $\phi_1, \ldots, \phi_n \to \psi$ represents that if $\phi_1, \ldots, \phi_n$ hold, then without exception it holds that ψ. It has a very general meaning; the unique restriction we will impose to prove our results is that we will assume throughout this section that every strict rule in an argumentation system is reasonable:

Definition 13 (Reasonable strict rules). *Let $(AS, \mathcal{K})$ be an argumentation theory and $AS = (\mathcal{L}, ^-, \mathcal{R}_s \cup \mathcal{R}_d, n)$ be an argumentation system. A strict rule $\phi_1, \ldots, \phi_n \to \psi \in \mathcal{R}_s$ is reasonable iff 1) for each ϕ_i $(1 \leq i \leq n)$ it holds $Atoms(\phi_i) \subseteq Atoms(\psi)$ or 2) $\forall\psi \in \mathcal{L}$, it holds $\{\phi_1, \ldots, \phi_n\} \vdash \psi$.*

A strict rule $\phi_1, \ldots, \phi_n \to \psi$ is reasonable if $\forall\phi_i$ $(1 \leq i \leq n)$, each atom in ϕ_i is also in ψ or $\{\phi_1, \ldots, \phi_n\}$ is inconsistent. Reasonable strict rules are very usual in many propositional logics. Now we define an inconsistency cleaned AF_2:

Definition 14 (Inconsistency-cleaned AF_2). *Let $\langle\mathcal{A}, \mathcal{D}, \mathcal{D}'\rangle$ be a AF_2 resulting from an argumentation theory AT. We define $\mathcal{A}_c = \{A \in \mathcal{A} \mid A \text{ is consistent}\}$, $\mathcal{D}_c = \mathcal{D} \cap (\mathcal{A}_c \times \mathcal{A}_c)$ and $\mathcal{D}'_c = \mathcal{D}' \cap (\mathcal{A}_c \times \mathcal{A}_c)$. We refer to $(\mathcal{A}_c, \mathcal{D}_c, \mathcal{D}'_c)$ as the inconsistency cleaned AF_2 resulting from an argumentation theory AT.*

By *inconsistency-cleaned version* of the $ASPIC^?$ system, we mean the $ASPIC^?$ system from which the inconsistency-cleaned AF_2 is constructed. The next concept is important to simplify the proofs of the results we will show in this section:

Definition 15 (Flat arguments). *Let A be an argument on the basis of an argumentation theory AT. We say that A is flat iff $\texttt{TopRule}(A)$ is strict, $A = [A_1, \ldots, A_n \to \alpha]$ and $\forall A_i$ $(1 \leq i \leq n)$, one of the following conditions holds:*

1. $\texttt{TopRule}(A_i)$ *is defeasible or*
2. $\texttt{Rules}(A_i) = \emptyset$.

Flat arguments have strict top rule and every of its strict subarguments comes from the set of premises. Henceforth, for any $AF_2\langle\mathcal{A}, \mathcal{D}, \mathcal{D}'\rangle$ resulting from argumentation theory AT, we will assume without loss of generality every $A \in \mathcal{A}$ with a strict top rule is flat. In order to prove some results in this section, we will need to identify the set of conclusions associated with a set of arguments:

Definition 16. *Let $\mathcal{A}$ be a set of arguments whose structure complies with Definition 4. We define* $\text{Concs}(\mathcal{A}) = \{\text{Conc}(A) \mid A \in \mathcal{A}\}$.

Next, we define the consequence function Cn_{sem}, such that $Cn_{sem}(AT)$ is a set of sets of conclusions under the argumentation semantics *sem*.

Definition 17. *Let $\mathfrak{AT}$ be the set of all argumentation theories that can be constructed from a language $\mathcal{L}$. Let $AT \in \mathfrak{AT}$ be an argumentation theory and $AF = (\mathcal{A}, \mathcal{D}, \mathcal{D}')$ be the resulting AF_2 from AT. We define $Cn_{sem} : \mathfrak{AT} \to 2^{2^{\text{Concs}(\mathcal{A})}}$ is a function s.t. $Cn_{sem}(AT) = \{\text{Concs}(E) \mid E \subseteq \mathcal{A} \text{ is an extension of } AF_2 \text{ under semantics sem}\}$, where $sem \in \{complete, grounded, preferred, stable, semi\text{-}stable\}$. We will use $Cn_c(AT)$ as the shortening of $Cn_{complete}(AT)$. For a set $\mathfrak{A}$ of propositional atoms, by $Cn_{sem}(AT)_{|\mathfrak{A}}$, we mean the set $\{\mathcal{F}_{|\mathfrak{A}} \mid \mathcal{F} \in Cn_{sem}(AT)\}$.*

In order to define the postulate for non-interference, we need to specify what the union of two argumentation theories looks like:

Definition 18 (Union of argumentation theories). *Let $AT_1 = (AS_1, \mathcal{K}_1)$ and $AT_2 = (AS_2, \mathcal{K}_2)$ be argumentation theories s.t. $\mathcal{K}_1 = \mathcal{K}_{n_1} \cup \mathcal{K}_{p_1}$, $\mathcal{K}_2 = \mathcal{K}_{n_2} \cup \mathcal{K}_{p_2}$, $AS_1 = (\mathcal{L}, ^-, \mathcal{R}_{s_1} \cup \mathcal{R}_{d_1}, n_1)$ and $AS_2 = (\mathcal{L}, ^-, \mathcal{R}_{s_2} \cup \mathcal{R}_{d_2}, n_2)$. Besides, we assume $n_1(r) = n_2(r)$ for any $r \in \mathcal{R}_{d_1} \cup \mathcal{R}_{d_2}$. We define $AT_1 \cup AT_2$ is an argumentation theory $AT = (AS, \mathcal{K})$ s.t. $\mathcal{K} = \mathcal{K}_n \cup \mathcal{K}_p$ with $\mathcal{K}_n = \mathcal{K}_{n_1} \cup \mathcal{K}_{n_2}$, and $\mathcal{K}_p = \mathcal{K}_{p_1} \cup \mathcal{K}_{p_2}$; $AS = (\mathcal{L}, ^-, \mathcal{R}_s \cup \mathcal{R}_d, n)$ with $\mathcal{R}_s = \mathcal{R}_{s_1} \cup \mathcal{R}_{s_2}$, $\mathcal{R}_d = \mathcal{R}_{d_1} \cup \mathcal{R}_{d_2}$ and $n(r) = n_1(r)$ if $r \in \mathcal{R}_{d_1}$; otherwise, $n(r) = n_2(r)$.*

Other very important notion to guarantee our results in this section is that of syntactically disjoint argumentation theories; it will be employed to characterise non-interference (Definition 22) and contamination (Definition 24).

Definition 19 (Syntactically disjoint argumentation theories). *Let AT_1 and AT_2 be argumentation theories. We say AT_1 and AT_2 are syntactically disjoint when $\text{Atoms}(AT_1) \cap \text{Atoms}(AT_2) = \emptyset$.*

The depth of an argument will be employed in the proof of Lemma 1:

Definition 20 (Depth of an argument). *Let A be an argument whose structure complies with Definition 4. The depth of A, denoted by $\text{depth}(A)$, is 1 if $\text{Rules}(A) = \emptyset$ or else $\text{depth}(A) = 1 + \max\{\text{depth}(A') \mid A' \in \text{Sub}(A)\}$.*

The following essential lemma states that for every argument A such that $\text{Conc}(A) \subseteq \text{Atoms}(AF)$ there exists an argument A' with the same conclusion, $\text{Atoms}(A') \subseteq \text{Atoms}(AF)$, and is not more vulnerable than A.

Lemma 1. *Let $AS_1 = (\mathcal{L}, ^-, \mathcal{R}_s \cup \mathcal{R}_{d_1}, n_1)$ and $AS_2 = (\mathcal{L}, ^-, \mathcal{R}_s \cup \mathcal{R}_{d_2}, n_2)$ be argumentation systems s.t. $\mathcal{R}_s$ is a set of strict rules and $\mathcal{R}_{d_1}$ and $\mathcal{R}_{d_2}$ are sets of defeasible rules. Let $AT = AT_1 \cup AT_2$ be an argumentation theory where $AT_1 = (AS_1, \mathcal{K}_1)$ and $AT_2 = (AS_2, \mathcal{K}_2)$ are syntactically disjoint, and $AF = (\mathcal{A}, \mathcal{D}, \mathcal{D}')$ and $AF_1 = (\mathcal{A}_1, \mathcal{D}_1, \mathcal{D}'_1)$ be respectively the inconsistency cleaned AF_2s resulting from AT and AT_1. For each argument $C \in \mathcal{A}$ such that $\text{Conc}(C) \subseteq \text{Atoms}(AT_1)$, $\exists C' \in \mathcal{A}_1$ with $\text{Conc}(C') = \text{Conc}(C)$, $C'^+ \subseteq C^+$ and $C'^- \subseteq C^-$.*

Proof. Let $\mathcal{K}_1 = \mathcal{K}_{n_1} \cup \mathcal{K}_{p_1}$ and $\mathcal{K}_2 = \mathcal{K}_{n_2} \cup \mathcal{K}_{p_2}$. We will prove by induction on depth(C) that $\forall C \in \mathcal{A}$, where Atoms(Conc($C$)) $\subseteq$ Atoms(AT_1), $\exists C' \in \mathcal{A}_1$ such that Conc(C') = Conc(C) and (1) Concs(Sub(C')) $\subseteq$ Concs(Sub(C)) and (2) DefR(C') $\subseteq$ DefR(C). Note $C'^+ = C^+$ follows directly from Conc(C') = Conc(C); condition (1) guarantees C' is consistent as C is consistent; condition (2) suffices to show $C'^- \subseteq C^-$ as AS_1 and AS_2 share the same set $\mathcal{R}_s$ of strict rules:

Suppose depth(C) = 1. There are two possibilities: $C \in \mathcal{K}_{n_1}$ or $C \in \mathcal{K}_{p_1}$. Thus, for $C' = C$, $C'^- = C^-$ and Conc(C') = Conc(C). Now assume conditions (1) and (2) hold for any argument C with depth(C) $\leq k$. We will show for any argument C with depth(C) = $k + 1$ they also hold. There are two possibilities:

- TopRule(C) $\in \mathcal{R}_d = \mathcal{R}_{d_1} \cup \mathcal{R}_{d_2}$. Note C is of the form $C_1, \ldots, C_n \Rightarrow$ Conc(C). As AT_1 and AT_2 are syntactically disjoint and Atoms(Conc(C)) $\subseteq$ Atoms(AT_1), TopRule(C) $\in \mathcal{R}_{d_1}$. It follows that for each $i \in \{1, \ldots, n\}$, Atoms(Conc(C_i)) $\subseteq$ Atoms(AT_1). As depth(C_i) $\leq k$, by induction hypothesis, there exists $C'_i \in \mathcal{A}_1$ such that Conc(C'_i) = Conc(C_i), DefR(C'_i) $\subseteq$ DefR(C'_i), and Concs(Sub(C'_i)) $\subseteq$ Concs(Sub(C_i)). Applying TopRule(C) we construct $C' = C'_1, \ldots, C'_n \Rightarrow$ Conc(C) from AT_1. Now we show that C' satisfies the requested properties.
  - $C' \in \mathcal{A}_1$ since Atoms(C') = Atoms(TopRule(C)) $\cup \bigcup_{i=1}^n$ Atoms(C'_i) $\subseteq$ Atoms(AT_1).
  - Conc(C') = Conc(C) since C and C' share the same top rule.
  - DefR(C') = $\{\{\text{DefR}(C)\} \cup \bigcup_{i=1}^n \text{DefR}(C'_i)\} \subseteq \{\{\text{TopRule}(C)\} \cup \bigcup_{i=1}^n \text{DefR}(C_i)\} = \text{DefR}(C)$.
  - Concs(Sub(C')) = $\{\{\text{Conc}(C)\} \cup \bigcup_{i=1}^n \text{Concs}(\text{Sub}(C'_i))\} \subseteq \{\{\text{Conc}(C)\} \cup \bigcup_{i=1}^n \text{Concs}(\text{Sub}(C_i))\} = \text{Concs}(\text{Sub}(C))$.
- TopRule(C) $\in \mathcal{R}_s$. Note C is of the form $C_1, \ldots, C_n \to$ Conc(C). As we have assumed C is flat, we can partition arguments C_i into two sets $\mathfrak{C}_p \cup \mathfrak{C}_d = \{1, \ldots, n\}$, in which $i \in \mathfrak{C}_p$ iff Conc(C_i) $\in \mathcal{K}_1 \cup \mathcal{K}_2$ and TopRule(C_i) = *undefined*, and $i \in \mathfrak{C}_d$ iff TopRule(C_i) $\in \mathcal{R}_{d_1} \cup \mathcal{R}_{d_2}$. Since AT_1 and AT_2 are syntactically disjoint, for $i \in \mathfrak{C}_p$, Atoms(Conc(C_i)) $\subseteq$ Atoms(AT_1) or Atoms(Conc(C_i)) $\subseteq$ Atoms(AT_2), and for $i \in \mathfrak{C}_d$, Atoms(TopRule(C_i)) $\subseteq$ Atoms(AT_1) or Atoms(TopRule(C_i)) $\subseteq$ Atoms(AT_2). We can partition the subarguments of C into two disjoint sets $\mathfrak{C}_1$ and $\mathfrak{C}_2$ such that for $i \in \mathfrak{C}_1$, Conc(C_i) $\subseteq$ Atoms(AT_1) and for $i \in \mathfrak{C}_2$, Conc(C_i) $\subseteq$ Atoms(AT_2). Let $\mathfrak{C}_p \cap \mathfrak{C}_1 = \{p_1, \ldots, p_k\}$, $\mathfrak{C}_d \cap \mathfrak{C}_1 = \{d_1, \ldots, d_m\}$, and $\mathfrak{C}_2 = \{b_1, \ldots, b_j\}$. For each $d_i \in \mathfrak{C}_d \cap \mathfrak{C}_1$, depth($C_{d_i}$) $\leq k$. By the induction hypothesis, for each $d_i \in \mathfrak{C}_d \cap \mathfrak{C}_1$, exists $C'_{d_i} \in \mathcal{A}_1$ s.t Conc(C'_{d_i}) = Conc(C_{d_i}), and DefR(C'_{d_i}) $\subseteq$ DefR(C_{d_i}). Note that Conc(C_{p_1}), ..., Conc(C_{p_k}), Conc(C'_{d_1}), ..., Conc(C'_{d_m}), Conc(C_{b_1}), ..., Conc(C_{b_j}) $\to$ Conc(C) corresponds to TopRule(C).
  Let $T = \text{Atoms}(\text{Conc}(C_{b_1})) \cup \ldots \cup \text{Atoms}(\text{Conc}(C_{b_j}))$. As $T \cap \text{Atoms}(\text{Conc}(C)) = \emptyset$, Conc($C_{p_1}$), ..., Conc($C_{p_n}$), Conc($C'_{d_1}$), ..., Conc($C'_{d_m}$) $\to$ Conc(C) $\in \mathcal{R}_s$; otherwise $\{\text{Conc}(C_{b_1}), \ldots, \text{Conc}(C_{b_j})\}$ is inconsistent (Definition 13), which cannot be true as C is consistent. Thus, we can construct an argument $C' = \text{Conc}(C_{p_1}), \ldots, \text{Conc}(C_{p_n}), \text{Conc}(C'_{d_1}), \ldots, \text{Conc}(C'_{d_m}) \to \text{Conc}(C)$ from AT_1 s.t. Conc(C') = Conc(C). Now we show C' satisfies the requested properties.

- $C' \in \mathcal{A}_1$ since $\texttt{Atoms}(C') = \texttt{Atoms}(\texttt{TopRule}(C)) \cup \bigcup_{(1 \leq i \leq n)} \texttt{Atoms}(C'_i) \subseteq \texttt{Atoms}(AT_1)$.
- $\texttt{Conc}(C') = \texttt{Conc}(C)$ since C and C' share the same top rule.
- In the construction of C' we resort to C_i for each $i \in \mathcal{C}_1 \cap \mathfrak{C}_p$ and for each $j \in \mathcal{C}_1 \cap \mathfrak{C}_d$, we resort to C'_j, obtained by induction hypothesis, s.t. $\texttt{DefR}(C'_j) \subseteq \texttt{DefR}(C_j)$ and $\texttt{Concs}(\texttt{Sub}(C'_j)) \subseteq \texttt{Concs}(\texttt{Sub}(C_j))$. It follows $\texttt{DefR}(C') \subseteq \texttt{DefR}(C)$ and $\texttt{Concs}(\texttt{Sub}(C')) \subseteq \texttt{Concs}(\texttt{Sub}(C))$. □

The notion of defense expresses when an argument C defends B from A:

Definition 21 (Defense). *Let $AF = (\mathcal{A}, \mathcal{D}, \mathcal{D}')$ be an AF_2 and $A, B, C \in \mathcal{A}$ such that $(A, B) \in \mathcal{D} \cup \mathcal{D}'$. An argument C defends B from A, denoted by $df(C, B, A)$, when 1) if $(A, B) \in \mathcal{D}$, then $(C, A) \in \mathcal{D}$; 2) if $(A, B) \in \mathcal{D}'$, then $(C, A) \in \mathcal{D}$ or $(C, A) \in \mathcal{D}'$.*

In addition to the fundamental result obtained by Lemma 1, we will use Lemmas 2, 3, 5 and 6 to prove Theorem 1, which is one of our main results. With the following lemma we establish a connection between complete extensions of AF with the arguments in $\mathcal{A}_1$:

Lemma 2. *Let $AT = AT_1 \cup AT_2$ for syntactically disjoint argumentation theories AT_1 and AT_2, and $AF = (\mathcal{A}, \mathcal{D}, \mathcal{D}')$ and $AF_1 = (\mathcal{A}_1, \mathcal{D}_1, \mathcal{D}'_1)$ be respectively the resulting inconsistency-cleaned AF_2s from AT and AT_1. For any complete extension $E \subseteq \mathcal{A}$ of AF, $\texttt{Concs}(E \cap \mathcal{A}_1) = \texttt{Concs}(E)_{|Atoms(AT_1)}$.*

Proof.

- If $\phi \in \texttt{Concs}(E \cap \mathcal{A}_1)$, then $\exists A \in E \cap \mathcal{A}_1$ such that $\texttt{Conc}(A) = \phi$. It follows $A \in E$ and $A \in \mathcal{A}_1$, and so $\phi \in \texttt{Concs}(E)$ and $\phi \in \texttt{Concs}(\mathcal{A}_1)$. As $\texttt{Atoms}(\texttt{Conc}(\phi)) \subseteq \texttt{Atoms}(AT_1)$, it holds $\phi \in \texttt{Concs}(E)_{|Atoms(AT_1)}$.
- If $\phi \in \texttt{Concs}(E)_{|Atoms(AT_1)}$, then $\exists A \in E$ s.t. $\texttt{Conc}(A) = \phi$ and $\texttt{Atoms}(\phi) \subseteq \texttt{Atoms}(AT_1)$. From Lemma 1, $\exists A' \in \mathcal{A}_1$ such that $\texttt{Conc}(A') = \phi$ and $A'^- \subseteq A^-$. As E is a complete extension of AF and $A \in E$, it must be $A' \in E$, and so $A' \in E \cap \mathcal{A}_1$. Thus, $\phi \in \texttt{Concs}(E \cap \mathcal{A}_1)$. □

Lemmas 3 and 4 will be used as intermediate steps in the demonstration of Lemma 5. Lemma 3 expresses the function f_{AF_1} associated with the subframework AF_1 in terms of the function f_{AF} associated with AF:

Lemma 3. *Let $AT = AT_1 \cup AT_2$ for syntactically disjoint argumentation theories AT_1 and AT_2, and $AF = (\mathcal{A}, \mathcal{D}, \mathcal{D}')$ and $AF_1 = (\mathcal{A}_1, \mathcal{D}_1, \mathcal{D}'_1)$ be respectively the resulting inconsistency-cleaned AF_2s from AT and AT_1. For any set of arguments $S \subseteq \mathcal{A}_1$, $f_{AF_1}(S) = f_{AF}(S) \cap \mathcal{A}_1$.*

Proof.

- $f_{AF_1}(S) \subseteq f_{AF}(S) \cap \mathcal{A}_1$. If $A \in f_{AF_1}(S)$, then $A \in \mathcal{A}_1$. It remains to prove $A \in f_{AF}(S)$: let $B \in \mathcal{A}$ s.t. $(B, A) \in \mathcal{D} \cup \mathcal{D}'$. It means $\texttt{Atoms}(\texttt{Conc}(B)) \subseteq \texttt{Atoms}(AT_1)$. By Lemma 1, $\exists B' \in \mathcal{A}_1$ s.t. $\texttt{Conc}(B') = \texttt{Conc}(B)$ and $B'^- \subseteq B^-$. Thus, $(B', A) \in \mathcal{D} \cup \mathcal{D}'$. As $A \in f_{AF_1}(S)$, $\exists C \in S$ s.t. $df(C, A, B')$. From $B'^- \subseteq B^-$, $(C, B) \in \mathcal{D} \cup \mathcal{D}'$. It follows $A \in f_{AF}(S)$. Thus $A \in f_{AF}(S) \cap \mathcal{A}_1$.

– $f_{AF}(S) \cap \mathcal{A}_1 \subseteq f_{AF_1}(S)$. Let $A \in f_{AF}(S) \cap \mathcal{A}_1$. It means that $\forall B \in \mathcal{A}$ such that $(B, A) \in \mathcal{D} \cup \mathcal{D}'$, $\exists C \in S$ such that $df(C, A, B)$. As $\mathcal{D}_1 \cup \mathcal{D}'_1 \subseteq \mathcal{D} \cup \mathcal{D}'$, it follows $\forall B \in \mathcal{A}_1$ such that $(B, A) \in \mathcal{D}_1 \cup \mathcal{D}'_1$, $\exists C \in S$ such that $df(C, A, B)$. Then $A \in f_{AF_1}(S)$. □

By Lemma 4 (employed in Lemma 5), if an argument A is attacked by a set of arguments that also attack arguments in a complete extension E, then $A \in E$.

Lemma 4. *Let $AF = (\mathcal{A}, \mathcal{D}, \mathcal{D}')$ be the AF_2 resulting from an argumentation theory AT. Let $AF_c = (\mathcal{A}_c, \mathcal{D}_c, \mathcal{D}'_c)$ be the inconsistency-cleaned AF_2. Let E be a complete extension of AF_c, $S \subseteq E$ and $A \in \mathcal{A}$. If A is consistent and $A^- \subseteq S^-$, then $A \in E$.*

Proof. Suppose A is consistent. It follows $A \in \mathcal{A}_c$. As E is a complete extension of AF, $\forall B \in S$, $B \in f_{AF}(E)$. As $A^- \subseteq S^-$, it follows $A \in f_{AF}(E) = E$. □

Lemmas 5 assures the complete extensions of AF when restricted to the arguments of its subframework AF_1 is a complete extension of AF_1:

Lemma 5. *Let $AT = AT_1 \cup AT_2$ for syntactically disjoint argumentation theories AT_1 and AT_2, and $AF = (\mathcal{A}, \mathcal{D}, \mathcal{D}')$ and $AF_1 = (\mathcal{A}_1, \mathcal{D}_1, \mathcal{D}'_1)$ be the inconsistency-cleaned AF_2s resulting from AT and AT_1 respectively. If E is a complete extension of AF, then $E \cap \mathcal{A}_1$ is a complete extension of AF_1.*

Proof. Let E be a complete extension of AF. Assume $E' = E \cap \mathcal{A}_1$. We will prove that E' is a complete extension of AF_1. Note E' is compatible in AF_1, since $E' \subseteq E$, E is compatible in AF and there are no new defeats in AF_1. Now we will show $E' = f_{AF_1}(E')$:

– $E' \subseteq f_{AF_1}(E')$. Let $A \in E'$. As also $A \in E$ and E is a complete extension of AF, it means $\forall B \in \mathcal{A}_1$ such that $(B, A) \in \mathcal{D}_1 \cup \mathcal{D}'_1$, $\exists C \in E$ such that $df(C, A, B)$. As $B \in \mathcal{A}_1$, $\mathtt{Atoms}(\mathtt{Conc}(C)) \subseteq \mathtt{Atoms}(AT_1)$. By Lemma 1, $\exists C' \in \mathcal{A}_1$, such that $\mathtt{Conc}(C') = \mathtt{Conc}(C)$ and $C'^- \subseteq C^-$. From Lemma 4, $C' \in E$, and so $C' \in E'$. Thus, $A \in f_{AF_1}(E')$.
– $f_{AF_1}(E') \subseteq E'$. As f is a monotony function, and E is a complete extension of AF, $f_{AF}(E') \subseteq f_{AF}(E) = E$, and so $f_{AF}(E') \cap \mathcal{A}_1 \subseteq E \cap \mathcal{A}_1$. From Lemma 3, $f_{AF_1}(E') = f_{AF}(E') \cap \mathcal{A}_1$. But then, we obtain $f_{AF_1}(E') \subseteq E \cap \mathcal{A}_1$. Thus, $f_{AF_1}(E') \subseteq E'$. □

Lemma 6 assures the admissible sets in AF_1 are also admissible sets in AF:

Lemma 6. *Let $AT = AT_1 \cup AT_2$ for syntactically disjoint argumentation theories AT_1 and AT_2, and $AF = (\mathcal{A}, \mathcal{D}, \mathcal{D}')$ and $AF_1 = (\mathcal{A}_1, \mathcal{D}_1, \mathcal{D}'_1)$ be respectively the resulting inconsistency-cleaned AF_2s from AT and AT_1. Let $S \subseteq \mathcal{A}_1$. If S is an admissible set in AF_1, then S is an admissible set if AF.*

Proof. As S is a compatible set in AF_1, S is also a compatible set in AF, since $\mathcal{A}_1 \subseteq \mathcal{A}$. It remains to show that $S \subseteq f_{AF}(S)$. Assume $B \in \mathcal{A}$ such that for some $A \in S$, $(B, A) \in \mathcal{D} \cup \mathcal{D}'$. It means $\mathtt{Atoms}(\mathtt{Conc}(B)) \subseteq \mathtt{Atoms}(AT_1)$. From

Lemma 1, $\exists B' \in \mathcal{A}_1$ such that $\text{Conc}(B') = \text{Conc}(B)$ and $B'^- \subseteq B^-$. Note $(B', A) \in \mathcal{D} \cup D'$ in AF_1. As S is admissible in AF_1, from Definition 11, $\exists C \in S$ s.t. $df(C, A, B')$. Given $B'^- \subseteq B^-$, $(C, B) \in \mathcal{D} \cup \mathcal{D}'$. It follows $S \subseteq f_{AF}(S)$.   □

In the sequel, we formally define the concept of non-interference.

Definition 22 (Non-interference). *The $ASPIC^?$ system satisfies non-interference under a semantics sem iff for every syntactically disjoint argumentation theories AT_1 and AT_2, it holds $Cn_{sem}(AT_1 \cup AT_2)_{|Atoms(AT_1)} = Cn_{sem}(AT_1)$.*

Non-interference means that, for disjoint argumentation theories AT_1 and AT_2, AT_1 does not influence the outcome with respect to the language of AT_2.

Theorem 1. *The inconsistency-cleaned version of the $ASPIC^?$ system satisfies non-interference under complete semantics.*

Proof. Let $AT = AT_1 \cup AT_2$ for syntactically disjoint argumentation theories AT_1 and AT_2 and $AF = (\mathcal{A}, \mathcal{D}, \mathcal{D}')$ and $AF_1 = (\mathcal{A}_1, \mathcal{D}_1, \mathcal{D}'_1)$ be respectively the resulting inconsistency-cleaned AF_2s from AT and AT_1. Let $\mathfrak{S}_1 = \{B_1, \ldots, B_n\}$ and $\mathfrak{S}_2 = \{S_1, \ldots, S_m\}$ be the set of complete extensions of AF and AF_1 respectively. We will prove that $L = R$, in which $L = Cn_c(AT)_{|Atoms(AT_1)} = \{\text{Concs}(B_1)_{|Atoms(AT_1)}, \ldots, \text{Concs}(B_n)_{|Atoms(AT_1)}\}$, and $R = Cn_c(AT_1)_{|Atoms(AT_1)} = \{\text{Concs}(S_1), \ldots, \text{Concs}(S_m)\}$.

From Lemma 2, $L = \{\text{Concs}(B_1 \cap \mathcal{A}_1), \ldots, \text{Concs}(B_n \cap \mathcal{A}_1)\}$. For each complete extension B of AF, $B \cap \mathcal{A}_1$ is a complete extension of AF_1 (Lemma 5). It remains to prove for any $S \in \mathfrak{S}_2$, $\exists B \in \mathfrak{S}_1$ with $B \cap \mathcal{A}_1 = S$. If S is a complete extension of AF_1 ($S = f_{AF_1}(S)$), S is an admissible set in AF (Lemma 6), i.e., $S \subseteq f_{AF}(S)$. Then $B = \bigcup_{n=1}^{\infty} f_{AF}^n(S)$ is a complete extension of AF as the least fixed point of f_{AF} contains S. We will prove that $B \cap \mathcal{A}_1 = S$. Intersecting both sides of $B = \bigcup_{n=1}^{\infty} f_{AF}^n(S)$ with $\mathcal{A}_1$, and applying Lemma 3, we get $B \cap \mathcal{A}_1 = (\bigcup_{n=1}^{\infty} f_{AF}^n(S)) \cap \mathcal{A}_1 = \bigcup_{n=1}^{\infty} f_{AF}^n(S) \cap \mathcal{A}_1 = \bigcup_{n=1}^{\infty} f_{AF_1}^n(S) = \bigcup_{n=1}^{\infty} S = S$.   □

$ASPIC^?$ is non trivial under semantics *sem* if the conclusions of an argumentation theory are never fully determined by the atoms.

Definition 23 (Non-trivial). *The $ASPIC^?$ system is non-trivial under semantics sem iff for each nonempty set $\mathfrak{A}$ of atoms, there are argumentation theories AT_1 and AT_2 such that $Atoms(AT_1) = Atoms(AT_2)$ and $Cn_{sem}(AT_1)_{|\mathfrak{A}} \neq Cn_{sem}(AT_2)_{|\mathfrak{A}}$.*

In the following theorem, we show that the inconsistency-cleaned version of the $ASPIC^?$ system satisfies non-triviality under complete semantics:

Theorem 2. *The inconsistency-cleaned version of the $ASPIC^?$ system satisfies non-triviality under complete semantics.*

Proof. Let $\mathfrak{A} = \{a_1, \ldots, a_n\}$ $(n \geq 1)$ a set of atoms. We will show that there are two inconsistency-cleaned argumentation frameworks $AF_1 = (\mathcal{A}_1, \mathcal{D}_1, \mathcal{D}'_1)$ and $AF_2 = (\mathcal{A}_2, \mathcal{D}_2, \mathcal{D}'_2)$ resulting from AT_1 and AT_2 respectively such that $\texttt{Atoms}(AT_1) = \texttt{Atoms}(AT_2)$ and $Cn_c(AT_1)_{|\mathfrak{A}} \neq Cn_c(AT_2)_{|\mathfrak{A}}$. Let $\mathcal{K}_{n_1} = \mathcal{K}_{n_2} = \mathcal{K}_{p_2} = \emptyset$, $\mathcal{K}_{p_1} = \{a_1, \ldots, a_n\}$, $\mathcal{R}_{d_2} = \{a_1 \Rightarrow a_1?; \ldots; a_n \Rightarrow a_n?\}$, $\mathcal{R}_{s_1} = \mathcal{R}_{s_2} = \mathcal{R}_{d_1} = \emptyset$. Thus, $Cn_c(AT_1) = \{\{a_1, \ldots, a_n\}\}$ and $Cn_c(AT_2) = \{\emptyset\}$, and so $Cn_c(AT_1) = Cn_c(AT_1)_{|\mathfrak{A}} \neq Cn_c(AT_2)_{|\mathfrak{A}} = Cn_c(AT_2)$. □

An argumentation theory AT_1 is contaminating when any other unrelated argumentation theory AT_2 becomes irrelevant when merged with AT_1:

Definition 24 (Contamination). *An argumentation theory AT_1 is contaminating under a semantics sem iff for every argumentation theory AT_2 s.t. AT_1 and AT_2 are syntactically disjoint, it holds $Cn_{sem}(AT_1) = Cn_{sem}(AT_1 \cup AT_2)$.*

Crash-resistance is strongly related to the concept of contamination:

Definition 25 (Crash-resistance). *We say that $ASPIC^?$ under a semantics sem satisfies crash-resistance iff there does not exists an argumentation theory AT that is contaminating under sem.*

The intuition behind crash-resistance is that one wants to avoid local problems having global effects.

Theorem 3. *If $ASPIC^?$ satisfies non-interference and non-triviality under complete semantics, then it also satisfies crash-resistance under complete semantics.*

Proof. (1) By absurd suppose the $ASPIC^?$ does not satisfy crash-resistance. Then there exists an argumentation theory AT_1 that is contaminating and $\texttt{Atoms}(AT_1) \subset \mathfrak{A}$. Let $\mathfrak{B} = \mathfrak{A} \backslash \texttt{Atoms}(AT_1)$. (2) By assumption $ASPIC^?$ is non-trivial. Thus, there are argumentation theories AT_2 and AT_3 such that $\texttt{Atoms}(AT_2) = \texttt{Atoms}(AT_3) \subseteq \mathfrak{B}$ and $Cn_c(AT_2)_{|\mathfrak{B}} \neq Cn_c(AT_3)_{|\mathfrak{B}}$. Note that both AT_2 and AT_3 are syntactically disjoint from AT_1. (3) By assumption $ASPIC^?$ satisfies non-interference, from which follows $Cn_c(AT_2)_{|\mathfrak{B}} = Cn_c(AT_2 \cup AT_1)_{|\mathfrak{B}}$ and $Cn_c(AT_3 \cup AT_1)_{|\mathfrak{B}} = Cn_c(AT_3)_{|\mathfrak{B}}$. (4) Given AT_1 is contaminating, $Cn_c(AT_1 \cup AT_2)_{|\mathfrak{B}} = Cn_c(AT_1)_{|\mathfrak{B}} = Cn_c(AT_1 \cup AT_3)_{|\mathfrak{B}}$. From (3) and (4), it follows that $Cn_c(AT_2)_{|\mathfrak{B}} = Cn_c(AT_3)_{|\mathfrak{B}}$. It is an absurd as from (2) we have $Cn_c(AT_2)_{|\mathfrak{B}} \neq Cn_c(AT_3)_{|\mathfrak{B}}$. □

Theorem 4. *The inconsistency-cleaned version of the $ASPIC^?$ system satisfies crash-resistance under complete semantics.*

Proof. It follows from Theorems 2, 1 and 3.

4   Conclusion and Future Works

In this work, we defined an argumentation framework, dubbed $ASPIC^?$, by introducing in $ASPIC^+$ [3] an interrogation mark ? as a plausibility operator to enhance any defeasible conclusion does not have the same status than an irrefutable one: In $ASPIC^?$, any defeasible rule have the form $\phi_1, \ldots, \phi_n \Rightarrow \phi?$.

As in [2], we distinguish strong contradictions from weak ones. We avoid the former and tolerate the latter. Then, we showed in $ASPIC^?$ conflicting arguments does not interfere with the acceptability of unrelated arguments. This is proved by combining solutions found in [6] and in [7,8] to show the postulates of Non-interference and Crash-Resistance hold in inconsistency-cleaned $ASPIC^?$: 1) as in [6], we resort to paraconsistent reasoning to tolerate conflicts; our differential is we tolerate only weak conflicts. 2) as in [7,8], we require for each argument, the set of conclusions of all its sub-arguments are consistent; our differential is that we eliminate only those arguments whose sets of conclusions lead to a strong conflict. Thus, our work paves the way to investigate in the context of structured argumentation alternative solutions to satisfy the postulates of Non-interference and Crash-resistance without having to delete all inconsistent arguments.

In the future we will study other ways to satisfy these postulates and which monotonic paraconsistent logics can be used as source of strict rules to avoid contaminating argumentation theories. We will also exploit the relation between $ASPIC^?$ and extended logic programas with paraconsistent semantics [13].

References

1. Carnielli, W., Marcos, J.: A taxonomy of C-systems. In: Paraconsistency, pp. 24–117. CRC Press (2002)
2. Pequeno, T., Buchsbaum, A.: The logic of epistemic inconsistency. In: Proceedings of the Second International Conference on Principles of Knowledge Representation and Reasoning, pp. 453–460 (1991)
3. Modgil, S., Prakken, H.: A general account of argumentation with preferences. Artif. Intell. **195**, 361–397 (2013)
4. Gorogiannis, N., Hunter, A.: Instantiating abstract argumentation with classical logic arguments: postulates and properties. Artif. Intell. **175**(9–10), 1479–1497 (2011)
5. Caminada, M., Modgil, S., Oren, N.: Preferences and unrestricted rebut. Computational Models of Argument (2014)
6. Grooters, D., Prakken, H.: Combining paraconsistent logic with argumentation. In: COMMA, pp. 301–312 (2014)
7. Wu, Y.: Between argument and conclusion-argument-based approaches to discussion, inference and uncertainty. Ph.D. thesis, University of Luxembourg (2012)
8. Wu, Y., Podlaszewski, M.: Implementing crash-resistance and non-interference in logic-based argumentation. J. Logic Comput. **25**(2), 303–333 (2015)
9. Rescher, N., Manor, R.: On inference from inconsistent premises. Theory Decis. **1**(2), 179–217 (1970)
10. Caminada, M.: Semi-stable semantics. OMMA **144**, 121–130 (2006)

11. Dung, P.: On the acceptability of arguments and its fundamental role in non-monotonic reasoning, logic programming and n-person games. Artif. Intell. **77**(2), 321–357 (1995)
12. Prakken, H.: An abstract framework for argumentation with structured arguments. Argument Comput. **1**(2), 93–124 (2010)
13. Damásio, C., Moniz Pereira, L.: A survey of paraconsistent semantics for logic programs. In: Besnard, P., Hunter, A. (eds.) Reasoning with Actual and Potential Contradictions. Handbook of Defeasible Reasoning and Uncertainty Management Systems, vol. 2, pp. 241–320. Springer, Cham (1998). https://doi.org/10.1007/978-94-017-1739-7_8

On the Refinement
of Compensation-Based Semantics
for Weighted Argumentation Frameworks

Henrique Viana[1](✉) and João Alcântara[2]

[1] Instituto Federal de Educação, Ciência e Tecnologia do Ceará (IFCE),
Rodovia CE-040, Km 137, 1 s/n, Aracati, CE, Brazil
henrique.viana@ifce.edu.br
[2] Departamento de Ciência da Computação, Universidade Federal do Ceará (UFC),
P.O. Box 12166, Fortaleza, CE 60455-760, Brazil
jnando@lia.ufc.br

Abstract. Acceptability semantics for the frameworks of weighted argumentation can satisfy up to one of the principles of (**Quality Precedence**), (**Cardinality Precedence**) or (**Compensation**), which are pairwise incompatible. In this paper we define two new principles: (**Quality Compensation**) and (**Cardinality Compensation**), which are weakened versions of (**Quality Precedence**) and (**Cardinality Precedence**), respectively. We show that these new principles are compatible with (**Compensation**) and propose two new semantics: a t-conorm-based, which can satisfy (**Quality Compensation**) and a cumulative sum-based semantics, which satisfies (**Cardinality Compensation**).

Keywords: Argumentation · T-conorms · Cumulative sum

1   Introduction

Argumentation is a reasoning process in which interacting arguments are built and evaluated. It is widely studied in Artificial Intelligence, namely for reasoning about making decisions [5,8,22] and modelling agents interactions [7,30]. An argumentation-based formalism or argumentation framework is generally defined as a set of arguments, attacks amongst the arguments, and a semantics for evaluating the arguments. A semantics assesses to what extent an argument is acceptable. Examples of semantics are those proposed by Dung, which compute extensions of arguments [6,11,21,26,29,32] and ranking semantics, which compute the overall strengths of each argument [2,10,12,13,18,31].

With respect to the ranking semantics, in the works of Amgoud and Ben-Naim [2], it was proposed a set of principles for it and consequently refined in [3]. Moreover, new principles were introduce to describe strategies that a semantics may use when it faces a conflict between the quality of attackers and their quantity [4]. The strategies are: i) privileging quality through the principle of (**Quality Precedence**), ii) privileging cardinality through the principle

© Springer Nature Switzerland AG 2021
A. Britto and K. Valdivia Delgado (Eds.): BRACIS 2021, LNAI 13073, pp. 344–358, 2021.
https://doi.org/10.1007/978-3-030-91702-9_23

of (**Cardinality Precedence**), or iii) simply allowing compensation between quality and cardinality through the principle of (**Compensation**).

It was pointed out some limitations in the literature about this subject. First, there was no semantics satisfying (**Cardinality Precendence**), which was unfortunate since (**Cardinality Precendence**) is a viable choice in (multiple criteria) decision making [17]. Second, there was only one semantics satisfying (**Quality Precedence**). Third, several other semantics satisfy (**Compensation**), however, none of them satisfies all the principles that are compatible with the compensation principle. With that in mind, Amgoud and Ben-Naim [4] provided three new semantics: a max-based, a cardinality-based and a sum-based one. A formal analysis and thorough comparison with other semantics were done to fill the previous gaps by introducing three semantics and show new semantics that enjoy more desirable properties than existing semantics.

Besides that, another important result pointed out is that some of the principles are incompatible. They cannot be satisfied all together by a semantics. This is particularly the case with the (**Quality Precedence**), (**Cardinality Precedence**) and (**Compensation**). From this perspective, in this work, we propose to explore further the relation among these three principles and show weaker principles that are intermediary of these three principles. Furthermore, we present two new semantics, one based on t-conorms operators [24], which present a mix of max-based semantics and the sum-based semantics; and the other semantics based on the notion of cumulative sum [25], which presents a trade off between the cardinality-based and the sum-based semantics. The novelty is both semantics satisfy not only all the basic principles, but also weakened versions of two incompatible principles. As far as we know, this is the first work to push forward the frontier of knowledge on the development of semantics aiming at satisfying weakened versions of incompatible principles.

The paper is organized as follows: in Sect. 2, we first recall some basic notions of Argumentation theory, introduce the notations used throughout the paper and show some examples of semantics and the principles that a semantics could satisfy. In Sect. 3, we then consider the contribution of the paper, with the introduction of two new semantics and two new principles. We provide a formal analysis and comparison of existing semantics and these new principles. Finally, in Sect. 4, we conclude the paper.

2   Foundations of Weighted Argumentation Frameworks

2.1   Basic Concepts

A weighted argumentation graph is a set of arguments and an attack relation between them. Each argument has a weight in the interval $[0, 1]$ representing its basic strength (the smaller the weight, the weaker the argument).

Definition 1 (WAG). *A weighted argumentation graph (WAG) is an ordered tuple $\mathbf{G} = \langle \mathcal{A}, w, \mathcal{R} \rangle$, where $\mathcal{A}$ is a non-empty finite set of arguments, w is a function from $\mathcal{A}$ to $[0, 1]$, and $\mathcal{R} \subseteq \mathcal{A} \times \mathcal{A}$.*

Intuitively, $w(a)$ is the basic strength of argument a, and $(a, b) \in \mathcal{R}$ (or $a\mathcal{R}b$) means argument a attacks argument b.

Example 1. Consider the WAG **G** below consisted of four arguments $a, b, c, \ldots, k$. For instance, the basic strength of a is $w(a) = 1$ and the basic strength of d is $w(d) = 0.6$. Besides that, argument d attacks argument a, which is depicted by a directed edge in the graph.

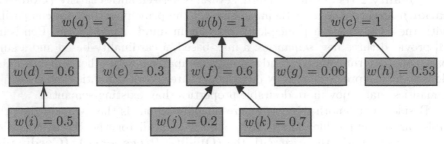

Definition 2 (Isomorphism). *Let* $\mathbf{G} = \langle \mathcal{A}, w, \mathcal{R} \rangle$ *and* $\mathbf{G}' = \langle \mathcal{A}', w', \mathcal{R}' \rangle$ *be two WAGs. An isomorphism from* **G** *to* **G**′ *is a bijective function* f *from* $\mathcal{A}$ *to* $\mathcal{A}'$ *such that: i)* $\forall a \in \mathcal{A}, w(a) = w'(f(a))$, *ii)* $\forall a, b \in \mathcal{A}, a\mathcal{R}b$ *iff* $f(a)\mathcal{R}'f(b)$.

An acceptability semantics is a function assigning a value, called *acceptability degree*, to every argument in a weighted argumentation graph. This value represents the *overall strength* of an argument, and is issued from the aggregation of the basic strength of the argument and the overall strengths of its attackers. The greater this value, the more acceptable the argument. Unlike extension semantics where arguments are either accepted or rejected, it is consider graded semantics, which may assign various acceptability degrees to arguments. Throughout the paper, we consider the scale $[0, 1]$.

Definition 3 (Semantics). *A semantics is a function* **S** *transforming any WAG* $\mathbf{G} = \langle \mathcal{A}, w, \mathcal{R} \rangle$ *into a vector* $Deg_{\mathbf{G}}^{\mathbf{S}}$ *in* $[0, 1]^n$, *where* $n = |\mathcal{A}|$. *For* $a \in \mathcal{A}$, $Deg_{\mathbf{G}}^{\mathbf{S}}(a)$ *is called acceptability degree of* a.

We present next the list of all notations used in the paper. Let $\mathbf{G} = \langle \mathcal{A}, w, \mathcal{R} \rangle$ be a WAG and $a \in \mathcal{A}$. $\mathrm{Att}_{\mathbf{G}}(a)$ denotes the set of all attackers of a in **G**, i.e. $\mathrm{Att}_{\mathbf{G}}(a) = \{b \in \mathcal{A} : b\mathcal{R}a\}$. For $\mathbf{G} = \langle \mathcal{A}, w, \mathcal{R} \rangle$ and $\mathbf{G}' = \langle \mathcal{A}', w', \mathcal{R}' \rangle$ such that $\mathcal{A} \cap \mathcal{A}' = \emptyset$, $\mathbf{G} \oplus \mathbf{G}'$ is the WAG $\langle \mathcal{A} \cup \mathcal{A}', w'', \mathcal{R} \cup \mathcal{R}' \rangle$ where for any $x \in \mathcal{A}$ (resp. $x \in \mathcal{A}'$), $w''(x) = w(x)$ (resp. $w''(x) = w'(x)$).

2.2   Examples of Weighted Semantics

The first semantics satisfies quality precedence, thus it favors the quality of attackers over their cardinality. It is based on a scoring function which follows a multiple steps process. At each step, the function assigns a score to each argument. In the initial step, the score of an argument is its basic strength. Then, in each step, the score is recomputed on the basis of the basic strength as well as the score of the strongest attacker of the argument at the previous step.

Definition 4 (Weighted Max-Based Semantics [4]). *For any WAG* $\mathbf{G} = \langle \mathcal{A}, w, \mathcal{R} \rangle$ *and* $a \in \mathcal{A}$, $Deg_{\mathbf{G}}^{\mathbf{Mbs}}(a) = \dfrac{w(a)}{1 + \max_{b \in Att_{\mathbf{G}}(a)} Deg_{\mathbf{G}}^{\mathbf{Mbs}}(b)}$. *By convention,* *if* $Att_{\mathbf{G}}(a) = \emptyset$, $\max_{b \in Att_{\mathbf{G}}(a)} Deg_{\mathbf{G}}^{\mathbf{Mbs}}(b) = 0$.

Example 2. Considering the WAG **G** from *Example 1*, we have that $\mathrm{Deg}_{\mathbf{G}}^{\mathbf{Mbs}}(a) = \dfrac{w(a)}{1 + \max_{b \in Att_{\mathbf{G}}(a)} \mathrm{Deg}_{\mathbf{G}}^{\mathbf{Mbs}}(b)} = \dfrac{w(a)}{1 + \max\{\mathrm{Deg}_{\mathbf{G}}^{\mathbf{Mbs}}(d), \mathrm{Deg}_{\mathbf{G}}^{\mathbf{Mbs}}(e)\}}$. Since the argument d is attacked by the argument i, we need to compute $\mathrm{Deg}_{\mathbf{G}}^{\mathbf{Mbs}}(d) = \dfrac{w(d)}{1 + \max\{\mathrm{Deg}_{\mathbf{G}}^{\mathbf{Mbs}}(i)\}} = \dfrac{0.6}{1 + 0.5} = 0.4$ (since there is no argument attacking i, $\mathrm{Deg}_{\mathbf{G}}^{\mathbf{Mbs}}(i) = w(i) = 0.5$). We also have that $\mathrm{Deg}_{\mathbf{G}}^{\mathbf{Mbs}}(e) = w(e) = 0.3$. Therefore, $\mathrm{Deg}_{\mathbf{G}}^{\mathbf{Mbs}}(a) = \dfrac{1}{1 + \max\{0.4, 0.3\}} = 0.71$.

For a matter of simplicity, in this paper we are considering examples with acyclic graphs, but the weighted semantics are also defined to deal with cyclic graphs. The details and proofs about this issue can be found in [4].

The second semantics, called weighted card-based, favors the number of attackers over their quality. It considers only arguments that have a basic strength greater than 0, called founded. This restriction is due to the fact that unfounded arguments are lifeless and their attacks are ineffective.

Definition 5 (Weighted Card-Based Semantics [4]). *Let* $\mathbf{G} = \langle \mathcal{A}, w, \mathcal{R} \rangle$ *be a WAG and* $a \in \mathcal{A}$. *The argument* a *is founded iff* $w(a) > 0$. *It is unfounded otherwise. Let* $AttF_{\mathbf{G}}(a)$ *denote the set of founded attackers of* a. *For any WAG* $\mathbf{G} = \langle \mathcal{A}, w, \mathcal{R} \rangle$ *and* $a \in \mathcal{A}$, $Deg_{\mathbf{G}}^{\mathbf{Cbs}}(a) = \dfrac{w(a)}{1 + |AttF_{\mathbf{G}}(a)| + \dfrac{\sum_{b \in AttF_{\mathbf{G}}(a)} Deg_{\mathbf{G}}^{\mathbf{Cbs}}(b)}{|AttF_{\mathbf{G}}(a)|}}$. *By convention, if* $Att_{\mathbf{G}}(a) = \emptyset$, $\sum_{b \in AttF_{\mathbf{G}}(a)} Deg_{\mathbf{G}}^{\mathbf{Cbs}}(b) = 0$.

Example 3. From the WAG **G** in *Example 1*, we have that $\mathrm{Deg}_{\mathbf{G}}^{\mathbf{Cbs}}(d) = 0.24$, $\mathrm{Deg}_{\mathbf{G}}^{\mathbf{Cbs}}(e) = 0.3$ and $\mathrm{Deg}_{\mathbf{G}}^{\mathbf{Cbs}}(a) = 0.3$.

The third semantics extends h-categorizer, initially proposed by Besnard and Hunter [12] for non-weighted and acyclic graphs. Then, it was extended to account for varying degrees of basic strengths, and any graph structure.

Definition 6 (Weighted h-Categorizer Semantics [4]). *For any WAG* $\mathbf{G} = \langle \mathcal{A}, w, \mathcal{R} \rangle$ *and* $a \in \mathcal{A}$, $Deg_{\mathbf{G}}^{\mathbf{Hbs}}(a) = \dfrac{w(a)}{1 + \sum_{b \in Att_{\mathbf{G}}(a)} Deg_{\mathbf{G}}^{\mathbf{Hbs}}(b)}$. *By convention, if* $Att_{\mathbf{G}}(a) = \emptyset$, $\sum_{b \in Att_{\mathbf{G}}(a)} Deg_{\mathbf{G}}^{\mathbf{Hbs}}(b) = 0$.

Example 4. From the WAG **G** in *Example 1*, we have that $\mathrm{Deg}_{\mathbf{G}}^{\mathbf{Hbs}}(d) = 0.4$, $\mathrm{Deg}_{\mathbf{G}}^{\mathbf{Hbs}}(e) = 0.3$ and $\mathrm{Deg}_{\mathbf{G}}^{\mathbf{Hbs}}(a) = 0.58$.

2.3   Principles and Properties

In the sequel, we present some principles, which are important for i) a better understanding of semantics, ii) the definition of reasonable semantics, iii) comparing semantics, iv) choosing suitable semantics for applications [4].

It was proposed 15 principles, which described the role and impact of attacks and basic strengths in the evaluation of arguments, and how these two elements are aggregated. The first principle, called (**Anonymity**), can be found in almost all axiomatic studies. In the argumentation literature, (**Anonymity**) is called *abstraction* in [2] and *language independence* in [9].

Principle 1 (Anonymity). *A semantics* **S** *satisfies anonymity iff for any two WAGs* $\mathbf{G} = \langle \mathcal{A}, w, \mathcal{R} \rangle$ *and* $\mathbf{G}' = \langle \mathcal{A}', w', \mathcal{R}' \rangle$, *for any isomorphism* f *from* $\mathbf{G}$ *to* $\mathbf{G}'$, *the following property holds:* $\forall a \in \mathcal{A}, Deg_{\mathbf{G}}^{\mathbf{S}}(a) = Deg_{\mathbf{G}'}^{\mathbf{S}}(f(a))$.

The second principle, called (**Independence**), states that the acceptability degree of an argument should be independent of any argument that is not connected to it.

Principle 2 (Independence). *A semantics* **S** *satisfies independence iff for any two WAGs* $\mathbf{G} = \langle \mathcal{A}, w, \mathcal{R} \rangle$ *and* $\mathbf{G}' = \langle \mathcal{A}', w', \mathcal{R}' \rangle$ *such that* $\mathcal{A} \cap \mathcal{A}' = \emptyset$, *the following holds:* $\forall a \in \mathcal{A}, Deg_{\mathbf{G}}^{\mathbf{S}}(a) = Deg_{\mathbf{G} \oplus \mathbf{G}'}^{\mathbf{S}}(a)$.

The next principle states that the acceptability degree of an argument a in a graph can depend on argument b only if there is a path from b to a, i.e., a finite non-empty sequence $\langle x_1, \ldots, x_n \rangle$ such that $x_1 = b, x_n = a$ and $\forall i < n, x_i \mathcal{R} x_{i+1}$.

Principle 3 (Directionality). *A semantics* **S** *satisfies directionality iff for any two WAGs* $\mathbf{G} = \langle \mathcal{A}, w, \mathcal{R} \rangle$ *and* $\mathbf{G}' = \langle \mathcal{A}', w', \mathcal{R}' \rangle$ *such that* $\mathcal{R}' = \mathcal{R} \cup \{(a, b)\}$, *it holds that:* $\forall x \in \mathcal{A}$, *if there is no path from* b *to* x, *then* $Deg_{\mathbf{G}}^{\mathbf{S}}(x) = Deg_{\mathbf{G}'}^{\mathbf{S}}(x)$.

The next principle, called (**Neutrality**), states that an argument, whose acceptability degree is 0, has no impact on the arguments it attacks.

Principle 4 (Neutrality). *A semantics* **S** *satisfies neutrality iff for any WAG* $\mathbf{G} = \langle \mathcal{A}, w, \mathcal{R} \rangle$, $\forall a, b \in \mathcal{A}$, *if i)* $w(a) = w(b)$, *and ii)* $Att_{\mathbf{G}}(b) = Att_{\mathbf{G}}(a) \cup \{x\}$ *with* $x \in A \backslash Att_{\mathbf{G}}(a)$ *and* $Deg_{\mathbf{G}}^{\mathbf{S}}(x) = 0$, *then* $Deg_{\mathbf{G}}^{\mathbf{S}}(a) = Deg_{\mathbf{G}}^{\mathbf{S}}(b)$.

The condition $w(a) = w(b)$ ensures that the attacks from $Att_{\mathbf{G}}(a)$ have the same effect on both arguments a and b. (**Equivalence**) principle ensures that the overall strength of an argument depends only on the basic strength of the argument and the overall strengths of its (direct) attackers.

Principle 5 (Equivalence). *A semantics* **S** *satisfies equivalence iff for any WAG* $\mathbf{G} = \langle \mathcal{A}, w, \mathcal{R} \rangle$, $\forall a, b \in \mathcal{A}$, *if i)* $w(a) = w(b)$, *and ii) there exists a bijective function* f *from* $Att_{\mathbf{G}}(a)$ *to* $Att_{\mathbf{G}}(b)$ *such that* $\forall x \in Att_{\mathbf{G}}(a), Deg_{\mathbf{G}}^{\mathbf{S}}(x) = Deg_{\mathbf{G}}^{\mathbf{S}}(f(x))$, *then* $Deg_{\mathbf{G}}^{\mathbf{S}}(a) = Deg_{\mathbf{G}}^{\mathbf{S}}(b)$.

(Maximality) principle states that an unattacked argument receives an acceptability degree equal to its basic strength.

Principle 6 (Maximality). *A semantics* **S** *satisfies maximality iff for any WAG* $\mathbf{G} = \langle \mathcal{A}, w, \mathcal{R} \rangle$, $\forall a \in \mathcal{A}$, *if* $Att_{\mathbf{G}}(a) = \emptyset$, *then* $Deg_{\mathbf{G}}^{\mathbf{S}}(a) = w(a)$.

The role of attacks is **(Weakening)** their targets. Indeed, when an argument receives an attack, its overall strength decreases whenever the attacker is "alive".

Principle 7 (Weakening). *A semantics* **S** *satisfies weakening iff for any WAG* $\mathbf{G} = \langle \mathcal{A}, w, \mathcal{R} \rangle$, $\forall a \in \mathcal{A}$, *if i)* $w(a) > 0$, *and ii)* $\exists b \in Att_{\mathbf{G}}(a)$ *such that* $Deg_{\mathbf{G}}^{\mathbf{S}}(b) > 0$, *then* $Deg_{\mathbf{G}}^{\mathbf{S}}(a) < w(a)$.

(Weakening) leads to strength loss as soon as an argument is attacked by at least one alive attacker. **(Counting)** principle states that each alive attacker has an impact on the overall strength of the argument. Thus, the more numerous the alive attackers of an argument, the weaker the argument.

Principle 8 (Counting). *A semantics* **S** *satisfies counting iff for any WAG* $\mathbf{G} = \langle \mathcal{A}, w, \mathcal{R} \rangle$, $\forall a, b \in \mathcal{A}$, *if i)* $w(a) = w(b)$, *ii)* $Deg_{\mathbf{G}}^{\mathbf{S}}(a) > 0$, *and iii)* $Att_{\mathbf{G}}(b) = Att_{\mathbf{G}}(a) \cup \{y\}$ *with* $y \in \mathcal{A} \backslash Att_{\mathbf{G}}(a)$ *and* $Deg_{\mathbf{G}}^{\mathbf{S}}(y) > 0$, *then* $Deg_{\mathbf{G}}^{\mathbf{S}}(a) > Deg_{\mathbf{G}}^{\mathbf{S}}(b)$.

(Weakening Soundness) principle goes further than weakening by stating that attacks are the only source of strength loss.

Principle 9 (Weakening Soundness). *A semantics* **S** *satisfies weakening soundness iff for any WAG* $\mathbf{G} = \langle \mathcal{A}, w, \mathcal{R} \rangle$, $\forall a \in A$ *such that* $w(a) > 0$, *if* $Deg_{\mathbf{G}}^{\mathbf{S}}(a) < w(a)$, *then* $\exists b \in Att_{\mathbf{G}}(a)$ *such that* $Deg_{\mathbf{G}}^{\mathbf{S}}(b) > 0$.

(Reinforcement) principle states that the stronger the source of an attack, the greater its intensity.

Principle 10 (Reinforcement). *A semantics* **S** *satisfies reinforcement iff for any WAG* $\mathbf{G} = \langle \mathcal{A}, w, \mathcal{R} \rangle$, $\forall a, b \in A$, *if i)* $w(a) = w(b)$, *ii)* $Deg_{\mathbf{G}}^{\mathbf{S}}(a) > 0$ *or* $Deg_{\mathbf{G}}^{\mathbf{S}}(b) > 0$, *iii)* $Att_{\mathbf{G}}(a) \backslash Att_{\mathbf{G}}(b) = \{x\}$, *iv)* $Att_{\mathbf{G}}(b) \backslash Att_{\mathbf{G}}(a) = \{y\}$, *and v)* $Deg_{\mathbf{G}}^{\mathbf{S}}(y) > Deg_{\mathbf{G}}^{\mathbf{S}}(x)$, *then* $Deg_{\mathbf{G}}^{\mathbf{S}}(a) > Deg_{\mathbf{G}}^{\mathbf{S}}(b)$.

(Resilience) principle states that an attack cannot completely kill an argument, i.e., to turn its acceptability degrees equal to 0.

Principle 11 (Resilience). *A semantics* **S** *satisfies resilience iff for any WAG* $\mathbf{G} = \langle \mathcal{A}, w, \mathcal{R} \rangle$, $\forall a \in A$, *if* $w(a) > 0$, *then* $Deg_{\mathbf{G}}^{\mathbf{S}}(a) > 0$.

(Proportionality) states that the stronger the target of an attack, the weaker its intensity.

Principle 12 (Proportionality). *A semantics* **S** *satisfies proportionality iff for any WAG* $\mathbf{G} = \langle \mathcal{A}, w, \mathcal{R} \rangle$, $\forall a, b \in A$ *such that i)* $Att_{\mathbf{G}}(a) = Att_{\mathbf{G}}(b)$, *ii)* $w(a) > w(b)$, *and iii)* $Deg_{\mathbf{G}}^{\mathbf{S}}(a) > 0$ *or* $Deg_{\mathbf{G}}^{\mathbf{S}}(b) > 0$, *then* $Deg_{\mathbf{G}}^{\mathbf{S}}(a) > Deg_{\mathbf{G}}^{\mathbf{S}}(b)$.

The three last principles concern possible choices offered to a semantics when it faces a conflict between the quality and the number of attackers. **(Quality Precedence)** principle gives more importance to the quality.

Principle 13 (Quality Precedence). *A semantics* **S** *satisfies quality precedence iff for any WAG* $\mathbf{G} = \langle \mathcal{A}, w, \mathcal{R} \rangle$, $\forall a, b \in A$, *if i)* $w(a) = w(b)$, *ii)* $Deg_{\mathbf{G}}^{\mathbf{S}}(a) > 0$, *and iii)* $\exists y \in Att_{\mathbf{G}}(b)$ *such that* $\forall x \in Att_{\mathbf{G}}(a), Deg_{\mathbf{G}}^{\mathbf{S}}(y) > Deg_{\mathbf{G}}^{\mathbf{S}}(x)$, *then* $Deg_{\mathbf{G}}^{\mathbf{S}}(a) > Deg_{\mathbf{G}}^{\mathbf{S}}(b)$.

(Cardinality Precedence) principle states that a great number of attackers has more effect on an argument than just few.

Principle 14 (Cardinality Precedence). *A semantics* **S** *satisfies cardinality precedence iff for any WAG* $\mathbf{G} = \langle \mathcal{A}, w, \mathcal{R} \rangle$, $\forall a, b \in A$, *if i)* $w(a) = w(b)$, *ii)* $Deg_{\mathbf{G}}^{\mathbf{S}}(a) > 0$, *and iii)* $|\{y \in Att_{\mathbf{G}}(b) : Deg_{\mathbf{G}}^{\mathbf{S}}(y) > 0\}| > |\{x \in Att_{\mathbf{G}}(a) : Deg_{\mathbf{G}}^{\mathbf{S}}(x) > 0\}|$, *then* $Deg_{\mathbf{G}}^{\mathbf{S}}(a) > Deg_{\mathbf{G}}^{\mathbf{S}}(b)$.

Finally, **(Compensation)** states that several weak attacks may compensate the quality of attacks overall.

Principle 15 (Compensation). *A semantics* **S** *satisfies compensation iff there exists a WAG* $\mathbf{G} = \langle \mathcal{A}, w, \mathcal{R} \rangle$, *such that for two arguments* $a, b \in A$, *i)* $w(a) = w(b)$, *ii)* $Deg_{\mathbf{G}}^{\mathbf{S}}(a) > 0$, *iii)* $|\{x \in Att_{\mathbf{G}}(a) : Deg_{\mathbf{G}}^{\mathbf{S}}(x) > 0\}| > |\{y \in Att_{\mathbf{G}}(b) : Deg_{\mathbf{G}}^{\mathbf{S}}(y) > 0\}|$, *iv)* $\exists y \in Att_{\mathbf{G}}(b)$ *such that* $\forall x \in Att_{\mathbf{G}}(a), Deg_{\mathbf{G}}^{\mathbf{S}}(y) > Deg_{\mathbf{G}}^{\mathbf{S}}(x)$ *and* $Deg_{\mathbf{G}}^{\mathbf{S}}(a) = Deg_{\mathbf{G}}^{\mathbf{S}}(b)$.

The results corresponding to the compatibility of the principles is stated below.

Proposition 1. *[4] The three following properties hold. i) **(Quality Precedence)**, **(Cardinality Precedence)** and **(Compensation)** are pairwise incompatible; ii) **(Independence)**, **(Directionality)**, **(Equivalence)**, **(Resilience)**, **(Reinforcement)**, **(Maximality)** and **(Quality Precedence)** are incompatible; iii) **(Cardinality Precedence)** (respectively **(Compensation)**) is compatible with all principles 1–12.*

It was shown that Weighted max-based semantics satisfies **(Quality Precedence)** as well as all the principles which are compatible with it [4]. It violates, however, **(Counting)** since by definition, this semantics focuses only on the strongest attacker of an argument, and neglects the remaining attackers.

Theorem 1. *[4] Weighted max-based semantics violates **(Cardinality Precedence)**, **(Compensation)**, **(Counting)** and **(Reinforcement)**. It satisfies all the remaining principles.*

Weighted card-based semantics satisfies **(Cardinality Precedence)** as well as all the principles that are compatible with it.

Theorem 2. *[4] Weighted card-based semantics satisfies all the principles except* **(Quality Precedence)** *and* **(Compensation)**.

Weighted h-categorizer semantics satisfies **(Compensation)** as well as all the principles that are compatible with it.

Theorem 3. *[4] Weighted h-categorizer semantics satisfies all the principles except* **(Quality Precedence)** *and* **(Cardinality Precedence)**.

3 Refinements of Compensation-Based Semantics

Fuzzy set theory has been shown to be a useful tool to describe situations in which the data are imprecise or vague. Fuzzy sets handle such situations by attributing a degree to which a certain object belongs to a set [15]. An important notion in fuzzy set theory is that of triangular norms and conorms: t-norms and t-conorms are used to define a generalized intersection and union of fuzzy sets [24]. Triangular norms and conorms serve as aggregation operators, which can be used, e.g., for querying databases [20], to compute the resulting degree of confidence of agents [33], in Approximate Reasoning [16], Information Retrieval [14], Neuro-symbolic Learning [19], Machine Learning [1], etc.

Definition 7 (T-conorm [24]). *A binary function* $\oplus : [0, 1] \times [0, 1] \rightarrow [0, 1]$ *is a t-conorm if it satisfies the following conditions:*

1. $\oplus\{a, b\} = \oplus\{b, a\}$ *(Commutativity);*
2. $\oplus\{a, \oplus\{b, c\}\} = \oplus\{\oplus\{a, b\}, c\}$ *(Associativity);*
3. $a \leq c$ *and* $b \leq d \Rightarrow \oplus\{a, b\} \leq \oplus\{c, d\}$ *(Monotonicity);*
4. $\oplus\{a, 0\} = a$ *(Neutral Element).*

A t-conorm acts as a disjunction in fuzzy logic or as a union in fuzzy set theory. When one of its arguments is 0, it returns its other argument; when one of its arguments is 1, it returns 1. It is both associative and commutative, and its partial derivatives with respect to its parameters are non-negative. T-conorms are a generalization of the usual two-valued logical disjunction (or the maximum operator), studied by classical logic, for fuzzy logics. The four basic t-conorms are described below:

Definition 8 (Basic T-conorms [24]). *The following are the four basic t-conorms:*

1. *Maximum t-conorm:* $\oplus_M\{x, y\} = max(x, y)$;
2. *Probabilistic sum t-conorm:* $\oplus_P\{x, y\} = x + y - x \cdot y$;
3. *Łukasiewicz t-conorm:* $\oplus_L\{x, y\} = min(x + y, 1)$;
4. *Drastic sum t-conorm:* $\oplus_D\{x, y\} = \begin{cases} 1, & if\ (x, y) \in\ ]0, 1] \times]0, 1]; \\ max(x, y), & otherwise. \end{cases}$

These four basic t-conorms are remarkable for several reasons. The drastic sum $\oplus_D$ and the maximum $\oplus_M$ are the largest and the smallest t-conorms, respectively (with respect to the pointwise order). The maximum $\oplus_M$ is the only t-conorm where each $x \in [0, 1]$ is an idempotent element (recall $x \in [0, 1]$ is called an idempotent element of $\oplus$ if $\oplus\{x, x\} = x$). The probabilistic sum $\oplus_P$ and the Łukasiewicz t-conorm $\oplus_L$ are examples of two important subclasses of t-conorms, namely, the classes of strict and nilpotent t-conorms, respectively (more details in [23]).

One way to compare t-conorms is using the notion of strength. Consider two t-conorms $\oplus_1$ and $\oplus_2$. If we have $\oplus_1\{x, y\} \leq \oplus_2\{x, y\}$ for all $x, y \in [0, 1]$, then we say that $\oplus_1$ is *weaker* than $\oplus_2$ or, equivalently, that $\oplus_2$ is *stronger* than $\oplus_1$, and we write in this case $\oplus_1 \leq \oplus_2$. We shall write $\oplus_1 < \oplus_2$ if $\oplus_1 \leq \oplus_2$ and $\oplus_1 \neq \oplus_2$. The drastic sum $\oplus_D$ is the strongest, and the Maximum $\oplus_M$ is the weakest t-conorm, i.e., for each t-conorm $\oplus$ we have $\oplus_M \leq \oplus \leq \oplus_D$. Between the four basic t-conorms we have these strict inequalities: $\oplus_M < \oplus_P < \oplus_L < \oplus_D$.

Example 5. Consider $x = 0.4$ and $y = 0.7$. We have that $\oplus_M\{0.4, 0.7\} = 0.7$, $\oplus_P\{0.4, 0.7\} = 0.82$, $\oplus_L\{0.4, 0.7\} = 1$ and $\oplus_D\{0.4, 0.7\} = 1$. We can state that the maximum t-conorm disconsiders all the values that are not the maximum, while the result of the probabilistic sum t-conorm takes into consideration the value of each argument. The Łukasiewicz t-conorm follows a similar idea, however, when the sum of the argument reaches a threshold, i.e. the value 1 which represents total membership, the result is equal to 1 (and all the excess is disconsidered). The Drastic sum t-conorm, as the name states, it is radical in the decision: if a argument x has a partial (i.e., $x \in (0, 1]$) or total membership (i.e., $x = 1$), the result of drastic t-conorm between x and any other element is equal to 1 (any partial membership is transformed in a total membership).

As the t-conorms are a natural generalization of the maximum operator, we can generalize the definition of Weighted Max-Based Semantics to a Weighted t-conorm Semantics.

Definition 9 (Weighted T-conorm Semantics). *For any WAG* $\mathbf{G} = \langle \mathcal{A}, w, \mathcal{R} \rangle$ *and* $a \in \mathcal{A}$, $Deg_{\mathbf{G}}^{\oplus}(a) = \dfrac{w(a)}{1 + \bigoplus_{b \in Att_{\mathbf{G}}(a)} Deg_{\mathbf{G}}^{\oplus}(b)}$. *By convention, if* $Att_{\mathbf{G}}(a) = \emptyset$, $\bigoplus_{b \in Att_{\mathbf{G}}(a)} Deg_{\mathbf{G}}^{\oplus}(b) = 0$.

Intuitively, we can think of the probabilistic sum semantics as the following idea: the degree of acceptability of an argument a is measured based on its weight and also the sum of pairs of acceptability degrees of the attacking arguments minus a rate (measured by the product between the attackers' acceptability degrees). It is a different approach to the maximum that only considers the largest value of an attacking argument and also different from the sum approach present in the h-categorizer semantics. In fact, the probabilistic sum semantics encompasses a bit of both worlds.

Łukasiewicz's semantics is much closer to h-categorizer semantics, and its big difference is that if the sum of the acceptability degrees of the attacking arguments exceeds 1, the weight of this attack is considered 1 (in the h-categorizer semantics this value can be greater than 1).

Drastic semantics is much less interesting when compared to the two previous ones, since it considers that if an argument is attacked by any other argument or a set of arguments with a degree of acceptability greater than 0, the total weight of the attack is always 1, regardless the values of the degrees of acceptability. In practice this means that when an argument is attacked, if the weight of any argument is greater than 0, the degree of acceptance of that argument attacked results in half its original weight.

Example 6. Considering the WAG **G** from *Example 1*, we have that $Deg_{\mathbf{G}}^{\oplus M}(a) = 0.71$, $Deg_{\mathbf{G}}^{\oplus P}(a) = 0.63$, $Deg_{\mathbf{G}}^{\oplus L}(a) = 0.58$ and $Deg_{\mathbf{G}}^{\oplus D}(a) = 0.5$. As said before, the drastic t-conorm is the strongest t-conorm and maximum t-conorm is the weakest. Consequently, the acceptability degree of an argument is higher for the maximum semantics and lower for the drastic semantics, when compared to the other t-conorms semantics.

With respect to the range of values of the acceptability degree for t-conorms semantics, we have the following results.

Proposition 2. *For any WAG* **G** $= \langle \mathcal{A}, w, \mathcal{R} \rangle$ *and for any* $a \in \mathcal{A}$, *we have that* $\{Deg_{\mathbf{G}}^{\oplus M}(a), Deg_{\mathbf{G}}^{\oplus P}(a), Deg_{\mathbf{G}}^{\oplus L}(a)\} \in [\frac{w(a)}{2}, w(a)]$ *and* $Deg_{\mathbf{G}}^{\oplus D}(a) = \frac{w(a)}{2}$ *or* $Deg_{\mathbf{G}}^{\oplus D}(a) = w(a)$.

Next, we will propose a weaker principle than **(Quality Precedence)**, named **(Quality Compensation)**. The idea of this principle is to prioritize, during a conflict of attacks on two arguments, the quality of attacks when the overall sum of attacks on these arguments is equal.

Principle 16 (Quality Compensation). *A semantics* **S** *satisfies quality compensation iff for any WAG* **G** $= \langle \mathcal{A}, w, \mathcal{R} \rangle$, $\forall a, b \in A$, *if i)* $w(a) = w(b)$, *ii)* $Deg_{\mathbf{G}}^{\mathbf{S}}(a) > 0$, *iii)* $\sum_{x \in Att_{\mathbf{G}}(a)} Deg_{\mathbf{G}}^{\mathbf{S}}(x) = \sum_{y \in Att_{\mathbf{G}}(b)} Deg_{\mathbf{G}}^{\mathbf{S}}(y)$, *and iv)* $\exists y \in Att_{\mathbf{G}}(b)$ *such that* $\forall x \in Att_{\mathbf{G}}(a), Deg_{\mathbf{G}}^{\mathbf{S}}(y) > Deg_{\mathbf{G}}^{\mathbf{S}}(x)$, *then* $Deg_{\mathbf{G}}^{\mathbf{S}}(a) > Deg_{\mathbf{G}}^{\mathbf{S}}(b)$.

This principle weakens **(Quality Precedence)** by introducing condition iii) $\sum_{x \in Att_{\mathbf{G}}(a)} Deg_{\mathbf{G}}^{\mathbf{S}}(x) = \sum_{y \in Att_{\mathbf{G}}(b)} Deg_{\mathbf{G}}^{\mathbf{S}}(y)$, where it considers equal the sum of the acceptability degrees of the attacking arguments. Therefore, if a semantics satisfies **(Quality Precedence)** it also satisfies **(Quality Compensation)**. Regarding Weighted t-conorms semantics, we have the following results according to their properties.

Theorem 4. *Considering the weighted t-conorm semantics:*

1. *Weighted Drastic t-conorm semantics violates only (Counting), (Reinforcement), (Quality Precedence), (Cardinality Precedence), (Compensation) and (Quality Compensation).*

2. *Weighted Łukasiewicz t-conorm violates only* **(Quality Precedence)**, **(Cardinality Precedence)** *and* **(Quality Compensation)**.

3. *Weighted Probabilistic sum t-conorm semantics violates only* **(Quality Precedence)** *and* **(Cardinality Precedence)**.

Weighted drastic t-conorm semantics behaves similar to weighted max-based semantics, except that it does not satisfy **(Quality Precendence)** or **(Quality Compensation)**. Although they are t-conorms, weighted Łukasiewicz and weighted probabilistic sum semantics go in a direction different from weighted max-based Semantics and satisfy **(Compensation)**, along with all the 1–12 principles. In special, weighted probabilistic sum also satisfies **(Quality Compensation)**, which is a weaker version of **(Quality Precedence)**, that is, it presents a balance between compensation and quality in its decisions.

Example 7. From the WAG **G** in *Example 1*, we have that $w(a) = w(b)$ and the sum of degrees of acceptability (for $\oplus_P$ semantics) of the attackers of a and b are, respectively, $\text{Deg}_{\mathbf{G}}^{\oplus_P}(d) + \text{Deg}_{\mathbf{G}}^{\oplus_P}(e) = 0.4 + 0.3 = 0.7$ and $\text{Deg}_{\mathbf{G}}^{\oplus_P}(e) + \text{Deg}_{\mathbf{G}}^{\oplus_P}(f) + \text{Deg}_{\mathbf{G}}^{\oplus_P}(g) = 0.3 + 0.34 + 0.06 = 0.7$. However, $\text{Deg}_{\mathbf{G}}^{\oplus_P}(a) = 0.63 < 0.64 = \text{Deg}_{\mathbf{G}}^{\oplus_P}(b)$, since the argument a has an attacker with the highest degree of acceptability ($\text{Deg}_{\mathbf{G}}^{\oplus_P}(d) = 0.4$). We can see that quality takes on importance when the total of values compared is indistinguishable. Otherwise, the semantics behaves like a compensating semantics.

The next semantics introduced in this paper is the Weighted CS-Based Semantics, based on the idea of Cumulative Sum. This operator has been applied in the areas of Outlier Detection [25], identifying rare items, events or observations which raise suspicions by differing significantly from the majority of the analyzed data. It is also studied in the area of Economy, from the notion of the Lorenz curve [27], which is most often used to represent economic inequality and it can also demonstrate unequal distribution in any system. Formally, a cumulative sum can be defined as follows.

Definition 10 (Cumulative Sum). *Consider the vectors* $L = (x_1, \ldots, x_n)$ *and* $L' = (x_{\sigma(1)}, \ldots, x_{\sigma(n)})$, *where* σ *is the permutation of* $\{1, \ldots, n\}$ *sorting the* x_i *in descending order. We define the vector of accumulated sum*

$$AS_L = (AS_{L'}^1, \ldots, A_{L'}^n), \text{ where } AS_{L'}^i = \sum_{x_k \in L', k=1}^{i} x_k.$$

The Cumulative Sum of L *is defined as* $CS_L = \sum AS_L$ *(the sum of its elements).*

A cumulative sum is a sequence of partial sums of a given sequence. For example, the cumulative sums of the sequence $a, b, c, \ldots$, are $a, a+b, a+b+c, \ldots$. After that, the sum of all these elements is performed.

Example 8. Consider the vector $V = (0.23, 0.26, 0.1)$ and $V' = (0.26, 0.23, 0.1)$ its ordered version. The cumulative sum of V is given by $CS_V = \sum AS_V = \sum (AS_{V'}^1, AS_{V'}^2, A_{V'}^3) = \sum (0.26, 0.49, 0.59) = 1.34$.

Definition 11 (Weighted CS-Based Semantics). *For any WAG* $G = \langle A, w, R \rangle$ *and* $a \in A$, $Deg_G^{CSbs}(a) = \dfrac{w(a)}{1 + CS_{AttS_G(a)}}$, *where* $AttS_G(a) = (Deg_G^{CSbs}(b_1), \ldots, Deg_G^{CSbs}(b_n))$ *and* $b_i \in Att_G(a)$, *for* $1 \le i \le n$. *By convention, if* $Att_G(a) = \emptyset$, *then* $CS_{AttS_G(a)} = 0$.

Weighted CS-Based Semantics extends the h-categorizer semantics by making the cumulative sum of the acceptability degrees of the attacking arguments.

Example 9. Considering the WAG G from *Example 1*, we have that $Deg_G^{Hbs}(a) = 0.58$ and $Deg_G^{CSbs}(a) = 0.47$. As it happened with the maximum t-conorm, which has a higher acceptability degree when compared with the other t-conorms, the acceptability degree of an argument is higher for the h-categorizer when compared to cumulative sum semantics.

As said previously, the cumulative sum is used in data analysis to detect variations and anomalies in a set of data, and this operator is sensible to the number of elements in the set. Thus, we propose a weaker principle than **(Cardinality Precedence)**, named **(Cardinality Compensation)**. The idea of this principle is to prioritize, during a conflict of attacks on two arguments, the quantity of attacks when the overall sum of attacks on these arguments is equal.

Principle 17 (Cardinality Compensation). *A semantics* **S** *satisfies cardinality compensation iff for any WAG* $G = \langle A, w, R \rangle$, $\forall a, b \in A$, *if i)* $w(a) = w(b)$, *ii)* $Deg_G^{S}(a) > 0$, *iii)* $\sum_{x \in Att_G(a)} Deg_G^{S}(x) = \sum_{y \in Att_G(b)} Deg_G^{S}(y)$, *and iv)* $|\{y \in Att_G(b) : Deg_G^{S}(y) > 0\}| > |\{x \in Att_G(a) : Deg_G^{S}(x) > 0\}|$, *then* $Deg_G^{S}(a) > Deg_G^{S}(b)$.

This principle weakens **(Cardinality Precedence)** by introducing condition iii) $\sum_{x \in Att_G(a)} Deg_G^{S}(x) = \sum_{y \in Att_G(b)} Deg_G^{S}(y)$, where it considers equal the sum of the acceptability degrees of the attacking arguments. Therefore, if a semantics satisfies **(Cardinality Precedence)** it also satisfies **(Cardinality Compensation)**. Regarding Weighted CS-Based semantics, we have the following results according to their properties.

Theorem 5. *Weighted CS-based semantics satisfies all the principles except* **(Cardinality Precedence)**, **(Quality Precedence)** *and* **(Quality Precedence)**. *Additionally, it satisfies* **(Cardinality Compensation)**.

Example 10. From the WAG G in *Example 1*, we have that $w(b) = w(c) = 1$ and the sum of degrees of acceptability (for cumulative sum semantics) of the attackers of b and c are, respectively, $Deg_G^{CSbs}(e) + Deg_G^{CSbs}(f) + Deg_G^{CSbs}(g) = 0.3 + 0.23 + 0.06 = 0.59$ and $Deg_G^{CSbs}(g) + Deg_G^{CSbs}(h) = 0.06 + 0.53 = 0.59$. However, $Deg_G^{CSbs}(b) = 0.41 < 0.47 = Deg_G^{CSbs}(c)$, since the argument b has more attackers than argument c. We can see that quantity takes on importance when the total of values compared is indistinguishable. Otherwise, the semantics behaves like a compensating semantics.

With respect to the range of values of the acceptability degree for CS-based semantics, they have the same result of the h-categorizes semantics.

Theorem 6. *For any WAG* $\mathbf{G} = \langle \mathcal{A}, w, \mathcal{R} \rangle$, *for any* $a \in \mathcal{A}$, $Deg_{\mathbf{G}}^{\mathbf{CS}}(a) \in (0, w(a)]$.

4 Conclusion

This paper introduced two new semantics for the weighted argumentation framework. The main objective is to show that there are semantics that have a hybrid behavior between the principles of **(Quality Precedence)**, **(Cardinality Precedence)** and **(Compensation)**. The semantics based on t-conorms, as the drastic t-conorms, Łukasiewicz and probabilistic sum seek an alternative to the maximum operator, which has the principle of **(Quality Precedence)** as a characteristic. It has been shown that the probabilistic sum t-conorm has the most interesting properties, because although it does not satisfy **(Quality Precedence)**, it satisfies all the basic principles, along with **(Compensation)** and **(Quality Compensation)**.

Table 1. The list of the principles satisfied (or violated) by the semantics.

	Mbs	Cbs	Hbs	$\oplus_{\mathbf{D}}$	$\oplus_{\mathbf{L}}$	$\oplus_{\mathbf{P}}$	CSbs
Anonymity	●	●	●	●	●	●	●
Independence	●	●	●	●	●	●	●
Directionality	●	●	●	●	●	●	●
Neutrality	●	●	●	●	●	●	●
Equivalence	●	●	●	●	●	●	●
Maximality	●	●	●	●	●	●	●
Weakening	●	●	●	●	●	●	●
Counting	×	●	●	×	●	●	●
Weakening soundness	●	●	●	●	●	●	●
Reinforcement	×	●	●	×	●	●	●
Resilience	●	●	●	●	●	●	●
Proportionality	●	●	●	●	●	●	●
Quality Precedence	●	×	×	×	×	×	×
Cardinality Precedence	×	●	×	×	×	×	×
Compensation	×	×	●	×	●	●	●
Quality Compensation	×	×	×	×	×	●	×
Cardinality Compensation	×	×	×	×	×	×	●

The semantics based on cumulative sum is intended to exhibit a hybrid behavior between a **(Compensation)** and **(Cardinality Precedence)**. Unlike the classical sum operator, presented in the h-categorizer semantics, the cumulative sum gives more weight to the total sum according to the number of elements: the more elements, the greater the sum value. As a result, we show that the cumulative sum semantics satisfies all the principles of the h-categorizer semantics plus a weak version of (Cardinality Precedence), called (Cardinality Compensation). Table 1 summarizes the results regarding the weighted argumentation framework operators and the satisfaction of all principles.

As future work, we intend to continue exploring other operators with intermediate characteristics between these three main incompatible principles. An alternative is to study t-conorm families. Besides the four basic t-conorms, it is possible to extend them into several families of t-conorms through parameters [28], resulting in several operators with different properties. Another point to be investigated are operators that exhibit intermediate behavior between quality and quantity. Furthermore, another possibility of research is to study other new principles in the weighted argumentation framework. A remarkable question remains open: is it possible to conceive a new semantics with a hybrid behavior involving not only two, but these three main incompatible principles?

References

1. Adeli, H., Hung, S.L.: Machine Learning: Neural Networks, Genetic Algorithms, and Fuzzy Systems. Wiley, Hoboken (1994)
2. Amgoud, L., Ben-Naim, J.: Ranking-based semantics for argumentation frameworks. In: Liu, W., Subrahmanian, V.S., Wijsen, J. (eds.) SUM 2013. LNCS (LNAI), vol. 8078, pp. 134–147. Springer, Heidelberg (2013). https://doi.org/10.1007/978-3-642-40381-1_11
3. Amgoud, L., Ben-Naim, J.: Axiomatic foundations of acceptability semantics. In: International Conference on Principles of Knowledge Representation and Reasoning (KR 2016), pp. pp-2 (2016)
4. Amgoud, L., Ben-Naim, J., Doder, D., Vesic, S.: Acceptability semantics for weighted argumentation frameworks. In: Twenty-Sixth International Joint Conference on Artificial Intelligence (2017)
5. Amgoud, L., Besnard, P.: Bridging the gap between abstract argumentation systems and logic. In: Godo, L., Pugliese, A. (eds.) SUM 2009. LNCS (LNAI), vol. 5785, pp. 12–27. Springer, Heidelberg (2009). https://doi.org/10.1007/978-3-642-04388-8_3
6. Amgoud, L., Cayrol, C.: A reasoning model based on the production of acceptable arguments. Ann. Math. Artif. Intell. 34(1), 197–215 (2002)
7. Amgoud, L., Maudet, N., Parsons, S.: Modelling dialogues using argumentation. In: Proceedings Fourth International Conference on MultiAgent Systems, pp. 31–38. IEEE (2000)
8. Amgoud, L., Prade, H.: Using arguments for making and explaining decisions. Artif. Intell. 173(3), 413–436 (2009)
9. Baroni, P., Giacomin, M.: On principle-based evaluation of extension-based argumentation semantics. Artif. Intell. 171(10–15), 675–700 (2007)
10. Baroni, P., Romano, M., Toni, F., Aurisicchio, M., Bertanza, G.: Automatic evaluation of design alternatives with quantitative argumentation. Argum. Comput. 6(1), 24–49 (2015)

11. Bench-Capon, T.J.: Persuasion in practical argument using value-based argumentation frameworks. J. Log. Comput. **13**(3), 429–448 (2003)
12. Besnard, P., Hunter, A.: A logic-based theory of deductive arguments. Artif. Intell. **128**(1–2), 203–235 (2001)
13. da Costa Pereira, C., Tettamanzi, A.G., Villata, S.: Changing one's mind: erase or rewind? possibilistic belief revision with fuzzy argumentation based on trust. In: IJCAI, International Joint Conference on Artificial Intelligence (2011)
14. De Baets, B., Kerre, E.: Fuzzy relations and applications. In: Advances in Electronics and Electron Physics, vol. 89, pp. 255–324. Elsevier (1994)
15. Deschrijver, G., Cornelis, C., Kerre, E.E.: On the representation of intuitionistic fuzzy t-norms and t-conorms. IEEE Trans. Fuzzy Syst. **12**(1), 45–61 (2004)
16. Deschrijver, G., Kerre, E.E.: On the composition of intuitionistic fuzzy relations. Fuzzy Sets Syst. **136**(3), 333–361 (2003)
17. Dubois, D., Fargier, H., Bonnefon, J.F.: On the qualitative comparison of decisions having positive and negative features. J. Artif. Intell. Res. **32**, 385–417 (2008)
18. Gabbay, D.M., Rodrigues, O.: Equilibrium states in numerical argumentation networks. Logica Universalis **9**(4), 411–473 (2015)
19. Giannini, F., Marra, G., Diligenti, M., Maggini, M., Gori, M.: Learning and t-norms theory. arXiv preprint arXiv:1907.11468 (2019)
20. Grzegorzewski, P., Mrówka, E.: Soft querying via intuitionistic fuzzy sets. In: Proceedings of the 9th International conference on Information Processing and management of Uncertainty in Knowledge-Based Systems IMPU 2002. Citeseer (2002)
21. Hunter, A.: A probabilistic approach to modelling uncertain logical arguments. Int. J. Approx. Reason. **54**(1), 47–81 (2013)
22. Kakas, A., Moraitis, P.: Argumentation based decision making for autonomous agents. In: Proceedings of the Second International Joint Conference on Autonomous Agents and Multiagent Systems, pp. 883–890 (2003)
23. Klement, E.P., Mesiar, R.: Logical, Algebraic, Analytic and Probabilistic Aspects of Triangular Norms. Elsevier Science B.V, Amsterdam (2005)
24. Klement, E.P., Mesiar, R., Pap, E.: Triangular Norms, 1st edn. Springer, Netherlands (2000). https://doi.org/10.1007/978-94-015-9540-7
25. Lazarevic, A., Kumar, V.: Feature bagging for outlier detection. In: Proceedings of the Eleventh ACM SIGKDD International Conference on Knowledge Discovery in Data Mining, pp. 157–166 (2005)
26. Li, H., Oren, N., Norman, T.J.: Probabilistic argumentation frameworks. In: Modgil, S., Oren, N., Toni, F. (eds.) TAFA 2011. LNCS (LNAI), vol. 7132, pp. 1–16. Springer, Heidelberg (2012). https://doi.org/10.1007/978-3-642-29184-5_1
27. Lorenz, M.O.: Methods of measuring the concentration of wealth. Publ. Am. Stat. Assoc. **9**(70), 209–219 (1905)
28. Mizumoto, M.: Pictorial representations of fuzzy connectives, part I: cases of t-norms, t-conorms and averaging operators. Fuzzy Sets Syst. **31**(2), 217–242 (1989)
29. Modgil, S.: Reasoning about preferences in argumentation frameworks. Artif. Intell. **173**(9–10), 901–934 (2009)
30. Prakken, H.: Formal systems for persuasion dialogue. Knowl. Eng. Rev. **21**(2), 163 (2006)
31. Rago, A., Toni, F., Aurisicchio, M., Baroni, P.: Discontinuity-free decision support with quantitative argumentation debates (2016)
32. Thimm, M.: A probabilistic semantics for abstract argumentation. In: ECAI, vol. 12, pp. 750–755 (2012)
33. Yager, R.R., Kreinovich, V.: Universal approximation theorem for uninorm-based fuzzy systems modeling. Fuzzy Sets Syst. **140**(2), 331–339 (2003)

Ontology Based Classification
of Electronic Health Records
to Support Value-Based Health Care

Avner Dal Bosco[1](✉) iD, Renata Vieira[2] iD, Bruna Zanotto[3] iD,
and Ana Paula Beck da Silva Etges[3] iD

[1] School of Technology, Pontifical Catholic University of Rio Grande do Sul,
PUCRS, Porto Alegre, Brazil
avner.bosco@edu.pucrs.br
[2] CIDEHUS, University of Évora, Évora, Portugal
renatav@uevora.pt
[3] Federal University of Rio Grande do Sul, Porto Alegre, Brazil

Abstract. Value-based health care management models require a precise accounting of health indexes such as risk events monitoring, clinical conditions, patient handling and outcomes. Currently this accounting is performed by manually reading and searching through electronic health records for these indexes. Our research proposes a way to make this an autonomous task that is performed by a computer using a Portuguese free-text concept classifier model based on ontologies. To validate our model we tested it on digital clinical records from 191 patients under ischemic stroke care. We have selected 30 management indexes to be identified in these texts. Our model reached, on average 56,8% of f1-score, varying from 5,83% to 94,78% f1-score across different management indexes.

Keywords: Ontologies · Electronic health records · Stroke · Health management

1 Introduction

Value-based healthcare systems (VBHC) allow fair rewards and proper recognition for health providers based on the quality and results of the service provided [3,5,11]. In this system, the responsible bodies for financing and rewarding these providers can be more confident about their investments, the main users get better services and results and, providers are encouraged to optimize their practice. The implementation of effective VBHC requires advances in computational

Financially supported by the Brazilian Coordination of Superior Level Staff Improvement (CAPES), the by Portuguese Foundation for Science and Technology (FCT)under the projects CEECIND/01997/2017, UIDB/00057/2020 and, the National Council for Scientific and Technological Development (CNPq) (project: 465518/2014-1).

© Springer Nature Switzerland AG 2021
A. Britto and K. Valdivia Delgado (Eds.): BRACIS 2021, LNAI 13073, pp. 359–371, 2021.
https://doi.org/10.1007/978-3-030-91702-9_24

intelligence to continually turn EHR data on information [7,10]. Through these records, services and results are evaluated. This is an exhausting manual task that has to be performed even on top of edge digital health records software; therefore it is essential to turn it into an automated task. This task is defined as the measure of service indexes such as risk events monitoring, clinical conditions, patient handling and outcomes. In order to measure these indexes it is necessary to find keywords and technical terms inside clinical records that are written as free-texts in the Portuguese language. Our research aims to automatize such task, and it is focused on patients under ischemic stroke care. For this context, 30 indicator indexes were chosen to measure the services grouped in: Clinical features, Evaluation measures and risk events, Clinical handling and Patient condition.

This challenge has been handled previously by Zanotto et al. (2021) [14] through machine learning techniques where good results were presented. However, these techniques often require vast amounts of annotated data and computational processing for model training. Our approach tries to avoid these issues by proposing a knowledge based system combined with natural language processing (NLP) techniques. We make use of the NLP methods to find keywords and match terms with an ontology which then uses axioms to classify a given text from the clinical records according to the management indexes.

The evaluation of our model was made on digital clinical records from 191 patients under stroke care. Our model was capable of to identify and to classify 28 of those indexes varying from 5,83% f1-score results and mcc score of 8,01% to 94,78% f1-score results and mcc-score of 94,78%. Considering all 30 indexes, our model reached, on average 56,8% of f1-score and a mcc-score of 57,97%.

2 Related Work

In Wang et al. (2003) [12] and Zhou and El-Gohary (2015) [15] we see the use of machine learning algorithms as part of the classification processes, either to learn the terms related to the domain or to make the classifications. A larger quantity of data was required to achieve good results. The domain application of Wang et al. (2003) were papers from the MEDLINE database and on Zhou and El-Gohary (2015) were construction regulation documents.

Other works follow a knowledge representation approach. The work of Allahyari et al. (2014) [1] uses an ontology for the classifications of English text from news on the web, they used graph projection for this purpose. In Chi et al. (2014) [4] an ontology was created in a semi-automated way, but due to data scarcity, quality and range of terms were not optimal. They worked with job hazards reports. In Schwertner et al. (2019) [9] the authors built an ontology based on domain knowledge from specialists, defining relations between concepts and sentences, they used the ontology as a classification tool. They also worked with clinical data information in the English language. In Gayathri and Kannan (2020) [6] the authors face a similar challenge that we had, their goal was to detect and identify health-related information on English text documents. On Yehia et al. (2019) [13] domain ontologies are used to classify sentences using rules, based on the relations between concepts. They worked with clinical data documents. In de Araujo et al. (2017) [2] an ontology that uses inference processes, based on linguistic rules defined by specialists, to classify texts is presented. In this paper the authors worked with documents and texts in Portuguese about judicial events.

These related works show that ontologies can be considered an alternative for text classification. Different approaches are applied to different domains and results were in general positive. We also see few movements towards techniques focused on the Portuguese language. Given the results of These related works, we considered that a model based on ontologies should be tested for our text classification task, in Portuguese language, to compare with previous work on the same problem that used machine learning in a study made by Zanotto et al. (2021) [14]. They present an evaluation of machine learning approaches for the same database used in the present work, the classification considers almost the same set of classes, they worked on 24 or the indexes. A comparison with this work is presented in the results section.

3 Available Data and Challenges

The main goal of this research was to verify the applicability and the performance of a computational model, based on ontologies, in the task of automatically detect and classify Portuguese free-texts from electronic health records. To accomplish our goal we focused on classifying clinical records from patients under ischemic stroke care. This decision is based on our proximity with a team of specialists of this domain. We also wanted to verify how this model compares against machine learning approaches on this same task and dataset [14].

The data available for this research contains Portuguese free texts clinical records of 191 patients that were treated for ischemic stroke incidents from 01/01/2019 up to 07/23/2019. Our challenge is to find terms and keywords in a given text from these records and given the detected words, classify the text in one the following 30 quality indexes that are shown in Table 1. This study was approved by the Hospital Ethics Committee (CAAE: 29694720000005330).

We aim to develop a computational model in a way that it permeates current Electronic Health Records software, allowing it to operate as first designed as it only would have to provide the data that is stored. With this approach, healthcare practitioners may keep using the same kind of software that they are used to, with no need to input any new data. As the model outputs only the indexes in which the texts were classified, practitioners would also be able keep the privacy and the details of their records and practices.

With the help of a team of domain specialists, we identified technical terms and keywords for the indexes that are to be found within the clinical records. We developed an ontology that plays three important roles in our model. The first one is to provide a list of terms (keywords) that are to be found in the texts. The second role is the definition of the relation of the indexes with the terms. Moreover, the last role is the classifier itself that reasons about the relation between texts and terms and classifies them into the appropriate index. To work along with our ontology, a term detecting algorithm was also developed.

Table 1. Management indexes and its sub-groups

Sub-group	Indexes			
Clinical Features	Coronary Disease	Previous Stroke Incident	Dyslipidemia	Drinker
	Atrial fibrillation	Systemic Arterial hypertension	Cancer	Diabetes
	Obesity	Smoker		
Clinical Handling and Care process	Location	Trombectomy	Thrombolysis	
Evaluation Measures and Risk Events	Intracranial Hemorrhage	Fall	Braden	Fall Risk
	Infection			
Results and Patient Status	Death	Pain	Feeding	Strength
	Paresis	Mobility	Mobility Level	Communication
	Cognitive Capacity	Rankin (mRs)	Self Care	NIHSS

3.1   Data Preprocessing and Annotation

All available clinical records were first anonymized. All records were split into sentences. After this step, 46.547 sentences were generated, in which the indexes would have to be detected. The sentence order was randomized to prevent annotation based on previous context, the idea was to analyse each sentence independently.

Two annotators, domain specialists, read the sentences and informed all the indexes that could be identified in each sentence. The results obtained from both annotators had the percent agreement between them measured by kappa, which was higher than 0.61. In the cases in which there were conflicts, both annotators would come to together to discuss and solve the conflict. No conflict was left unsolved. At the conclusion of these step it was identified that only 17.471 out of the 46.547 sentences were related to one or more of the indicators. The sentences occurrences of each indicator is detailed in Table 8.

4   Methods

4.1   Ontology Based Classification Algorithm

Note that our ontology is a task ontology, developed for classifying sentences containing terms into 30 different indexes. Thus our ontology has three main concepts: the *'Terms'* concept, in which all the sub-concepts are keywords that are to be found in the texts; the *'Sentences'* concept that contains all the sentences processed by the text-detection algorithm; and the *'Index'* concept that contains the description of all the indexes in which the sentences should be classified into. An object property relation *'contain'*, expresses the presence of a given term in a sentence. To the *'Terms'* concept we added a few subsets of concepts: The *'Values'* sub-concept aims to specify the terms that require a numeric value in order to compose a classification; The *'Negations'* sub-set contains expressions that would indicate the negation of the occurrence of an index; the *'PastTense'* sub-set contains terms that would signal the occurrence of an index in some point of the past, these terms often do so by appearing at some point after the mention of an index, and the *'PastTenseRetroactive'* is also for identification of the past, but this specifies the terms that appear before the mention of the index in the sentence. Whenever the terms from these subsets were found, they create the appropriate relation, for instance, *'sentence negates index'*, instead of *'sentence contains index'*.

Our ontology plays the following roles: It serves as knowledge model of the terminology of this domain and it plays as a text classifier trough its inference capabilities.

The inference process is based in assertions in a logical form that together comprise the overall theory that the ontology describes in this domain, this assertions are calles axioms and in this ontology it refers to which terms a sentence must contain in order to be classified as an element of a given index. Table 2 shows us a few examples of these axioms. This ontology was built in OWL language using the *Protégé* tool.

The term detection algorithm receives the instances of the class *Terms*. The algorithm then runs through each sentence and tries to match the given terms to the words in the sentences. Whenever a match is found, the algorithm registers it in the ontology composing the triple *sentence, relation, term*.

Our approach also uses *'word embedding'* models. The *'word embedding'* used was developed on the basis of electronic health records of a brazilian hospital[1] [8], which was trained using 21 million sentences culminating in a model with 63 thousand words. We chose this language model, since this is based on Portuguese EHR. The application of this models has two main goals: To circumvent grammar errors that often occur in these clinical records and; To expand the vocabulary list of terms that were defined initially in the ontology, which means that it brings new rellated words, for instance the medical term *'coronarina'* is not defined in the ontology, but this model relates it to term *'coronaria'* covering this terminology gap. Hence our algorithm uses these models to search for similar words. Using the cosine similarity, the top 10 words are retrieved to expand the list of terms given by the ontology.

To optimize running time two parallel lists are kept by the algorithm: the first one is in charge of storing all words from the texts that do not have a match in the ontology; the second list keeps track of all the words from the sentence that do not match with the terms in the ontology, but some of its word embedding similar do. After all the sentences have been processed and put under the *'Sentences'* set in the ontology, the reasoning process starts and the classification is made. The results are then evaluated by comparing them with the annotated data.

---

[1] https://www.inf.pucrs.br/linatural/wordpress/recursos-e-ferramentas/word-embed dings-para-saude/.

Table 2. Axioms examples

Indicator	Classification	Axioms
Thrombolysis	1	(contain some delta) and (contain some reperfusão) and (contain some terapia) SubClassOf trombólise1
		(contain some reperfusão) and (contain some terapia) and (sem some contraindicação) SubClassOf trombólise1
	0	(contain some reperfusão) and (contain some terapia) and (sem some indicação) SubClassOf trombólise0
Rankin (mRs)	0–6	rankin some xsd:decimal[>= 0 , <= 6] SubClassOf rankin
Mobility Level	11	(contain some auxiliar) and (contain some deambula) and (contain some um) SubClassOf NívelMobilidade11
		mobilidade some xsd:decimal[>= 11 , < 12] SubClassOf NívelMobilidade11
	12	NívelMobilidade11 and (contain some prontidão) SubClassOf NívelMobilidade12
Strength	3	(contain some contra) and (contain some gravidade) and (contain some movimento) SubClassOf ForçaNível3
	4	ForçaNível3 and (contain some resistência) SubClassOf ForçaNível4
		((contain some força) and (contain some perda)) and ((contain some leve) or (contain some sutil)) SubClassOf ForçaNível4
	5	(MobilidadeSemAjuda or NívelMobilidade12 or NívelMobilidade13 or NívelMobilidade14 or NívelMobilidade15) SubClassOf ForçaNível5
		(contain some maior) and (contain some resistência) and (contain some supera) SubClassOf ForçaNível5

4.2 Example

To better understand how our model works let us take the set of sentences shown in Table 3. Our model receives these sentences as input and classifies them. Consider sentence number 6, the first step is to tokenize it. Table 4 shows the result of this step. Next, every word is treated to remove characters that are not either alphabets or numbers, and changed to lowercase, as seen in Table 5. The first word from our sample is *'após'*, our model uses the *word embedding model* to expand this word and get new similar words. Table 6 shows us all the similar words found for this term. After that they are compared with our defined terms. In this example, the term *'após'* and all the similar terms are searched in the list of defined terms in the ontology. For this instance the term *'após'* is not defined and hence no relation between *'sentence #6'* and this term is created.

Our model stores this term in a list of unmatched terms, for the next time it appears it will not be evaluated again. Next, we have the term *'trombolise'*. This one is also expanded with our word-embedding model and the similar terms are shown in Table 7. Again, the term and all the similar words are matched against the ontology.

Table 3. Randomized set of sentences

#	Sentence
1	Cardiologia - início acompanhamento a pedido da Dr.nome_do_médico Sr.nome_do_paciente, 78 anos
2	[Sentença de outra evolução] Nega tabagismo
3	[Sentença de outra evolução] AVC prévio em 2017
4	# Fibrilação atrial em 2008–2009 (uso de amiodarona até fev/2017) recorrência de FA documentada desde março/2019 - reiniciou amiodarona e usou até abril/19
5	# IMC = 35.
6	#Após trombolis, vomitos hipotensao e dor

Table 4. Set of words from sentence #6

#	Words
1	#Após
2	trombolis
3	vomitos
4	e
5	hipotensão
6	nega
7	dor

Table 5. Set of treated words from sentence #6

#	Words
1	após
2	trombolis
3	vomitos
4	e
5	hipotensão
6	nega
7	dor

Table 6. Set of similar words given by the word-embedding model for the term *'após'*

#	Words	
1	após	(`após0'), (`apósOo'), (`apósck'), (`apósjá'), (`apósi'),~
		(`apapós'), (`apósqt'), (`apósc'), (`apóso'), (`apósa')

In this scenario we notice that the term given by the sentence is *'trombolise'*, whereas the definition contains *'trombólise'*. So as expected the term itself is not found in the ontology list as it is defined with proper spelling. However the word embedding model captures this misspelling. As one of the similar words is *'trombólise'* then our model correctly creates the relation between *'sentence #6'* and the term *'trombólise'*. Our algorithm then writes in the ontology the relation *'Sentence#6 contain trombólise'*. As this relation is specified in the axiom *'(contain some alteplase) or (contain some trombolisada) or (contain some trombolítico) or (contain some trombólise) SubClassOf thrombolysis'* once the reasoning process is complete this sentence would then be classified as an elements of the set *'thrombolysis'* which is the set of all sentences that tell us that the index *Thrombolysis* is present.

These steps are then taken to every word in the sentence, and at the end, the relations found are stored in the ontology that will next reason about them.

5   Results

For evaluation of the proposed model the 46.547 sentences were processed and the output was compared to the manual annotation to measure 'precision', 'recall', 'mcc-score' and 'f1-score'. The running time for the whole task was also measured. Table 8 shows the total occurrences of sentences annotated for each index and the results obtained.

The model took 532,43 s to process all the 46.547 sentences achieving, in average, 'f1-score' of 56,8%, 'mcc-score' of 57,97%, 'precision' of 64,89% and 'recall' of 57,97%. Some indexes, such as *'Thrombolysis'* and *'Atrial fibrillation'*, achieved results over 80%; however others, such as *'Pain'* and *'Mobility'*, did not reach over 20%.

Table 7. Set of similar words given by the word-embedding model for the term *'trombolise'*

#	Words	Expansions
2	trombolis	(`trombíolise'),~(`trombolisec'),~(`trombolisdo'),~(`00trombolise'),
		(`trombolíse'),
		(`trombolisetc'), (`trombolize'),~(`trombolisada'), (`trombólise'),
		(`trombse')

Table 8. Ontology based model results

Running time	532,43 s				
Index	Occurrences	f1 (%)	mcc (%)	precision	recall
Braden	260	94,78	94,78	97,12	92,55
Dyslipidemia	143	94,07	94,22	100	88,81
Smoker	283	92,24	92,24	94,96	89,68
Systemic Arterial Hyp.	589	91,26	91,15	90,99	91,15
Diabetes	354	89,89	89,86	92,63	87,31
Location	1512	89,01	88,67	92,29	87,74
Thrombolysis	499	87,75	87,66	88,94	86,45
Rankin (mRs)	189	87,61	87,69	92,36	83,33
Atrial fibrillation	292	83,65	83,84	76,44	92,68
Drinker	109	82,61	82,59	80,85	84,44
Obesity	86	81,36	81,36	82,76	80
NIHSS	320	79,33	79,22	81,23	77,52
Coronary Disease	316	78,5	78,37	77,06	80
Trombectomy	236	64,97	68,82	97,46	48,73
Paresis	510	64,33	64,43	72,77	57,65
Feeding	1576	61,83	64,54	90,8	46,68
Fall Risk	447	61,77	63,23	80,16	50,25
Communication	1134	61,54	62,77	81,06	49,49
Mobility Level	845	50,75	53,01	75,91	39,01
Fall	22	36,89	44,97	23,46	86,36
Cognitive Capacity	759	33,8	33,28	28,59	41,03
Strength	690	27,48	27,63	37,15	22,5
Previous Stroke Incident	238	26,76	32,05	56,14	14,48
Infection	1247	25,58	30,2	56,9	16,5
Mobility	845	19,64	24,87	51,15	11,59
Cancer	247	17,23	17,06	15,14	20
Pain	636	13,37	13,46	10,12	19,69
Intracranial Hemorrhage	216	5,83	8,01	19,35	3,43
Death	2335	0,2	−0,24	2,86	0,11
Self Care	482	0	−0,74	0	0
Total/Average	17417	56,80	57,97	64,89	54,97
Weighted Average		48	75,21	80,61	66,97

These results were compared to the ones obtained by Zanotto et al. (2021) [8] in which a machine learning approach was used for the same challenge. Several supervised computational machine learning (ML) methods, including recent neural and non-neural methods were evaluated on the basis of a 5-fold cross-validation procedure. The best results were achieved by the *W+C+SVM* model, which is based on word-TFIDF and character-TFIDF for input representation and SVM for the classification.

Table 9 presents the *f1-score* results of our ontology based approach and the machine learning approach. Recall that we have machine learning evaluation results for only some of the indexes (24) and can only compare those.

Table 9. *'F1-score'* results comparison between the ontology based approach and machine learning for 24 indexes. When the difference is 5 points or more, the higher score is in bold.

Indexes	Occurrences	Ontology	W+C+SVM
Dyslipidemia	143	**94,07**	83,2
Smoker	283	**92,24**	82,1
Systemic Arterial Hyp.	589	**91,26**	86
Diabetes	354	89,89	89
Location	1512	89,01	88,9
Thrombolysis	499	87,75	85,8
Rankin (mRs)	189	**87,61**	26,9
Atrial fibrillation	292	**83,65**	71,3
Drinker	109	**82,61**	38,6
Obesity	86	81,36	81,7
NIHSS	320	**79,33**	12,4
Coronary Disease	316	**78,5**	61,2
Trombectomy	236	64,97	**72,6**
Paresis	510	64,33	**88,7**
Feeding	1576	61,83	**89,5**
Fall Risk	447	61,77	**89,6**
Communication	1134	61,54	**74,4**
Mobility level	845	**50,75**	40,5
Previous Stroke Incident	238	26,76	**67,1**
Infection	1247	25,58	**79,9**
Mobility	845	19,64	**75,7**
Pain	636	13,37	**52**
Intracranial hemorrhage	216	5,83	**66,4**
Death	2335	0,2	**89,5**
Average		62,24	**70,54**

Our approach performs well on indexes that the machine learning model doesn't and vice-versa. This signals that, by delegating a given a indicator to the most adequate classifier, a combined model could perform well on a larger range of indexes. For instance classifications made by the ontology would benefit the analysis when it comes to the indicator *'Dyslipidemia'* whereas the indicator *'Infection'* would benefit from the machine learn approach. This decision could also be modeled in the ontology and reasoned by it. Further efforts should be put into this matter as a collaborative future work.

5.1 Errors Analysis

In Table 10 we present the most common cases of *'false negatives'* and *'false positives'*. In general the performance of the model is related to the coverage of terms in the ontology. Misspellings and technical slang played a big part on classification errors, the word-embedding model does not cover all the possible variations of terms.

Cases 3, 12 and 13 are related to differences in spelling which were not captured in the most similar words according to the WE model. Examples 1, 2 and 9 are cases in which the Index is present in the sentences, but the specific terms in these sentences were not defined in our ontology.

Table 10. Most common erros

Index	#	Confusion matrix	Annot.	Class	Sentence
Death	1	False Negative	0.0	−1	Condição Ventilatória: ar ambiente, eupneico
	2	False Negative	0.0	−1	Ambiente - Na poltrona, estável, colaborativo, sem queixas, acompanhado da filha.
Previous Stroke	3	False Negative	1.0	−1	D # AVC isquêmico previo - sem sequelas aparentes -mrankin (mRs) previo: 3 # Demência de Alzheimer - Tem vida de relação, corversa, caminha, alimenta-se
	4	False Negative	1.0	−1	Paciente com história de AVC isquêmico em out/18 e dezembro de 2018
	5	False Negative	1.0	−1	#Atual: AVCI #Prévio: AVC/DM2/HAS/DPOC
	6	False Positive	0.0	1.0	# Nega AVC ou Infarto prévio
Cancer	7	False Positive	0.0	1.0	# CA bexiga em 2012 #
	8	False Positive	0.0	1.0	# 2016 - Ca de células claras rim direito - Nefrectomia parcial Dir # Descolamento de retina há 1 ano - Olho Esq # Adenocarcinoma com células em anel de sinete do esôfago distal - QT até junho/19 com progressão da doença + prótese esofágica # medicações em uso
Intracranial Hemorrhage	9	False Negative	1.0	−1	> transformação hemorrágica
	10	False Positive	0.0	1.0	Não há evidência de lesão expansiva, hemorragia intracraniana ou desvios da linha média
Pain	11	False Positive	0.0	1.0	Sem queixas de dor ou desconforto
Location	12	False Negative	1.0	−1	Emergencia/Enfermagem
	13	False Negative	1.0	−1	> EMG HMV->

Past tense and negation are also common source of errors, as expected, since these elements in language require specific sophisticated solutions on their own. In examples 4, 7, 8 the past is indicated through the date of the event, neither our ontology nor our algorithm was prepared for this. Errors in 6, 10 and 11 show us cases in which one *'negation'* term is negating more than one Index. Our model expects one negation term per event.

6   Conclusion

In this paper, we proposed a Portuguese text classifier based on ontologies. Our results show that this approach achieved good results for at least 18 out of the 30 indexes. We believe that this research demonstrates how ontologies is a good alternative for Portuguese medical texts classifications and, because of that, can contribute to the implementation of VBHC programs, contributing to the transformation of health care systems. An advantage of this approach is direct explainability.

For future work we plan to evaluate the model on EHR from different institutions to validate the generality of the results shown here. We also plan to expand the coverage of terms and for that end, we plan to create another *word embedding* model, more tailored to the stroke patients context. As a continuation of the project we plan to align our task ontology with stroke domain ontologies.

References

1. Allahyari, M., Kochut, K.J., Janik, M.: Ontology-based text classification into dynamically defined topics. In: Proceedings of the 8th IEEE International Conference on Semantic Computing, Newport Beach, Estados Unidos da America, pp. 273–278. IEEE (2014)
2. de Araujo, D.A., Rigo, S.J., Barbosa, J.L.V.: Ontology-based information extraction for juridical events with case studies in Brazilian legal realm. Artif. Intell. Law **25**, 379–396 (2017)
3. Bessa, R.d.O.: Análise dos modelos de remuneração médica no setor de saúde suplementar brasileiro. Ph.D. thesis, FGV (2011)
4. Chi, N.W., Lin, K.Y., Hsieh, S.H.: Using ontology-based text classification to assist job hazard analysis. Adv. Eng. Inf. **28**, 381–394 (2014)
5. Engel, G.L.: The clinical application of the biopsychosocial model. J. Med. Philos. Forum Bioeth. Philos. Med. **6**(2), 101–124 (1981). https://doi.org/10.1093/jmp/6.2.101
6. Gayathri, M., Kannan, R.: Ontology based concept extraction and classification of ayurvedic documents. Procedia Comput. Sci. **172**, 511–516 (2020)
7. Gonçalves, F.N.R.: Optimizing patients' pathways in international cooperation, by doing value based healthcare (VBHC). Acta medica portuguesa **32**(2), 167–168 (2019)
8. Dias Pereira dos Santos, H., D. P. S. Ulbrich, A.H., Woloszyn, V., Vieira, R.: An initial investigation of the charlson comorbidity index regression based on clinical notes. In: 2018 IEEE 31st International Symposium on Computer-Based Medical Systems (CBMS), pp. 6–11 (2018). https://doi.org/10.1109/CBMS.2018.00009

9. Schwertner, M.A., Rigo, S.J., Araújo, D.A., Silva, A.B., Eskofier, B.: Fostering natural language question answering over knowledge bases in oncology EHR. In: Proceedings of the 32nd International Symposium on Computer-Based Medical Systems, Cordoba, Espanha, pp. 501–506. IEEE (2019)

10. da Silva Etges, A.P.B., Ruschel, K.B., Polanczyk, C.A., Urman, R.D.: Advances in value-based healthcare by the application of time-driven activity-based costing for inpatient management: a systematic review. Value Health **23**(6), 812–823 (2020). https://doi.org/10.1016/j.jval.2020.02.004. https://www.sciencedirect.com/science/article/pii/S1098301520301303

11. Uzuelli, F.H.d.P., Costa, A.C.D.d., Guedes, B., Sabiá, C.F., Batista, S.R.R.: Reforma da atenção hospitalar para modelo de saúde baseada em valor e especialidades multifocais. Ciência Saúde Coletiva **24**, 2147–2154 (2019)

12. Wang, B.B., Mckay, R.I.B., Abbass, H.A., Barlow, M.: A comparative study for domain ontology guided feature extraction. In: Proceedings of the 26th Australasian Computer Science Conference, pp. 69–78. Australian Computer Society Inc., Adelaide, Austrália (2003)

13. Yehia, E., Boshnak, H., Abdelgaber, S., Abdo, A., Elzanfaly, D.: Ontology-based clinical information extraction from physician's free-text notes. J. Biomed. Inform. **98**, 103–117 (2019)

14. Zanotto, B., et al.: Automatic classification of electronic health records for a value-based program through machine learning. Value Health (2021, to appear). Accepted for publication in Virtual ISPOR 2021

15. Zhou, P., El-Gohary, N.: Ontology-based multilabel text classification of construction regulatory documents. J. Comput. Civ. Eng. **30**, 40–54 (2015)

Machine Learning and Data Mining

A Co-occurrence Based Approach
for Mining Overlapped Co-clusters
in Binary Data

Yuri Santa Rosa Nassar dos Santos[1,2]([✉]) [ID], Rafael Santiago[1,2] [ID],
Raffaele Perego[3] [ID], Matheus Henrique Schaly[1], Luis Otávio Alvares[1] [ID],
Chiara Renso[3] [ID], and Vania Bogorny[1,2] [ID]

[1] Universidade Federal de Santa Catarina, Florianópolis, Brazil
yuri.nassar@posgrad.ufsc.br
[2] INE, Programa de Pós-Graduação em Ciência da Computação, Florianópolis, Brazil
[3] ISTI-CNR, Pisa, Italy

Abstract. Co-clustering is a specific type of clustering that addresses
the problem of simultaneously clustering objects and attributes of a data
matrix. Although general clustering techniques find non-overlapping co-
clusters, finding possible overlaps between co-clusters can reveal embedded patterns in the data that the disjoint clusters cannot discover. The
overlapping co-clustering approaches proposed in the literature focus on
finding *global* overlapped co-clusters and they might overlook interesting
local patterns that are not necessarily identified as global co-clusters.
Discovering such *local* co-clusters increases the granularity of the analysis, and therefore more specific patterns can be captured. This is the
objective of the present paper, which proposes the new Overlapped
Co-Clustering (OCoClus) method for finding overlapped co-clusters on
binary data, including both *global* and *local* patterns. This is a non-
exhaustive method based on the co-occurrence of attributes and objects
in the data. Another novelty of this method is that it is driven by an
objective cost function that can automatically determine the number
of co-clusters. We evaluate the proposed approach on publicly available
datasets, both real and synthetic data, and compare the results with a
number of baselines. Our approach shows better results than the baseline
methods on synthetic data and demonstrates its efficacy in real data.

Keywords: Co-clustering · Overlapped co-clusters · Binary data

1 Introduction

Over the years, the task of clustering complex data has become more challenging
since a high number of attributes can increase computational complexity and
affect cluster consistency [17]. One way to deal with this complexity is to use
the co-clustering approach, which simultaneously clusters objects (rows) and
attributes (columns) in matrix data [5]. The focus of these methods relies on

© Springer Nature Switzerland AG 2021
A. Britto and K. Valdivia Delgado (Eds.): BRACIS 2021, LNAI 13073, pp. 375–389, 2021.
https://doi.org/10.1007/978-3-030-91702-9_25

finding co-clusters, where each co-cluster is formed by a subset of objects and attributes that can represent a submatrix of a given matrix.

Co-clustering approaches use, in general, a non-overlapping strategy, which means that an *element* of a co-cluster can belong to only one co-cluster [1, 14]. However, in many real situations, an *element* can participate simultaneously in more than one co-cluster. For example, a movie could be both *thriller* and *science-fiction*, a song can be both *rock* and *high-energy*, etc. Therefore, an overlapping strategy is important because it identifies *intersections* between co-clusters, revealing patterns that could be lost when using disjoint co-clustering. Besides, the detection of overlapping co-clusters has proven to be challenging since it is not trivial to evaluate the clustering quality [8].

We notice that most of the overlapping co-clustering approaches proposed in the literature have two characteristics: (1) they discover *global* clusters, and (2) they tend to fit a specific application. Examples can be found in text mining [2], bioinformatics [13], recommendation systems [15], and social network analysis [18], to name a few. In contrast, the works of Fu et al. [4], Li [7], Whang et al. [16], and Zhu et al. [19], not only focus on *global* clusters on binary datasets, but they were designed for generic purposes. The limitation of these works is that they do not detect *local* co-clusters, i.e., refined groups formed by objects and attributes that identify an overlap pattern on *global* co-clusters.

In this paper, we propose a new co-clustering method that combines simplicity of use with the capacity to extract overlapping global and local co-clusters. This method is named Overlapped Co-Clustering (OcoClus) and it is based on the co-occurrence of attributes and objects. The main novelty of this method is that, unlike the traditional overlapped co-clustering, it is able to infer a new type of patterns called *local* co-clusters. Furthermore, OCoClus is driven by an objective cost function that does not require the user-defined parameter of the number of co-clusters.

In summary, we make the following contributions: (i) propose an incremental co-clustering approach that is application-independent and an algorithm that can find both overlapped and non-overlapped *global* and *local* co-clusters; (ii) use a cost function that, finds the number of co-clusters automatically, that ranks the co-clusters from the most relevant to the less relevant, and that finds overlapped co-clusters.

The remainder of this work is organized as follows. The basic concepts definition of our work are presented in Sect. 2. Section 3 presents the works that are related to our proposal. Section 4 presents the details of our method. Section 5 presents the evaluation of the method with synthetic and real data. Finally, the conclusion and further research directions of our work are presented in Sect. 6.

2   Basic Concepts

In this section we present the basic concepts to guide the reader throughout this paper.

Let D be a *binary* matrix with N rows (objects) and M columns (attributes). The element d_{ij} of D, where i and j are integers that $1 \leq i \leq N$ and $1 \leq j \leq M$, is equal to 1 if the j-th attribute occurs in the i-th object (*true element*); otherwise, it is 0. Co-clustering is the grouping task of finding K (*global*) co-clusters in D where each co-cluster is formed by a subset of objects and attributes [11]. The subset of objects I can be represented as a binary vector of length N, where $I_i = 1$ indicates that the i-th object is present in I. Similar to that, a subset of attributes J with $J_j = 1$ indicates that the j-attribute is present in J with length M. More formally, a co-cluster can be defined as follows:

Definition 1. ***Co-cluster:*** *Let D be a binary matrix, I be the subset of objects, J be the subset of attributes; a co-cluster C is defined as $C = \langle I, J \rangle$. The elements c_{ij} of co-cluster C are formed by the outer product of its subsets I and J ($C \in \{0, 1\}^{|I| \times |J|}$). Thus, a co-cluster C can represent a submatrix of D.*

Such (*global*) co-cluster C can be formed with only the *true elements* in D or mixed with true elements and *noise elements* ($d_{ij} = 0$). In this paper, the terms *global* co-cluster and co-cluster are interchangeably used. The co-occurrence between objects and attributes can form a co-cluster C which can be simplified by searching elements $d_{ij} = 1$ in the matrix D. Furthermore, it can reduce the search space once the goal is to identify true co-occurrences. Inspired by [9], we adapted four concepts for co-clustering problem: cost function $\mathcal{F}_P$, *pure* co-cluster PC, noise thresholds ϵ_I and ϵ_J, and expanded *pure* co-cluster EC. The cost function $\mathcal{F}_P$ can be used to evaluate the process of forming a co-cluster. More formally, we can define the cost function $\mathcal{F}_P$ as follows:

Definition 2. ***Cost Function:*** *Let C^* be a co-cluster, $\prod$ be a set of global co-clusters, D be an input matrix, ρ be a weight of importance for the co-clusters cost, $\mathcal{N}$ be a noise matrix, γ_{C^*} and $\gamma_{\mathcal{N}}$ be user-defined functions for measuring the costs of co-clusters and noise; a cost function $\mathcal{F}_P$ is defined as $\mathcal{F}_P(\prod, D) = \rho \times \sum_{C^* \in \prod} \gamma_{C^*}(C^*) + \gamma_{\mathcal{N}}(\mathcal{N})$.*

The objective is to minimize $\mathcal{F}_P$ regarding ρ, $\gamma_{C^*}(C^*)$ and $\gamma_{\mathcal{N}}(\mathcal{N})$. Such noise matrix $\mathcal{N}$ used by [9] takes into account the *false positives, false negatives,* and the already covered elements in D. Regarding the set of co-clusters $\prod$ and matrix D, the *false positives* are elements $d_{ij} = 0$ covered by some pattern in $\prod$, while *false negatives* are elements $d_{ij} = 1$ not covered by any pattern in $\prod$. The concept of *pure* co-cluster simplifies the identification of a *global* co-cluster, which identifies a disjoint *global* co-cluster that contains only *true elements*. Thus, we can define a *Pure Co-cluster PC* as follows:

Definition 3. ***Pure Co-cluster:*** *Let D be a binary matrix, d_{ij} be an element of D; a pure co-cluster $PC = \langle PC_J, PC_I \rangle$ is formed by a subset of objects PC_I and a subset of attributes PC_J that identifies only the true elements. Thus, PC can represent a submatrix of D, which contains only the true elements $d_{ij} = 1$.*

A *pure* co-cluster PC can be expanded with noisy objects and attributes. We use two thresholds to control the amount of noise in a co-cluster: ϵ_I for objects and ϵ_J for attributes. Thus, the noise thresholds ϵ_I and ϵ_J can be defined as follows:

Definition 4. *Noise Thresholds:* Let C^* be a co-cluster, C_J^* and C_I^* be the subsets of attributes/objects that define C^*; a maximum noise threshold for objects ϵ_I and attributes ϵ_J limit the amount of noise that can be included in C^*. Thus, each new object must be included in at least $(1 - \epsilon_I) \times ||C_J^*||$ attributes of C^*, while each new attribute must be included in at least $(1 - \epsilon_J) \times ||C_I^*||$ objects of C^*.

The noise threshold value can range from $[0, 1]$, where 0 does not allow any noise while 1 allows the maximum amount. The number of objects and attributes of a given subset is measured by the L^1-norm $|| \cdot ||$ (or Hamming norm), which simply counts the number of bits 1 in the vector. From that, the expanded *pure* co-cluster EC represents a expanded version of PC with noise data. More formally, an *Expanded Pure Co-cluster EC* can be defined as follows:

Definition 5. *Expanded Pure Co-cluster:* Let PC be the pure co-cluster, EC_I be a subset of objects, EC_J be a subset of attributes, ϵ_I be the noise object threshold, ϵ_J be the noise attribute threshold; an expanded pure co-cluster is defined as $EC = \langle EC_J, EC_I \rangle$, where EC_J and EC_I can contain new attributes and objects not present in PC regarding the noise thresholds ϵ_I and ϵ_J.

3   Related Works

Because of the difficulty in finding co-clusters, there is no method widely accepted as the state-of-the-art; instead, there are algorithms that perform better in certain types of data than others. Since a complete review is out of the scope of this paper, we shall briefly discuss some reference algorithms. For a comprehensive review of co-clustering algorithms, we refer to [12].

Dhillon [3] used the matrix decomposition using the eigenvectors combined with bipartite graph to find *global* co-clusters in a real-valued matrix. It needs to know the number of co-clusters *a priori* and the order of the discovered co-clusters is not important. Furthermore, it uses a support matrix to include some attributes as noise data; however, it does not have any noise-parameter to control the number of objects or attributes as noise data. Kluger et al. [6] extended the Dhillon [3] approach by using the singular value decomposition to find *global* co-clusters. It assumes that the data have a checkerboard structure in the matrix. This approach includes each element of a matrix into one co-cluster without overlap; therefore, it cannot control the noise data.

Fu et al. [4] proposed a Bayesian-based overlapping co-clustering approach based on a multivariate distribution to find *global* co-clusters in a binary data matrix. It assumes that the number of co-clusters is known *a priori*. This approach does not indicate that the order of the discovered co-clusters is important. It tolerates noisy elements in the co-clusters; however, it does not have a noise-parameter to control the number of objects or attributes included in the co-cluster as noise. Zhu et al. [19] proposed an overlapping co-clustering approach to approximate a binary data matrix with the sum of identified *global* co-clusters. This method needs to have the number of co-clusters a priori. Furthermore, it does not deal with noise data and does not associate any importance for the co-cluster that explains the discovered order.

Lucchese et al. [9] proposed a frequent pattern mining method for binary datasets. The patterns are formed by sets of attributes and objects, where they can represent a non-overlapped *global* co-cluster. It uses a generalized cost function to drive the mining process to find the number of patterns automatically. The discovered order of the patterns is relevant regarding the cost function; therefore, it can be seen as a ranking. Finally, two noise thresholds control the number of noisy attributes and objects included in a pattern.

Whang et al. [16] modeled the input data as a bipartite graph to find *global* overlapping co-clusters in binary data. This method allows to include noise objects in the co-clusters with a probability distribution function that models the noise. However, it does not define noise thresholds to control the number of noise objects and attributes. It can automatically infer the number of co-clusters, besides that, the method does not consider that the order of the discovered co-clusters is relevant in the process.

Li [7] presented a generalized overlapped co-clustering approach that uses singular value decomposition on the binary data matrix to identify *global* co-clusters. This method does not include noise automatically or by a user-defined noise threshold; it searches for homogeneous co-clusters without noise. Furthermore, it can infer the number of co-clusters; however, it is not guaranteed to converge to the optimum number. Finally, the method does not show that the discovered order of these co-clusters is relevant to it.

4 The Overlapped Co-clustering Approach

In this section we present a new method called *OCoClus* (Overlapped Co-Clustering) for finding overlapped co-clusters in a binary dataset by identifying both *global* and *local* co-clusters. OCoClus searches for the co-occurrences between attributes and objects to identify co-clusters where a cost function drives the co-clustering process. In the following we present the method definitions in Sect. 4.1 and the proposal details in Sect. 4.2.

4.1 Method Definitions

Local co-clusters are patterns in the data related to specific characteristics that are overlooked by *global* co-clusters since it finds clusters which are in the intersection of objects and attributes of the *global* clusters. Thus, we formally define a *local* co-cluster LC as follows:

Definition 6. *Local Co-cluster:* *Let $\prod$ be the set of co-clusters, LC_I be a subset of objects, LC_J be a subset of attributes, C be the co-cluster in $\prod$; a local co-cluster is defined as $LC = \langle LC_I, LC_J \rangle$, where the object intersections of co-cluster C with the co-clusters in $\prod$ forms LC_I, and the union of the attributes between C and the intersected co-clusters in $\prod$ forms LC_J.*

We propose a new cost function $\mathcal{F}$ designed to make the overlapping and non-overlapping co-clusters equally important including both *global* and *local* co-cluster. The difference between the new cost function in Definition 7 and the cost function given in Definition 2 is that the new cost function considers just the size of the co-cluster and the quantity of noise that can be included in the co-cluster. However, the cost function in Definition 2 weights the relevance of the patterns regarding its size, it penalizes the patterns that cover an element already covered and does not include an element into the expected pattern. From that, we define the new cost function $\mathcal{F}$ as follows:

Definition 7. Cost Function: *Let $\prod$ be the set of co-clusters, D be the binary matrix, C^* be the co-cluster, $||C_J^*||$ and $||C_I^*||$ be the size of the subsets of attributes/objects that define C^*, $\mathcal{N}$ be the number of noise elements included in C^* ($d_{ij} = 0$), and H be the part that does not consider noise data; a cost function $\mathcal{F}$ is defined as $\mathcal{F}(\prod, D) = H + \mathcal{N}$, where $H = \sum_{C^* \in \prod} (||C_I^*|| + ||C_J^*||) - (||C_I^*|| \times ||C_J^*||)$. Thus, the objective is to minimize $\mathcal{F}$ regarding C_I^*, C_J^* and $\mathcal{N}$.*

Regarding the new cost function $\mathcal{F}$, part H contributes to the cost function evaluating co-clusters without noise, while part $\mathcal{N}$ contributes by allowing some noise data regarding the maximum noise thresholds. Once we have already formalized the main definitions, it is simple to define the overlapped co-cluster used to represent the *global* and *local* patterns as follows:

Definition 8. Overlapped Co-cluster: *Let $\prod$ be the set of co-clusters, X be the co-cluster $\in \prod$ with its subset of attributes X_J and objects X_I, Op be the set of co-clusters $\in \prod$ that intersect X_I, and Op_J and Op_I be the subset of attributes and objects of Op; an overlapped co-cluster is formally defined as $OC = \langle X_J \cup Op_J, X_I \cap Op_I \rangle$.*

Considering the Definition 8, the subset of objects of OC_I is formed by the nested intersection of objects between X_I and Op_I ($X_I \cap Op_I$), and the subset of attributes OC_J by joining the attributes of X_J with Op_J ($X_J \cup Op_J$).

4.2   Method Description

Algorithm 1 is the main algorithm that organizes our approach. It receives four input parameters: the matrix D, the number of co-clusters K, the object noise threshold ϵ_I and the attribute noise threshold ϵ_J. As a result, it outputs a set of co-clusters Φ which contain K co-clusters that can overlap.

In Algorithm 1, the set of co-clusters $\prod$ is set as empty (line 1), and the residual matrix D_r is initiated with D, which is used to find uncovered co-clusters in D (line 2). The algorithm iterates over *findPureCocluster* (line 4) and *expandPure-Cocluster* (line 5) methods at most K times, where K is the maximum number of co-clusters (line 3). In *findPureCocluster* method, the attributes in D_r are sorted in descending order (from the most frequent to the least) and stored in a list S to maximize the probability of forming large co-clusters. Therefore, the attributes

Algorithm 1. OCoClus

Input: D: input matrix
              K: max number of clusters {optional}
              ϵ_I: max object noise threshold {optional}
              ϵ_J: max attribute noise threshold {optional}
Output: Φ: set of disjoint and overlapped co-clusters
             OCoClus $(K, D, \epsilon_J, \epsilon_I)$
1: $\prod \leftarrow \emptyset$
2: $D_r \leftarrow D$ {residual matrix}
3: **for** $i = 1, \ldots, K$ **do**
4:      $PC, E \leftarrow$ findPureCocluster$(D, D_r, \prod)$ {Definition 3}
5:      $EC \leftarrow$ expandPureCocluster$(PC, E, \prod, D, \epsilon_I, \epsilon_J)$ {Definition 5}
6:      **if** $\mathcal{F}(\prod, D) < \mathcal{F}(\prod \cup EC, D)$ **then**
7:          **break**
8:      **end if**
9:      $\prod \leftarrow \prod \cup EC$
10:     $D_r(i, j) \leftarrow 0 \ \forall i, j$ where $EC_I(i) = 1 \wedge EC_J(j) = 1$
11: **end for**
12: $\Phi \leftarrow$ findOverlap$(\prod)$ {Definition 8}
13: **return** Φ

in S are evaluated for being added to a co-cluster without backtracking reducing the search space. Only the true elements in D forms the *pure* co-cluster PC regarding the attributes in S. With this, the number of objects and attributes that co-occur are used in the cost function $\mathcal{F}$ stated in Definition 7 to evaluate if the tested subsets of objects and attributes can minimize the cost function. The PC grows in the number of objects and attributes as long as the cost function $\mathcal{F}$ is minimized. Besides, some attributes cannot be used to form the PC, then these attributes are stored in an extension list E. The output is a *pure* co-cluster PC and an extension list of attributes E.

In *expandPureCocluster* method, OCoClus expands PC with new objects and attributes that allows noise data (line 5). With this, the *expanded* co-cluster EC is initiated with PC identified at line 4. Then, the process is similar to *findPureCocluster*; however, at this part, the method checks if the inclusion does not exceed the maximum noise thresholds (Definition 4) and improves the cost function $\mathcal{F}$. This inclusion occurs in two steps. First, the method tries to include new objects that are not present in EC and does not modify the current attributes. Second, it does not modify the current objects and tries to include the attributes stored in the extension list E one at a time without backtracking. If an attribute is included in EC, the process goes back to the first step and repeats both steps. We remark that each new object and attribute is included in EC if such inclusions respect both Definition 4 and Definition 7. This process is repeated while E is not empty. The *expandPureCocluster* returns an *expanded* co-cluster EC as the output.

Given the output of the *expandedPureCocluster*, if the new co-cluster EC minimizes the cost function $\mathcal{F}$ of the model (line 6), it is added to $\prod$ (line 9). However, if EC does not minimize the cost function $\mathcal{F}$ of the model, even though the parameter K does not reach its maximum value, the algorithm stops searching for new co-clusters (line 7). Besides, if the cost function $\mathcal{F}$ is improved, the residual matrix D_r is then updated with EC (line 10). The updated residual matrix D_r is used in the next iteration to find new patterns in D that are not covered by any previous co-cluster. OCoClus can find the number of co-clusters automatically whenever K is not given. However, if the user misspecify the value of K, then the true number of co-clusters may not be discovered.

So far, $\prod$ covers non-overlapped patterns in the data (line 9); hence, it cannot show which co-clusters *share* characteristics. Therefore, OCoClus refines these non-overlapped co-clusters to identify *global* and *local* overlapped co-clusters as stated in Definition 8. With this, the *findOverlap* method (line 12) iterates over $\prod$ to identify possible overlapped co-clusters by taking the nested object intersections between the co-clusters in $\prod$. The nested intersection considers what is shared among all intersected co-clusters instead of a common intersection between pairs of co-clusters. If such a co-cluster intersection exists, the attributes of the co-clusters involved in the intersection are joined. The next step is to delete the redundant co-clusters, i.e., co-cluster totally covered by another co-cluster. From that, the *findOverlap* is a simple and effective method that allows OCoClus to find both overlapped co-cluster structures. Its simplicity and effectiveness become possible by exploring the nested intersections of objects and joining attributes separately regarding the co-clusters in $\prod$. This process looks simple once the cost function $\mathcal{F}$ already evaluated the disjoint co-clusters in the previous methods, but it effectively identifies overlapped co-clusters. At the end, *findOverlap* returns the set of non-overlapped and overlapped (if exist) co-clusters Φ. Finally, Algorithm 1 returns this set of non-overlapped and overlapped patterns including both *global* and *local* co-clusters (line 13).

Proposition 1. *Let K be the maximum number of non-overlapped co-clusters, N the total number of objects, M the total number of attributes, and P the number of overlapped co-clusters. The computational complexity of findPureCocluster method is $O(MN)$, expandPureCocluster method is $O(M(MN+N)) = O(M^2 N)$, and findOverlap method is $O(K^2+P^2)$. Regarding the overall complexity of our algorithm, OCoClus calls findPureCocluster and expandPureCocluster methods, then builds D_r for each of the K (or less) non-overlapped co-clusters and finalizes with the findOverlap method. Thus, the computational complexity of the OCoClus Algorithm is $O(KM^2 N + (K^2+P^2))$.*

5   Experimental Evaluation

We compare OCoClus[1] with four publicly available methods presented in the related works to use as the baseline methods, which are: Li [7], Lucchese et al. [9], Kluger et al. [6], and Dhillon [3]. We include the works of Dhillon and Kluger et al. because they are consolidated approaches in the literature and publicly available as a package by *scikit learn*[2]. Considering their stable implementation, we selected these works once the overlapping baseline methods fail to find the embedded overlapped co-clusters. Therefore, we include those non-overlapped

---

[1] https://github.com/bigdata-ufsc/ococlus.
[2] https://scikit-learn.org/stable/modules/biclustering.html.

Table 1. Datasets description.

Dataset	Number of objects	Number of attributes	Sparsity (%)	Number of co-clusters
Synthetic-1	100	100	76.62	7
Synthetic-2	600	1000	77.97	10
Synthetic-3	100	100	68	4
CAL500	502	103	76.6	–
CV-19	5729	567	98.18	–

methods in the baseline to compare such a co-clustering result. We used three synthetic datasets named *synthetic-1* and *synthetic-2*, and *synthetic-3*, where we artificially embedded the co-clusters (patterns) to create the ground-truth datasets. Furthermore, we also evaluated OCoClus on two real-life datasets, named CAL500[3] and CV-19[4], to show its efficacy in the real application scenario. All the experiments data and source code are made public.

We performed the experiments in a machine with a processor Intel i7-7700 3.6 GHz, 16 GB of memory, and OS Windows 10 64bits. Furthermore, we ran 15 independent simulations for all methods on each synthetic dataset to compute the average and standard deviation of the evaluation metrics score. Table 1 shows the main characteristics of the datasets used in the experiments. It shows the total number of objects and attributes, the sparsity in the data (percentage of zeros), and the number of co-clusters for the synthetic datasets.

We use four evaluation metrics to assess the quality of the OCoClus. First, we use the reconstruction error matrix to measure the difference between data input and found co-clusters given by $\text{Rec}_{error} = ||X \veebar Y||$ similar to [7]. In short, we take the sum of the element-wise *xor* ($\veebar$) between the input data matrix X (ground-truth) and the reconstructed matrix Y regarding the found co-clusters to measure the quantity of *false positives* and *false negatives*. The clustering quality is better when the result of the Rec_{error} is equal or close to zero. The other three metrics are *Omega* index (overlapped version of ARI measure), *Overlapped Normalized Mutual Information* (ONMI) and *overlapped F1* measure (F_{score}) [10]. For these three last measures, the clustering quality is better when the result is equal or close to one, where one is the maximum score. We decided to use these measures since our approach focuses on the overlapping problem and therefore the traditional measures like for example *Adjusted Rand Index* (ARI), *Normalized Mutual Information* (NMI), and F_{score} are not suitable to capture the overlapping behaviour.

5.1   Evaluation of OCoClus with Synthetic Data

To be fair with all methods, we set the number of co-clusters K according to the ground-truth shown in Table 1. Regarding the noise control used by

---

[3] http://mulan.sourceforge.net/datasets-mlc.html.
[4] https://ti.saude.rs.gov.br/covid19/; just passed away people data were used.

Table 2. Score of the evaluation metrics for the synthetic datasets.

Dataset	Work	Rec$_{error}$	Omega	ONMI	F$_{score}$
Synthetic-1	Li	2086.27 ∓ 371.33	−0.0110 ∓ 0.0228	0.0060 ∓ 0.0103	0.0717 ∓ 0.1051
	Lucchese$_{t_1}$	0 ∓ 0	0.6640 ∓ 0	0.5419 ∓ 0	0.6617 ∓ 0
	Lucchese$_{t_2}$	160 ∓ 0	0.6125 ∓ 0	0.509 ∓ 0	0.6359 ∓ 0
	Lucchese$_{t_3}$	160 ∓ 0	0.6125 ∓ 0	0.509 ∓ 0	0.6359 ∓ 0
	Dhillon	1404 ∓ 0	0.1818 ∓ 0	0.1648 ∓ 0	0.4868 ∓ 0
	Kluger	3759 ∓ 0	0.2107 ∓ 0	0.1824 ∓ 0	0.3155 ∓ 0
	OCoClus	**0 ∓ 0**	**1 ∓ 0**	**1∓ 0**	**1 ∓ 0**
Synthetic-2	Li	19528.33 ∓ 1568.05	0.748 ∓ 0.022	0.2624 ∓ 0.0092	0.4968 ∓ 0.0177
	Lucchese$_{t_1}$	353 ∓ 0	0.7644 ∓ 0	0.5985 ∓ 0	0.711 ∓ 0
	Lucchese$_{t_2}$	40078 ∓ 0	0.7483 ∓ 0	0.326 ∓ 0	0.5425 ∓ 0
	Lucchese$_{t_3}$	40078 ∓ 0	0.7483 ∓ 0	0.326 ∓ 0	0.5425 ∓ 0
	Dhillon	31426 ∓ 0	0.9106 ∓ 0	0.3627 ∓ 0	0.5001 ∓ 0
	Kluger	34382 ∓ 0	0.9653 ∓ 0.001	0.1744 ∓ 0.004	0.2828 ∓ 0.0036
	OCoClus	**0 ∓ 0**	**1 ∓ 0**	**1 ∓ 0**	**1 ∓ 0**
Synthetic-3	Li	1530.4 ∓ 923.1	0.0152 ∓ 0.0554	0.0395 ∓ 0.0355	0.3466 ∓ 0.1857
	Lucchese$_{t_1}$	0 ∓ 0	0.1813 ∓ 0	0.4031 ∓ 0	0.6462 ∓ 0
	Lucchese$_{t_2}$	1000 ∓ 0	0 ∓ 0	0.0003 ∓ 0	0.4767 ∓ 0
	Lucchese$_{t_3}$	1000 ∓ 0	0 ∓ 0	0.0003 ∓ 0	0.4767 ∓ 0
	Dhillon	2200 ∓ 0	−0.0123 ∓ 0	0.1565 ∓ 0	0.3856 ∓ 0
	Kluger	1000 ∓ 0	0.0835 ∓ 0	0.2935 ∓ 0	0.4576 ∓ 0
	OCoClus	**0 ∓ 0**	**1 ∓ 0**	**1 ∓ 0**	**1 ∓ 0**

Lucchese et al. [9], we use three configurations (t_1, t_2 and t_3) of noise threshold parameters to assess its clustering result when the noise values change. The configuration t_1 uses the object noise threshold $\epsilon_I = 0$ and attribute noise threshold $\epsilon_J = 0$. For the last two configurations t_2 and t_3, the noise values are the same used in Lucchese et al. [9]. The configuration t_2 uses $\epsilon_I = 0.5$ and $\epsilon_J = 0.8$, while configuration t_3 uses $\epsilon_I = 1$ and $\epsilon_J = 1$.

Table 2 shows respectively the average and standard deviation from the evaluation metrics for each method and synthetic dataset. It can be seen that OCoClus obtained the best score result in all synthetic datasets; hence, it finds the embedded overlapped co-clusters. Lucchese$_{t_1}$ obtained the second best result while the other two configurations obtained the same score values because they identified the same co-clusters. The method proposed by Li [7] obtained the worst result among the methods. This happens because the method sometimes does not converge to any co-cluster which makes its overall result worse than or close to the non-overlapped methods. Besides, it can be seen in Table 2 that the baseline methods do not find the real number of co-clusters once their evaluation scores do not reach the best value. Regarding the non-overlapped methods, the method proposed by Kluger et al. [6] shows the worse overall co-clustering result. Meanwhile, the method of Li [7] improved slightly its overall clustering result in *synthetic-2* dataset compared to *synthetic-1* and *synthetic-3*. However, it does not overcome the clustering results of the non-overlapped approaches at all.

In summary, it can be seen in Table 2 that OCuClus outperformed the baseline methods in all evaluation metrics. Such a result occurs because OCoClus identifies all *global* and *local* co-clusters while the baselines fail to find both co-cluster structures correctly in the data. The baselines focus on the *global* non-overlapped and overlapped structures. Regarding the baseline methods, using the constraint t_1 in the work of Lucchese et al. [9], this configuration generated the best clustering result. However, considering the other two constraints, we notice that they do not improve the clustering result. Furthermore, the method proposed by Li [7] shows an overall worse clustering result, even though it improved its performance in the synthetic-2 dataset but not enough to overcome all methods. The methods of Kluger et al. [6] and Dhillon [3], in general, obtained stable results in comparison with Li [7].

5.2 Real Application Scenario

In this section, we used OCoClus on two real datasets to demonstrate its general utility. We set the noise thresholds ϵ_I and ϵ_J to the minimum value, and this means that we are not allowing any attribute or object to be added as noise in the co-clusters. With this parameter control, it is possible to have a better understanding of the co-cluster structure. In fact, OCoClus finds *pure* co-clusters when the noise thresholds are set to the minimum value (zero); this means that all attributes that occur in all objects do not include the presence of noise.

Music Annotation. The left side of Fig. 1 shows the bitmap of the CAL500 dataset, and the right side shows the bitmap of the OCoClus result. Similar to Li [7], the question is to identify song sets that share similar annotations. Moreover, we are interested in finding which are the common annotations that distinguish each song set. This task can enhance our perception of the relationship between songs and annotations and therefore it can be applied to music retrieval and recommendation system. We used OCoClus in the processed dataset and the main co-clusters are highlighted in the right side of Fig. 1.

OCoClus identified three main levels which are within the red lines and four main co-clusters. The two larger *global* co-clusters have the size 150 × 13 and 100 × 12. Further, looking into these two co-clusters we found such annotations as "NOT-Song-Fast_Tempo", "NOT-Emotion-Angry-Aggressive", "NOT-Song-Heavy_Beat" and "NOT-Emotion-Bizarre-Weird" for the first co-cluster, and the "Song-Fast_Tempo", "NOT-Emotion-Angry-Aggressive", "Song-Heavy- _Beat" and "Song-High_Energy" for the second co-cluster. The first co-cluster includes songs such as "For you and I" by 10cc, "Three little birds" by Bob Marley and The Weilers, and "I'll be your baby tonight" by Bob Dylan. In the second co-cluster includes songs such as "Trapped" by 2pac, "Dirty deeds done dirt cheap" by AC/DC and "Livin on a prayer" by Bon Jovi. Considering the song attributes in the first cluster, it can be seen that songs are formed by a slow rhythm and without strong beats. The second cluster characterized songs with strong beats and a fast rhythm. Therefore, the method identified clusters with opposite characteristics, showing two groups of users with different preferences.

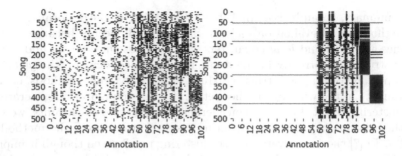

Fig. 1. CAL500 Music dataset. The left side shows the bitmap of the binary annotation matrix where objects represent songs and attributes represent annotations (black = presence; white = absence). The right side shows the bitmap of the identified co-clusters. Objects and attributes are ordered in the same way in both figures just for visualization purpose. (best seen in color) (Color figure online)

Considering the *local* patterns, for instance, OCoClus finds a co-cluster of songs as "Summertime" by Dj Jazzy Jeff and The fresh prince, "Sunset 138 bpm remix" by Dj Markitos and "Teenage shutdown" by Electric Frankenstein to name few songs that share some annotations as "Song-Texture_Electric", "Song-Fast_Tempo" and "Song-High_Energy". The fourth *local* co-cluster consists of 45 songs and 7 music annotations. It characterizes a group formed by songs with drums, men on vocals, with electronic and acoustic parts. We identified the music genres like Rock, Pop music, Pop rock, and Alternative rock in this co-cluster. For instance, to name a few songs, this cluster has the "Soul and Fire" by Sebadoh, "Clocks" by Coldplay, "Tubthumping" by Chumbawamba, "Last Goodbye" by Jeff Buckley, "November Rain" by Guns N 'Roses, and "Wonderful Tonight" by Eric Clapton. Thus, it can be seen that OCoClus is useful for finding overlapped and non-overlapped co-clusters that identify the relationship between songs based on the semantic annotations.

Coronavirus Information. Table 3 shows the co-clusters with its attributes and the number of objects. The union of clusters G8 and G9 show the cluster of older people with a total of 1137. Attribute *Senior_3* aggregates people 80 years old or above, and *Senior_2* aggregates people from 70 to 79 years old. Once the *Senior_3* and heart disease attributes appear in G8, they are relevant to form this cluster with 913 people and no other attributes have improved the cost function for it. The G9 cluster has 224 people who died in August, which also happened in the G6 cluster (467 people) during July. These two months mark the peak of the winter season in the region where these people lived.

The clusters G2, G3, G4, G7, and G10 can be seen as a group of people who had at least one main symptom of Covid-19 associated with some comorbidity. The non-overlapped co-clusters are the first 10 groups in Table 3. We notice that the Male attribute is present in two clusters (G1 and G3) regarding the top

Table 3. Description of the clusters in the CV-19 dataset.

Clusters	Attributes	Number of objects
G1	Dyspnea, Cough, Male, Fever	1174
G2	Dyspnea, Other_Comorbidities, Female	1066
G3	Heart_Disease, Dyspnea, Male	1272
G4	Cough, Diabetes	1307
G5	Fever, PORTO ALEGRE - R10, PORTO ALEGRE	631
G6	Other_Comorbidities, Infected_July, Death_July	467
G7	Other_Symptoms, Dyspnea, Female	569
G8	Senior_3, Heart_Disease	913
G9	Senior_2, Death_August, Infected_August	224
G10	Fever, Other_Comorbidities, Male	769
G11	Dyspnea, Heart_Disease, Cough, Fever, Male	539
G12	Dyspnea, Cough, Fever, Diabetes, Male	410
G13	Dyspnea, Cough, Diabetes, Other_Comorbidities, Female	258
G14	Dyspnea, Other_Comorbidities, Other_Symptoms, Female	365
G15	Dyspnea, Cough, Fever, Heart_Disease, Diabetes, PORTO ALEGRE - R10, PORTO ALEGRE, Male	39
G16	Dyspnea, Cough, Fever, Heart_Disease, Diabetes, Male	241
G17	Dyspnea, Cough, Fever, Heart_Disease, Diabetes, Senior_3, Male	41
G18	Dyspnea, Cough, Fever, Heart_Disease, Senior_3, Male	135

four. Cluster G1 identifies 1174 men who experienced the three main symptoms of Covid-19. Meanwhile, the Female attribute is present in one cluster regarding the top four. It identifies 1066 women who presented Dyspnea and other comorbidities as main attributes for this group.

The identified overlapped co-clusters show details that are overlooked in disjoint co-clusters and these co-clusters are the last 8 groups (G11–G18) in Table 3. For instance, cluster G11 identifies a group of 539 men that felt symptoms as dyspnea, fever, and cough and had heart disease problem. In the same way, cluster G12 identifies 410 men with symptoms as in G1, but now it has those with diabetic issues. In comparison, the overlapped cluster G13 identifies a group of 258 women with cough symptoms in combination with diabetes and other comorbidities. Cluster G14 represents another pattern since it identifies 365 women with dyspnea and other symptoms in combination with other comorbidites.

Groups G11 and G12 are examples of *global* overlapped co-clusters, while the groups G15 and G17 are two different examples of *local* co-clusters. Cluster G15 represents a group of 39 men from Porto Alegre region and lived in the capital, where they all felt the main symptoms of covid-19 and who had heart disease and diabetes problems. For group G17, characterizes a group of 41 older men over 80 years of age who had the main symptoms and who had heart disease and diabetes. Regarding the *local* co-clusters, it can be seen that such co-clusters identify detailed patterns that are overlooked by *global* co-clusters. The experiment with the Covid-19 dataset is an example of a real problem related to data analysis complexity. Thus, it can be seen that each co-cluster reveals a meaning pattern according to the information granularity.

6   Conclusion and Future Works

We proposed *OCoClus*, a new non-exhaustive overlapped co-clustering method for binary data, designed for general purpose analysis. OCoClus is based on the detection of co-occurrence of objects and attributes, to identify *global* and *local* co-clusters that overlap. Besides that, when there are no overlapped patterns in the dataset, OCoClus can identify the non-overlapped co-clusters. Furthermore, it is driven by a cost function to automatically identify the number of co-clusters. We performed experiments on synthetic and real data that demonstrates the efficacy and utility of our proposed method.

OCoClus found all embedded co-clusters in the synthetic datasets used as ground-truth, proven by the fact that OCoClus obtained the maximum score in the evaluation metrics. Such a result shows that OCoClus outperformed the limitations of the baseline methods. Nevertheless, we prove the usefulness of our method in two real datasets where we show that OCoClus identified co-clusters that can represent meaningful patterns. We highlight the fact that the obtained results are interesting to propose new specialized systems that use the identified co-clusters as input to decision support systems.

Like any work in the literature, our approach also has space for improvements as future research. First, the number of co-clusters is driven by a cost function regarding the number of objects and attributes. Then, the method tends to find rectangular clusters which may generate patterns with few attributes for big data mining. Second, an interesting research direction is to adapt the method to deal with heterogeneous data. Third, it may be interesting to set the noise thresholds ϵ_I and ϵ_J in a data-driven way Finally, identifying uncorrelated co-clusters in the data matrix is another interesting direction to improve the method.

Acknowledgements. This work has been partially supported by CAPES (Finance code 001), CNPQ, FAPESC (Project Match - co-financing of H2020 Projects - Grant 2018TR 1266), and the European Union's Horizon 2020 research and innovation programme under GA N. 777695 (MASTER). The views and opinions expressed in this paper are the sole responsibility of the author and do not necessarily reflect the views of the European Commission.

References

1. Affeldt, S., Labiod, L., Nadif, M.: Ensemble block co-clustering: a unified framework for text data. In: Proceedings of the 29th ACM International Conference on Information & Knowledge Management, pp. 5–14 (2020)

2. Brunialti, L.F., Peres, S.M., da Silva, V.F., de Moraes Lima, C.A.: The BinOvN-MTF algorithm: overlapping columns co-clustering based on non-negative matrix tri-factorization. In: 2017 Brazilian Conference on Intelligent Systems (BRACIS), pp. 330–335. IEEE, Uberlandia, Brazil (2017)
3. Dhillon, I.S.: Co-clustering documents and words using bipartite spectral graph partitioning. In: Proceedings of the Seventh ACM SIGKDD International Conference on Knowledge Discovery and Data Mining, pp. 269–274. ACM, Association for Computing Machinery, New York (2001)
4. Fu, Q., Banerjee, A.: Bayesian overlapping subspace clustering. In: 2009 Ninth IEEE International Conference on Data Mining, pp. 776–781. IEEE (2009)
5. Hartigan, J.A.: Direct clustering of a data matrix. J. Am. Stat. Assoc. **67**(337), 123–129 (1972)
6. Kluger, Y., Basri, R., Chang, J.T., Gerstein, M.: Spectral biclustering of microarray data: coclustering genes and conditions. Genome Res. **13**(4), 703–716 (2003)
7. Li, G.: Generalized co-clustering analysis via regularized alternating least squares. Comput. Stat. Data Anal. **150**, 106989 (2020)
8. Lucchese, C., Orlando, S., Perego, R.: A generative pattern model for mining binary datasets. In: Proceedings of the 2010 ACM Symposium on Applied Computing, pp. 1109–1110. ACM (2010)
9. Lucchese, C., Orlando, S., Perego, R.: A unifying framework for mining approximate top-k binary patterns. IEEE Trans. Knowl. Data Eng. **26**(12), 2900–2913 (2013)
10. Lutov, A., Khayati, M., Cudré-Mauroux, P.: Accuracy evaluation of overlapping and multi-resolution clustering algorithms on large datasets. In: 2019 IEEE International Conference on Big Data and Smart Computing (BigComp), Kyoto, Japan. IEEE (2019)
11. Madeira, S.C., Oliveira, A.L.: Biclustering algorithms for biological data analysis: a survey. IEEE/ACM Trans. Comput. Biol. Bioinform. **1**(1), 24–45 (2004)
12. Padilha, V.A., Campello, R.J.: A systematic comparative evaluation of biclustering techniques. BMC Bioinform. **18**(1), 1–25 (2017)
13. Pio, G., Ceci, M., D'Elia, D., Loglisci, C., Malerba, D.: A novel biclustering algorithm for the discovery of meaningful biological correlations between micrornas and their target genes. BMC Bioinform. **14**(S7), S8 (2013)
14. Role, F., Morbieu, S., Nadif, M.: CoClust: a python package for co-clustering. J. Stat. Softw. **88**(1), 1–29 (2019)
15. Vlachos, M., Dünner, C., Heckel, R., Vassiliadis, V.G., Parnell, T., Atasu, K.: Addressing interpretability and cold-start in matrix factorization for recommender systems. IEEE Trans. Knowl. Data Eng. **31**(7), 1253–1266 (2018)
16. Whang, J.J., Rai, P., Dhillon, I.S.: Stochastic blockmodel with cluster overlap, relevance selection, and similarity-based smoothing. In: 2013 IEEE 13th International Conference on Data Mining. IEEE (2013)
17. Xu, R., Wunsch, D.: Survey of clustering algorithms. IEEE Trans. Neural Netw. **16**(3), 645–678 (2005)
18. Zheng, et al.: Clustering social audiences in business information networks. Pattern Recognit. **100**, 107126 (2020)
19. Zhu, H., Mateos, G., Giannakis, G.B., Sidiropoulos, N.D., Banerjee, A.: Sparsity-cognizant overlapping co-clustering for behavior inference in social networks. In: 2010 IEEE International Conference on Acoustics, Speech and Signal Processing, pp. 3534–3537. IEEE (2010)

A Comparative Study on Concept Drift Detectors for Regression

Marília Lima[1]([envelope])[iD], Telmo Silva Filho[2][iD],
and Roberta Andrade de A. Fagundes[1][iD]

[1] University of Pernambuco, Pernambuco, Brazil
mncal@ecomp.poli.br, roberta.fagundes@upe.br
[2] Federal University of Paraiba, Paraiba, Brazil
telmo@de.ufpb.br

Abstract. Context: in the field of machine learning models are trained to learn from data, however often the context at which a model is deployed changes, degrading the performances of trained models and giving rise to a problem called Concept Drift (CD), which is a change in data distribution. **Motivation:** CD has attracted attention in machine learning literature, with works proposing modification to well-known algorithms' structures, ensembles, online learning and drift detection, but most of the CD literature regards classification, while regression drift is still poorly explored. **Objective:** The goal of this work is to perform a comparative study of CD detectors in the context of regression. **Results:** we found that (i) PH, KSWIN and EDDM showed higher detection averages; (ii) the base learner has a strong impact in CD detection and (iii) the rate at which CD happens also affects the detection process. **Conclusion:** our experiments were executed in a framework that can easily be extended to include new CD detectors and base learners, allowing future studies to use it.

Keywords: Concept drift detection · Regression · Comparative study

1 Introduction

Machine learning is a sub-field of artificial intelligence which develops models that are able to learn from data. Learning usually happens statically with fixed labels during the process. In the context of regression with continuous dependent variables, many models have been proposed with useful results in many scenarios. However, there has been a growing interest in models that are able to process streams of data and can learn incrementally [10].

While working with data streams, models often face changes in the probability distribution of input data, which can negatively affect performance if the model was trained on data sampled from a significantly different distribution. This problem is most commonly referred to as concept drift (CD). CD can happen in many different scenarios, such as weather forecasting, stock price prediction and fake news identification [17]. CD might happen in the conditional

© Springer Nature Switzerland AG 2021
A. Britto and K. Valdivia Delgado (Eds.): BRACIS 2021, LNAI 13073, pp. 390–405, 2021.
https://doi.org/10.1007/978-3-030-91702-9_26

distribution of the dependent variable given the predictors, which is called real drift, or it can affect only the independent variable distribution, which is called virtual drift [12].

Due to its significance, CD has been widely studied in recent years with approaches involving structural changes applied to well-known models, ensembles and drift detection. However, most of these works targeted the task of classification. In this context, Gonçalves et al. (2014) [9] ran a large comparative study of CD detectors using Naive Bayes as the base learner. Their work showed that there is no single best CD detector and performance depends on the pair (dataset, CD detector), i.e. "there is no free lunch", as expected in machine learning. These results motivate a similar study in the context of regression with continuous targets.

Therefore, this paper's main contribution is to perform a wide comparative study of CD detectors for regression. We used seven different CD detectors together with 10 regression models. The 70 detector-base learner combinations were applied to four synthetic and four real datasets with virtual CD. Thus, our study presents significant variation of base learners, detectors and datasets. Experiments were evaluated according to prediction error and number of detected drift points. As another contribution, we develop our work such that future research can easily expand it by adding new detectors, base learners and datasets. In addition, this study focuses on continuous outputs in the dependent variable, which is still scarcely explored in the literature.

The rest of this paper is organized as follows: Sect. 2 presents the drift detection methods. Section 3 describes the parameters used in the drift detectors, the datasets used in the experiments, and the adopted evaluation methodology. The results obtained in the experiments are analyzed in Sect. 4. Finally, Sect. 5 presents our conclusions.

2   Background

CD detection methods use a base learner (regressor/classifier) to identify if an input is a drift point, i.e. for each instance the method outputs a drift prediction based on the base learner prediction and the observed target value. Most detectors perform this analysis in groups of instances at each time.

Many detectors have been proposed for regression. One of the first was the drift detection method (DDM) [8], which is based on the idea that the base learner's error rate decreases as the number of samples increases, as long as the data distribution is stationary. If DDM finds an increased error rate above a calculated threshold, it detects that CD has happened. The threshold is given by $p_{min} + s_{min}$ is minimum. Where, p_{min} is the minimum recorded error rate and s_{min} is the minimum recorded standard deviation.

DDM was later extended to monitor the average distance between two errors instead of only the error rate. This new method was called early drift detection method (EDDM) [2].

A different approach, called adaptive windowing (ADWIN) [4], is a sliding window algorithm which detects drifts and records updated statistics about data streams. The size of the sliding window is decided based on statistics calculated at different points of the stream. If the difference between the calculated statistics is higher than a predefined threshold, ADWIN detects the CD and discards all data from the stream up to the detected drift point.

Another early CD detector, called Page-Hinkley (PH) [14], computes observed values and their average up until the current point in a data stream. PH then detects the drift if the observed average is higher than a predefined threshold.

Frías-Blanco et al. (2014) [7] proposed two detectors. The first, called HDDM_A, is based on Hoeffding's inequality and used the average of the continuous values in a data stream to determine if the data contain CD. The second method, HDDM_W, uses McDiarmid's bounds and the exponentially weighted moving average (EWMA) statistic to estimate if an instance represents CD or not.

The Kolmogorov-Smirnov Windowing (KSWIN) method [16] uses the Kolmogorov-Smirnov (KS) statistical test to monitor data distributions without assuming any particular distributions. KSWIN keeps a fixed window size and compares the cumulative distribution of the current window to the previous one, detecting CD if the KS test rejects the null hypothesis that the distributions are the same.

3   Experiment Configuration

In this section we describe the datasets and the parametrization used for the drift detectors as well as the evaluation methodology used in the experiments.

3.1   Datasets

The datasets chosen for the experiments have all been previously used to study the concept drift problem. To analyze the performance of the methods, four synthetic datasets and four real-world datasets were used. The datasets allow us to analyze how the detectors identify the points of deviation in the data and false points detected. We use the synthetic data sets from the work of Almeida et al. (2019) [1] following the same methodological process of adding deviation. For each data set 5000 samples are created. The domain of each attribute is divided into ten equal-sized parts. The first 2000 samples correspond to the first seven parts of the domain of each variable. 1000 new instances are added and the domain is expanded until the 5000 samples are completed. Table 1 presents synthetic data.

The real-world datasets used are Bike [12], FCCU1 (gasoline concentration), FCCU2 (LDO concentration) and FCCU3 (LPG concentration) [18]. We performed an analysis on the target variable using the interquartile range method (IQR), $IQR = Q_3 - Q_1$, where $= Q_3$ is the third quartile and $= Q_1$ is the first quartile. We perform a variability estimate to calculate lower $L_{inf} = \bar{y} - 1.5 * IQR$ and higher $L_{sup} = \bar{y} + 1.5 * IQR$ bounds for drift identification, where $\bar{y}$ is the mean of the target variable. At the end of the analysis we

Table 1. Synthetic datasets

Dataset	Function	Domain
3d_Mex.hat	$y = \frac{sin\sqrt{x_1{}^2 + x_2{}^2}}{\sqrt{x_1{}^2 + x_2{}^2}} + \epsilon$	$x_i \sim U[-4\pi, 4\pi]$
Friedman#1	$y = 10sin(\pi x_1 x_2) + 20(x_3 - 0.5)^2 + 10x_4 + 5x_5 + \epsilon$	$x_i \sim U[0, 1]$
Friedman#3	$y = tan^{-1}\frac{x_2 x_3 - \frac{1}{x_2 x_4}}{x_1} + \epsilon$	$x_1 \sim U[0, 1]$ $x_2 \sim U[40\pi, 560\pi]$ $x_3 \sim U[0, 1]$ $x_4 \sim U[0, 11]$
Mult	$y = 0.79 + 1.27x_1 x_2 + 1.56x_1 x_4 + 3.42x_2 x_5 + 2.06x_3 x_4 x_5 + \epsilon$	$x_i \sim U[0, 1]$

noticed that none of the data points were out of bounds. Thus, we artificially added drift along the data. Figure 1 shows the distributions of y and y_{Drift} (y with artificial drift).

To add the deviation we add the value of the dependent variable (y) by multiplying the standard deviation and a number that leaves the value of y greater than L_{sup}. The added drift in the Bike dataset ((a)) has the characteristic of recurring and abrupt speed, as it appears in the data in a certain period of time, returns to the original concept, and appears again, in addition to abruptly changing the concept. We add two offsets by multiplying by the numbers three and four respectively. In the FCCU1, FCCU2, FCCU3 dataset we add three abrupt deviations. Which we multiply: by three, seven and, twelve (For FCCU1); three, six, three (For FCCU2); three, six, nine (For FCCU3).

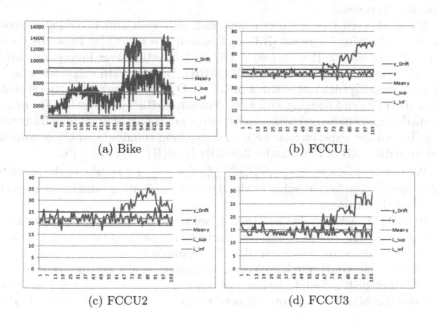

(a) Bike (b) FCCU1

(c) FCCU2 (d) FCCU3

Fig. 1. Datasets

Table 2 presents a summary of the main characteristics of each dataset, such as number of predictive attributes, number of samples and amount of drift.

Table 2. Dataset features

Name	Dataset	# Samples	# Attributes	Drift quantity	Test batch size
3-D Mex.Hat (Mex)	Synthetic	5000	2	3	250
Friedman #1 (Fried1)	Synthetic	5000	5	3	250
Friedman #3 (Fried3)	Synthetic	5000	4	3	250
Multi (Multi)	Synthetic	5000	5	3	250
Bike	Real-word	731	8	2	25
FCCU1	Real-world	104	6	3	13
FCUU2	Real-world	104	6	3	13
FCCU3	Real-world	104	6	3	13

3.2 Evaluation Setup

To evaluate the drift detection methods using the presented datasets, we use a methodology similar to the one described in [1]. For each time step, a specified amount of instances of the training set are read and used to train the regressor. Next, other examples from the test set are used to test the regressor. If the detection method identifies the occurrence of CD, the test data are used for training, which corresponds to the popular train-test-train approach. This procedure is repeated 30 times.

We used seven base learners paired with all tested drift detection methods. They are: Boosting Regressor (BR) [6], Decision Trees (DT) [3], Lasso Regression (LR) [20], Hoeffding Tree Regressor (HTR) [13], Multilayer Perception (MLP) [19], Bagging with meta-estimator BR (B-BR), Bagging with meta-estimator LR (B-LR), Bagging with meta-estimator DT (B-DT), Bagging with meta-estimator HTR (B-HTR) and Bagging with meta-estimator MLP (B-MLP) [5]. The experiments described here used Scikit-multiflow: A multi-output streaming framework [11]. To run the experiments, we used a machine running a 64-bit operational system with 4 GB of RAM and a 2.20 GHz Intel(R) i5-5200U CPU.

We use $RelMAE = \frac{|y\_true - y\_pred|}{mean\_absolute\_error(y\_true, y\_pred)}$ to evaluate the predicted value against the actual value of the data set, indicating whether it correctly classified the instance.

If RelMAE gets a value less than 1, we treated the learner's prediction as correct, otherwise it was considered wrong. This value is passed to a parameterized drift detector, which will flag the example for no drift or true drift. If true drift is identified, the base learner is trained on the instance that has just arrived. If no drift is found, the instance is not used for training. Scikit-multiflow provides the following concept drift detection methods: ADWIN, DDM, EDDM,

HDDM_A, HDDM_W, KSWIN and PageHinkley, which were used with their default settings for the paper. The prequential evaluation method or interleaved test-then-train method strategy is used for the evaluation of the learning models with concept drift and we performed thirty iterations [1]. We chose the Mean Square Error (MSE) for error metric. In addition, we carry out statistical tests to analyze the predictive performance of the learning models in relation to the MSE error. The Friedman test is applied with $\alpha = 0.05$ [9]. We present our statistical analysis using critical difference diagrams.

We initially performed experiments to identify the best parameter settings for each of the base learners. Specifically, we conducted a grid search to choose hyperparameters from the base learners using the initial training data. After identifying the parameters that most influence performance we conducted the actual experiments. All models are from sklearn [15] and scikit-multiflow [11].

4   Experimental Results

In this section we present the results of the experiments with the concept drift detection methods.

4.1   Predictive Accuracy and Drift Identification

We performed a broad base learner study together with the detectors. Therefore, in addition to the detectors, we compare the base learner without updating with test data (called No Partial) and updating whenever new test data arrives, regardless of detection (called Partial). Figure 2 shows the heat map of the mean MSE value of all CD detectors and base learners across the 30 iterations with distinct seeds. Errors are normalized between 0 and 1.

We analyzed the average points of detection for all different scenarios. Figure 3 presents the mean error along the data stream for dataset 3d_Mex.hat. Only the LR, HTR, B-LR and B-HTR base learners showed lower errors as the new test data arrived, even in cases when no CD was detected by ADWIN, DDM, HDDM_A and HDDM_W. A higher number of wrong detections can lead to larger errors, because models are retrained with data from the detected new concept. Additionally, due to the small number of training data of each new concept, base learners that are not able to learn incrementally may show worse performances.

Differently from the previous scenario, Fig. 4 shows that for dataset *Friedman*#1 most models have increasing errors along the data stream. However, base learner HTR together with detector PH have decreased errors beginning with point 2000. This happened due to successful drift point detections followed by base learner updates. Base learners MLP and B-MLP also showed lower errors when paired with the ADWIN and PH detectors.

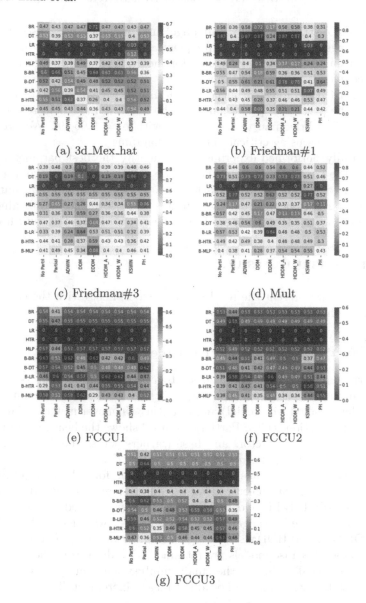

Fig. 2. Heatmap of normalized MSE errors

Figure 5 presents results for dataset *Friedman#*3. Most base learners, except MLP and B-MLP, had lower errors at point 2000. For dataset Mult, Fig. 6 shows variation in the error value, with increasing errors starting at point 2000 in most cases. In such cases, the error increases because it is necessary to retrain the base learners given the new concept. Thus, we can see that the base learners were not updated correctly.

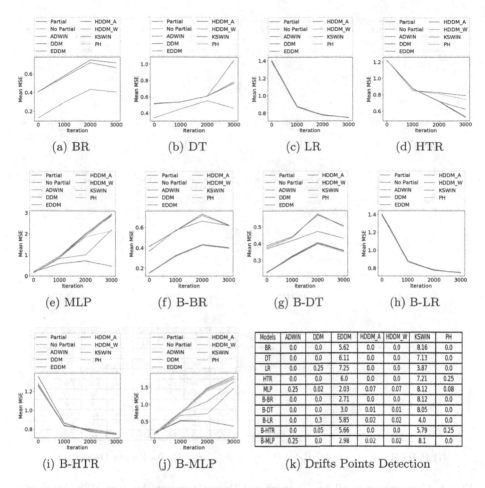

Fig. 3. Comparison of all base learner and Detectors using Average MSE Accuracies in dataset 3d_Mex.hat, and Average Drifts Points Detection.

Regarding synthetic data, the best average detection performances were obtained by EDDM, KSWIN and PH, considering the different base learners. These indicates that in these scenarios where concept drift happens gradually, these methods are able to detect small variations in data distribution.

For the Bike dataset (Fig. 7), only ADWIN and PH were not able to detect any concept drift in the data. The base learners varied significantly according in their error values. Additionally, we observed that all pairs of base learners and detectors increased their errors around point 200, with worse performances coming from ADWIN and PH. These detectors did not adjust well to the concept changes in this dataset.

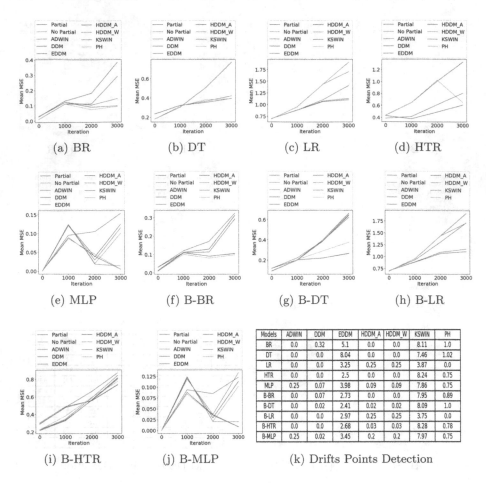

(a) BR   (b) DT   (c) LR   (d) HTR

(e) MLP   (f) B-BR   (g) B-DT   (h) B-LR

(i) B-HTR   (j) B-MLP   (k) Drifts Points Detection

Fig. 4. Comparison of all base learner and Detectors using Average MSE Accuracies in dataset Friedman#1, and Average Drifts Points Detection.

Due to small changes in the average errors for datasets FCCU1, FCCU2 and FCCU3, Fig. 8 only shows the average detection points. Only KSWIN was able to find drift points. This is likely due to lack of enough data to train the base learners, which would allow better precision for the detectors. For FCCU1 and FCCU3, only the Partial model, which always retrains with the new test data, obtained significantly lower errors. For FCCU2, all models behaved similarly, with slightly decreasing errors.

Our results show that, in addition to the base learner and to the detector, the pace at which the drift happens and the number of instances available to update the base learner all contribute significantly to the overall performance.

4.2 Statistical Analysis

We used critical difference diagrams and Nemenyi test to compare all detectors across all 8 datasets for the different base learners. The resulting diagrams can be seen in Fig. 9. This analysis shows that the detectors have statistically similar performances for all base learners.

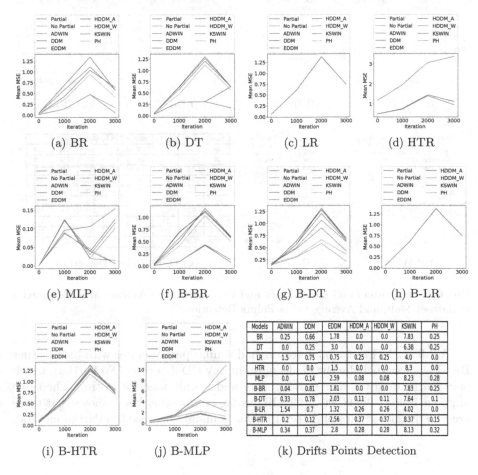

(a) BR   (b) DT   (c) LR   (d) HTR

(e) MLP   (f) B-BR   (g) B-DT   (h) B-LR

(i) B-HTR   (j) B-MLP   (k) Drifts Points Detection

Models	ADWIN	DDM	EDDM	HDDM_A	HDDM_W	KSWIN	PH
BR	0.25	0.66	1.78	0.0	0.0	7.83	0.25
DT	0.0	0.25	3.0	0.0	0.0	6.38	0.25
LR	1.5	0.75	0.75	0.25	0.25	4.0	0.0
HTR	0.0	0.0	1.5	0.0	0.0	8.3	0.0
MLP	0.0	0.14	2.59	0.08	0.08	8.23	0.28
B-BR	0.04	0.81	1.81	0.0	0.0	7.83	0.25
B-DT	0.33	0.78	2.03	0.11	0.11	7.64	0.1
B-LR	1.54	0.7	1.32	0.26	0.26	4.02	0.0
B-HTR	0.2	0.12	2.56	0.37	0.37	8.37	0.15
B-MLP	0.34	0.37	2.8	0.28	0.28	8.13	0.32

Fig. 5. Comparison of all base learner and Detectors using Average MSE Accuracies in datase Friedman#3, and Average Drifts Points Detection.

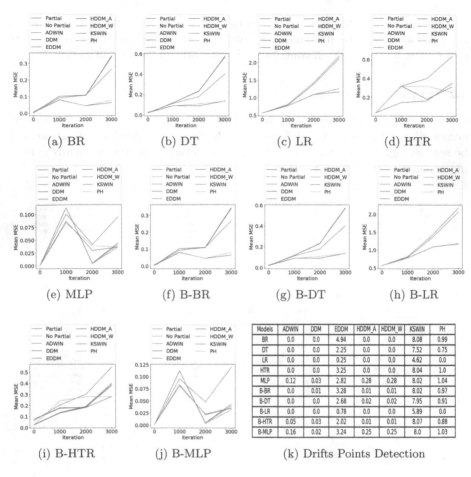

Fig. 6. Comparison of all base learner and Detectors using Average MSE Accuracies in dataset Mult, and Average Drifts Points Detection.

As shown in Fig. 9a, all detectors had similar performances with base learner BR. This even includes the model without any CD detectors (No Partial), which only learns from the first training data and is only used for testing. The same happens in Figs. 9b, 9f, 9i and 9g. Figure 9c, on the other hand, shows that there are two distinct groups of detectors. This is mainly because the Partial method,

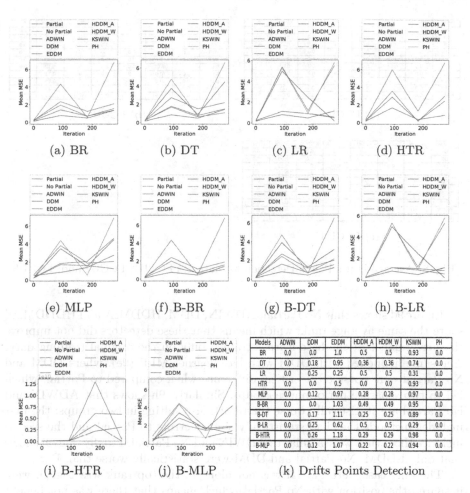

Fig. 7. Comparison of all base learner and Detectors using Average MSE Accuracies in dataset Bike, and Average Drifts Points Detection.

which always updates with every new block of test data in the stream, significantly outperformed every detector, except for EDDM and KSWIN. HDDM_A, HDDM_W, ADWIN, PH and DDM had worse errors than No Partial on average, although these results were not statistically significant. Finally, DDM was statistically worse than Partial, because their distance is larger than the critical difference.

Models	ADWIN	DDM	EDDM	HDDM_A	HDDM_W	KSWIN	PH
BR	0.0	0.0	0.0	0.0	0.0	0.02	0.0
DT	0.0	0.0	0.0	0.0	0.0	0.01	0.0
LR	0.0	0.0	0.0	0.0	0.0	0.25	0.0
HTR	0.0	0.0	0.0	0.0	0.0	0.25	0.0
MLP	0.0	0.0	0.0	0.0	0.0	0.01	0.0
B-BR	0.0	0.02	0.0	0.0	0.0	0.02	0.0
B-DT	0.0	0.0	0.0	0.0	0.0	0.02	0.0
B-LR	0.0	0.0	0.0	0.0	0.0	0.25	0.0
B-HTR	0.0	0.0	0.0	0.0	0.0	0.25	0.0
B-MLP	0.0	0.0	0.0	0.0	0.0	0.01	0.0

(a) FCCU1

Models	ADWIN	DDM	EDDM	HDDM_A	HDDM_W	KSWIN	PH
BR	0.0	0.0	0.0	0.0	0.0	0.1	0.0
DT	0.0	0.02	0.0	0.0	0.0	0.11	0.0
LR	0.0	0.0	0.0	0.0	0.0	0.25	0.0
HTR	0.0	0.0	0.0	0.0	0.0	0.25	0.0
MLP	0.0	0.0	0.0	0.0	0.0	0.08	0.0
B-BR	0.0	0.01	0.0	0.0	0.0	0.12	0.0
B-DT	0.0	0.0	0.0	0.0	0.0	0.21	0.0
B-LR	0.0	0.0	0.0	0.0	0.0	0.25	0.0
B-HTR	0.0	0.0	0.0	0.0	0.0	0.25	0.0
B-MLP	0.0	0.0	0.0	0.0	0.0	0.06	0.0

(b) FCCU2

Models	ADWIN	DDM	EDDM	HDDM_A	HDDM_W	KSWIN	PH
BR	0.0	0.0	0.0	0.0	0.0	0.08	0.0
DT	0.0	0.0	0.0	0.0	0.0	0.09	0.0
LR	0.0	0.0	0.0	0.0	0.0	0.25	0.0
HTR	0.0	0.0	0.0	0.0	0.0	0.25	0.0
MLP	0.0	0.0	0.0	0.0	0.0	0.04	0.0
B-BR	0.0	0.0	0.0	0.0	0.0	0.03	0.0
B-DT	0.0	0.0	0.0	0.0	0.0	0.1	0.0
B-LR	0.0	0.0	0.0	0.0	0.0	0.25	0.0
B-HTR	0.0	0.0	0.0	0.0	0.0	0.25	0.0
B-MLP	0.0	0.0	0.0	0.0	0.0	0.06	0.0

(c) FCCU3

Fig. 8. Average Drifts Points Detection for datasets FCCU1, FCCU2 and, FCCU3

Figure 9d shows that No Partial, ADWIN, DDM, HDDM_A and HDDDM_W share the same average rank, which means that these detectors did not improve the performance of the base learner trained only on the original training data. The same happened with DDM with base learner MLP (9e), where DDM and No Partial were outside the critical difference when compared to Partial, which led to two separate performance groups. Similarly, 9h shows that ADWIN and No Partial are statistically worse than Partial, resulting in two groups: the first includes methods that had statistically similar errors to Partial and the second includes those that were similar to ADWIN and No Partial. Finally, Fig. 9j shows that only EDDM, No Partial and DDM were statistically worse than Partial.

The only case were Partial was not alone in the top rank was Fig. 9i, were it shared the position with No Partial, which means that there was not benefit in retraining the model. It is important to point out that it is actually a good result when a detector is statistically similar to Partial, because it means that, as a result of using the detector, the base learner performed just as well as if it was always retrained, even though it only retrains when the detector identifies a drift.

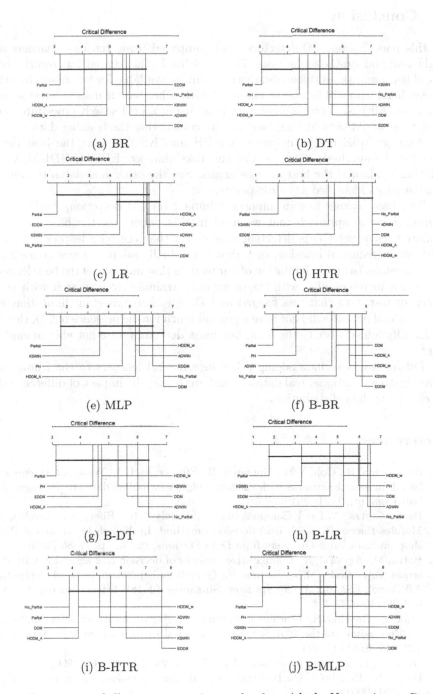

Fig. 9. Comparison of all regressors against each other with the Nemenyi test. Groups of regressors that are not significantly different (at α 0:05) are connected

5   Conclusion

In this paper, seven CD detectors were compared using ten base learners and eight real and synthetic datasets. The resulting Python-based framework[1] used the sklearn and skmultiflow libraries and will be available for future use by other researchers, who will be able to extend it with other base learners, detectors and datasets. All base learners were optimized using a grid search procedure, such that they would present their best results considering the training data.

Average MSE results indicate that PH and KSWIN were the best detectors for the synthetic datasets. For the Bike dataset, EDDM, HDDM_A and HDDM_W showed the best average errors. For the other real datasets, none of the detectors identified any drift points.

The base learner has an important impact on CD detection. Additionally, considering the approach that we used in this paper, i.e. the base learner is updated whenever CD is detected, it is useful that the base learner is able to perform incremental learning, such that it gradually adapts to new concepts.

Therefore, the main limitation of our work is that only one of the base learners can learn incrementally, with the other ones training from scratch with every batch of test data that was flagged as CD. Another important limitation was that our real datasets did not have a ground truth for the presence of CD, thus we artificially added drift to the data, but most detectors were not able to capture that.

Future works include adding other incremental learners to the framework, investigating additional real datasets and analyzing the impact of different rates of change in data distribution.

References

1. de Almeida, R., Goh, Y.M., Monfared, R., Steiner, M.T.A., West, A.: An ensemble based on neural networks with random weights for online data stream regression. Soft Comput., 1–21 (2019)
2. Baena-García, M., del Campo-Ávila, J., Fidalgo, R., Bifet, A., Gavalda, R., Morales-Bueno, R.: Early drift detection method. In: Fourth International Workshop on Knowledge Discovery from Data Streams, vol. 6, pp. 77–86 (2006)
3. Batra, M., Agrawal, R.: Comparative analysis of decision tree algorithms. In: Panigrahi, B.K., Hoda, M.N., Sharma, V., Goel, S. (eds.) Nature Inspired Computing. AISC, vol. 652, pp. 31–36. Springer, Singapore (2018). https://doi.org/10.1007/978-981-10-6747-1_4
4. Bifet, A., Gavalda, R.: Learning from time-changing data with adaptive windowing. In: Proceedings of the 2007 SIAM International Conference on Data Mining, pp. 443–448. SIAM (2007)
5. Breiman, L.: Bagging predictors. Mach. Learn. **24**(2), 123–140 (1996)
6. Duffy, N., Helmbold, D.: Boosting methods for regression. Mach. Learn. **47**(2), 153–200 (2002)

---

[1] https://github.com/Marilia-Lima/Framework-Detectors-Concept-Drift.

7. Frias-Blanco, I., del Campo-Ávila, J., Ramos-Jimenez, G., Morales-Bueno, R., Ortiz-Diaz, A., Caballero-Mota, Y.: Online and non-parametric drift detection methods based on Hoeffding's bounds. IEEE Trans. Knowl. Data Eng. **27**(3), 810–823 (2014)
8. Gama, J., Medas, P., Castillo, G., Rodrigues, P.: Learning with drift detection. In: Bazzan, A.L.C., Labidi, S. (eds.) SBIA 2004. LNCS (LNAI), vol. 3171, pp. 286–295. Springer, Heidelberg (2004). https://doi.org/10.1007/978-3-540-28645-5_29
9. Gonçalves Jr., P.M., de Carvalho Santos, S.G., Barros, R.S., Vieira, D.C.: A comparative study on concept drift detectors. Expert Syst. Appl. **41**(18), 8144–8156 (2014)
10. Mastelini, S.M., de Leon Ferreira de Carvalho, A.C.P.: 2CS: correlation-guided split candidate selection in Hoeffding tree regressors. In: Cerri, R., Prati, R.C. (eds.) BRACIS 2020. LNCS (LNAI), vol. 12320, pp. 337–351. Springer, Cham (2020). https://doi.org/10.1007/978-3-030-61380-8_23
11. Montiel, J., Read, J., Bifet, A., Abdessalem, T.: Scikit-multiflow: a multi-output streaming framework. J. Mach. Learn. Res. **19**(72), 1–5 (2018). http://jmlr.org/papers/v19/18-251.html
12. Oikarinen, E., Tiittanen, H., Henelius, A., Puolamäki, K.: Detecting virtual concept drift of regressors without ground truth values. Data Min. Knowl. Discov. **35**(3), 726–747 (2021). https://doi.org/10.1007/s10618-021-00739-7
13. Osojnik, A., Panov, P., Džeroski, S.: Tree-based methods for online multi-target regression. J. Intell. Inf. Syst. **50**(2), 315–339 (2017). https://doi.org/10.1007/s10844-017-0462-7
14. Page, E.S.: Continuous inspection schemes. Biometrika **41**(1/2), 100–115 (1954)
15. Pedregosa, F., et al.: Scikit-learn: machine learning in Python. J. Mach. Learn. Res. **12**, 2825–2830 (2011)
16. Raab, C., Heusinger, M., Schleif, F.M.: Reactive soft prototype computing for concept drift streams. Neurocomputing **416**, 340–351 (2020)
17. dos Santos, V.M.G., de Mello, R.F., Nogueira, T., Rios, R.A.: Quantifying temporal novelty in social networks using time-varying graphs and concept drift detection. In: Cerri, R., Prati, R.C. (eds.) BRACIS 2020. LNCS (LNAI), vol. 12320, pp. 650–664. Springer, Cham (2020). https://doi.org/10.1007/978-3-030-61380-8_44
18. Soares, S.G., Araújo, R.: An on-line weighted ensemble of regressor models to handle concept drifts. Eng. Appl. Artif. Intell. **37**, 392–406 (2015)
19. Valença, M.: Fundamentos das redes neurais: exemplos em java. Olinda, Pernambuco: Editora Livro Rápido (2010)
20. Xu, H., Caramanis, C., Mannor, S.: Robust regression and lasso. IEEE Trans. Inf. Theory **56**(7), 3561–3574 (2010)

A Kullback-Leibler Divergence-Based Locally Linear Embedding Method: A Novel Parametric Approach for Cluster Analysis

Alexandre L. M. Levada[1]([⊠])(iD) and Michel F. C. Haddad[2,3](iD)

[1] Computing Department, Federal University of São Carlos, São Carlos, SP, Brazil
[2] Department of Land Economy, University of Cambridge, Cambridge, UK
mfch2@cam.ac.uk
[3] School of Business and Management,
Queen Mary University of London, London, UK

Abstract. Numerous problems in machine learning require some type of dimensionality reduction. Unsupervised metric learning deals with the definition of intrinsic and adaptive distance functions of a dataset. Locally linear embedding (LLE) consists of a widely used manifold learning algorithm that applies dimensionality reduction to find a more compact and meaningful representation of the observed data through the capture of the local geometry of the patches. In order to overcome relevant limitations of the LLE approach, we introduce the LLE Kullback-Leibler (LLE-KL) method. Our objective with such a methodological modification is to increase the robustness of the LLE to the presence of noise or outliers in the data. The proposed method employs the KL divergence between patches of the KNN graph instead of the pointwise Euclidean metric. Our empirical results using several real-world datasets indicate that the proposed method delivers a superior clustering allocation compared to state-of-the-art methods of dimensionality reduction-based metric learning.

Keywords: Locally linear embedding · Metric learning · KL divergence

1 Introduction

The main objective of dimensionality reduction unsupervised metric learning is to produce a lower dimensional representation that preserves the intrinsic local geometry of the data. The locally linear embedding (LLE) consists of one of the pioneering algorithms of such a class, which has been successfully applied to non-linear feature extraction for pattern classification tasks [9]. Although more efficient than linear methods, the LLE still has some important limitations.

Firstly, it is rather sensitive to noise or outliers [14]. Secondly, in datasets that are not represented by smooth manifolds, the LLE frequently fails to produce reasonable results. It is worth mentioning that variations of the LLE have

© Springer Nature Switzerland AG 2021
A. Britto and K. Valdivia Delgado (Eds.): BRACIS 2021, LNAI 13073, pp. 406–420, 2021.
https://doi.org/10.1007/978-3-030-91702-9_27

been proposed to overcome such limitations, with varying degrees of success. In Hessian eigenmaps, the local covariance matrix used in the computation of the optimal reconstruction weights is replaced by the Hessian matrix (i.e., the second order derivatives matrix), which encodes relevant curvature information [1,13,15]. A further extension is known as the local tangent space alignment (LTSA).

The difference is that the LTSA builds the locally linear patch with an approximation of the tangent space through the application of the principal component analysis (PCA) over a linear patch. Subsequently, the local representations are aligned to the point in which all tangent spaces become aligned in the unfolded manifold [16]. In addition, the modified LLE (MLLE) introduces several linearly independent weight vectors for each neighborhood of the KNN graph. Some works in the relevant literature describe that the local geometry of MLLE is more stable in comparison with the LLE [17]. One remaining issue still present in such recent developments of the LLE method is the adoption of Euclidean distance as the similarity measure within a linear patch.

In the proposed LLE-KL method, we incorporate the KL divergence into the estimation of the optimal reconstruction weights. The main contribution of the proposed method is that, through the replacement of the pointwise Euclidean distance by a patch-based information-theoretic distance (i.e., KL divergence), the LLE becomes less sensitive to the presence of noise or outliers. Our empirical clustering analysis performed subsequently to the dimensionality reduction for several real-world datasets popularly used in the machine learning literature, indicate that the proposed LLE-KL method is capable of generating more reasonable clusters in terms of silhouette coefficients compared to state-of-the-art manifold learning algorithms, such as the Isomap [12] and UMAP [6].

The remainder of the paper is organized as follows. Section 2 discusses previous relevant work, with focus on the relative entropy and the regular LLE algorithm. Section 3 details the proposed LLE-KL algorithm. Section 4 presents the data used as well as computational experiments and empirical results. Section 5 concludes with our final remarks and suggestions for future research.

2   Related Work

In this section, we discuss the relative entropy (KL divergence) between probability density functions as a similarity measure among random variables, exploring its computation in the Gaussian case. Moreover, we discuss the LLE algorithm and detail its mathematical derivation.

2.1   Kullback-Leibler Divergence

In machine learning applications, the problem of quantifying similarity levels between different objects or clusters consists of a challenging task, especially in cases where the standard Euclidean distance is not a reasonable choice. Many studies on feature selection adopt statistical divergences to select the set of features that should maximize some measure of separation between classes. Part of their success is due to the fact that most dissimilarity measures are related to distance metrics. In such a context, information theory provides a solid mathematical background for metric learning in pattern classification. The entropy of a continuous random vector x is given by:

$$H(p) = - \int p(x)[log\ p(x)]dx = -E\ [log\ p(x)] \tag{1}$$

where $p(x)$ is the probability density function (pdf). In a similar fashion, we may define the cross-entropy between two probability density functions $p(x)$ and $q(x)$:

$$H(p,q) = - \int p(x)[log\ q(y)]dx \tag{2}$$

The KL divergence is the difference between cross-entropy and entropy [4]:

$$D_{KL}(p,q) = H(p,q) - H(p) = \int p(x)log\left(\frac{p(x)}{q(x)}\right)dx = E_p\left[log\left(\frac{p(x)}{q(x)}\right)\right] \tag{3}$$

An important property is that the relative entropy is non-negative, therefore, $D_{KL}(p,q) \geq 0$.

2.2   The LLE Algorithm

The LLE consists of a local method, thus, the new coordinates of any $\vec{x}_i \in R^m$ depends only on the neighborhood of the respective point. The main hypothesis behind the LLE is that for a sufficiently high density of samples, it is expected that a vector $\vec{x}_i$ and its neighbors define a linear patch (i.e., they all belong to an Euclidean subspace [9]). Hence, it is possible to characterize the local geometry by linear coefficients:

$$\hat{\vec{x}}_i \approx \sum_j w_{ij}\vec{x}_j \qquad \text{for} \qquad \vec{x}_j \in N(\vec{x}_i) \tag{4}$$

consequently, we can reconstruct a vector as a linear combination of its neighbors.

The LLE algorithm requires as input an $n \times m$ data matrix X, with rows $\vec{x}_i$, a defined number of dimensions $d < m$, and an integer $k > d+1$ for finding local neighborhoods. The output is an $n \times d$ matrix Y, with rows $\vec{y}_i$. The LLE algorithm may be divided into three main steps [9,10], as follows:

1. From each $\vec{x}_i \in R^m$ find its k nearest neighbors;
2. Find the weight matrix W that minimizes the reconstruction error for each data point $\vec{x}_i \in R^m$;
3. Find the coordinates Y which minimize the reconstruction error using the optimum weights.

In the following subsections, we describe how to obtain the solution to each step of the LLE algorithm.

Finding Local Linear Neighborhoods. A relevant aspect of the LLE is that this algorithm is capable of recovering embeddings which intrinsic dimensionality d is smaller than the number of neighbors, k. Moreover, the assumption of a linear patch imposes the existence of an upper bound on k. For instance, in highly curved datasets, it is not reasonable to have a large k, otherwise such an assumption would be violated. In the uncommon situation where $k > m$, it has been shown that each sample may be perfectly be reconstructed from its neighbors, from which another problem emerges: the reconstruction weights are not anymore unique.

To overcome this limitation, some regularization is necessary in order to break the degeneracy [10]. Lastly, a further concern in the LLE algorithm refers to the connectivity of the KNN graph. In the case the graph contains multiple connected components, then the LLE should be applied separately on each one of them, otherwise the neighborhood selection process should be modified to assure global connectivity [10].

Least-Squares Estimation of the Weights. The second step of the LLE is to reconstruct each data point from its nearest neighbors. The optimal reconstruction weights may be computed in closed form. Without loss of generality, we may express the local reconstruction error at point $\vec{x}_i$ as follows:

$$E(\vec{w}) = \left\| \sum_j w_j (\vec{x}_i - \vec{x}_j) \right\|^2 = \sum_j \sum_k w_j w_k (\vec{x}_i - \vec{x}_j)^T (\vec{x}_i - \vec{x}_k) \quad (5)$$

Defining the local covariance matrix C as:

$$C_{jk} = (\vec{x}_i - \vec{x}_j)^T (\vec{x}_i - \vec{x}_k) \quad (6)$$

we then have the following expression for the local reconstruction error:

$$E(\vec{w}) = \sum_j \sum_k w_j C_{jk} w_k = \vec{w}^T C \vec{w} \quad (7)$$

Regarding the constraint $\sum_j w_j = 1$, that may be interpreted in two different manners, namely geometrically and probabilistically. From a geometric point of view, it provides invariance under translation, thus, by adding a given constant vector $\vec{c}$ to $\vec{x}_i$ and all of its neighbors, the reconstruction error remains unchanged. In terms of probability, enforcing the weights to sum to one leads W to become a stochastic transition matrix [10] directly related to Markov chains and diffusion maps [11]. As detailed below, in the minimization of the squared error the solution is found through an eigenvalue problem. In fact, the estimation of the matrix W reduces to n eigenvalue problems. Considering that there are no constraints across the rows of W, we may then find the optimal weights for each sample $\vec{x}_i$ separately, drastically simplifying the respective computations. Therefore, we have n independent constrained optimization problems given by:

$$\arg\min_{\vec{w}_i} \ \vec{w}_i^T C_i \vec{w}_i \quad \text{s.t.} \quad \vec{1}^T \vec{w}_i = 1 \quad (8)$$

for $i = 1, 2, ..., n$. Using Lagrange multipliers, we express the Lagrangian function as follows:

$$L(\vec{w}_i, \lambda) = \vec{w}_i^T C_i \vec{w}_i - \lambda(\vec{1}^T \vec{w}_i - 1) \tag{9}$$

Taking the derivatives with relation to $\vec{w}_i$ as follows:

$$\frac{\partial}{\partial \vec{w}_i} L(\vec{w}_i, \lambda) = 2C_i \vec{w}_i - \lambda \vec{1} = 0 \tag{10}$$

leads to

$$C_i \vec{w}_i = \frac{\lambda}{2} \vec{1} \tag{11}$$

Which is equivalent to solving the following linear system:

$$C_i \vec{w}_i = \vec{1} \tag{12}$$

and then normalizing the solution to guarantee that $\sum_j w_i(j) = 1$ by dividing each coefficient of the vector $\vec{w}_i$ by the sum of all the coefficients:

$$w_i(j) = \frac{w_i(j)}{\sum_j w_i(j)} \quad \text{for} \quad j = 1, 2, ..., m \tag{13}$$

In the case that k (i.e., number of neighbors) is greater than m (i.e., number of features) then, in general, the space spanned by k distinct vectors consists of the whole space. This means that $\vec{x}_i$ may be expressed exactly as a linear combination of its k-nearest neighbors. In fact, if $k > m$, there are generally infinitely many solutions to $\vec{x}_i = \sum_j w_j \vec{x}_j$, due to the fact that there would be more unknowns (k) than equations (m). In such a case, the optimization problem is ill-posed and, thus, regularization is required. A common regularization technique refers to the Tikonov regularization, which adds a penalization term to the least squares problem, as follows:

$$\left\| \vec{x}_i - \sum_j w_j \vec{x}_j \right\|^2 + \alpha \sum_j w_j^2 \tag{14}$$

where α controls the degree of regularization. In other words, it selects the weights which minimize a combination of reconstruction error as well as the sum of the squared weights. In the case that $\alpha \to 0$, then there is a least-squares problem. However, in the opposite limit (i.e., $\alpha \to \infty$), the squared-error term becomes negligible, allowing the minimization of the Euclidean norm of the weight vector $\vec{w}$. Typically, α is set to be a small but non-zero value. In this case, the n independent constrained optimization problems are:

$$\arg\min_{\vec{w}_i} \vec{w}_i^T C_i \vec{w}_i + \alpha \vec{w}_i^T \vec{w}_i \quad \text{s.t.} \quad \vec{1}^T \vec{w}_i = 1 \tag{15}$$

for $i = 1, 2, ..., n$. The Lagrangian function is defined by:

$$L(\vec{w}_i, \lambda) = \vec{w}_i^T C_i \vec{w}_i + \alpha \vec{w}_i^T \vec{w}_i - \lambda(\vec{1}^T \vec{w}_i - 1) \tag{16}$$

Taking the derivative with respect to $\vec{w}_i$ and setting the result to zero:

$$2C_i \vec{w}_i + 2\alpha \vec{w}_i = \lambda \vec{1} \tag{17}$$

$$(C_i + \alpha I)\vec{w}_i = \frac{\lambda}{2} \vec{1} \tag{18}$$

$$\vec{w}_i = \frac{\lambda}{2}(C_i + \alpha I)^{-1} \vec{1} \tag{19}$$

where λ is selected to properly normalize $\vec{w}_i$, which is equivalent to solving the following linear system:

$$(C_i + \alpha I)\vec{w}_i = \vec{1} \tag{20}$$

and then normalize the solution. In other words, to regularize the problem it is necessary to add a small perturbation into the main diagonal of the matrix C_i. In all experiments reported in the present paper, we use the regularization parameter $\alpha = 10^{-4}$.

Finding the Coordinates. The main idea in the last stage of the LLE algorithm is to use the optimal reconstruction weights estimated by least-squares as the proper weights on the manifold and then solve for the local manifold coordinates. Thus, fixing the weight matrix W, the goal is to solve another quadratic minimization problem to minimize the following:

$$\Phi(Y) = \sum_{i=1}^{n} \left\| \vec{y}_i - \sum_j w_{ij} \vec{y}_j \right\|^2 \tag{21}$$

Hence, the following question needs to be addressed: what are the coordinates $\vec{y}_i \in R^d$ (approximately on the manifold), that such weights (W) reconstruct? In order to avoid degeneracy, we have to impose two constraints, as follows:

1. The mean of the data in the transformed space is zero, otherwise we would have an infinite number of solutions;
2. The covariance matrix of the transformed data is the identity matrix, therefore, there is no correlation between the components of $\vec{y} \in R^d$ (this is a statistical constraint to assess that the output space is the Euclidean one, defined by an orthogonal basis).

However, unlikely the estimation of the weights W, finding the coordinates does not simplify into n independent problems, because each row of Y appears in Φ multiple times, as the central vector y_i and also as one of the neighbors of other vectors. Thus, firstly, we rewrite equation (21) in a more meaningful manner using matrices, as follows:

$$\Phi(Y) = \sum_{i=1}^{n} \left(\vec{y}_i - \sum_j w_{ij} \vec{y}_j \right)^T \left(\vec{y}_i - \sum_j w_{ij} \vec{y}_j \right) \tag{22}$$

Applying the distributive law and expanding the summation, we then obtain:

$$\Phi(Y) = \sum_{i=1}^{n} \vec{y}_i^T \vec{y}_i - \sum_{i=1}^{n} \sum_j \vec{y}_i^T w_{ij} \vec{y}_j$$

$$- \sum_{i=1}^{n} \sum_j \vec{y}_j^T w_{ji} \vec{y}_i + \sum_{i=1}^{n} \sum_j \sum_k \vec{y}_j^T w_{ji} w_{ik} \vec{y}_k \tag{23}$$

Denoting by Y the $d \times n$ matrix in which each column $\vec{y}_i$ for $i = 1, 2, ..., n$ stores the coordinates of the i-th sample in the manifold and acknowledging that $\vec{w}_i(j) = 0$, unless $\vec{y}_j$ is one of the neighbors of $\vec{y}_i$, we may express $\Phi(Y)$ as follows:

$$\Phi(Y) = Tr(Y^T Y) - Tr(Y^T W Y)$$
$$- Tr(Y^T W^T Y) + Tr(Y^T W^T W Y)$$
$$= Tr(Y^T Y) - Tr(Y^T (WY))$$
$$- Tr((WY)^T Y) + Tr((WY)^T (WY))$$
$$= Tr(Y^T (Y - WY) - (WY)^T (Y - WY))$$
$$= Tr((Y - WY)^T (Y - WY))$$
$$= Tr(((I - W)Y)^T ((I - W)Y))$$
$$= Tr(Y^T (I - W)^T (I - W)Y) \tag{24}$$

Defining the $n \times n$ matrix M as:

$$M = (I - W)^T (I - W) \tag{25}$$

we get the following optimization problem:

$$\underset{Y}{\arg\min} \ Tr(Y^T M Y) \quad \text{subject to} \quad \frac{1}{n} Y^T Y = I \tag{26}$$

Thus, the Lagrangian function is given by:

$$L(Y, \lambda) = Tr(Y^T M Y) - \lambda \left(\frac{1}{n} Y^T Y - I \right) \tag{27}$$

Differentiating and setting the result to zero, finally leads to:

$$MY = \beta Y \tag{28}$$

where $\beta = \frac{\lambda}{n}$, showing that the Y must be composed by the eigenvectors of the matrix M. Since there is a minimization problem to be solved, it is then required to select Y to compose the d eigenvectors associated to the d smallest eigenvalues. Note that M being an $n \times n$ matrix, it contains n eigenvalues and n orthogonal eigenvectors. Although the eigenvalues are real and non-negative values, its smallest value is always zero, with the constant eigenvector $\vec{1}$. Such a bottom eigenvector corresponds to the mean of Y and should be discarded to impose the constraint that $\sum_{i=1}^{n} \vec{y}_i = 0$ [7]. Therefore, to get $\vec{y}_i \in R^d$, where $d < m$, it is necessary to select the $d + 1$ smallest eigenvectors and discard the constant eigenvector with zero eigenvalue. In other words, one must select the d eigenvectors associated to the bottom non-zero eigenvalues.

3   The KL Divergence-Based LLE Method

The main motivation to introduce the LLE-KL method is to find a surrogate for the local matrix C_i for each sample of the dataset. Recall that, originally, $C_i(j, k)$ is computed as the inner product between $\vec{x}_i - \vec{x}_j$ and $\vec{x}_i - \vec{x}_k$, which means that we employ the Euclidean geometry in the estimation of the optimal reconstruction weights. In the definition of such matrix it is used a non-linear distance function, namely the relative entropy between Gaussian densities estimated within different patches of the KNN graph. Our inspiration is the parametric PCA, an information-theoretic extension of the PCA method that applies the KL divergence to compute a surrogate for the covariance matrix (i.e., entropic covariance matrix) [5].

Let $X = \{\vec{x}_1, \vec{x}_2, \ldots, \vec{x}_n\}$, with $\vec{x}_i \in R^m$, be our data matrix. The first step in the proposed method consists of building the KNN graph from X. At this early stage, we employ the extrinsic Euclidean distance to compute the nearest neighbors of each sample $\vec{x}_i$. Denoting by η_i the neighborhood system of $\vec{x}_i$, a patch P_i is defined as the set $\{\vec{x}_i \cup \eta_i\}$. It is worth noticing that the number of elements of P_i is $K + 1$, for $i = 1, 2, \ldots, n$. In other words, a patch P_i is given by an $m \times (k+1)$ matrix:

$$P_i = \begin{bmatrix} x_i(1) & x_{i1}(1) & \ldots & x_{ik}(1) \\ x_i(2) & x_{i1}(2) & \ldots & x_{ik}(2) \\ \vdots & \vdots & \ddots & \vdots \\ \vdots & \vdots & \ldots & \vdots \\ x_i(m) & x_{i1}(m) & \ldots & x_{ik}(m) \end{bmatrix} \tag{29}$$

The rationale of the proposed method is considering each row of the matrix P_i as a sample of a univariate Gaussian random variable, and then estimating the parameters μ (mean) and σ^2 (variance) of each row, leading to an m-dimensional vector of tuples, as follows:

$$\vec{p}_i = [(\mu_i(1), \sigma_i^2(1)), ..., (\mu_i(m), \sigma_i^2(m))] \tag{30}$$

Let $p(x)$ and $q(x)$ be univariate Gaussian densities, $N(\mu_1, \sigma_1^2)$ and $N(\mu_2, \sigma_2^2)$, respectively. Then, the relative entropy becomes:

$$D_{KL}(p, q) = log\left(\frac{\sigma_2}{\sigma_1}\right) + \frac{1}{2\sigma_2^2} E_p[(x - \mu_2)^2] - \frac{1}{2\sigma_1^2} E_p[(x - \mu_1)^2] \tag{31}$$

By the definition of central moments, we then have:

$$E_p[(x - \mu_1)^2] = \sigma_1^2 \tag{32}$$

$$E_p[(x - \mu_2)^2] = E[x^2] - 2E[x]\mu_2 + \mu_2^2 \tag{33}$$

$$E[x^2] = Var[x] + E^2[x] = \sigma_1^2 + \mu_1^2 \tag{34}$$

which leads to the following closed-form equation:

$$D_{KL}(p, q) = log\left(\frac{\sigma_2}{\sigma_1}\right) + \frac{\sigma_1^2 + (\mu_1 - \mu_2)^2}{2\sigma_2^2} - \frac{1}{2} \tag{35}$$

As the relative entropy is not symmetric, it is possible to compute its symmetrized counterpart as follows:

$$D_{KL}^{sym}(p, q) = \frac{1}{2}[D_{KL}(p, q) + D_{KL}(q, p)] \tag{36}$$

$$= \frac{1}{4}\left[\frac{\sigma_1^2 + (\mu_1 - \mu_2)^2}{\sigma_2^2} + \frac{\sigma_2^2 + (\mu_1 - \mu_2)^2}{\sigma_1^2} - 2\right]$$

$$= \frac{1}{4\sigma_1^2 \sigma_2^2}\left[\left(\sigma_1^2 - \sigma_2^2\right)^2 + (\mu_1 - \mu_2)^2 \left(\sigma_1^2 + \sigma_2^2\right)\right]$$

Let $\vec{d}_{ij}$ be the m-dimensional vector of symmetrized relative entropies between the components of $\vec{p}_i$ and $\vec{p}_j$, computed by the direct application of equation (36) in each pair of parameter vectors. Then, in the proposed method, the entropic matrix C_i is computed as follows:

$$C_i = \vec{d}_{ij}^T \vec{d}_{ik} \tag{37}$$

where the column vector $\vec{d}_{ij}$ involves the patches $\vec{p}_i$ and $\vec{p}_j$, while the row vector $\vec{d}_{ik}^T$ involves the patches $\vec{p}_i$ and $\vec{p}_k$. It is worth noticing that unlike the established LLE method - which employs the pairwise Euclidean distance - the proposed LLE-KL method employs a patch-based distance (relative entropy) instead, becoming less sensitive to the presence of noise and outliers in the observed data.

The remaining steps of the algorithm are precisely the same as in the established LLE. The set of logical steps comprising the LLE-KL method is detailed in Algorithm 1.

Algorithm 1. KL divergence based LLE

1: **function** LLE-KL(X, K, d)
2:     From input data $X_{m \times n}$, build a KNN graph.
3:     **for** $\vec{x}_i \in X^T$ **do**
4:         Compute the $K \times K$ matrix C_i as:

$$C_i = \vec{d}_{ij}^T \vec{d}_{ik} \tag{38}$$

5:         Solve $C_i \vec{w}_i = \vec{1}$ to estimate $\vec{w}_i \in R^K$.
6:             Normalize the weights in $\vec{w}_i$ so that $\sum_j \vec{w}_i(j) = 1$.
7:     **end for**
8:     Construct the $n \times n$ matrix W using $\vec{w}_i$'s.
9:     Compute $M = (I - W)^T (I - W)$.
10:     Find the eigenvalues and eigenvectors of M.
11:     Select the bottom d non-zero eigenvectors of M and define the matrix Y, where each column is an eigenvector.
12:     **return** Y
13: **end function**

In the subsequent sections, we present some computational experiments to compare the performance of the proposed method against several manifold learning algorithms.

4   Data, Experiments and Results

In the present section is presented the data used, experiments performed, and a discussion on the respective empirical results.

4.1   Data

In the present study, a total of 25 datasets are used as input data. Such datasets are widely used in machine learning estimations and applications. The datasets used in the present study were collected from openML.org, which contains detailed information regarding the number of instances, features, and classes for each one of datasets.

4.2   Experiments and Results

In order to test and evaluate the proposed method, a series of computational experiments is performed to quantitatively compare clustering results obtained subsequently to the dimensionality reduction into 2-D spaces using the silhouette coefficient (SC) - i.e., a measure of goodness-of-fit into a low-dimensional representation [8]. All results are reported in Table 1. It is worth noticing that in 23 out of 25 datasets (i.e., 92% of the cases), the LLE-KL method obtains the best performance in terms of the SC.

Regarding the means and medians, one may realize that the proposed method performs better in comparison with the established LLE as well as two of its variations (i.e., Hessian LLE and LTSA). In addition, the proposed method also achieves a superior performance compared to the state-of-the-art algorithm UMAP. It is worth noticing that, commonly, the UMAP requires a reasonably large sample size to perform well, which is not always the case in the empirical analysis of the present paper.

To check if the results obtained by the proposed method are statistically superior to the competing methods, we perform a Friedman test - i.e. non-parametric test for paired data in case of more than two groups [2]. For a significant level $\alpha = 0.01$, we conclude that there is strong evidence to reject the null hypothesis that all groups are identical ($p = 1.12 \times 10^{-11}$). In order to analyze which groups are significantly different, we then perform a Nemenyi post-hoc test [3]. According to this test, there is strong evidence that the LLE-KL method produces significantly higher SCs compared to the PCA ($p < 10^{-3}$), Isomap ($p < 10^{-3}$), LLE ($p < 10^{-3}$), Hessian LLE ($p < 10^{-3}$), LTSA ($p < 10^{-3}$), and UMAP ($p < 10^{-3}$).

The proposed method also has some limitations. One caveat of the LLE-KL - which is common to other manifold learning algorithms, consists of the out-of-sample problem. It is not clear how to properly evaluate new samples that are not part of the training set. Another drawback refers to the definition of the parameter K (i.e., number of neighbors) that controls the patch size. Previous computational experiments reveal that the SC is rather sensitive to changes in such a parameter. Our strategy in this case is then as follows: for each dataset, we build the KKN graphs for all values of K in the interval [2, 40].

We select the best model as the one that maximizes the SC among all values of K. Although we are using the class labels to perform model selection, in fact, the LLE-KL method performs unsupervised metric learning, in the sense that we do not employ the class labels in the dimensionality reduction. A data visualization comparison of the clusters obtained through the application of the LLE and LLE-KL methods for two distinct datasets (i.e., tic-tac-toe and corral) are depicted in Figs. 1 and 2. It is worth noticing that the discrimination between the two classes is more evident in the proposed method than in the established version of the LLE, since there is visibly less overlap between the blue and red cluster.

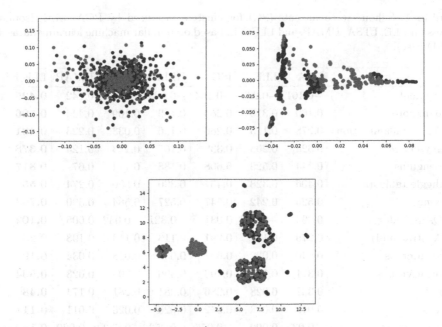

Fig. 1. Comparison between clusters generated subsequently to the application of the LLE, LLE-KL (with the number of neighbors set as K = 9), and UMAP for the tic-tac-toe dataset.

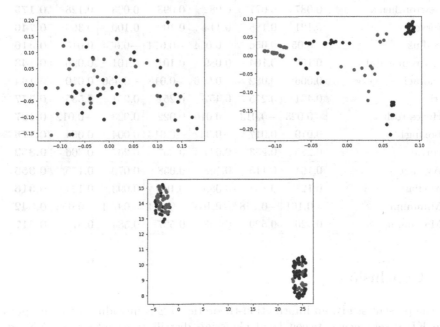

Fig. 2. Comparison between clusters generated subsequently to the application of the LLE, LLE-KL (with the number of neighbors set as K = 14), and UMAP for the corral dataset.

Table 1. Silhouette coefficients (SC) for clusters produced by PCA, LLE, Isomap, Hessian LLE, LTSA, UMAP, and LLE-KL, based on popular machine learning datasets (2-D case).

	PCA	LLE	ISO	HLLE	LTSA	UMAP	LLE-KL
segment	−0.161	−0.248	−0.164	−0.623	−0.068	0.142	**0.425**
banknote	0.160	0.315	0.284	0.109	0.109	0.444	**0.596**
Acute-inflammations	0.278	0.113	0.266	0.156	0.033	0.223	**0.491**
geyser1	0.328	0.205	0.339	0.165	0.173	0.124	**0.378**
penguins	0.344	0.529	0.608	0.188	0.114	0.67	**0.817**
diggle_table_a2	0.406	0.328	0.45	0.390	0.390	0.274	**0.55**
wine	0.526	0.242	0.547	0.527	0.584	0.590	**0.763**
Qsar-biodeg	0.094	−0.062	0.031	−0.321	−0.032	0.005	**0.102**
Australian(4)	0.279	0.13	0.291	0.113	0.113	0.193	**0.23**
prnn_crabs	0.040	0.022	0.037	0.037	0.038	0.034	**0.19**
prnn_viruses	0.371	0.112	0.496	0.396	0.396	0.023	**0.594**
bolts	0.337	0.028	0.286	0.281	0.281	0.174	**0.48**
kidney	0.019	0.018	0.042	0.026	0.026	0.011	**0.111**
attendence	−0.034	0.000	−0.076	−0.053	−0.053	−0.020	**0.182**
parity5	−0.062	−0.051	−0.048	−0.043	−0.044	−0.052	**0.042**
wildcat	0.151	0.149	0.125	−0.020	−0.021	0.026	**0.166**
newton_hema	0.087	0.077	0.082	0.093	0.093	0.128	**0.179**
servo	0.121	0.185	0.114	0.100	0.100	0.366	**0.746**
collins	−0.05	0.056	−0.054	−0.674	−0.674	0.015	**0.316**
environmental	0.105	0.104	0.089	0.101	0.101	0.096	**0.143**
thoracic_surgery	0.006	0.082	−0.006	−0.018	−0.450	0.010	**0.194**
iris	0.401	0.275	0.452	0.251	0.251	0.535	**0.577**
Hayes-roth	−0.023	−0.013	−0.010	0.022	0.023	−0.012	**0.107**
boxing1	0.019	0.017	−0.03	−0.013	0.004	0.064	**0.176**
corral	0.285	0.253	0.041	0.267	0.267	0.306	**0.332**
Average	0.161	0.115	0.168	0.058	0.070	0.175	**0.355**
Median	0.121	0.104	0.089	0.100	0.093	0.124	**0.316**
Minimum	−0.161	−0.248	−0.164	−0.674	−0.674	−0.052	**0.042**
Maximum	0.526	0.529	0.608	0.527	0.584	0.67	**0.817**

5　Conclusion

In the present study, an information-theoretic LLE is introduced to incorporate the KL divergence between local Gaussian distributions into the KNN adjacency graph. The rationale of the proposed method is to replace the pointwise Euclidean distance by a more appropriate patch-based distance. Such a methodological modification results in a more robust method against the presence of noise or outliers in the data.

Our claim is that the proposed LLE-KL is a promising alternative to the existing manifold learning algorithms due to the proposition of its theoretical methodological improvements and considering our computational experiments that support two main points. Firstly, the quality of the clusters produced by the LLE-KL method indicates a superior performance than those obtained by competing state-of-the-art manifold learning algorithms. Secondly, the LLE-KL non-linear features may be more discriminative in supervised classification than features obtained by state-of-the-art manifold learning algorithms.

Suggestions of future work include the use of other information-theoretic distances, such as the Bhattacharyya and Hellinger distances, as well as geodesic distances based on the Fisher information matrix. Another possibility is the non-parametric estimation of the local densities using kernel density estimation techniques (KDE). In this case, non-parametric versions of the information-theoretic distances may be employed to compute a distance function between the patches of the KNN graph. The ϵ-neighborhood rule may also be used for building the adjacency relations that define the discrete approximation for the manifold, leading to non-regular graphs. Furthermore, a supervised dimensionality reduction based metric learning approach may be created by removing the edges of the KNN graph for which the endpoints belong to different classes as a manner to impose that the optimal reconstruction weights use only the neighbors that belong to the same class within the sample.

Acknowledgments. This study was partially financed the Coordenação de Aperfeiçoamento de Pessoal de Nível Superior - Brasil (CAPES) - Finance Code 001.

References

1. Donoho, D.L., Grimes, C.: Hessian eigenmaps: locally linear embedding techniques for high-dimensional data. Proc. Natl. Acad. Sci. **100**(10), 5591–5596 (2003)
2. Friedman, M.: The use of ranks to avoid the assumption of normality implicit in the analysis of variance. J. Am. Stat. Assoc. **32**(200), 675–701 (1937)
3. Hollander, M., Wolfe, D.A., Chicken, E.: Nonparametric Statistical Methods, 3 edn. Wiley, New York (2015)
4. Kullback, S., Leibler, R.A.: On information and sufficiency. Ann. Math. Stat. **22**(1), 79–86 (1951)
5. Levada, A.L.M.: Parametric PCA for unsupervised metric learning. Pattern Recognit. Lett. **135**, 425–430 (2020)
6. McInnes, L., Healy, J., Melville, J.: UMAP: uniform manifold approximation and projection for dimension reduction. arXiv 1802.03426 (2020)
7. de Ridder, D., Duin, R.P.: Locally linear embedding for classification. Technical report, Delft University of Technology (2002)
8. Rousseeuw, P.J.: Silhouettes: a graphical aid to the interpretation and validation of cluster analysis. J. Comput. Appl. Math. **20**, 53–65 (1987)
9. Roweis, S., Saul, L.: Nonlinear dimensionality reduction by locally linear embedding. Science **290**, 2323–2326 (2000)

10. Saul, L., Roweis, S.: Think globally, fit locally: unsupervised learning of low dimensional manifolds. J. Mach. Learn. Res. **4**, 119–155 (2003)
11. Talmon, R., Cohen, I., Gannot, S., Coifman, R.R.: Diffusion maps for signal processing: a deeper look at manifold-learning techniques based on kernels and graphs. IEEE Signal Process. Mag. **30**(4), 75–86 (2013)
12. Tenenbaum, J.B., de Silva, V., Langford, J.C.: A global geometric framework for nonlinear dimensionality reduction. Science **290**, 2319–2323 (2000)
13. Wang, J.: Hessian locally linear embedding. In: Wang, J. (ed.) Geometric Structure of High-Dimensional Data and Dimensionality Reduction, pp. 249–265. Springer, Heidelberg (2012). https://doi.org/10.1007/978-3-642-27497-8_13
14. Wang, J., Wong, R.K., Lee, T.C.: Locally linear embedding with additive noise. Pattern Recognit. Lett. **123**, 47–52 (2019)
15. Xing, X., Du, S., Wang, K.: Robust hessian locally linear embedding techniques for high-dimensional data. Algorithms **9**, 36 (2016)
16. Zhang, Z.Y., Zha, H.Y.: Principal manifolds and nonlinear dimensionality reduction via tangent space aligment. SIAM J. Sci. Comput. **26**(1), 313–338 (2004)
17. Zhang, Z., Wang, J.: MLLE: modified locally linear embedding using multiple weights. In: Schölkopf, B., Platt, J.C., Hofmann, T. (eds.) Advances in Neural Information Processing Systems 19, Proceedings of the 20th Conference on Neural Information Processing Systems, Vancouver, British Columbia, pp. 1593–1600 (2006)

Classifying Potentially Unbounded Hierarchical Data Streams with Incremental Gaussian Naive Bayes

Eduardo Tieppo[1,2]([✉]) [iD], Jean Paul Barddal[1] [iD], and Júlio Cesar Nievola[1] [iD]

[1] Pós-Graduação em Informática (PPGIa), Pontifícia Universidade Católica do Paraná (PUCPR), Curitiba, Brazil
{eduardo.tieppo,jean.barddal,nievola}@ppgia.pucpr.br
[2] Instituto Federal do Paraná (IFPR), Pinhais, Brazil
eduardo.tieppo@ifpr.edu.br

Abstract. Hierarchical data stream classification inherits the properties and constraints of hierarchical classification and data stream classification concomitantly. Therefore, it requires novel approaches that (i) can handle class hierarchies, (ii) can be updated over time, and (iii) are computationally light-weighted regarding processing time and memory usage. In this study, we propose the *Gaussian Naive Bayes for Hierarchical Data Streams* (GNB-hDS) method: an incremental Gaussian Naive Bayes for classifying potentially unbounded hierarchical data streams. GNB-hDS uses statistical summaries of the data stream instead of storing actual instances. These statistical summaries allow more efficient data storage, keep constant computational time and memory, and calculate the probability of an instance belonging to a specific class via the Bayes' Theorem. We compare our method against a technique that stores raw instances, and results show that our method obtains equivalent prediction rates while being significantly faster.

Keywords: Hierarchical classification · Data stream classification · Gaussian Naive Bayes · Incremental learning

1 Introduction

Hierarchical classification is required on problems where instances are labeled with classes that are related to one another in a hierarchy, such as in recognition of music genres and subgenres [12], computer-aided diagnosis where diseases are categorized by their etiology [43], recognition of animals, which are organized in a taxonomy [28,42], and, recently, even helping in COVID-19 identification using the hierarchical etiology of pneumonia [29].

However, classification techniques often assume that data samples of a particular problem are static and fully available to a learning model in a well-defined

Supported by the Coordenação de Aperfeiçoamento de Pessoal de Nível Superior - Brasil (CAPES) - Finance Code 001.

© Springer Nature Switzerland AG 2021
A. Britto and K. Valdivia Delgado (Eds.): BRACIS 2021, LNAI 13073, pp. 421–436, 2021.
https://doi.org/10.1007/978-3-030-91702-9_28

training step [26]. This assumption does not reflect many of the real-world scenarios in which classification is applied. The ever-increasing volume of data from
diverse sources such as the Internet, wireless sensors, mobile devices, or social
networks produces massive large-scale data streams [23, 27, 32].

Data streams are potentially unbounded over time and hence cannot be
stored in memory. Also, as the time component is intrinsic in data streams,
these are expected to be transient, i.e., the underlying data distribution is ever-
changing, thus resulting in variations in the target concept, a phenomenon named
concept drift [10, 19, 21, 39].

When merged, hierarchical classification and data stream classification areas
combine their properties and introduce new challenges in a roughly unexplored
area: the hierarchical classification of data streams. Consequently, novel algorithms for hierarchical data stream classification must: (i) handle class hierarchies, (ii) be updatable over time, (iii) detect and adapt to changes in data
behavior, and (iv) be computationally light-weighted regarding processing time
and memory consumption [10, 19, 31].

In this study, we propose the *GNB-hDS* method: an Incremental Gaussian
Naive Bayes for classifying potentially unbounded hierarchical data streams.
GNB-hDS uses statistical summaries of the data stream instead of storing raw
instances.

Despite the relevant application of Bayesian classifiers in hierarchical and
data stream classification tasks separately, they have not been adapted yet to
their intersection task. Therefore, to the best of our knowledge, this is the first
method that combines incremental Bayesian learning with hierarchical classification. These statistical summaries allow a more efficient data storage, holding
constant computational time and memory usage, and permit the calculation of
the probability of a given instance belonging to a specific class via the Bayes'
Theorem.

The novel contributions of this work are as follows:

- We qualify Gaussian Naive Bayes, a well-known classification technique [11],
  to work with potentially unbounded hierarchical data streams and in an incremental fashion by using updatable statistical summaries related to a class
  hierarchy.
- We propose *GNB-hDS*, a method for the hierarchical classification of data
  streams using summarization techniques. The model is incremental and handles potentially unbounded data streams with constant memory usage.

Furthermore, as a byproduct of this research, we make the source code for
the proposed method, as well as the datasets used in the experiments, available
for reproducibility.

The remainder of this paper is organized as follows. Section 2 describes the
problem of hierarchical classification of data streams and Sect. 3 brings forward related works. Section 4 describes the proposed incremental Gaussian Naive
Bayes for the hierarchical classification of data streams. Section 5 comprises the
experimental protocol and the discussion of the results obtained. Finally, Sect. 6
concludes this paper and states envisioned future works.

2   Problem Statement

As mentioned above, in this paper, we are particularly interested in hierarchical data stream classification. This specific task combines characteristics and challenges from two different areas, and thus, it differs from classical classification in two key aspects.

First, concerning hierarchical classification, instances of a problem are assigned to a label path that belongs to a hierarchically structured set of classes instead of one single independent label [35]. Figure 1 compares a general approach of (a) flat (non-hierarchical) classification, and (b) hierarchical classification in an illustrative problem. In flat classification, the decision must be made while considering all the classes of the problem (all the possible song genres). Meanwhile, hierarchical classification concerns an existing class taxonomy, which can be used to make first smaller and generic decisions about the problem (in the example, decide first between Rock and R&B genres), and then more specific ones.

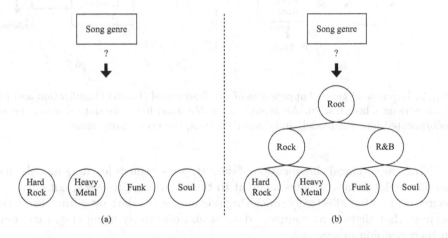

Fig. 1. Example of general approaches of (a) flat and (b) hierarchical classification in a hypothetical music genre problem. The class taxonomy can be used to lead smaller specific decisions about the classes by splitting the context complexity.

Second, concerning data stream classification, there is not the concept of a complete and fully available dataset; instead, instances of a problem are provided to the model sequentially over time [19]. Figure 2 compares (a) a traditional classification process and (b) a data stream classification process. In traditional (or batch) classification, the dataset is assumed to be static and completely available to the model at the training step. Next, the dataset is divided into training and test subsets; the training data is submitted to the learning model that reviews them as many times as necessary until obtaining a single satisfactory test model. This final model is then applied to the subset of test data and provides predictions and, consequently, accuracy estimates.

In contrast, in streaming scenarios, data is made available sequentially over time, and even a single instance can be provided to the model at a time. Each instance is tested by the model, resulting in a prediction, and only after that it is incorporated into the model (being used as training data). This process, entitled 'test-then-train', is repeated for each instance, or chunk of instances, that is gathered from the stream. Any processed instance needs to be eventually discarded to maintain the model stable to process new instances since the data stream is potentially unbounded.

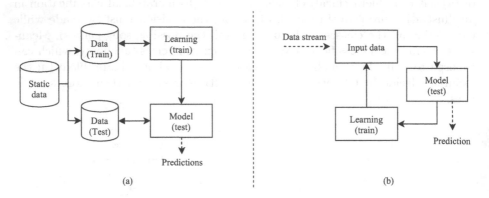

(a)   (b)

Fig. 2. Example of general approaches of (a) Traditional (batch) classification and (b) Data Stream Classification. An input data is obtained from the data stream, tested, incorporated into the model, and discarded; then, the cycle starts again.

Thus, hierarchical classification of data streams regards learning models that use hierarchical data streams as input to their learning processes, not only as a source of data but effectively processing portions of the data over time, using the premise that there is no complete dataset and effectively using class taxonomy in their decision processes.

More formally, we let hDS define a hierarchical data stream in the $[(\vec{x}^t, \vec{y}^t)]|_{t=0}^{\infty}$ format providing instances $(\vec{x}^t, \vec{y}^t)$ on a specific timestamp t, where $\vec{x}^t$ represents a d-dimensional features set and its values, and $\vec{y}^t$ represents the corresponding ground-truth label path (hierarchically structured classes).

These hierarchically structured classes compose a regular concept hierarchy arranged on a partially ordered set $(Y, \succ)$, where Y is a finite set containing all label paths and the relationship $\succ$ is defined as an asymmetric, anti-reflexive, and transitive subsumption (is-a) relation [35]. Finally, the classification of hierarchical data streams can be formally defined as $f^t : \vec{x}^t \mapsto \vec{y}^t$, where the function f^t is continuously updated by mapping features $\vec{x}$ to the corresponding label paths $\vec{y}^t$ accurately.

As the data streams are potentially infinite due to their time component, learning models are restrained by finite computational resources and must work with bounded memory and time, analyzing each instance only once according to their arrival and then discarding it. The processing time of an incoming instance from the data stream must not surpass the ratio in which new instances become available. Otherwise, the learning model will need to discard new instances without analyzing them [5, 10].

3   Related Work

Machine learning models based on the Bayes' Theorem have been widely used in classification since their outputs are human-readable, they can naturally handle missing values, and are relatively easy to implement [22, 36].

In hierarchical classification, Bayesian classifiers were used with different levels of adaptation. The authors in [14] used Bayesian probabilities attached to each node in the hierarchy using a Local Classifier per Node approach [35] and a top-down strategy to analyze the binary predictions along with the hierarchical structure. Similarly, the authors in [13] used binary classifiers for each class in the hierarchy considering both the parent and child classes of the current class.

In the works of [7, 45], the authors also used Bayes-based classifiers within a Local Classifier per Node approach but to perform hierarchical multilabel classification. Finally, the authors in [36] proposed a Naive Bayes fitted to the hierarchical classification using a global approach [35].

A Bayesian classifier fitted to handle hierarchical classification needs to be adapted, at least, to consider the relationship between the hierarchically structured classes in the calculation of probabilities [36].

In data stream classification, incremental adaptations of Bayesian classifiers have been widely studied and are also widely applied in state-of-the-art algorithms. Data stream classification can be handled by a Naive Bayes classifier in a straightforward manner, since the learning model only needs to incrementally store summaries of data that allow the probabilities calculations as new instances are provided from the data stream [25].

The authors in [3] introduced the idea of recalculating probabilities for each instance provided to a model and this idea was later reinforced by the authors in [2, 25]. In the work [33], the authors proposed an incremental Bayes Tree based on statistical summaries of data which are updated with each incoming instance. The authors in [9] used Naive Bayes classifiers ensembled with other tree-based classifiers to improve specific leaf node predictions. Finally, the authors in [4] also used incremental statistical summaries to restrain a Naive Bayes classifier and cope with limited computational resources.

It is noteworthy, nonetheless, to highlight the work of [28], where the authors proposed an incremental k-Nearest Neighbors (kNN) [1] approach for the hierarchical classification of data streams. This can be considered a seminal work of the area, yet, it does depict drawbacks such as kNN relies on distance computations, which are computationally intensive and can put in jeopardy time and memory usage constraints required by streaming scenarios [27, 38]. In this sense, in Sect. 5, we compare our proposal (GNB-hDS) against the one proposed in [28] and show that GNB-hDS uses Bayes probabilities to obtain competitive prediction correctness with better computational performance.

4   Proposed Method

Our proposal, hereafter referred to as Gaussian Naive Bayes for Hierarchical Data Streams (*GNB-hDS*), is an incremental method for the hierarchical classification of data streams based on the Naive Bayes technique [11,18].

The main idea behind *GNB-hDS* is the use of incremental data summaries, specifically the mean, standard deviation, and the number of data instances, that allow the calculation of probabilities used in the Bayes' Theorem [11,22]. These incremental data summaries are attached to nodes of the hierarchy and are updated as new instances are gathered from the data stream. We implemented two key adaptations in the traditional Naive Bayes classifier to make it handle hierarchical data streams:

- Regarding the hierarchical data structure, the original algorithm was modified to consider not only one class but all related classes of a given instance. As the hierarchical data structure represents a subsumption relation, any new instance provided from the data stream also belongs to its ancestors. Thus, we traverse the hierarchy to update all data summaries of parent nodes recursively until the root node of the hierarchy.
- Regarding the streaming input data, the algorithm must store incremental statistical descriptors instead of the actual instances. Thus, we need to compute the mean, the standard deviation, and the count of data instances assigned to each class incrementally, discarding the instance after it is analyzed.

Regarding the stated problem approach, *GNB-hDS* represents the class taxonomy in a tree structure using local classifiers at each parent node and assigns leaf node classes as the last class of one predicted label path $\vec{y}^t$ (mandatory leaf-node and single path prediction) [35].

We point out that although the GNB-hDS method has been implemented here in a more specific way regarding the stated problem, GNB-hDS also supports direct acyclic graphs and non-mandatory leaf node prediction in its concept. To that, the data structure of a given node in the hierarchy should allow links with more than one parent node, and the top-down strategy used in the prediction step must consider some stopping criteria (e.g., a probability threshold) resulting in partial depth label paths.

Figure 3 illustrates the process performed by *GNB-hDS*. Circles represent classes, and dashed squares enclose classifiers. The method represents the class taxonomy in a tree structure, where R stands for the root node of the hierarchy and classes are related with each other (as described in Sect. 2).

When receiving an incoming instance for prediction, the method tackles the hierarchy using a Local Classifier per Parent Node (LCPN) approach [35], thus analyzing the current parent node and predicting between its child nodes by using probabilities obtained with the Bayes' Theorem. This process is repeated until a leaf node is reached.

Each node in the tree stores the count of instances (n), a d-dimensional incremental mean $(\bar{x}_n)$ and a d-dimensional incremental standard deviation (σ_n) of the class represented (as shown in class 2). After the incoming instance processing, the statistical descriptors $(n, \bar{x}_n \sigma_n)$ are updated incrementally with the instance feature values on all the classes through the hierarchy regarding the ground-truth label path of that instance.

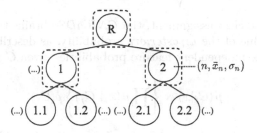

Fig. 3. Illustration of GNB-hDS method.

As before introduced, the instances are represented by data summaries comprising three statistical descriptors stored incrementally: (i) the count of class instances, (ii) the d-dimensional mean instance, and (iii) the d-dimensional standard deviation of the instances of a given class.

The number of instances assigned to a class C is stored in an attached counter. When an instance is retrieved from the stream, the C-th class counter is incremented alongside the counters of C's ancestors.

The incremental mean ($\bar{x}_n$) and the incremental standard deviation (σ_n) considering each attribute from a d−dimensional x_n instance are obtained, respectively, from Eqs. 1 and 2, where n stands for the number of instances observed so far assigned to C [15, 40].

Also, it is important to reinforce that the incremental mean and the incremental standard deviation are d-dimensional descriptors as the features set and its values from the d-dimensional x_n instance. Note that Eqs. 1 and 2 support only continuous feature sets and the current mean and the standard deviation of the previous observed instances assigned to C are represented by $\bar{x}_{n-1}$ and σ_{n-1}.

$$\bar{x}_n = \frac{(\bar{x}_{n-1}(n-1)) + x_n)}{n} \tag{1}$$

$$\sigma_n = \sqrt{\frac{(n-2)\sigma_{n-1}^2 + (n-1)(\bar{x}_{n-1} - \bar{x}_n)^2 + (x_n - \bar{x}_n)^2}{n-1}} \tag{2}$$

The prediction of the class to be assigned to an incoming instance provided from the data stream is performed in three steps: (i) computation of the *a priori* probabilities based on the count of class instances, (ii) computation of likelihood probabilities based on the Bayes' Theorem for each attribute of the incoming instance, and (iii) calculation of the maximum value of the *a posteriori* probability from the product of the independent feature probabilities given a class C.

The calculation of the likelihood probability is described in Eq. 3, where i represents a feature index and j a class index [11].

$$p\left(x_i \mid C_j\right) = \frac{1}{\sqrt{2\pi\sigma_{i,j}^2}} exp\left\{-\frac{1}{2}\left(\frac{x_i - \bar{x}_{i,j}}{\sigma_{i,j}}\right)^2\right\} \tag{3}$$

To perform the class assignment, the *GNB-hDS* obtains the class label with the maximum value of the *a posteriori* probability, as described in Eq. 4, from the product of the independent feature probabilities given C [11].

$$p\left(C_j \mid x\right) \propto \left\{\prod_i p\left(x_i \mid C_j\right)\right\} p(C_j) \tag{4}$$

Moreover, these three steps are performed from the top of the hierarchy data structure and repeated until a leaf node is reached, resulting in the union of the class assignments made from Eq. 4 and representing the final label path assigned to the incoming instance.

Algorithm 1 shows the proposed Gaussian Naive Bayes for Hierarchical Data Streams (*GNB-hDS*). It receives a hierarchical data stream hDS supplying instances $(\vec{x}, \vec{y})$ over time and, if required, outputs a set of predicted labels (a label path) $\widehat{\vec{y}_i}$ for each given instance $(\vec{x}, \vec{y})$, where $\vec{x}$ represents a d-dimensional features set and its values, and $\vec{y}$ represents the corresponding ground-truth label path of that instance.

The algorithm starts by understanding and representing the class taxonomy from the hierarchical data stream. The first loop (line 2 onwards) receives an

---

Algorithm 1:
GNB-hDS - Gaussian Naive Bayes for Hierarchical Data Streams

---

input   : a hierarchical data stream hDS providing instances $(\vec{x}, \vec{y})$
output: a predicted label path $\widehat{\vec{y}_i}$ for the input instance

---

1 Tree ← classTaxonomy(hDS);
2 **foreach** ($\vec{x} \in hDS$) **do**
3 | predictedNode ← Tree.root;
4 | **while** $\neg(predictedNode.isLeaf)$ **do**
5 | | **foreach** ($childNode \in predictedNode.children$) **do**
6 | | | priors ← priorProbability(childNode.Class);
7 | | **end**
8 | | likelihood ← likelihoodProbability($\vec{x}$,priors);
9 | | posterior ← posteriorProbability(likelihood,priors);
10 | | predictedNode ← argmax(posterior);
11 | | $\widehat{\vec{y}_i} \leftarrow \widehat{\vec{y}_i} \cup \{predictedNode.label\}$;
12 | **end**
13 | UpdateStatisticalDescriptors($\vec{y}_i$);
14 **end**

incoming instance from the hierarchical data stream. The following loop (lines 4–12) handles the hierarchy using the LCPN approach by predicting one of the children labels possible for that parent node.

The *a priori* probabilities are calculated in line 6 using the counts of class instances. The likelihood and posterior probabilities are calculated in lines 8 and 9 by the application of Eqs. 3 and 4, respectively. The predicted node for the evaluated parent is obtained in line 10, and the respective single label is appended to a partial label path $\widehat{y_i}$ (line 11). This process is repeated until a leaf node is reached and the label path $\widehat{y_i}$ is complete and ready to be output by the algorithm.

Finally, the algorithm updates the statistical descriptors (the count n of class instances, the incremental mean instance $\bar{x}_n$, and the incremental standard deviation σ_n) of all classes contained in $\widehat{y_i}$, from the leaf to the root class.

5   Analysis

In this section, we report the experimental analysis conducted to compare our proposal against existing works in hierarchical data stream classification. First, we provide the experimental protocol adopted. Next, we discuss the results in terms of prediction and performance.

5.1   Experimental Protocol

Table 1 depicts the 14 hierarchically labeled datasets used in our testbed, listing their number of instances, features, and classes, the number of labels per level in the hierarchy (from top-level to leaf level), and references. These datasets contain different features, instances, and domains, thus assessing how our proposal behaves in different scenarios.

Table 1. Datasets used in the experiment.

Dataset	Instances	Features	Classes	Labels per level	Reference
Entomology	21,722	33	14	4, 6, 9, 10	[28]
Ichthyology	22,444	15	15	2, 6, 12, 15	[28]
Insects-a-b	52,848	33	6	1, 1, 2, 6	[37]
Insects-a-i	355,275	33	6	1, 1, 2, 6	[37]
Insects-i-a-r-b	79,986	33	6	1, 1, 2, 6	[37]
Insects-i-a-r-i	452,044	33	6	1, 1, 2, 6	[37]
Insects-i-b	57,018	33	6	1, 1, 2, 6	[37]
Insects-i-g-b	24,15	33	6	1, 1, 2, 6	[37]
Insects-i-g-i	143,323	33	6	1, 1, 2, 6	[37]
Insects-i-i	452,044	33	6	1, 1, 2, 6	[37]
Insects-i-r-b	79,986	33	6	1, 1, 2, 6	[37]
Insects-i-r-i	452,044	33	6	1, 1, 2, 6	[37]
Insects-o-o-c	905,145	33	24	4, 10, 14, 24	[37]
Instruments	9,419	30	31	5, 10, 31	[28]

During the experiments, classifiers were assessed in terms of hierarchical F-measure [24]. Like traditional classification metrics, the hierarchical F-Measure (hF) relies on hierarchical precision and recall components, but instances are associated with a path of labels, and the entire path is evaluated.

The hierarchical F-Measure is depicted in Eq. 5, while its precision (hP) and recall (hR) components are described in Eqs. 6 and 7, respectively. In both precision and recall metrics, $\widehat{\vec{y}_i}$ is the set of labels predicted for the i-th instance, and $\vec{y}_i$ is its corresponding ground-truth label set.

$$hF = \frac{2 \times hP \times hR}{hP + hR} \qquad (5)$$

$$hP = \frac{\sum_i \left| \widehat{\vec{y}_i} \cap \vec{y}_i \right|}{\sum_i \left| \widehat{\vec{y}_i} \right|} \qquad (6)$$

$$hR = \frac{\sum_i \left| \widehat{\vec{y}_i} \cap \vec{y}_i \right|}{\sum_i \left| \vec{y}_i \right|} \qquad (7)$$

We report the hF metric using the prequential test-then-train [10,20] validation method, where each instance is used to test the model before it is used for training and updating [10,21].

Furthermore, we measured the time performance by calculating the number of instances that a classifier can process per second.

We compared our proposed *GNB-hDS* to the hierarchical kNN described in Sect. 3 proposed in [28], hereafter referred to as *kNN-hDS*. We set up *kNN-hDS* with $k \in \{1, 3, 5\}$ and n (buffer size) $\in \{5, 10, 15, 20\}$. The method *GNB-hDS* does not require setting parameters.

Finally, the results obtained by both methods were assessed using Wilcoxon hypothesis tests [41] with a 95% confidence level according to the protocol provided in [16] to verify significant differences in the hF and instances processed per second rates obtained by both methods.

The experiments in this paper were performed using Python 3.7. The proposed script containing the *GNB-hDS* method, as well as the datasets, are freely available for download[1].

5.2   Results

Table 2 shows the Hierarchical F-measure (hF) and Instances per second rates obtained by *kNN-hDS* and *GNB-hDS* in the datasets (greater values are highlighted in bold). In the *kNN-hDS method*, rates represent the best hF results obtained in the parameters configuration (as described in Sect. 5.1).

In terms of predictive performance assessment, the *GNB-hDS* method obtained better hF rates in 10 out of the 14 datasets. However, hF values are

---

[1] http://www.ppgia.pucpr.br/~jean.barddal/datasets/GNB-hDS.zip.

Table 2. Hierarchical F-measure (hF) and Instances per second rates obtained during experiments.

Dataset	hF (%)		Instances per second	
	kNN-hDS	GNB-hDS	kNN-hDS	GNB-hDS
Entomology	**51.51**	48.64	127	**379**
Ichthyology	40.55	**46.82**	157	**395**
Insects-a-b	80.95	**81.11**	151	**489**
Insects-a-i	79.14	**80.88**	153	**494**
Insects-i-a-r-b	79.49	**81.42**	153	**495**
Insects-i-a-r-i	78.52	**81.57**	153	**491**
Insects-i-b	79.78	**80.55**	148	**500**
Insects-i-g-b	**83.29**	81.53	158	**483**
Insects-i-g-i	78.94	**80.40**	154	**495**
Insects-i-i	78.63	**80.90**	152	**497**
Insects-i-r-b	**80.14**	78.57	153	**491**
Insects-i-r-i	78.60	**81.61**	153	**494**
Insects-o-o-c	55.24	**64.14**	75	**282**
Instruments	**65.42**	48.31	79	**262**
Average	72.16	**72.60**	140.43	**446.21**

similar across both methods, such that the average difference between them is 0.44% while favoring *GNB-hDS*. Despite the improvements, the Wilcoxon test showed no statistical difference between hF rates obtained by the methods (p-$value = 0.2209$).

Concerning processing speed comparison, the *GNB-hDS* method was able to process more instances per second across all datasets, with an average rate of 446.21 instances against 140.43 of the *kNN-hDS* method. Thus, on average, our method was able to process 3.2 times more instances than the *kNN-hDS* method.

A one-tailed Wilcoxon test indicated a statistical difference between instances per second rates obtained by both methods (p-$value = 0.0005$) and confirmed that *GNB-hDS* is significantly faster when compared to *kNN-hDS* method.

Considering predictive performance and processing speed rates, *GNB-hDS* can obtain computational performance improvements without significant threats to the predictive performance by using statistical summaries of data combined with the class hierarchy information.

As aforementioned, the *GNB-hDS* method uses the premise of a Gaussian (normal) data distribution to deal with instance representation in the learning model [30]. In this sense, in addition to the previously described analysis, we investigated if *GNB-hDS* could use its premise to obtain better hF rates when data is normally distributed.

Thus, the *GNB-hDS* method, in addition to its speed, would present an additional advantage to the *kNN-hDS* method (or to any other method that does not use the premise of data normality) since it would be more adapted to classify normally distributed data. This advantage can be even more noticeable when we consider the data stream context, where data are potentially unbounded and statistical descriptors, such as mean and standard deviation, are more likely to obtain better representations of the population.

To examine this claim, we perform a Shapiro-Wilk test of normality [34] in all the datasets and applied the Yeo-Johnson power transformation technique [44] in all data to improve their adherence to a more normal distribution. The data normality was measured by Shapiro-Wilk test before and after the application of the Yeo-Johnson transformation.

Table 3 depicts the Shapiro-Wilk W statistic before (raw data) and after (transformed data) the Yeo-Johnson transformation. W statistic is bounded by 1, and closer values to this upper bound represent data more fitted to a normal distribution.

In addition, Table 3 depicts the Hierarchical F-measure (hF) obtained by both GNB-hDS and kNN-hDS methods when applied in both raw and transformed datasets. One can note that the predictive performance of GNB-hDS was improved when using transformed data in 13 out 14 datasets. Likewise, the average hF increased 2.35%, with noticeable increases in some datasets, such as Entomology (5.23%) and Insects-o-o-c (5.32%). Oppositely, kNN-hDS could not achieve the same improvements. In fact, kNN-hDS obtained lower hF rates with the transformed data resulting in a decrease of 1.6% in the average hF from 72.16% to 70.56%.

Table 3. Shapiro-Wilk W statistic of datasets and Hierarchical F-measure (hF) obtained by GNB-hDS with raw and transformed data.

Dataset	Shapiro-Wilk W statistic		hF (%) obtained by GNB-hDS		hF (%) obtained by Local kNN-hDS	
	Raw data	Transformed data	Raw data	Transformed data	Raw data	Transformed data
Entomology	0.7489	0.9517	48.64	**53.87**	**51.51**	51.07
Ichthyology	0.9028	0.9839	46.82	**50.27**	**40.55**	35.86
Insects-a-b	0.7236	0.9240	81.11	**81.90**	**80.95**	78.78
Insects-a-i	0.7248	0.9272	80.88	**84.05**	**79.14**	76.96
Insects-i-a-r-b	0.7268	0.9273	81.42	**83.48**	**79.49**	78.48
Insects-i-a-r-i	0.7234	0.9269	81.57	**83.40**	**78.52**	76.60
Insects-i-b	0.7239	0.9239	80.55	**82.28**	**79.78**	77.90
Insects-i-g-b	0.7280	0.9273	**81.53**	81.42	**83.29**	81.66
Insects-i-g-i	0.7227	0.9288	80.40	**83.16**	**78.94**	77.02
Insects-i-i	0.7234	0.9269	80.90	**83.05**	**78.63**	76.58
Insects-i-r-b	0.7250	0.9252	78.57	**79.58**	**80.14**	79.03
Insects-i-r-i	0.7234	0.9269	81.61	**83.45**	**78.60**	76.60
Insects-o-o-c	0.7416	0.9468	64.14	**69.46**	**55.24**	55.03
Instruments	0.9689	0.9868	48.31	**49.93**	65.42	**66.25**
Average	0.7577	0.9381	72.60	**74.95**	**72.16**	70.56

Finally, we performed one-tailed Wilcoxon tests to verify if the results obtained with the transformed datasets are significantly higher than with raw data for both *GNB-hDS* and *kNN-hDS* methods.

On *kNN-hDS*, the test indicated a statistical difference between performances with both data (*p-value* = 0.0009) favoring raw data, i.e., *kNN-hDS* does not benefit from a more normal distributed data. In contrast, on *GNB-hDS* the test indicated a statistical difference between performances with both data (*p-value* = 0.0006) favoring the transformed data and has confirmed that *GNB-hDS* can take advantage of a more normal distribution-like data, thus corroborating our claims.

6 Conclusion

In this paper, we proposed *GNB-hDS*, an algorithm for hierarchical classification of data streams using data summaries to represent data. Our proposal is incremental and handles potentially unbounded data streams with constant memory consumption. Consequently, the proposed method processes more instances per second without dreadful impacts in prediction rates when compared to existing kNN-based techniques. To the best of our knowledge, our method extends the state-of-the-art being the first incremental method based on Bayes' Theorem tailored for hierarchical data streams classification.

The resulting source code and all the datasets used in the experiments are freely available for download to be used as a baseline to further research on the hierarchical classification of data streams, such as data preprocessing, computational resources analysis, and concept drift detection and adaptation.

In future works, we are interested in designing and applying other data summaries and different window types to maintain more than one *a priori* probabilities per class to allow *a posteriori* probabilities calculation weighted by data newness. Also, we are interested in applying existing drift detectors [6,8,17] to increase the responsiveness to changes in the data distribution.

References

1. Aha, D.W., Kibler, D., Albert, M.K.: Instance-based learning algorithms. Mach. Learn. **6**(1), 37–66 (1991)
2. Alcobé, J.: Incremental learning of tree augmented Naive Bayes classifiers. In: Garijo, F.J., Riquelme, J.C., Toro, M. (eds.) IBERAMIA 2002. LNCS (LNAI), vol. 2527, pp. 32–41. Springer, Heidelberg (2002). https://doi.org/10.1007/3-540-36131-6_4
3. Anderson, J.R., Matessa, M.: Explorations of an incremental, Bayesian algorithm for categorization. Mach. Learn. **9**(4), 275–308 (1992)
4. Bahri, M., Maniu, S., Bifet, A.: A sketch-based Naive Bayes algorithms for evolving data streams. In: 2018 IEEE International Conference on Big Data (Big Data), pp. 604–613. IEEE (2018)

5. Barddal, J.P., Gomes, H.M., Enembreck, F., Pfahringer, B., Bifet, A.: On dynamic feature weighting for feature drifting data streams. In: Frasconi, P., Landwehr, N., Manco, G., Vreeken, J. (eds.) ECML PKDD 2016. LNCS (LNAI), vol. 9852, pp. 129–144. Springer, Cham (2016). https://doi.org/10.1007/978-3-319-46227-1_9

6. Barros, R.S., Cabral, D.R., Gonçalves Jr., P.M., Santos, S.G.: RDDM: reactive drift detection method. Expert Syst. Appl. 90, 344–355 (2017)

7. Bi, W., Kwok, J.T.: Bayes-optimal hierarchical multilabel classification. IEEE Trans. Knowl. Data Eng. 27(11), 2907–2918 (2015)

8. Bifet, A., Gavalda, R.: Learning from time-changing data with adaptive windowing. In: Proceedings of the 2007 SIAM International Conference on Data Mining, pp. 443–448. SIAM (2007)

9. Bifet, A., Holmes, G., Pfahringer, B., Kirkby, R., Gavalda, R.: New ensemble methods for evolving data streams. In: Proceedings of the 15th ACM SIGKDD International Conference on Knowledge Discovery and Data Mining, pp. 139–148 (2009)

10. Bifet, A., Kirkby, R.: Data stream mining a practical approach (2009)

11. Bishop, C.M.: Pattern Recognition and Machine Learning. springer, Heidelberg (2006)

12. Burred, J.J., Lerch, A.: A hierarchical approach to automatic musical genre classification. In: Proceedings of the 6th International Conference on Digital Audio Effects, pp. 8–11. Citeseer (2003)

13. de Campos Merschmann, L.H., Freitas, A.A.: An extended local hierarchical classifier for prediction of protein and gene functions. In: Bellatreche, L., Mohania, M.K. (eds.) DaWaK 2013. LNCS, vol. 8057, pp. 159–171. Springer, Heidelberg (2013). https://doi.org/10.1007/978-3-642-40131-2_14

14. Cesa-Bianchi, N., Gentile, C., Zaniboni, L.: Incremental algorithms for hierarchical classification. J. Mach. Learn. Res. 7, 31–54 (2006)

15. Chan, T.F., Golub, G.H., LeVeque, R.J.: Algorithms for computing the sample variance: analysis and recommendations. Am. Stat. 37(3), 242–247 (1983)

16. Demšar, J.: Statistical comparisons of classifiers over multiple data sets. J. Mach. Learn. Res. 7, 1–30 (2006)

17. Frías-Blanco, I., del Campo-Ávila, J., Ramos-Jimenez, G., Morales-Bueno, R., Ortiz-Díaz, A., Caballero-Mota, Y.: Online and non-parametric drift detection methods based on Hoeffding's bounds. IEEE Trans. Knowl. Data Eng. 27(3), 810–823 (2014)

18. Friedman, N., Geiger, D., Goldszmidt, M.: Bayesian network classifiers. Mach. Learn. 29(2), 131–163 (1997)

19. Gama, J.: Knowledge Discovery from Data Streams. Chapman and Hall/CRC (2010)

20. Gama, J., Sebastião, R., Rodrigues, P.P.: On evaluating stream learning algorithms. Mach. Learn. 90(3), 317–346 (2013)

21. Gama, J., Žliobaitė, I., Bifet, A., Pechenizkiy, M., Bouchachia, A.: A survey on concept drift adaptation. ACM Comput. Surv. (CSUR) 46(4), 44 (2014)

22. Han, J., Pei, J., Kamber, M.: Data Mining: Concepts and Techniques. Elsevier, Amsterdam (2011)

23. Hesabi, Z.R., Tari, Z., Goscinski, A., Fahad, A., Khalil, I., Queiroz, C.: Data summarization techniques for big data—a survey. In: Khan, S.U., Zomaya, A.Y. (eds.) Handbook on Data Centers, pp. 1109–1152. Springer, New York (2015). https://doi.org/10.1007/978-1-4939-2092-1_38

24. Kiritchenko, S., Famili, F.: Functional annotation of genes using hierarchical text categorization. In: Proceedings of BioLink SIG, ISMB, January 2005

25. Klawonn, F., Angelov, P.: Evolving extended Naive Bayes classifiers. In: Sixth IEEE International Conference on Data Mining-Workshops (ICDMW 2006), pp. 643–647. IEEE (2006)

26. Kotsiantis, S.B., Zaharakis, I., Pintelas, P.: Supervised machine learning: a review of classification techniques. Emerg. Artif. Intell. Appl. Comput. Eng. **160**, 3–24 (2007)

27. Nguyen, H.L., Woon, Y.K., Ng, W.K.: A survey on data stream clustering and classification. Knowl. Inf. Syst. **45**(3), 535–569 (2015)

28. Parmezan, A.R.S., Souza, V.M.A., Batista, G.E.A.P.A.: Towards hierarchical classification of data streams. In: Vera-Rodriguez, R., Fierrez, J., Morales, A. (eds.) CIARP 2018. LNCS, vol. 11401, pp. 314–322. Springer, Cham (2019). https://doi.org/10.1007/978-3-030-13469-3_37

29. Pereira, R.M., Bertolini, D., Teixeira, L.O., Silla Jr., C.N., Costa, Y.M.: COVID-19 identification in chest x-ray images on flat and hierarchical classification scenarios. Comput. Methods Programs Biomed. **194**, 105532 (2020)

30. Pontes, E.A.S.: A brief historical overview of the Gaussian curve: from Abraham de Moivre to Johann Carl Friedrich Gauss. Int. J. Eng. Sci. Invent. (IJESI), 28–34 (2018)

31. Prasad, B.R., Agarwal, S.: Stream data mining: platforms, algorithms, performance evaluators and research trends. Int. J. Database Theory Appl. **9**(9), 201–218 (2016)

32. Quiñonero-Candela, J., Sugiyama, M., Schwaighofer, A., Lawrence, N.D.: Dataset Shift in Machine Learning. The MIT Press, Cambridge (2009)

33. Seidl, T., Assent, I., Kranen, P., Krieger, R., Herrmann, J.: Indexing density models for incremental learning and anytime classification on data streams. In: Proceedings of the 12th International Conference on Extending Database Technology: Advances in Database Technology, pp. 311–322 (2009)

34. Shapiro, S.S., Wilk, M.B.: An analysis of variance test for normality (complete samples). Biometrika **52**(3/4), 591–611 (1965)

35. Silla, C.N., Freitas, A.A.: A survey of hierarchical classification across different application domains. Data Min. Knowl. Discov. **22**(1–2), 31–72 (2011)

36. Silla Jr., C.N., Freitas, A.A.: A global-model Naive Bayes approach to the hierarchical prediction of protein functions. In: 2009 Ninth IEEE International Conference on Data Mining, pp. 992–997. IEEE (2009)

37. Souza, V.M.A., Reis, D.M., Maletzke, A.G., Batista, G.E.A.P.A.: Challenges in benchmarking stream learning algorithms with real-world data. Data Min. Knowl. Discov., 1–54 (2020). https://doi.org/10.1007/s10618-020-00698-5

38. Steinbach, M., Ertöz, L., Kumar, V.: The Challenges of clustering high dimensional data. In: Wille, L.T. (ed.) New Directions in Statistical Physics, pp. 273–309. Springer, Heidelberg (2004). https://doi.org/10.1007/978-3-662-08968-2_16

39. Tsymbal, A.: The problem of concept drift: definitions and related work. Comput. Sci. Dep. Trinity Coll. Dublin **106**(2), 58 (2004)

40. West, D.: Updating mean and variance estimates: an improved method. Commun. ACM **22**(9), 532–535 (1979)

41. Wilcoxon, F.: Individual comparisons by ranking methods. In: Kotz, S., Johnson, N.L. (eds.) Breakthroughs in Statistics. Springer Series in Statistics (Perspectives in Statistics), pp. 196–202. Springer, New York (1992). https://doi.org/10.1007/978-1-4612-4380-9_16

42. Wu, F., Zhang, J., Honavar, V.: Learning classifiers using hierarchically structured class taxonomies. In: Zucker, J.-D., Saitta, L. (eds.) SARA 2005. LNCS (LNAI), vol. 3607, pp. 313–320. Springer, Heidelberg (2005). https://doi.org/10.1007/11527862_24

43. Yassin, N.I., Omran, S., El Houby, E.M., Allam, H.: Machine learning techniques for breast cancer computer aided diagnosis using different image modalities: a systematic review. Comput. Methods Progr. Biomed. **156**, 25–45 (2018)
44. Yeo, I.K., Johnson, R.A.: A new family of power transformations to improve normality or symmetry. Biometrika **87**(4), 954–959 (2000)
45. Zaragoza, J.C., Sucar, E., Morales, E., Bielza, C., Larranaga, P.: Bayesian chain classifiers for multidimensional classification. In: Twenty-Second International Joint Conference on Artificial Intelligence. Citeseer (2011)

Coarsening Algorithm via
Semi-synchronous Label Propagation
for Bipartite Networks

Alan Demétrius Baria Valejo[1]([✉]), Paulo Eduardo Althoff[2],
Thiago de Paulo Faleiros[2], Maria Lígia Chuerubim[3], Jianglong Yan[4],
Weiguang Liu[4], and Liang Zhao[5]

[1] Department of Computing, Federal University of São Carlos (UFSCar),
São Carlos, SP, Brazil
alanvalejo@ufscar.br
[2] University of Brasília (Unb), Brasília, DF, Brazil
thiagodepaulo@unb.br
[3] Federal University of Uberlândia (UFU), Uberlândia, Brazil
marialigia@ufu.br
[4] School of Computer Science, Zhongyuan University of Technology,
Zhengzhou, China
jianglong.yan@hotmail.com, weiguang.liu@zut.edu.cn
[5] Department of Computing and Mathematics (DCM), FFCLRP,
University of São Paulo (USP), Ribeirão Preto, SP, Brazil
zhao@usp.br

Abstract. Several coarsening algorithms have been developed as a powerful strategy to deal with difficult machine learning problems represented by large-scale networks, including, network visualization, trajectory mining, community detection and dimension reduction. It iteratively reduces the original network into a hierarchy of gradually smaller informative representations. However, few of these algorithms have been specifically designed to deal with bipartite networks and they still face theoretical limitations that need to be explored. Specifically, a recently introduced algorithm, called MLPb, is based on a synchronous label propagation strategy. In spite of an interesting approach, it presents the following two problems: 1) A high-cost search strategy in dense networks and 2) the cyclic oscillation problem yielded by the synchronous propagation scheme. In this paper, we address these issues and propose a novel fast coarsening algorithm more suitable for large-scale bipartite networks. Our proposal introduces a semi-synchronous strategy via cross-propagation, which allows a time-effective implementation and deeply

This work was carried out at the Center for Artificial Intelligence (C4AI-USP), with support by the São Paulo Research Foundation (FAPESP) under grant number: 2019/07665-4 and by the IBM Corporation. This work is also supported in part by FAPESP under grant numbers 2013/07375-0, 2015/50122-0 and 2019/14429-5, the Brazilian National Council for Scientific and Technological Development (CNPq) under grant number 303199/2019-9, and the Ministry of Science and Technology of China under grant number: G20200226015.

© Springer Nature Switzerland AG 2021
A. Britto and K. Valdivia Delgado (Eds.): BRACIS 2021, LNAI 13073, pp. 437–452, 2021.
https://doi.org/10.1007/978-3-030-91702-9_29

reduces the oscillation phenomenon. The empirical analysis in both synthetic networks and real-world networks shows that our coarsening strategy outperforms previous approaches regarding accuracy and runtime.

Keywords: Complex networks · Bipartite networks · Multilevel method · Multiscaling analysis

1   Introduction

Bipartite networks are a broadly pervasive class of networks, also known as two-layer networks, where the set of nodes is split into two disjoint subsets called "layers" and links can connect only nodes of different layers. These networks are widely used in science and technology to represent pairwise relationship between categories of entities, e.g. documents and terms, patient and gene expression (or clinical variables) or scientific papers and their authors [17,19]. Over the last years, there has been a growing scientific interest in bipartite networks given their occurrence in many data analytic problems, such as community detection and text classification.

Several coarsening algorithms have been proposed as a scalable strategy to address hard machine learning problems in networks, including network visualization [29], trajectory mining [15], community detection [30] and dimensionality reduction [21]. These algorithms build a hierarchy of reduced networks from an initial problem instance, yielding multiple levels-of-detail, Fig. 1. It is commonly used for generating multiscale networks and, most notably, as a step of the well-known multilevel method.

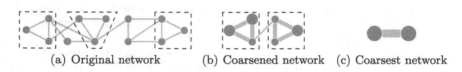

(a) Original network   (b) Coarsened network   (c) Coarsest network

Fig. 1. Coarsening process. In (a), group of nodes are matched; in (b), the original network is coarsened, i.e., matched nodes are collapsed into a super-node and links incident in matched nodes are collapsed into super-edges; the coarsest network is illustrated in (c). The coarsening process is repeated, level by level, until the desired network size is reached.

However, only a few coarsening algorithms have been specifically designed to deal with bipartite networks, as showed in a recent survey [28], and they still face theoretical limitations that open for scientific investigation. Specifically, a recently introduced algorithm, proposed in [24], called MLPb, is based on a label propagation strategy that uses the diffusion of information to build a hierarchy of informative simplified representations of the original network. It implements a high time-cost strategy that searches the whole two-hop neighborhood of each node, which limits its use to sparse networks with a low link-density. As an additional limitation, MLPb uses a synchronous strategy which is known to yield a cyclic oscillation problem in some topological structures, as bipartite components.

To overcome these issues, we propose a novel fast coarsening algorithm based on the cross-propagation concept suitable for large-scale bipartite networks. Specifically, two-fold contribution:

- We design a novel coarsening algorithm that uses a semi-synchronous strategy via cross-propagation, which only considers the direct neighbors of nodes, which implies a cost-efficient implementation and can deeply reduce the oscillation phenomenon.
- We improve the classical cross-propagation strategy using the multilevel process by adding two restrictions: The first defines the minimum number of labels at the algorithm convergence and the second enforces size constraints to groups of nodes belonging to the same label. These restrictions increase the potential and adaptability of cross-propagation to foster novel applications in bipartite networks and can foster future research.

The empirical analysis, considering a set of thousands of networks (both synthetic and real-world networks), demonstrated that our coarsening strategy outperforms previous approaches regarding accuracy and runtime.

The remainder of the paper is organized as follows: First, we introduce the basic concepts and notations. Then, we present the proposed coarsening strategy. Finally, we report results and summarize our findings and discuss future work.

2   Background

A network $\mathcal{G} = (\mathcal{V}, \mathcal{E}, \sigma, \omega)$ is bipartite (or two-layer) if its set of nodes $\mathcal{V}$ and $\mathcal{V}^1 \cap \mathcal{V}^2 = \emptyset$. $\mathcal{E}$ is the set of links, wherein $\mathcal{E} \subseteq \mathcal{V}^1 \times \mathcal{V}^2$. A link (u, v) may have an associated weight, denoted as $\omega(u, v)$ with $\omega : \mathcal{V}^1 \times \mathcal{V}^2 \to \mathbb{R}$; and a node u may have an associated weight, denoted as $\sigma(u)$ with $\sigma : V \to \mathbb{R}$. The degree of a node $u \in \mathcal{V}$ is denoted by $\kappa(u) = \sum_{v \in \mathcal{V}} w(u, v)$. The h-hop neighborhood of u, denoted by $\Gamma_h(u)$, is formally defined as the nodes in set $\Gamma_h(u) = \{v \mid$ there is a path of length h between u and $v\}$. Thus, $\Gamma_1(u)$ is the set of nodes adjacent to u; $\Gamma_2(u)$ is the set of nodes 2-hops away from u, and so forth.

2.1   Label Propagation

The label propagation (LP) algorithm is a popular, simple and time-effective algorithm, commonly used in community detection [20]. Every node is initially assigned to a unique label, then, at each iteration, each node label is updated with the most frequent label in its neighborhood, following the rule:

$$l'_u = \arg\max_{l \in \mathcal{L}} \sum_{v \in \Gamma_1(u)} \delta(l_v, l), \tag{1}$$

wherein l_v is the current label of v, l'_u is the new label of u, $\mathcal{L}$ is the label set for all nodes and δ is the Kronecker's delta. Intuitively, groups of densely connected nodes will converge to a single dominant label.

LP has been widely studied, extended and enhanced. The authors in [1] proposed a modified algorithm that maximizes the network modularity. The authors in [11] presented a study of LP in bipartite networks. The authors in [12], improved in [2], introduced a novel algorithm that maximizes the modularity through LP in bipartite networks. The authors in [7] presented a variation of this concept to k-partite networks. In the multilevel context, the authors in [14] proposed a coarsening algorithm based on LP and, recently, the authors in [24] extended this concept to handle bipartite networks.

Synchronous LP formulation can yield cyclic oscillation of labels in some topological structures, as bipartite, nearly bipartite, ring, star-like components and other topological structures within them. Specifically, after an arbitrary step, labels values indefinitely oscillate between them, i.e. a node exchanges its label with a neighbor and, in a future iteration, this exchange is reversed. This problem is illustrated in Fig. 2.

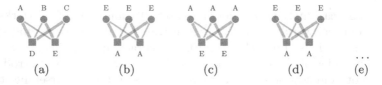

Fig. 2. Oscillation phenomenon. In (a), labels are randomly assigned to nodes; in (b), a propagation process updates the labels; in the subsequent iterations, labels values indefinitely oscillate between them.

To suppress this problem, it is used the asynchronous [20] or semi-synchronous [4] strategy, in which a node or a group of nodes is updated at a time, respectively. For bipartite networks are common to apply the *cross-propagation concept*, a semi-synchronous strategy [11], in which nodes in a selected layer are set as propagators and nodes in the other layer are set as receivers. The process is initially performed from the propagator to the receivers, then it is performed in the reverse direction, as illustrated in Fig. 3.

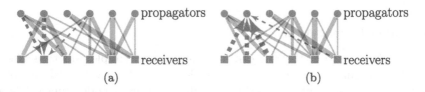

Fig. 3. Cross-propagation in bipartite networks. In (a), labels are propagated from top layer to bottom layer; in (b), the process is performed in the reverse direction.

2.2   Coarsening in Bipartite Networks

A popular strategy to solve large-scale network problems (or data-intensive machine learning problems) is through a multiscale analysis of the original problem instance, which can involve a coarsening process that builds a sequence of networks at different levels of scale. Coarsening algorithms are commonly used as a step of the multilevel method, whose aim is to reduce the computational cost of a target algorithm (or a task) by applying it on the coarsest network. It operates in three phases [25]:

Coarsening phase: Original network $\mathcal{G}_0$ is iteratively coarsened into hierarchy of smaller networks $\{\mathcal{G}_1, \mathcal{G}_2, \cdots, \mathcal{G}_\mathcal{H}\}$, wherein $G_\mathcal{H}$ is the *coarsest network*. The process implies in collapsing nodes and links into single entities, referred to as super-node and super-link, respectively.

Solution finding phase: The *target algorithm* or a task is applied or evaluated in the coarsest representation $G_\mathcal{H}$.

Uncoarsening phase: The solution obtained in $G_\mathcal{H}$ is projected back, through the intermediate levels $\{G_{\mathcal{H}-1}, G_{\mathcal{H}-2}, \cdots, G_1\}$, until G_0.

It is notable that the coarsening is the key component of the multilevel method, since it is problem-independent, in contrast to the other two phases that are designed according to the target task [25]. Therefore, many algorithms have been developed and, recently, some strategies able to handle bipartite networks have gained notoriety.

One of the first, proposed in [22,23], called OPM_{hem} (one-mode projection-based matching algorithm), decomposes the bipartite structure into two unipartite networks, one for each layer. Although it increases the range of analysis options available (as classic and already established algorithms), this decomposition can lead to loss of information, reflecting in the performance of the algorithm.

Later, the authors in [27] introduced two coarsening algorithms, called $RGMb$ (random greedy matching) and GMb (greedy matching), that uses directly the bipartite structure to select a pairwise set of nodes. They use the well-known and useful concept of a two-hop neighborhood. As a drawback, performing this search on large-scale bipartite networks with a high link-density can be computationally impractical.

Recently, the authors in [24] proposed a coarsening based on label propagation through the two-hop neighborhood. Despite its accuracy, it uses a standard and synchronous propagation strategy that can lead to instability and it does not guarantee the convergence.

The growing interest in coarsening algorithms for bipartite networks is recent and current strategies faces several theoretical limitations that remain mostly unexplored, consequently, open to scientific exploration. To overcome these issues, in the next section, we present a novel coarsening strategy.

3 Coarsening via Semi-synchronous Label Propagation for Bipartite Networks

We design a coarsening strategy via semi-synchronous label propagation for bipartite networks (CLPb). We use the cross-propagation concept to diffuse labels between layers. After the convergence, nodes in the *same layer* that belong to the *same label* will be collapsed into a single super-node.

3.1 Algorithm

A label is defined as a tuple $\mathcal{L}_u(l, \beta)$, wherein l is the current label and $\beta \in [0, 1] \subset \mathbb{R}^+$ is its score. At first, each node $u \in V$ is initialized with a starting label $\mathcal{L}_u = (u, 1.0/\sqrt{\kappa(u)})$, i.e. the initial $\mathcal{L}_u$ is denoted by its *id* (or *name*) with a maximum score, i.e. $\beta = 1.0$. To reduce the influence of *hub nodes*[1], in all iteration, β must be normalized by its node degree, as follows:

$$\mathcal{L}'_u = \left(l_i, \frac{\beta_i}{\sqrt{\kappa(u)}} \right) \tag{2}$$

Each step propagates a new label to a receiver node u selecting the label with maximum β from the union of its neighbors' labels, i.e. $\mathcal{L}_u = \cup \mathcal{L}_v \; \forall \; v \in \Gamma_1(u)$, according to the following filtering rules:

1. Equal labels $\mathcal{L}^{eq} \subseteq \mathcal{L}u$ are merged and the new β' is composed by the sum of its belonging scores:

$$\beta' = \sum_{(l,\beta) \in \mathcal{L}^{eq}} \beta, \tag{3}$$

2. The belonging scores of the remaining labels are normalized, i.e.:

$$\mathcal{L}_u = \{(l_1, \frac{\beta_1}{\beta^{sum}}), (l_2, \frac{\beta_2}{\beta^{sum}}), \dots, (l_\gamma, \frac{\beta_\gamma}{\beta^{sum}})\}, \tag{4}$$

$$\beta^{sum} = \sum_{i=1}^{\gamma} \beta_i, \tag{5}$$

where γ is the number of remaining labels.
3. The label with the largest β is selected:

$$\mathcal{L}'_u = \arg\max_{(l,\beta) \in \mathcal{L}_u} \mathcal{L}_u. \tag{6}$$

4. The size of the coarsest network is naturally controlled by the user, i.e. require defining a number of reduction levels, a reduction rate or any other parameter to fit a desired network size. Here, the minimum number of labels η for each layer is a user-defined parameter. A node $u \in \mathcal{V}^i$, with $i \in \{1, 2\}$ define a

---

[1] The highest-degree nodes are often called *hubs*.

layer, is only allowed to update its label if, and only if, the number of labels in the layer $|\mathcal{L}^i|$ remains equal to or greater than η^i, i.e.:

$$|\mathcal{L}^i| \leq \eta^i. \tag{7}$$

5. At last, a classical issue in the multilevel context is that super-nodes tend to be highly unbalanced at each level [25]. Therefore, is common to constrain the size of the super-nodes from an *upper-bound* $\mu \in [0,1] \subset \mathbb{R}^+$, which limits the maximum size of a group of labels in each layer:

$$S^i = \frac{1.0 + (\mu * (\eta^i - 1)) * |\mathcal{V}^i|}{\eta^i}, \tag{8}$$

wherein $\mu = 1.0$ and $\mu = 0$ implies highly imbalanced and balanced groups of nodes, respectively. Therefore, a node u with weight $\sigma(v)$ can update its current label l to a new label l' if, and only if:

$$\sigma(u) + \sigma(l') \leq S^i \quad \text{and} \quad \sigma(l') = \sum_{v \in l'} \sigma(v). \tag{9}$$

If restrictions 3 or 4 are not attained, the algorithm returns to step 3; the label with the maximum β is removed and a new ordered label is selected. The process is repeated until a label that satisfies the restrictions 3 and 4 is obtained. Figure 4 shows one step of CLPb in a bipartite network using the previously defined strategy. The propagation process is repeated $\mathcal{T}$ (user-defined parameter) times until the convergence or stops when there are no label changes.

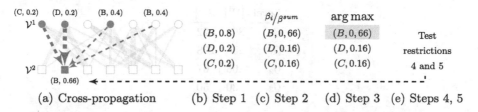

(a) Cross-propagation   (b) Step 1 (c) Step 2   (d) Step 3   (e) Steps 4, 5

Fig. 4. One step of the CLPb algorithm in a bipartite network. In (a), the process is performed from the top layer, considering the propagators nodes $\in \mathcal{V}^1$, to the bottom layer, considering the receiver nodes $\in \mathcal{V}^2$. At first, represented in (b), equal labels are merged. In (c), second step, remaining labels are normalized. In third step, the label B is selected, as showed in (d). In (e), the restriction 4 and 5 are tested. Finally, label B is propagated to the node in the bottom layer, as illustrated by the black dashed line.

After the cross-propagation convergence, the algorithm collapses each group of matched nodes (i.e. nodes with same label) into a single *super-node*. Links that are incident to matched nodes are collapsed into the so-called *super-links*. Figure 5 illustrates this process.

Note, the CLPb process does not guarantee that the desired minimum number of labels η will be reached at the current level, i.e. the algorithm can stop with a number of labels greater than the desired one. However, the multilevel process naturally mitigates this problem, since CLPb is performed, level by level, in the subsequent coarsened networks, until the desired number of nodes are reached. I.e., the original network $\mathcal{G}_0$ is iteratively coarsened into a hierarchy of smaller networks $\{\mathcal{G}_1, \mathcal{G}_2, \cdots, \mathcal{G}_{\mathcal{H}}, \cdots\}$, wherein $\mathcal{G}_{\mathcal{H}}$ is an arbitrary level. Table 1 summarizes three levels automatically achieved by CLPb when evaluated in the UCForum [10].

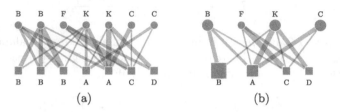

(a)   (b)

Fig. 5. Contraction process. In (a), group of nodes are matched using CLPb algorithm; in (b), the original network is coarsened, i.e., nodes that share labels are collapsed into a super-node and links incident to matched nodes are collapsed into super-edges.

Table 1. UCForum: contains 899 users and 522 posts on forums. Considering $\eta^1 = 30$ and $\eta^2 = 50$ as an input user-parameter, CLPb automatically builds three levels to reach the desired network size.

Level	$|\mathcal{V}^1|$	$|\mathcal{V}^2|$
$\mathcal{G}_0$	899	522
$\mathcal{G}_1$	258	258
$\mathcal{G}_2$	68	67
$\mathcal{G}_3$	30	50

Naturally, users can control the maximum number of levels and the reduction factor ρ for each layer, rather than input the desired number of nodes in the coarsest network. In this case, the desired number of nodes for each layer and in each level can be defined as exemplified in Eq. 10. Alternatively, users can stop the algorithm in an arbitrary level. However, this is a technical decision and the stop-criterion in Eq. 10 is commonly used in the literature [25].

$$\eta_i = (1 - \rho^i) * |\mathcal{V}^i| \tag{10}$$

3.2 Complexity

The computational complexity of the LP is near-linear regarding the number of links, i.e., $\mathcal{O}(|\mathcal{V}|+|\mathcal{E}|)$ steps are needed at each iteration. If a constant number of $\mathcal{T}$ iterations are considered, then $\mathcal{O}(\mathcal{T}(|\mathcal{V}|+|\mathcal{E}|))$. The contraction process (illustrated in Fig. 5), first, iterates over all matched nodes $\in \mathcal{V}_\mathcal{H}$ to create super-nodes $\in \mathcal{V}_{\mathcal{H}+1}$, then, each link in $\mathcal{E}_\mathcal{H}$ is selected to create super-links $\in \mathcal{E}_{\mathcal{H}+1}$, therefore, $\mathcal{O}(|\mathcal{V}| + |\mathcal{E}|)$. These complexities are well-known in the literature, and the expanded discussion can be found in [25] and [20]. Based on these considerations, the CLPb complexity is $\mathcal{O}(\mathcal{T}(|\mathcal{V}| + |\mathcal{E}|)) + \mathcal{O}(|\mathcal{V}| + |\mathcal{E}|)$ at each level.

4 Experiments

We compared the performance of CLPb with four state-of-the-art coarsening algorithms, namely MLPb, OPM_{hem}, RGMb and GMb (discussed in Sect. 2 and presented in the survey [28]). First, we conducted an experiment in a set of thousands of synthetic networks and, then, we test the performance of the algorithms in a set of well-known real networks.

A common and practical approach to verify the quality of a coarsened representation is mapping each super-node as a group (community or cluster) and evaluate them using quality measures. This type of analysis is considered a benchmark approach in the literature, as discussed in the recent surveys [25, 28] and in other studies, as [6,8,9,18,24]. Therefore, it is natural to use this analysis in our empirical evaluation.

The following two measures were considered: normalized mutual information (NMI) [13], which quantifies the quality of the disjoint clusters comparing the solution found by a selected algorithm with the baseline (or ground truth), and Murata's Modularity [16], which quantifies the strength of division of a network into communities. Experiments were executed on a Linux machine with an 6-core processor with 2.60 GHz and 16 GB main memory.

4.1 Synthetic Networks

The benchmark analysis was conducted on thousands of synthetic networks obtained employing a network generation tool called BNOC, proposed in [26]. Each network configuration was replicated 10 times to obtain the average and standard deviation. Default parameters are presented in [26].

First, we evaluated the sensibility of the algorithms regarding the noise level in the networks. The noise level is a disturbance or error in the dataset (the proportion of links wrongly inserted), e.g., 0.5 means that half of the links are not what they should be. Noise can negatively affect the algorithm's performance in terms of accuracy.

A set of 1000 synthetic bipartite networks with distinct noise level was generated, as follows: $|\mathcal{V}| = 2,000$ with $|\mathcal{V}^1| = |\mathcal{V}^2|$, noise within the range $[0.0, 1.0]$ and 20 communities for each layer. Figure 6(a) depicts the NMI values for the evaluated algorithm as a function of the amount of noise. The algorithms exhibit distinct behaviors. MLPb and CLPb obtained high NMI values with a low level of noise, however, NMI values for MLPb decrease quickly after 0.22 noise level, whereas CLPb decreases slowly. Therefore, MLPb revealed a high noise sensibility. Although GMb, RGMb and OPM_{hem} algorithms obtained the lowest NMI values, mainly, within the range $[0.0, 0.4]$, their performances decrease slowly compared with MLPb.

We also evaluated the sensibility of the algorithms regarding the number of communities in the networks. A set of 1000 synthetic bipartite networks with distinct number of communities was generated, as follows: $|\mathcal{V}| = 2,000$ with $|\mathcal{V}^1| = |\mathcal{V}^2|$, communities within the range $[1, 500]$ and 0.3 of noise level. Figure 6(b) depicts the NMI values for the evaluated algorithm as a function of the number of communities. GMb, RGMb and OPM_{hem} presented a high sensibility to a low number of communities in the network, specifically, within the range $[1, 100]$, in contrast, CLPb and MLPb obtained high NMI values in the same range. Within the range $[200, 500]$, all algorithms obtained close NMI values.

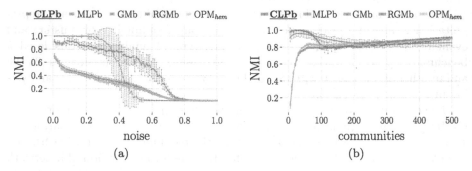

Fig. 6. NMI of the algorithms in relation to the noise level (a) and number of communities and (b) in 2,000 synthetic networks.

A Nemenyi post-hoc test [5] was applied to the results depicted in Figs. 6(a) and 6(b) to detect statistical differences in the performances of the algorithms. The critical difference (CD) is indicated at the top of each diagram and algorithms' average ranks are placed in the horizontal axis, with the best ranked to the left. A black line connects algorithms for which there is no significant performance difference. According to the Nemenyi statistics, the critical value for comparing the mean-ranking of two different algorithms at 95 percentiles is 0.04, i.e. significant differences are above this value. CLPb was ranked first followed by GMb, the pair MLPb and RGMb and then, in last, OPM_{hem}. Furthermore, CLPb performs statistically better than MLPb, RGMb and OPM_{hem} algorithms.

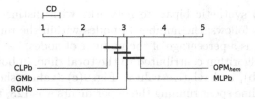

Fig. 7. Nemenyi post-hoc test applied to the results depicted in Figs. 6(a) and 6(b).

We assessed the scalability of the algorithms in terms of the absolute and relative total time spent. First, a set of 1000 synthetic networks with distinct link-density was generated, as follows: link-density within the range $[0.01, 0.99]$, wherein 0.01 indicates very sparse networks and 0.99 indicates very dense networks with $m \approx n^2$; $|V| = 5,000$ with $|V^1| = |V^2|$ and 20 communities at each layer. Figure 8 shows how each algorithm contributed to the total time, in both absolute values, Figs. 8(a) and 8(b), and relative values, Fig. 8(c) (values shown on top of the bars). The total time spent running the experiments was $419,857.968$ s, or nearly 116 h. CLPb spent $4,351.2$ s, which is nearly 1.0% of the total time, furthermore, CLPb ran 18 to 35 times faster than the other algorithms. GMb and OPM_{hem} were the most expensive algorithms.

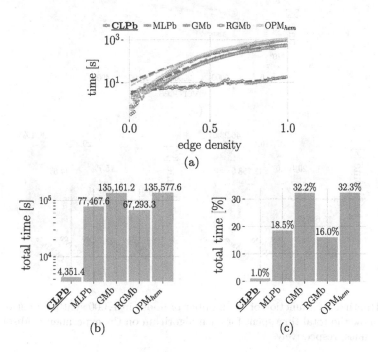

Fig. 8. Runtime as a function of link-density in $1,000$ synthetic networks: (b) and (c) shows the total time spent for each algorithm on the experiments, absolute and relative values, respectively.

A set of 1000 synthetic bipartite networks with distinct number of nodes was generated, as follows: the number of nodes within the range $[1,000, 40,000]$ and communities as a percentage of the number of nodes, i.e. $|V| * 0.01$. Figure 9 shows how each algorithm contributed to the total time, in both absolute values, Figs. 9(a) and 9(b), and relative values, Fig. 9(c) (values shown on top of the bars). The total time spent running the experiments was $128,151.711\,s$, or nearly 35 h. CLPb spent $5,644.2\,s$, which is nearly 4.4% of the total time, furthermore, CLPb ran $3,3$ to $8,2$ times faster than the other algorithms. GMb and OPM$_{hem}$ were the most expensive algorithms.

4.2   Real-World Networks

We considered six real-world bipartite networks available at KONECT (the Koblenz Network Collection) [10]. We took the largest connected component of each network. Network properties are detailed in Table 2(a). Murata's modularity was used to obtain the accuracy of the algorithms by reducing the networks to 30%, 50% and 80% of its original sizes.

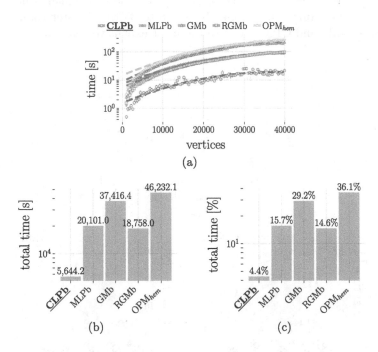

Fig. 9. Runtime as a function of the number of nodes in $1,000$ synthetic networks: (b) and (c) show the total time spent for each algorithm on the experiments, absolute and relative values, respectively.

Considering 30%, 50% and 80% of network reduction, summarized in Tables 2(b), 2(c) and 2(d), CLPb yielded the best values on 4, 5 and 3 networks, respectively; and MLPb yielded the best values on 2 networks at each of the three cases. The pair GMb and OPM_{hem} obtained the best value on one network with 80% of network reduction.

A Nemenyi post-hoc test was applied to the results, shown in Figs. 10(a), 10(b) and 10(c), obtained from 30%, 50% and 80% of network reduction, respectively. Figure 10(d) summarizes the overall results. According to the Nemenyi statistics, the critical value for comparing the average-ranking of two algorithms at the 95 percentile is 1.11. According to Fig. 10(d), CLPb was ranked best, followed by MLPb, the pair GMb and OPM and, in last, RGMb. Furthermore, CLPb performs statistically better than GMb, RGMb and OPM_{hem}.

The empirical investigation showed that CLPb yielded more accurate and stable results compared to the standard algorithms and requires considerably lower execution time. It is a strong indicator of its performance on large-enough problem sizes and must foster the development of novel scalable solutions defined in bipartite networks, including network visualization, trajectory mining, community detection or graph partitioning, data dimension reduction and optimization of high-complexity algorithms [3,15,27].

Table 2. Modularity scores of the algorithms: (a) summaries the properties of the networks; (b), (c) and (c) presented modularity scores of the algorithms considering 30%, 50% and 80% of network reduction, respectively.

(a)

Dataset	$|\mathcal{V}^1|$	$|\mathcal{V}^2|$	$|\mathcal{E}|$
Ucforum	248	610	1,249
MCrime	754	509	1,377
N-reactome	8,788	15,433	41,087
Condmat	13,861	19,466	53,628
Movielens	3,919	2,378	8,868
Dbpedia	54,909	19,866	98,895

(b)

Dataset	CLPb	MLPb	GMb	RGMb	OPM_{hem}
Ucforum	0.139	**0.181**	0.135	0.120	0.135
MCrime	**0.589**	0.564	0.556	0.541	0.556
N-reactome	**0.453**	0.438	0.431	0.391	0.431
Condmat	0.455	**0.463**	0.454	0.409	0.454
Movielens	**0.293**	0.273	0.242	0.233	0.242
Dbpedia	**0.520**	0.507	0.475	0.449	0.475

(c)

Dataset	CLPb	MLPb	GMb	RGMb	OPM_{hem}
Ucforum	0.165	**0.264**	0.145	0.119	0.145
MCrime	**0.651**	0.624	0.630	0.583	0.630
N-reactome	**0.540**	0.517	0.467	0.427	0.467
Condmat	**0.563**	0.557	0.530	0.454	0.530
Movielens	**0.330**	0.325	0.248	0.234	0.248
Dbpedia	**0.583**	**0.583**	0.517	0.480	0.517

(d)

Dataset	CLPb	MLPb	GMb	RGMb	OPM_{hem}
Ucforum	**0.291**	0.282	0.129	0.109	0.129
MCrime	0.696	0.724	**0.773**	0.663	**0.773**
N-reactome	**0.622**	0.567	0.496	0.437	0.496
Condmat	0.585	**0.700**	0.611	0.536	0.611
Movielens	**0.382**	0.364	0.254	0.220	0.254
Dbpedia	0.645	**0.691**	0.542	0.524	0.542

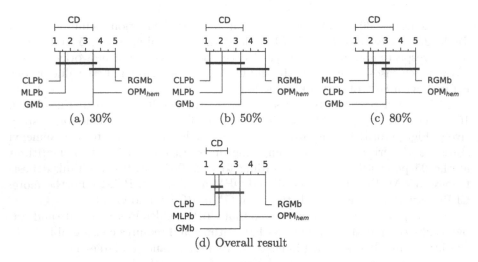

Fig. 10. Nemenyi post-hoc test: (a), (b) and (c) represent the results depicted in Tables (b), (d) and (d); alternatively, (d) summarizes the overall results.

5 Conclusion

We have proposed a novel time-effective semi-synchronous coarsening algorithm to handle large-scale bipartite networks, called CLPb. We introduce the cross-propagation concept in the model to overcome unstable issues, as the cyclic oscillations found state-of-the-art algorithms. Furthermore, CLPb employs a search strategy that only visits the immediate neighborhood of each node, which is more suitable to evaluate large-scale networks even with high link-density. Consequently, the algorithm has low computational complexity.

Empirical analysis on thousands of networks provided compelling evidence that CLPb outperforms the state-of-the-art algorithms regarding accuracy and demands considerably shorter execution times, specifically, CLPb was 4 to 35 times faster as compared to the established algorithms.

Note, this study intended to present our algorithm and validate them from an empirical approach following the state-of-the-art guidelines [28]. We now intend to employ the CLPb in real applications, e.g., in future work, we plan to extend the CLPb to dimension reduction and semi-supervised classification tasks. Another issue that deserves further attention is its application to network visualization.

References

1. Barber, M.J., Clark, J.W.: Detecting network communities by propagating labels under constraints. Phys. Rev. E **80**, 026129 (2009)
2. Beckett, S.J.: Improved community detection in weighted bipartite networks. R. Soc. Open Sci. **3**(1), 140536 (2016)

3. Cintra, D., Valejo, A., Lopes, A., Oliveira, M.: Visualization to assist interpretation of the multilevel paradigm in bipartite graphs. In: 15th International Joint Conference on Computer Vision, Imaging and Computer Graphics Theory and Applications, pp. 133–140 (2019)
4. Cordasco, G., Gargano, L.: Label propagation algorithm: a semi-synchronous approach. Int. J. Soc. Netw. Min. **1**(1), 3–26 (2012)
5. Demšar, J.: Statistical comparisons of classifiers over multiple data sets. J. Mach. Learn. Res. **7**, 1–30 (2006)
6. Dias, M.D., Mansour, M.R., Dias, F., Petronetto, F., Silva, C.T., Nonato, L.G.: A hierarchical network simplification via non-negative matrix factorization. In: Proceedings of the Conference on Graphics, Patterns and Images (SIBGRAPI), pp. 119–126 (2017)
7. Ding, C., Li, T., Wang, D.: Label propagation on k-partite graphs. In: 2009 International Conference on Machine Learning and Applications, pp. 273–278. IEEE (2009)
8. Karypis, G., Kumar, V.: Metis - unstructured graph partinioning and sparse matrix ordering system. Technical report, University of Minnesota, Department of Computer Science (1995)
9. Karypis, G., Kumar, V.: A fast and high quality multilevel scheme for partitioning irregular graphs. SIAM J. Sci. Comput. **20**(1), 359–392 (1998)
10. Kunegis, J.: KONECT: the Koblenz network collection. In: Proceedings of the 22nd International Conference on World Wide Web, pp. 1343–1350 (2013)
11. Liu, X., Murata, T.: How does label propagation algorithm work in bipartite networks? In: 2009 IEEE/WIC/ACM International Joint Conference on Web Intelligence and Intelligent Agent Technology, vol. 3, pp. 5–8. IEEE (2009)
12. Liu, X., Murata, T.: An efficient algorithm for optimizing bipartite modularity in bipartite networks. J. Adv. Comput. Intell. Intell. Inform. **14**(4), 408–415 (2010)
13. Meilă, M.: Comparing clusterings—an information based distance. J. Multivar. Anal. **98**(5), 873–895 (2007)
14. Meyerhenke, H., Sanders, P., Schulz, C.: Partitioning complex networks via size-constrained clustering. In: Gudmundsson, J., Katajainen, J. (eds.) SEA 2014. LNCS, vol. 8504, pp. 351–363. Springer, Cham (2014). https://doi.org/10.1007/978-3-319-07959-2_30
15. Minatel, D., Valejo, A., Lopes, A.: Trajectory network assessment based on analysis of stay points cluster. In: Brazilian Conference on Intelligent Systems (BRACIS), pp. 564–569 (2018)
16. Murata, T.: Modularities for bipartite networks. In: Proceedings of the 20th ACM Conference on Hypertext and Hypermedia, pp. 245–250 (2009)
17. Newman, M.E.J.: Scientific collaboration networks. II. Shortest paths, weighted networks, and centrality. Phys. Rev. E Stat. Nonlinear Soft Matter Phys. **64**(1), 016132 (2001)
18. Noack, A., Rotta, R.: Multi-level algorithms for modularity clustering. In: Vahrenhold, J. (ed.) SEA 2009. LNCS, vol. 5526, pp. 257–268. Springer, Heidelberg (2009). https://doi.org/10.1007/978-3-642-02011-7_24
19. de Paulo Faleiros, T., Valejo, A., de Andrade Lopes, A.: Unsupervised learning of textual pattern based on propagation in bipartite graph. Intell. Data Anal. **24**(3), 543–565 (2020)
20. Raghavan, N., Albert, R., Kumara, S.: Near linear time algorithm to detect community structures in large-scale networks. Phys. Rev. E Stat. Nonlinear Soft Matter Phys. **76**, 036106 (2007)

21. Sakellaridi, S., Fang, H.R., Saad, Y.: Graph-based multilevel dimensionality reduction with applications to eigenfaces and latent semantic indexing. In: Proceedings of the International Conference on Machine Learning and Applications (ICMLA), pp. 194–200 (2008)

22. Valejo, A., Ferreira, V., de Oliveira, M.C.F., de Andrade Lopes, A.: Community detection in bipartite network: a modified coarsening approach. In: Lossio-Ventura, J.A., Alatrista-Salas, H. (eds.) SIMBig 2017. CCIS, vol. 795, pp. 123–136. Springer, Cham (2018). https://doi.org/10.1007/978-3-319-90596-9_9

23. Valejo, A., Lopes, A., Filho, G., Oliveira, M., Ferreira, V.: One-mode projection-based multilevel approach for community detection in bipartite networks. In: International Symposium on Information Management and Big Data (SIMBig), Track on Social Network and Media Analysis and Mining (SNMAN), pp. 101–108 (2017)

24. Valejo, A., Faleiros, T.P., Oliveira, M.C.R.F., Lopes, A.: A coarsening method for bipartite networks via weight-constrained label propagation. Knowl. Based Syst. **195**, 105678 (2020)

25. Valejo, A., Ferreira, V., Fabbri, R., Oliveira, M.C.R.F., Lopes, A.: A critical survey of the multilevel method in complex networks. ACM Comput. Surv. **53**(2), 35 (2020)

26. Valejo, A., Goes, F., Romanetto, L.M., Oliveira, M.C.F., Lopes, A.A.: A benchmarking tool for the generation of bipartite network models with overlapping communities. Knowl. Inf. Syst. **62**, 1641–1669 (2019)

27. Valejo, A., Oliveira, M.C.R.F., Filho, G.P., Lopes, A.A.: Multilevel approach for combinatorial optimization in bipartite network. Knowl. Based Syst. **151**, 45–61 (2018)

28. Valejo, A.D.B., de Oliveira dos Santos, W., Naldi, M.C., Zhao, L.: A review and comparative analysis of coarsening algorithms on bipartite networks. Eur. Phys. J. Spec. Top. (4), 1–11 (2021). https://doi.org/10.1140/epjs/s11734-021-00159-0

29. Walshaw, C.: A multilevel algorithm for force-directed graph drawing. In: Proceedings of the International Symposium on Graph Drawing, vol. 1984, pp. 171–182 (2001)

30. Zhu, M., Meng, F., Zhou, Y., Yuan, G.: An approximate spectral clustering for community detection based on coarsening networks. Int. J. Adv. Comput. Technol. **4**(4), 235–243 (2012)

Evaluating Clustering Meta-features
for Classifier Recommendation

Luís P. F. Garcia[1]([✉]) [iD], Felipe Campelo[2] [iD], Guilherme N. Ramos[1] [iD],
Adriano Rivolli[3] [iD], and André C. P. de L. F. de Carvalho[4] [iD]

[1] Department of Computer Science, University of Brasília, Brasília, Brazil
{luis.garcia,gnramos}@unb.br
[2] College of Engineering and Physical Sciences, Aston University, Birmingham, UK
f.campelo@aston.ac.uk
[3] Computing Department, Technological University of Paraná,
Cornélio Procópio, Brazil
rivolli@utfpr.edu.br
[4] Institute of Mathematical and Computer Sciences, University of São Paulo,
São Carlos, Brazil
andre@icmc.usp.br

Abstract. Data availability in a wide variety of domains has boosted
the use of Machine Learning techniques for knowledge discovery and classification. The performance of a technique in a given classification task
is significantly impacted by specific characteristics of the dataset, which
makes the problem of choosing the most adequate approach a challenging
one. Meta-Learning approaches, which learn from meta-features calculated from the dataset, have been successfully used to suggest the most
suitable classification algorithms for specific datasets. This work proposes
the adaptation of clustering measures based on internal indices for supervised problems as additional meta-features in the process of learning a
recommendation system for classification tasks. The gains in performance
due to Meta-Learning and the additional meta-features are investigated
with experiments based on 400 datasets, representing diverse application
contexts and domains. Results suggest that (*i*) meta-learning is a viable
solution for recommending a classifier, (*ii*) the use of clustering features
can contribute to the performance of the recommendation system, and
(*iii*) the computational cost of Meta-Learning is substantially smaller
than that of running all candidate classifiers in order to select the best.

Keywords: Meta-learning · Meta-features · Characterization
measures · Clustering problems

1 Introduction

Data analysis is a pervasive activity in several areas of industrial and scientific
activities, often involving Machine Learning (ML) approaches, which have been
successfully applied in a wide variety of domains [35]. However, the variability

© Springer Nature Switzerland AG 2021
A. Britto and K. Valdivia Delgado (Eds.): BRACIS 2021, LNAI 13073, pp. 453–467, 2021.
https://doi.org/10.1007/978-3-030-91702-9_30

of dataset characteristics makes it challenging to accurately select the most adequate ML approach to handle new data. This issue has also been tackled from a ML perspective, where the task is learning, based on meta-data extracted from several datasets related to a given application or domain, a model to predict which ML approach may be better suited for exploring that particular domain. This approach, commonly referred to as Meta-Learning (MtL), has shown promising results in predicting classifier performances for a problem, based on statistical or geometrical characteristics of a dataset [16].

MtL approaches consist of the data-driven selection of techniques, through knowledge extracted from previous tasks, which may be then applied by a recommendation system to predict the preferred approach for a new, previously unseen problem [4]. This depends on the construction of a *meta-dataset* from the information on a group of datasets in the class of the problem. For each dataset in the group, its descriptive characteristics (*meta-features*) are extracted and combined into one or more *meta-examples*, which are then labeled according to the observed performances of different ML algorithms or an order-preserving transformation such as their rank [4], composing the target feature to be predicted. From the meta-dataset, a *meta-model* can be induced by a learning algorithm and then used in a recommendation system to predict which algorithm is expected to have the best performance when a new problem needs to be addressed [34].

Despite their success, most MtL studies still lack an in-depth analysis of the meta-features [29] which are known to be crucial in the successful use of MtL [3]. Several works propose new meta-features [15–17], but only a few present important details such as their asymptotic computational cost, the degree of information presented, or the importance of the meta-features for the investigated problems [16, 24].

This paper presents the use of clustering measures as meta-features in an MtL framework, to learn a recommendation system for classifiers. These clustering measures are based on internal indices which extract information, such as compactness or separation, to evaluate the goodness of a clustering structure. Although some of these measures have been used in unsupervised MtL scenarios [26, 37], this work proposes the use of those measures on classification problems. The proposed approach uses class labels as cluster indicators rather than the numerical results of a clustering algorithm, and is expected to extract informative measures for quantifying statistical or geometrical characteristics of datasets.

The main goal of this paper is to investigate whether clustering measures can be applied in this context, and whether they influence the choices of the recommendation system. The experimental results suggest that including these measures as meta-features contributes to more accurate recommendations in the problem of classifier selection. Additionally, an initial evaluation of computational costs indicates that this MtL approach is substantially less computationally expensive than testing all classifiers on the dataset for the selection of the best one.

2   Background

This section presents the background information necessary to describe the proposed approach: Sect. 2.1 introduces the Meta-Learning framework, including the process of building a meta-dataset and how to recommend algorithms. Section 2.2 elaborates on clustering meta-features, a subset of internal indices based on compactness and separation of the goodness of a clustering structure.

2.1   Meta-learning

Different algorithms have distinct learning strategies, thus the essence of learning can only be captured by considering different learning algorithms with diverse biases to acquire domain specific information [4]. This concept was initially introduced to make the algorithm selection problem systematic [30], and the goal is to predict the best algorithm to solve a specific given problem, when more than one option is available.

The components of this model are the space of problem instances ($\mathcal{P}$); the space of instance meta-features ($\mathcal{F}$); the space of algorithms ($\mathcal{A}$); and the space of evaluation measures ($\mathcal{Y}$). From these components, an MtL system can obtain a model capable of mapping a dataset or problem $p \in \mathcal{P}$, described by meta-features $f \in \mathcal{F}$, into one algorithm $\alpha \in \mathcal{A}$ able to solve the problem with a good predictive performance according to measure $y \in \mathcal{Y}$. The meta-learner recommendation would be the algorithm with the best expected $y(\alpha(p))$. This can be further improved, for instance, by the inclusion of components which may guide theoretical support to refine the recommendation system [34].

A crucial component of these previous approaches is the definition of the meta-features ($\mathcal{F}$) used to describe general properties of datasets [3]. They must be able to provide evidence on the future performance of the algorithms in $\mathcal{A}$ and to discriminate, with an acceptable computational cost, the performance of a group of algorithms. The main meta-features used in the MtL literature can be divided into six main groups [31], called *standard* meta-features in this work. They represent meta-features based general high-level summaries of the data, statistical and information theory properties of the data, properties of Decision Trees (DTs) induced from the data and the performance of simple and fast learning algorithms [9,25,29].

Another concern is the definition of the set of problem instances ($\mathcal{P}$), since the use of a large number of diverse datasets is recommended to induce a reliable meta-model [4]. Attempts to reduce the bias in this choice include using datasets from different data repositories, such as UCI [13] and OpenML [36]. The importance of problem diversity is based on the underlying assumption that the meta-model is expected to generalize the acquired knowledge when faced with new problem instances without explicit constraints in terms of expected problem characteristics.

The selected algorithms $\mathcal{A}$ represent a set of candidate algorithms to be recommended in the algorithm selection process. Ideally, these algorithms should also be sufficiently different from each other and represent all regions in the algorithm space [34]. The models induced by the algorithms can be evaluated by different measures. Quality measures $\mathcal{Y}$ for assessing the models depend on the nature of the problem. Classifiers, for example, can be evaluated by different measures such as accuracy, F_β, area under the ROC curve (AUC) or the Kappa coefficient, among others.

The step following the extraction of meta-features from the datasets and training the set of algorithms is the labeling of meta-examples in the meta-base. The three properties most frequently used in this task are [4]: (i) the best performing algorithm on the meta-example's dataset (a meta-classification problem); (ii) the ranking of the algorithm according to its performance (a meta-ranking problem); and (iii) the raw performance value of each algorithm on the dataset (a meta-regression problem).

Differently from previous works on MtL applied on clustering data [26,37], this work presents a MtL regression task based on the use of clustering meta-features using internal indices, to extract information like separation and compactness on a supervised scenario using the class labels. In this case, the main objective is to improve the algorithm recommendation using internal indices, motivated by their generally low computational cost and high degree of information. The standard meta-features [31] are considered to provide a comparison baseline and to allow an objective analysis of the variation in performance resulting from using the new clustering meta-features. The evaluation of the recommender system includes assessing the performance in the base- and meta-levels and the cost in execution time.

2.2   Clustering Meta-features

A few definitions must be presented before the description of the clustering measures. Let $X \in \mathbb{R}^{N \times q}$ be a matrix of N observations, each represented by a q-dimensional vector $\mathbf{x}_\ell \in \mathbb{R}^q$; and $U_K(X) = \{X_1, X_2, \ldots, X_K\}$ be an exhaustive partition of X into K mutually exclusive clusters X_k with sizes $n_k > 0$, $k = 1, \ldots, K$. In all definitions below, the clustering meta-features are functions of a partition $U_K(X)$, which in the case of this paper is given by the class labels. The $U_K(X)$ is kept implicit in the definitions in order to keep the notation cleaner, but the reader should keep this relationship in mind.

Let $\bar{\mathbf{x}}_k$ denote the mean point of all observations belonging to cluster X_k, and $\bar{\bar{\mathbf{x}}}$ denote the grand mean of all observations in X. Also, let $dist(\cdot, \cdot)$ denote the distance (e.g., Euclidean) between two points. Then,

$$\delta_{ij} \triangleq \min_{\substack{\mathbf{x} \in X_i \\ \mathbf{y} \in X_j}} dist(\mathbf{x}, \mathbf{y}) \tag{1}$$

denotes the single-linkage distance between the two clusters X_i, X_j, and

$$\Delta_k \triangleq \max_{\mathbf{x}, \mathbf{y} \in X_k} dist(\mathbf{x}, \mathbf{y}) \tag{2}$$

represents the diameter of a cluster X_k.

Additionally, consider some vectors of distances. Let $\mathbf{d}^+ \in \mathbb{R}^{N_W}$ be a vector of all N_W within-cluster distances (i.e., all distances between pairs of points having equal class labels in the data), $\mathbf{d}^- \in \mathbb{R}^{N_B}$ be a vector containing all N_B between-cluster distances (between pairs of points having different class labels), and $\mathbf{d}^\bullet \in \mathbb{R}^{N_T}$ denote a vector containing the distances between all N_T pairs of points in X, with $N_T = N_W + N_B$.

Finally, let the following quantities be defined: the within-groups sum of squares,

$$WGSS = \sum_{k=1}^{K} \sum_{\mathbf{x}_i \in X_k} \left[dist(\mathbf{x}_i, \bar{\mathbf{x}}_k)\right]^2, \tag{3}$$

and the between-groups sum of squares,

$$BGSS = \sum_{k=1}^{K} n_k \left[dist(\bar{\mathbf{x}}_k, \bar{\mathbf{x}})\right]^2. \tag{4}$$

Given the preceding definitions, the clustering meta-features used in this work are formalized as follows:

– Dunn's separation index (VDU) [14]:

$$VDU = \min_{\substack{i,j \in [1,K] \\ i \neq j}} \frac{\delta_{ij}}{\max_{k \in [1,K]} \Delta_k}. \tag{5}$$

– Davies-Bouldin index (VDB) [11]:

$$VDB = \frac{1}{K} \sum_{i=1}^{K} \max_{j \neq i} \frac{\Delta_i + \Delta_j}{dist(\bar{\mathbf{x}}_i, \bar{\mathbf{x}}_j)}. \tag{6}$$

– Baker-Hubert index (Γ) [1]: let

$$s^+ = \sum_{\forall d_i \in \mathbf{d}^+} \sum_{\forall d_j \in \mathbf{d}^-} One(d_i < d_j) \quad \text{and} \quad s^- = \sum_{\forall d_i \in \mathbf{d}^+} \sum_{\forall d_j \in \mathbf{d}^-} One(d_i > d_j) \tag{7}$$

where $One(condition)$ is a function that returns 1 if the condition is true and 0 otherwise; then:

$$\Gamma = \frac{s^+ - s^-}{s^+ + s^-}. \tag{8}$$

– Tie-corrected Kendall tau (τ) [12]:

$$\tau = \frac{s^+ - s^-}{\sqrt{N_W N_B N_T (N_T - 1)/2}}. \tag{9}$$

- Ray-Turi index (ν) [28]:

$$\nu = \frac{1}{N} \frac{WGSS}{\min_{i \neq j} \delta_{ij}^2}. \tag{10}$$

- Mean inter-centroid distance (INT) [2]:

$$INT = \frac{2}{K(K-1)} \sum_{k=1}^{K-1} dist(\bar{\mathbf{x}}_k, \bar{\mathbf{x}}_{k+1}). \tag{11}$$

- Global silhouette index (SIL) [32]: for a given point $\mathbf{x} \in X_k$, let $a(\mathbf{x})$ denote the mean distance between this point and all other points belonging to the same cluster, X_k; $\eth(\mathbf{x}, k')$ denote the mean distance between this point and all points belonging to a distinct cluster $X_{k'}$ ($k' \neq k$); and

$$b(\mathbf{x}) = \min_{k' \neq k} \eth(\mathbf{x}, k'). \tag{12}$$

The silhouette width of point $\mathbf{x}$ is then calculated as

$$\mathcal{S}(\mathbf{x}) = \frac{b(\mathbf{x}) - a(\mathbf{x})}{\max(b(\mathbf{x}), a(\mathbf{x}))}, \tag{13}$$

and the global silhouette index can be calculated as

$$SIL = \frac{1}{K} \sum_{k=1}^{K} \sum_{\forall \mathbf{x} \in X_k} \frac{\mathcal{S}(\mathbf{x})}{n_k}. \tag{14}$$

- Point biserial index (PB) [12]:

$$PB = \left(\frac{|\mathbf{d}^+|_1}{N_W} - \frac{|\mathbf{d}^-|_1}{N_B} \right) \sqrt{\frac{N_W N_B}{N_T^2}}. \tag{15}$$

with $|\cdot|_1$ denoting the ℓ_1 norm of a vector.
- Calinski-Harabasz index (CH) [8]:

$$CH = \frac{(N-K)\,BGSS}{(K-1)\,WGSS}. \tag{16}$$

- Xie-Beni index (XB) [38]:

$$XB = \frac{1}{N} \frac{WGSS}{\min_{i \neq j} \delta_{ij}}. \tag{17}$$

- Normalized Relative Entropy (NRE) [26]:

$$NRE = \sum_{k=1}^{K} \frac{n_k}{N} \log_2 \left(\frac{n_k}{N} \right). \tag{18}$$

– C index (C) [21]: let s_{min} denote the sum of the N_W smallest elements of $\mathbf{d}^\bullet$ and s_{max} the sum of its N_W largest elements. Then:

$$C = \frac{|\mathbf{d}^+|_1 - s_{min}}{s_{max} - s_{min}}. \tag{19}$$

– Mean of distances to cluster centroids (CM):

$$CM = \frac{1}{K} \sum_{k=1}^{K} \sum_{\forall \mathbf{x}_i \in X_k} dist(\mathbf{x}_i, \bar{\mathbf{x}}_k) \tag{20}$$

– Connectivity (CN) [7,19]:

$$CN = \sum_{i=1}^{N} \sum_{j=1}^{L} \mathcal{I}(\mathbf{x}_i, \eta_j(\mathbf{x}_i)), \tag{21}$$

where $\eta_j(\mathbf{x}_i)$ denotes the j-th nearest-neighbor to point $\mathbf{x}_i$, and $\mathcal{I}(\mathbf{x}_i, \eta_j(\mathbf{x}_i))$ is an indicator function that receives the value of $1/j$ if $\mathbf{x}_i$ and $\eta_j(\mathbf{x}_i)$ belong to the same cluster, zero otherwise.

– Average scattering for clusters (SD_{scat}) [18]:

$$SD_{scat} = \frac{\sum_{k=1}^{K} |\hat{\boldsymbol{\sigma}}_k^2|_2}{K |\hat{\boldsymbol{\sigma}}_\bullet^2|_2}, \tag{22}$$

where $\hat{\boldsymbol{\sigma}}_\bullet^2$ denotes the vector of variance estimates for all attributes in all clusters, and $\hat{\boldsymbol{\sigma}}_k^2$ is the vector of variance estimates for all attributes considering only the observations in the k-th cluster.

– Total separation between clusters (SD_{dis}) [18]:

$$SD_{dis} = \frac{\kappa_{max}}{\kappa_{min}} \sum_{k=1}^{K} \left(\sum_{\substack{k'=1 \\ k' \neq k}}^{K} dist(\bar{\mathbf{x}}_k, \bar{\mathbf{x}}_{k'}) \right)^{-1} \tag{23}$$

where:

$$\kappa_{max} = \max_{k' \neq k} dist(\bar{\mathbf{x}}_k, \bar{\mathbf{x}}_{k'}) \quad \text{and} \quad \kappa_{min} = \min_{k' \neq k} dist(\bar{\mathbf{x}}_k, \bar{\mathbf{x}}_{k'}). \tag{24}$$

– Akaike's Information Criterion (AIC) [33]: The AIC of a first-order multiple linear regression model of class labels on each attribute of the dataset. The linear model is given by

$$class(\mathbf{x}_i) = \hat{\beta}_0 + \hat{\boldsymbol{\beta}}\mathbf{x}_i + e_i, \tag{25}$$

with $class(\mathbf{x}_i)$ denoting the numerically-encoded class of point $\mathbf{x}_i \in X$, $\hat{\beta}_0$ and $\hat{\boldsymbol{\beta}}$ representing the fitted coefficients of the model, and e_i the residual related to the i-th observation in the dataset. The AIC of the model is then given as:

$$AIC = -2\ln(L) + 2q \tag{26}$$

where $\ln L$ is the log-likelihood value estimated for the model (25).

- Bayesian Information Criterion (BIC) [33]: The BIC of the regression model described in (25), which is calculated as:

$$BIC = -2\ln(L) + q\ln(N) \tag{27}$$

3   Methodology

This works aims to investigate the use of clustering measures as meta-features for learning a recommendation system for classification tasks. More specifically, the standard and clustering meta-features are used in the MtL setup designed to predict the accuracy of some popular classification techniques for a given dataset. The objectives are: (i) to determine whether the MtL approach results in an improved recommendation system; (ii) to investigate whether the clustering meta-features contribute to the performance of this recommendation system; and (iii) to characterize the execution times required to extract each set of meta-features.

To train and assess the meta-learner, a meta-dataset was populated with the meta-feature values (both clustering and standard meta-features) for a collection of problem instances, labeled with the performances of known classification techniques. The meta-models were induced by regression techniques and evaluated in the base-level, measuring the predictive performance of the classifiers, and the and meta-level, measuring the impact of recommending the best one. The computational costs for theses processes was recorded for evaluation.

Four hundred datasets from the OpenML repository [36], representing diverse application contexts and domains, were used in this experiment. They were selected considering a maximum number of $10,000$ observations, 500 features and 10 classes, to constrain the computational costs of the process. For each dataset, both standard and clustering meta-features, as described in Sect. 2.2, were computed. The averages of 10-fold cross-validated predictive accuracies achieved by each of five classification techniques, for each dataset, were also calculated for labeling the meta-examples.

The classification approaches used were: C4.5 decision tree [27] with pruning based on subtree raising; k-Nearest Neighbors (kNN) model [22] with $k = 3$; Multilayer Perceptron (MLP) [20] with learning rate of 0.3, momentum of 0.5 and a single hidden layer; Random Forest (RF) [5] with 500 trees; and Support Vector Machine (SVM) [10] with radial basis kernel. These hyper-parameter values were defined following the standard configurations of the implementations used.

This process resulted in a meta-dataset containing 130 meta-features (112 standard and 18 clustering) and 400 samples for each classifier, labeled by the mean accuracy of the classification method. This meta-dataset was then used to train regression models to estimate the expected accuracy of each classifier, as a function of the meta-features. Five regression techniques known to have different biases were tested: Classification And Regression Trees (CART) [6] algorithm with pruning; Distance-weighted k-Nearest Neighbor (DWNN) [22] with $k = 3$ and Gaussian kernel; Multilayer Perceptron Regression (MLPR) [20] with learning rate of 0.3, momentum of 0.5 and a single hidden layer; Random Forest Regressor (RFR) [5] with 500 trees; and Support Vector Regression (SVR) [10] using radial basis kernel. As with the classifiers, the regressor hyper-parameters were also set as the default values of the implementations used without any problem-specific tuning.

To train and evaluate the regression models, 10-fold cross-validation rounds were also executed. The models obtained were evaluated for quality, considering a comparison to two simple baselines: Random (RD) and Default (DF). The RD baseline represents the observed performance of a randomly chosen classifier in the meta-base. The DF baseline was set as the accuracy of the classifier that most often presented the best classification performance across all datasets.

The final step is the analysis of the trade-off between the computational cost of each set of meta-features and that of evaluating all classifiers, in a cross-validation setup, through direct comparison of single threaded experiments. The experiments were done in a cluster node with two Intel Xeon E5-2680v2 processors and 128 GB DDR3. The standard meta-features, as well as some of the clustering ones, are provided by the `mfe` package[1].

4   Experimental Results

Since the main goal of algorithm recommender systems using MtL is to suggest, within the known options, the algorithm with the most appropriate bias to a particular dataset [4], the first analysis of this experiment focuses on such objective. Figure 1 presents the results of two analyses on the meta-base involving the classifiers on the selected 400 datasets.

The distribution of accuracy values, summarized by the boxplots in Fig. 1a, suggests that all approaches have generally similar distributions of performance values across the datasets employed, with reasonably high median accuracies for most problems. Random Forest has a slightly higher median value than the others, while kNN has the lowest median and the largest variability in performance values. Figure 1b shows how many times each algorithm was the "winner", i.e. the number of datasets for which each presented the best performance. These results imply that the set of choices is considered adequate for MtL since each one was the best performing in a non-empty subset of the dataset. It is clear from the figures that RF was most frequently the best classifier in the dataset, which means it will be the most probable choice for the random baseline recommender system and the fixed choice for the default baseline one.

To investigate their quality on the task of learning to recommend a classifier, the selected regressors (MLPR, CART, DWNN, SVR, and RFR) and the

---

[1] https://github.com/rivolli/mfe.

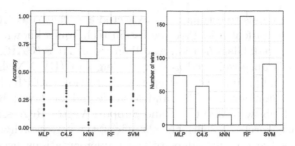

(a) Distribution of accuracies per classifier.

(b) Number of times each
classifier was ranked as
best.

Fig. 1. Performance of classifiers over the 400 datasets.

two baselines regressors (RD and DF) were applied to the meta-dataset considering two sets of meta-features: the 112 standard meta-features (described in
Sect. 2.1) and the full set of 130 meta-features (incorporating the 18 clustering
measures from Sect. 2.2). Figure 2 shows the average normalized mean squared
error (NMSE) of each meta-regressor on predicting the expected accuracy of the
classifiers for each set of meta-features, estimated by 10-fold cross-validation.

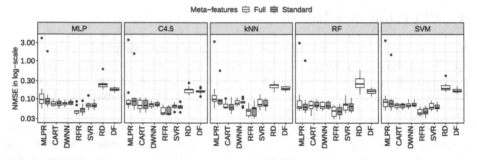

Fig. 2. The log-scaled NMSE for each combination of meta-feature set, classification
and regression models.

The boxplots indicate that all meta-regressors provided better predictions
than the baseline approaches (with the exception of a single observation for the
MLPR regressor). RFR seems to have a slight advantage when compared to the
other regressors, for both sets of features. MLPR, CART, DWNN and SVR show
a similar performance distribution in most cases, with MLPR having a higher
interquartile range and a few outliers.

Although visual inspection does not necessarily provide a clear winner
between the two sets of meta-features, it is sufficient to indicate that: (*i*) they
are clearly less error-prone than the baselines, and (*ii*) their NMSE values show

a relatively low variance. More objectively, an ANOVA model followed by Dunnett pairwise comparisons indicates a statistically significant difference at the 95% confidence level between the MtL algorithms (in aggregate) against the two baselines ($p < 10^{-10}$ for both the MtL×DF and MtL×RD comparisons). The effects of different meta-feature sets, classifier methods and distinct replicates were removed by blocking [23].

A second analysis was conducted to investigate the effect of adding the clustering meta-features on the predictive ability of the meta-regressors. This analysis was performed by removing the two baseline methods and fitting a blocked ANOVA model [23] on rank-transformed data (required in this case to meet the ANOVA assumptions), with the meta-feature sets as the experimental factor and the classifier methods, meta-regressors, and replicates as blocking factors. This test reveal a statistically significant positive effect of adding the clustering ones on the performance of the best regressor, RFR, across all classifiers (paired t-test for standard×full meta-feature sets, $p = 0.036$).

Figure 3 shows the results of a direct comparison between the recommended classifier. The x-axis lists the meta-regressors and the y-axis shows the percent differences in accuracy, averaged over all datasets. As seen in the analysis of Fig. 2 there is a noticeable advantage of using the MtL approaches over the random baseline (RD), as well as against the fixed choice of RF, which is best overall classifier (DF), in the case of the three most successful regressors. The magnitude of the gains is also substantial, with 40–50% gains over RD and 10–20% gains over DF in the case of DWNN, RFR and SVR.

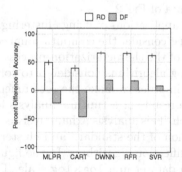

Fig. 3. Differences between base-classifier accuracies over baselines. Vertical lines represent standard errors.

The impact of the meta-features can be further investigated via the analysis of the RFR model. Figure 4 shows the most relevant meta-features, according to their individual contribution to the reduction of the mean squared error, MSE (measured as the increase in MSE when that meta-feature was omitted), as well as their corresponding groups. The x-axis lists the meta-features in decreasing order of relevance, with the mean change in MSE shown in the y-axis. Vertical bars indicate standard errors of estimation.

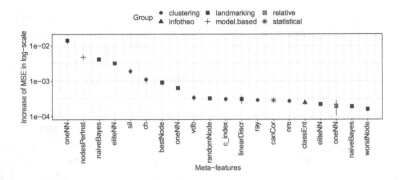

Fig. 4. Top-ranked meta-features selected by the RFRs in log-scale, based on the average increase in MSE when the meta-feature is omitted.

The standard meta-features are more strongly represented, as expected due to their informative value [31]. Some meta-features appear more than once due to distinct summarization functions being used. Landmarking measures, which mainly relate to the performance of simple meta-models induced by the kNN, Naïve Bayes and simple node DT algorithms, represent over half of the top features. Interestingly, six of the 18 clustering features are ranked in the 20 most relevant for the RFR model, in this order: *sil*, *ch*, *ray*, *vdb*, c_index and *nre*. This suggests that these features are relevant to the MtL task, at least for the case of the RFR meta-regressor, which would corroborate the result of the hypothesis test performed on the data of Fig. 2.

Finally, the practical applications of using clustering meta-features for a MtL recommender system must consider the computational costs involved. The trade-off between the runtime of the characterization process and the evaluation of all alternative data modeling algorithms considered to solve the task under study should favor the former in order to make the recommendation system useful.

Figure 5 presents the results of a runtime analysis to evaluate this trade-off for the proposed approach. This analysis exhaustively compared the *single-thread* runtime cost for extraction of the standard and clustering meta-features to the cost for running all classification algorithms. In the figure, each point represents the cost for processing a dataset in a `log×log` scale. The *x*-axis shows the cost for running all classifiers while the *y*-axis shows the cost for extracting the meta-features. This enables a straightforward visual analysis: if a point is above the line, the *y* value is less than the *x* value, thus running the classifiers is more expensive than extracting the features.

The standard meta-features are cheaper than running all classification algorithms in around 89% of the cases. The clustering features are considerably cheaper than that running all classification algorithms. This low cost in computation, associated with the informative nature of some of the clustering meta-features (seen in Fig. 4), suggests their usefulness in, and their potential for, improving MtL recommender systems.

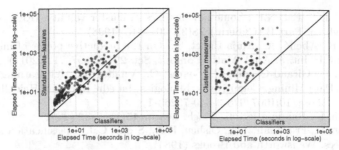

(a) Runtime of the standard (b) Runtime of the clustering
meta-features.   meta-features.

Fig. 5. Execution times for computing meta-features compared to applying all classifiers.

5 Conclusions and Future Works

This work investigated both the gains in performance associated with MtL in general, and the use of clustering meta-features in a recommender system for classification algorithms. Experimental results showed that the selected classifiers were adequate for a recommender system and that using the recommended model resulted in improvements over a random or fixed (best expected) algorithm choice.

The results presented in this paper suggest two tentative conclusions: (i) that meta-learning, as an approach to recommend classifiers for unseen problems, has the potential to provide good choices with a reduced computational budget; and (ii) that clustering-based meta-features are suitable to enhance this MtL task, which may indicate that they are able to capture relevant properties from datasets to describe the performance of classifiers.

Future works include investigating the impact of the underlying grouping structure of the datasets in the adequacy of the clustering meta-features, as well as investigating the correlations of the meta-features to enable the selection of a more efficient, parsimonious subset of meta-feature. The incorporation of a more diverse set of classification algorithms for the recommender system and the incorporation of hyper-parameter tuning for the classifiers and regressors also represent interesting new areas for research.

Acknowledgment. Research carried out using the computational resources of the Center for Mathematical Sciences Applied to Industry (CeMEAI) funded by FAPESP (grant 2013/07375-0).

References

1. Baker, F.B., Hubert, L.J.: Measuring the power of hierarchical cluster analysis. J. Am. Stat. Assoc. **70**(349), 31–38 (1975)

2. Bezdek, J.C., Pal, N.R.: Some new indexes of cluster validity. IEEE Trans. Syst. Man Cybern. Part B (Cybern.) **28**(3), 301–315 (1998)
3. Bilalli, B., Abelló, A., Aluja-Banet, T.: On the predictive power of meta-features in OpenML. Int. J. Appl. Math. Comput. Sci. **27**(4), 697–712 (2017)
4. Brazdil, P., Giraud-Carrier, C., Soares, C., Vilalta, R.: Metalearning - Applications to Data Mining. Cognitive Technologies, 1st edn. Springer, Heidelberg (2009). https://doi.org/10.1007/978-3-540-73263-1
5. Breiman, L.: Random forests. Mach. Learn. **45**(1), 5–32 (2001)
6. Breiman, L., Friedman, J.H., Olshen, R.A., Stone, C.J.: Classification and Regression Trees. Wadsworth and Brooks (1984)
7. Brock, G., Pihur, V., Datta, S., Datta, S.: clValid: An R package for cluster validation. J. Stat. Softw. **25**(4), 1–22 (2008). http://www.jstatsoft.org/v25/i04/
8. Caliński, T., Harabasz, J.: A dendrite method for cluster analysis. Commun. Stat. Theory Methods **3**(1), 1–27 (1974)
9. Castiello, C., Castellano, G., Fanelli, A.M.: Meta-data: characterization of input features for meta-learning. In: Torra, V., Narukawa, Y., Miyamoto, S. (eds.) MDAI 2005. LNCS (LNAI), vol. 3558, pp. 457–468. Springer, Heidelberg (2005). https://doi.org/10.1007/11526018_45
10. Cristianini, N., Shawe-Taylor, J.: An Introduction to Support Vector Machines and Other Kernel-based Learning Methods. Cambridge University Press, Cambridge (2000)
11. Davies, D.L., Bouldin, D.W.: A cluster separation measure. IEEE Trans. Pattern Anal. Mach. Intell. **PAMI-1**(2), 224–227 (1979)
12. Desgraupes, B.: clusterCrit Vignette (2018). https://CRAN.R-project.org/package=clusterCrit/vignettes/clusterCrit.pdf
13. Dua, D., Graff, C.: UCI machine learning repository (2017). http://archive.ics.uci.edu/ml
14. Dunn, J.C.: Well-separated clusters and optimal fuzzy partitions. J. Cybern. **4**(1), 95–104 (1974)
15. Filchenkov, A., Pendryak, A.: Datasets meta-feature description for recommending feature selection algorithm. In: Artificial Intelligence and Natural Language and Information Extraction, Social Media and Web Search FRUCT Conference (AINL-ISMW FRUCT), vol. 7, pp. 11–18 (2015)
16. Garcia, L.P.F., Lorena, A.C., de Souto, M.C.P., Ho, T.K.: Classifier recommendation using data complexity measures. In: 24th International Conference on Pattern Recognition (ICPR), pp. 874–879 (2018)
17. Garcia, L.P.F., Rivolli, A., Alcobaça, E., Lorena, A.C., de Carvalho, A.C.P.L.F.: Boosting meta-learning with simulated data complexity measures. Intell. Data Anal. **24**(5), 1011–1028 (2020)
18. Halkidi, M., Batistakis, Y., Vazirgiannis, M.: On clustering validation techniques. J. Intell. Inf. Syst. **17**(2–3), 107–145 (2001)
19. Handl, J., Knowles, J.: Exploiting the trade-off — the benefits of multiple objectives in data clustering. In: Coello Coello, C.A., Hernández Aguirre, A., Zitzler, E. (eds.) EMO 2005. LNCS, vol. 3410, pp. 547–560. Springer, Heidelberg (2005). https://doi.org/10.1007/978-3-540-31880-4_38
20. Haykin, S.S.: Neural Networks: A Comprehensive Foundation. Prentice Hall, Hoboken (1999)
21. Hubert, L., Schultz, J.: Quadratic assignment as a general data analysis strategy. Br. J. Math. Stat. Psychol. **29**(2), 190–241 (1976)
22. Mitchell, T.M.: Machine Learning. McGraw Hill Series in Computer Science. McGraw Hill, New York (1997)

23. Montgomery, D.C.: Design and Analysis of Experiments, 5th edn. Wiley, Hoboken (2000)
24. Muñoz, M.A., Villanova, L., Baatar, D., Smith-Miles, K.: Instance spaces for machine learning classification. Mach. Learn. **107**(1), 109–147 (2018). https://doi.org/10.1007/s10994-017-5629-5
25. Pfahringer, B., Bensusan, H., Giraud-Carrier, C.G.: Meta-learning by landmarking various learning algorithms. In: 17th International Conference on Machine Learning (ICML), pp. 743–750 (2000)
26. Pimentel, B.A., de Carvalho, A.C.P.L.F.: A new data characterization for selecting clustering algorithms using meta-learning. Inf. Sci. **477**, 203–219 (2019)
27. Quinlan, J.R.: Induction of decision trees. Mach. Learn. **1**(1), 81–106 (1986)
28. Ray, S., Turi, R.H.: Determination of number of clusters in k-means clustering and application in colour segmentation. In: 4th International Conference on Advances in Pattern Recognition and Digital Techniques (ICAPRDT), pp. 137–143 (1999)
29. Reif, M., Shafait, F., Goldstein, M., Breuel, T., Dengel, A.: Automatic classifier selection for non-experts. Pattern Anal. Appl. **17**(1), 83–96 (2014)
30. Rice, J.R.: The algorithm selection problem. Adv. Comput. **15**, 65–118 (1976)
31. Rivolli, A., Garcia, L.P.F., Soares, C., Vanschoren, J., de Carvalho, A.C.P.L.F.: Characterizing classification datasets: a study of meta-features for meta-learning. CoRR abs/1808.10406, 1–49 (2019)
32. Rousseeuw, P.J.: Silhouettes: a graphical aid to the interpretation and validation of cluster analysis. J. Comput. Appl. Math. **20**, 53–65 (1987)
33. Sakamoto, Y., Ishiguro, M., Kitagawa, G.: Akaike Information Criterion Statistics. Springer, Netherlands (1986)
34. Smith-Miles, K.A.: Cross-disciplinary perspectives on meta-learning for algorithm selection. ACM Comput. Surv. **41**(1), 1–25 (2008)
35. Stephenson, N., et al.: Survey of machine learning techniques in drug discovery. Curr. Drug metab. **20**(3), 185–193 (2019)
36. Van Rijn, J.N., et al.: OpenML: a collaborative science platform. In: European Conference on Machine Learning and Knowledge Discovery in Databases (ECML/PKDD), pp. 645–649 (2013)
37. Vukicevic, M., Radovanovic, S., Delibasic, B., Suknovic, M.: Extending meta-learning framework for clustering gene expression data with component-based algorithm design and internal evaluation measures. Int. J. Data Min. Bioinform. (IJDMB) **14**(2), 101–119 (2016)
38. Xie, X.L., Beni, G.: A validity measure for fuzzy clustering. IEEE Trans. Pattern Anal. Mach. Intell. **13**(8), 841–847 (1991)

Fast Movelet Extraction and Dimensionality Reduction for Robust Multiple Aspect Trajectory Classification

Tarlis Tortelli Portela$^{(\boxtimes)}$ [iD], Camila Leite da Silva[iD], Jonata Tyska Carvalho[iD], and Vania Bogorny[iD]

Universidade Federal Santa Catarina, Florianópolis, Santa Catarina, Brazil
{tarlis.portela,camila.leite.ls}@posgrad.ufsc.br,
{jonata.tyska,Vania.Bogorny}@ufsc.br

Abstract. Mobility data analysis has received significant attention in the last few years. Enriching spatial-temporal trajectory data with semantic information, which is the definition of Multiple Aspect Trajectories, presents lots of opportunities, but also many challenges. Regarding trajectory classification, the state-of-the-art method called MASTER-Movelets has shown to have the best classification accuracy over several datasets. Indeed, this method generates interpretable patterns called movelets which are the most discriminant sequences of points. Despite its increased performance, the method is computationally expensive and does not scale well, which makes its application unfeasible for large datasets. In this paper we propose a pivot based approach to reduce the search space, selecting only most promising trajectory points to extract movelets. We additionally provide a method to define a limited number of semantic dimensions for movelets. Experiments show that the proposed method is at least 50% faster for extracting the movelets, and shows a average drop of 82% of input to the classification models while keeping a similar classification accuracy level. Additionally, our scalability analysis with respect to computation time shows that the proposed method scales better than the other methods as the dataset grows in number of points, trajectories and dimensions.

Keywords: Trajectory classification · Multiple aspect trajectories · Data mining · Movelets

1 Introduction

Mobility data analysis is important to different purposes and applications. The movement of people, vehicles, ships, and hurricanes are examples of mobility data. This data are represented as a sequence of points located in space and time, called *moving object trajectories*.

In 2016 emerged the concept of multiple aspect trajectories [3,10], a broader concept in which spatio-temporal points can be enriched with several semantic

© Springer Nature Switzerland AG 2021
A. Britto and K. Valdivia Delgado (Eds.): BRACIS 2021, LNAI 13073, pp. 468–483, 2021.
https://doi.org/10.1007/978-3-030-91702-9_31

Fig. 1. Example of multiple aspect trajectory.

aspects, as shown in Fig. 1. The meaning of semantic aspect is any type of information that is neither spatial nor temporal. This information represents any aspect as the name of a visited place or Point of Interest (POI), the price or rate of the place, the weather condition, the transportation mode, the individual's mood, etc. This new trajectory representation poses new challenges on trajectory data mining, specially in classification, which is the problem we focus in this paper.

Trajectory classification is the task of finding the class label of the moving object based on its trajectories [8]. It is important for identifying the strength level of a hurricane [8], the transportation mode of a moving object [2], the type of a vessel (cargo, fish, tourism, etc.) [8], the user that is the owner of a trajectory [5], etc. The great challenge related to multiple aspect trajectory classification is the large number and the heterogeneity of the dimensions associated to each trajectory point. Most works for trajectory classification have developed methods for a specific problem or application, and considered only space and time information, as summarized in [9]. On the contrary of traditional classification literature that propose new classifiers, trajectory classification relies on developing new methods for feature extraction to feed a classifier.

A recent method designed for robust trajectory classification and that has been specifically developed for multiple aspect trajectories is MASTERMovelets [5]. It extracts the subtrajectories that better discriminate each class, which are called *movelets*, and that are used as input to classification algorithms. MASTERMovelets automatically explores all possible dimension combinations in each subtrajectory of any size in the dataset, while seeking for the best subtrajectories for representing the classes. The *movelets* represent the behavior of a class, and they are normally *frequent* subtrajectories inside a class, i.e., they are movement subsequences that are recurrent patterns of a class and not common to other classes. A subtrajectory may have any length in terms of number of points, and MASTERMovelets explores all possible subtrajectory sizes (e.g. one point, two points, etc.). Indeed, it explores all dimension combinations (e.g. space; space and time; space, time and POI category, etc.) in order to choose the most discriminant ones, what makes the method very robust, but also very time consuming. Therefore, to use it on real world datasets, more efficient strategies must be developed for extracting movelets.

In this work we propose a new method for discovering movelets, that is based on [5], and uses a pivot strategy, called *SUPERMovelets*. The pivots are the trajectory parts (or subtrajectories) inside a class that will potentially generate the best movelet candidates. By introducing this concept, initially our approach reduces the search space for movelet discovery by computing the distance of one subtrajectory only to subtrajectories within the same class, finding the pivots. Moreover, instead of using all dimensions, our method automatically selects the number of dimensions that better characterize each class. As a consequence, the proposed method contributes to a significant dimensionality reduction in terms of movelets that will be used by the classifier for reducing the classification training time. In summary, we make the following contributions: (i) we propose a new and more efficient method for discovering movelets, maintaining similar accuracy levels; (ii) propose a smart strategy that automatically selects the best number of dimensions for each class; (iii) mitigate the curse of dimensionality problem, by reducing the number of movelets; (iv) present a robust experimental evaluation to show the computational time reduction and scalability.

The rest of the paper is organized as follows: Sect. 2 presents the main works related to multiple aspect trajectory classification, Sect. 3 presents the main concepts for understanding our approach, Sect. 4 describes our method, Sect. 5 presents the experimental evaluation, and Sect. 6 concludes the paper.

2   Related Works

Existing methods for trajectory classification do not propose new classifiers. The focus is usually on discovering a set of discriminant features that better characterize a trajectory class [4,8,9]. In this paper we limit the state of the art to works that consider semantic aspects of trajectories. A complete list of works for both raw and multiple aspect trajectory classification can be found in [9].

The first works to use some semantic information for classification are reported in [14,15]. These works collected data from smartphones such as latitude, longitude, altitude, and date, plus deriving other features like speed, and matching bus or metro lines to classify the user movement in walking, running, or driving. The works of [6,7,19] consider only semantic dimensions. In [7] the authors use the semantics of the roads to segment trajectories for classification of vehicles. The methods presented in [6] and [19] use the POI identifier to classify the moving object.

The algorithms Movelets [4] and MASTERMovelets [5] discover *relevant subtrajectories*, which are called movelets, without the need of extracting other features. However, the processing time and computational cost are extremely high. The main difference between these methods is that Movelets encapsulates the distances of all trajectory dimensions in a single distance value, while MASTERMovelets (that has a higher complexity) keeps the distance of each dimension in a vector of distances, and therefore achieves a much better accuracy than Movelets when using all data dimensions. MASTERMovelets [5] was specifically developed for multiple aspect trajectory classification, and it largely outperformed state-of-the-art methods in terms of accuracy. Furthermore, the resulting *movelets* are interpretable, giving insights about the data, so the classification results are explainable.

The work of [16] performs much faster than Movelets [4], providing a similar classification accuracy level, but for a single dimension. When combining several dimensions, the user must manually test and select the best dimension combinations, while MASTERMovelets is able to automatically select the best dimension combinations and generate movelets with heterogeneous and different numbers and types of dimensions. Another recent work is MARC [12], which uses word embeddings and encapsulates all trajectory dimensions including space, time and semantics to feed a neural network classifier. On the spatial dimension this is the first work that uses the geoHash [11] combined to other dimensions. It reaches a very high accuracy, outperforming the Movelets [4] when using all dimensions, but as the classifier is limited to neural networks, the resulting patterns are not interpretable.

3 Basic Concepts

This section presents the main concepts that are necessary for better understanding this work, and are based on [5], as they share the same structures. We start with the concept of trajectory in Definition 1:

Definition 1. *Multiple Aspect Trajectory: a multiple aspect trajectory T_i is a sequence of m elements $T_i = \langle e_1, e_2, ..., e_m \rangle$, where each element is characterized by a set of l dimensions $D = \{d_1, d_2, ..., d_l\}$, also called aspects.*

In order to simplify the problem, we assume that all trajectory elements have the same number of aspects. Hereafter we will refer to trajectory aspects as dimensions.

The behavior patterns of moving objects are normally characterized by a trajectory part, and not an entire trajectory. A trajectory part is called subtrajectory. The subtrajectory concept is detailed in Definition 2:

Definition 2. *Subtrajectory: given a trajectory T_i of size m, a subtrajectory $s_{a,b} = \langle e_a, e_{a+1}..., e_b \rangle$ is a contiguous subsequence of T_i starting at element e_a and ending at element e_b, where $1 \leq a \leq b \leq m$. The subtrajectory $s_{a,b}$ can be represented by all the dimensions D or a subset of dimensions $D' \subseteq D$. The length of the subtrajectory is defined as $w = |s_{a,b}|$. In addition, the set of all subtrajectories of length w in T_i is represented by $S_{T_i}^w$, and the set of all subtrajectories of all lengths in T_i is $S_{T_i}^*$.*

To generate every possible subtrajectory from each trajectory of the dataset, it is necessary to consider all possible subtrajectory sizes and combination of dimensions. Since an element has multiple and heterogeneous dimensions, first the distances between all elements in the trajectories are calculated, as defined in Definition 3. The dimensions often store different data types, e.g., categorical, numerical and so on. Hence, using a vector of distances instead of one aggregated value enable the analysis of different dimensions combinations, and the use custom distance functions for each dimension.

Definition 3. *Distance vector between two multidimensional elements: the distance between two multidimensional elements e_i and e_j represented by c dimensions is $dist\_e(e_i, e_j)$, that returns a distance vector $V = (v_1, v_2, ..., v_c)$, where each $v_k = dist_k\_e(e_i, e_j)$ is the distance between the two elements at dimension k, and respects the property of symmetry $dist_k\_e(e_i, e_j) = dist_k\_e(e_j, e_i)$.*

In order to perform trajectory classification, it is necessary to seek for similarities among trajectories of the same class, and compare subtrajectories of the same length. One way is to seek a position with the minimum distance vector between a subtrajectory and a trajectory, this position is called *best alignment*. Furthermore, the best alignment implies pairing a subtrajectory to a trajectory either in a subset or all dimensions together. It begins with calculating distances between the elements of the subtrajectory and a trajectory T_i. For subtrajectories of two or more elements, each alignment position use a representative vector of distances that is a sum of element distances for each dimension. Then, the best alignment consists in ranking the distances vector in each dimension for all possible positions, and getting the average rank of distance with the lowest value. As an example of pairing a subtrajectory into a trajectory, Fig. 2 (left) presents the best alignment of a subtrajectory (movelet candidate) in the trajectory T_i, in the spatial dimension. We can see that the best alignment, highlighted with a rectangle, is the position where the distance between subtrajectory of $\mathcal{M}_1$ and the trajectory is minimal.

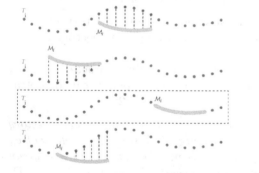 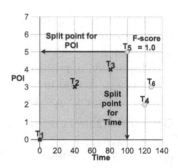

Fig. 2. (left) Example of subtrajectory best alignment. (right) Example of split point selection for a movelet candidate with the dimensions time and POI rating.

The combination of subtrajectories with a subset of the trajectory dimensions are called *movelet candidates*, and they are evaluated and pruned in order to identify the most representative subtrajectories of each class. The *movelet candidates* are described in Definition 4:

Definition 4. *Movelet Candidate: a movelet candidate $\mathcal{M}$ from a subtrajectory $s_{start,end}$ is a tuple $\mathcal{M} = (T_i, start, w, C, \mathbb{W}, quality, sp)$, where T_i is a trajectory of the dataset T; start is the position in T_i where the subtrajectory begins, and*

w is the subtrajectory length ($w = |s_{start,end}|$); C is a subset of dimensions such that $C \subseteq D$; $\mathbb{W}$ is a set of pairs ($W^s_{T_j}, class_{T_j}$), where $W^s_{T_j}$ is a distance vector of the subtrajectory $s_{(start,end)}$ to each trajectory T_j in $\boldsymbol{T}$. The distances are calculated using the best alignment between $s_{(start,end)}$ and each trajectory T_j; quality is a relevance score given to the candidate $\mathcal{M}$; sp is a set of distance values, called split points, that better divide the classes used to measure the candidate relevance.

The set $\mathbb{W}$ is the *movelet candidate* best alignments, it represents the distances (for each dimension) to all trajectories in the dataset, and one best alignment is selected as the split point vector sp. The split point sp is a *given* point that divides the multiple dimensional space. Figure 2 (right) presents an example of split point selection for a given movelet candidate extracted from the class represented by the **x** class. This split point is the best alignment in a trajectory of an opposite class **o** that best separates the classes considering dimensions together, which is a similar concept used in Support Vector Machines (SVM). In the example, we evaluate each trajectory from the **o** class (t_4, t_5, t_6) using the values of their best alignment to separate the classes. The split point is selected by calculating the F-score value for each trajectory that is not the target class. The chosen split point is the one point that represents the alignment between the movelet candidate and the trajectory t_5, once it gives the best F-score value. After qualifying each *movelet candidate*, only those with the best *quality*, without point overlapping in the trajectory T_i from where it was extracted, are kept and are called *movelets*.

Definition 5. *Movelet: given a trajectory T_i, and a movelet candidate $\mathcal{M}_x$ containing a subtrajectory s with $s_{a,b} \subseteq T_i$, $\mathcal{M}_x$ is a movelet if for each movelet candidate $\mathcal{M}_y$ containing a subtrajectory $u_{f,g}$ with $u_{f,g} \subseteq T_i$ that overlaps $s_{a,b}$ in at least one element, $\mathcal{M}_x.quality > \mathcal{M}_y.quality$.*

The output is a table $|\boldsymbol{T}| \times |\boldsymbol{M}|$ that will be used as input for training the classification algorithms. Each row of the table is a trajectory, each column is a movelet from the set of movelets $\boldsymbol{M}$, and the table values are 1 and 0, in order to represent the presence or absence of a movelet in a trajectory. A movelet is present in a trajectory when its best alignment distances to that trajectory is lower than the split point values. It means that the movelet *covers* the trajectory considering the movelet dimensions C and sp values.

4   SUPERMovelets

In this Section we propose SUPERMovelets[1], a new method for reducing the search space and the computational cost for finding *movelets*. Our method is inspired by [18] and [13], and finds *pivots*, which are the relevant trajectory parts for extracting the movelets. SUPERMovelets also automatically finds a

---

[1] Source code will be published if accepted for publication.

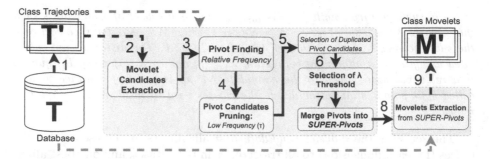

Fig. 3. Overview of our proposed method.

threshold λ for limiting the maximum number of trajectory dimensions required in each movelet candidate, as an alternative for not exploring all dimension combinations. The pivots and maximum number of dimensions λ are extracted for each class individually. The SUPERMovelets can be divided in three main steps: (i) pivot finding, (ii) the automatic selection of λ and (iii) *movelets* extraction.

Figure 3 presents an overview of our method. Given a trajectory dataset **T**, the algorithm starts selecting the trajectories of one class **T'** (step 1) for extracting the pivots. For each trajectory T_i in the class trajectory set, SUPER-Movelets extracts all movelet candidates (step 2). The method calculates the relative frequency to each movelet candidate, with respect to the trajectories of the same class as T_i (step 3) and a threshold τ, the most frequent ones will become the pivot candidates. After finding the candidates, SUPERMovelets performs the pivot candidate pruning, where the ones with relative frequency less than τ are removed (step 4). This step is detailed in Sect. 4.1. Next, it selects only the pivot candidates with higher frequency that repeats at least once in the class trajectories (step 5). We suppose that the most discriminant trajectory parts are the ones that repeat in total or partially in the trajectories of the same class. It calculates the λ threshold based on the common number of dimensions in the pivots (step 6), detailed in Sect. 4.2. The pivot points are compared and merged into *SUPER-Pivots* when they are neighbors or have overlapping points (step 7). For last (step 8), the *SUPER-Pivots* and the maximum number of dimensions λ will be used for movelet extraction following the same steps as the MASTERMovelets method, which for each trajectory outputs the *movelets* and aggregates the result to the class set of movelets **M'** (step 9).

4.1   Pivot Finding

In order to reduce the number of subtrajectories tested for generating movelets, SUPERMovelets identifies the most promising ones of each class, called *pivots* (Definition 6). The pivots are defined by the frequency they appear in each trajectory, assuming that most frequent subtrajectories are more relevant for movelet extraction. By doing so, the movelets search space is reduced, and so is the computational cost, making trajectory classification using movelets faster and scalable. Assuming that moving objects have recurrent movement patterns, and that these patterns are more frequent in trajectories of each individual and less frequent in trajectories of other individuals, and considering several data dimensions, we hypothesize that frequent subtrajectories in a class are very discriminant and can lead to similar classification accuracy results as MASTER-Movelets while reducing its computational cost.

Definition 6. *Pivot Candidate: a pivot candidate is a tuple* $\mathcal{P} = (T_i, s_{(start,end)}, C, \mathbb{W}, quality_{piv})$, *where* T_i *is a trajectory from* $\boldsymbol{T}$; $s_{start,end}$ *is a subtrajectory extracted from* T_i, C *is a subset with the candidate dimensions,* $C \subseteq D$; $\mathbb{W}$ *is a set of pairs* $(W_{T_j}^s, class_{T_j})$, *that contains the distances of the best alignment of* $\mathcal{P}$ *to every trajectory* $\boldsymbol{T'}$ *that belongs to the same class as* T_i. *The* $W_{T_j}^s$ *is a distance vector of the subtrajectory* $s_{(start,end)}$ *to each trajectory* T_j *in* $\boldsymbol{T'}$. *The* $quality_{piv}$ *is the relative frequency of the* $\mathcal{P}$ *in its class.*

The pivot candidates represent subtrajectories of different sizes and with any dimension combination from trajectories of a given class, and their quality is based on the relative frequency that they appear in the class. The quality of the pivots is measured considering all dimensions that are present in the subtrajectory that originates the pivot candidate. Equation (1) describes this quality function for a *pivot candidate* ($\mathcal{P}$) of a class:

$$quality_{piv} = \frac{\sum_{d=1}^{d=|C|} freq_{piv}(\mathcal{P}, d, \mathbf{T'})}{|C|} \tag{1}$$

Where C is the set of trajectory dimensions of $\mathcal{P}$, and the quality is the average proportion that $\mathcal{P}$ occurred in trajectories of the class in each dimension d of C. As there are different combinations of dimensions in each $\mathcal{P}$, we measure the relative frequency as the average count that a $\mathcal{P}$ occurred in each dimension, as described in (2).

$$freq_{piv}(\mathcal{P}, d, \mathbf{T'}) = \frac{\sum_{i=1}^{i=|\mathbf{T'}|} \begin{cases} 1, & W_{T_i}^s[d] = 0 \\ 0, & \text{otherwise.} \end{cases}}{|\mathbf{T'}|} \tag{2}$$

Considering that $W_{T_i}^s[d]$ is the distance of the *pivot candidate* for dimension d, and that $|\mathbf{T'}|$ is the number of trajectories of a given class. The relative frequency of $freq_{piv}(\mathcal{P}, d, \mathbf{T'})$ is the average occurrence of $\mathcal{P}$ in each trajectory of its class. Figure 4 (a) presents an example of the quality given to two pivot candidates with a single dimension each, both extracted from trajectories of *Class 4* in the figure, which has five trajectories in total (T_7 to T_{11}). The first pivot candidate piv_a (Fig. 4a) has two points, and was extracted from trajectory T_7. It occurred in trajectories T_7, T_8, T_9 and T_{11}, which represent 80% of the trajectories from *Class 4*, thus the quality of this pivot candidate is 0.8. The second pivot candidate piv_b has three points, and was extracted from trajectory T_9. It occurred in trajectories T_9 and T_{11}, which represent 40% of the trajectories of *Class 4*, thus the quality of this pivot candidate is 0.4.

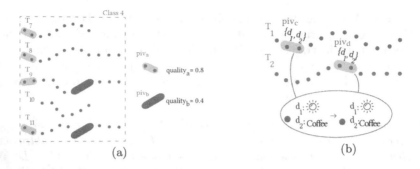

Fig. 4. Filtering by: (a) quality proportion of the *pivot candidates*; (b) redundant *pivot candidates*, both of the same size and trajectory dimensions.

After qualifying each *pivot candidate*, they are filtered by **low quality** and **redundancy**. The first filter consists in discarding *pivot candidates* that have quality lower than τ. The second filter consists in the application of a *duplicate search function* over the *pivot candidates* that passed from the first filter, i.e., keeping only pivot candidates that occur at least two times. Figure 4 (b) presents an example of the redundancy that is tackled by the second filter. The pivot candidates piv_c and piv_d are considered redundant when both have the same size (two trajectory points) and the same trajectory dimensions with the same values (equal POI names and equal weather conditions).

Finally, the Pivot Finding merges the resultant *pivots* for identifying in each trajectory which are the *SUPER-Pivots*. The main objective of this merging step is to identify consecutive pivots or the ones that overlap. If two pivots are consecutive in the trajectory or they share the same point (they have overlapping points), then the SUPERMovelets ensembles them into one *SUPER-Pivot*.

4.2   Selection of λ Threshold

To combine all trajectory dimensions to generate all possible subtrajectories, seeking for the best dimension combination in each movelet candidate, means that for each subtrajectory there is an explosion of dimension combinations. To limit this combinatorial explosion, the SUPERMovelets automatically selects the maximum number of dimensions that a movelet candidate can have based on the set of **pivots**, by counting the absolute number of dimensions that each pivot has, as given in (3). The mode of the number of dimensions present in the set of **pivots** of a class is chosen as the threshold λ, given in (4). It means that the movelet candidates extracted from the trajectories of that class will not have more dimensions than λ.

$$freq_{\mathbf{P}}(dim) = \sum_{i=1}^{i=|\mathbf{P}|} \begin{cases} 1, & \mathcal{P}_i.|C| = dim \\ 0, & \text{otherwise.} \end{cases} \tag{3}$$

$$\lambda = \underset{dim}{\text{argmax}}\, freq_{\mathbf{P}}(dim) \tag{4}$$

4.3   *Movelets* Extraction

The last step of SUPERMovelets is to extract the *movelets*. This process follows the same steps as described in Sect. 3 and MASTERMovelets [5], with the difference that the *movelet candidates* are only extracted from the SUPER-Pivots, and the dimension combination is limited to λ. With this, we aim to make the trajectory classification faster, scalable and to maintain the same accuracy level obtained when used the original, and larger, dataset.

5   Experimental Evaluation

We used four datasets publicly available and commonly used by state-of-the-art methods. They are check-in trajectories enriched with semantic dimensions from three different Location-Based Social Networks (LBSN). These datasets were also used in [5] to evaluate MASTERMovelets, which makes fairer and easier to compare the results. The trajectories were split in weeks in order to have many trajectory examples for each user as necessary for classification tasks. The class label is the user which is the owner of the trajectories. Table 1 shows the characteristics of each dataset, with the number of trajectories, the size of the smallest and the longest trajectory, the attributes of the dataset and their classes. All datasets have a large number of classes, ranging from 193 to 300.

Table 1. Summary of the used datasets.

Dataset	Description		Dataset	Description	
Brightkite	Traj. size	10–50	Gowalla	Traj. size	10–50
	Trajectories:	7,911		Trajectories:	5,329
	Attributes	Lat, Lon, POI, Time, Weekday		Attributes	Lat, Lon, POI, Time, Weekday
Foursquare NY	Traj. size	10–144	Foursquare Generic	Traj. size	10–144
	Trajectories:	3,079		Trajectories:	3,079
		Lat, Lon, POI, POI Category			
	Attributes:	Time, Weekday, Weather,		Attributes:	POI Category, Time, Weekday,
		Price, Rating			Weather, Price, Rating

Brightkite: is a social media [1] that provides the anonymized user that made the check-in, the POI semantic reference and the spatio-temporal information of where and when the check-in was made. We used a total of 300 random users for analysis, with a filter of a minimum of 10 points and maximum of 50 points per trajectory to guarantee consistency. Trajectories were enriched with the semantic information of the weekday of each check-in, and the resultant dataset has a total of 7, 911 trajectories, with trajectory sizes varying from 10 to 50 points.

Gowalla: has users around the world [1] and has the same dimensions as the Brightkite. From this dataset we also extracted a total of 300 random users, while limiting the trajectory sizes for a minimum of 10 and maximum of 50 points, also enriched with the semantic information of the weekday. The resultant dataset has a total of 5, 329 trajectories, with sizes ranging from 10 to 50 points.

Foursquare: dataset from the Foursquare social media with multiple aspect trajectories in New York, USA [17]. It provides the anonymized user that made the check-in, the POI and the spatial position in time when the check-in was made. We considered trajectories with at least 10 check-ins, resulting in 193 users for this dataset. The points were enriched with the semantic information of the weekday, the POI category from Foursquare API[2], the numerical information of the price and rating of the POIs, and the Weather from Wunderground API[3]. The resultant dataset has a total of 3079 trajectories, with trajectory sizes varying from 10 to 144. The **Foursquare Generic** is composed by the same trajectories from the Foursquare NY, but it consists of a harder problem, where we removed the specific information of spatial position (lat, lon), and the specific semantics of the POI, keeping only the POI Category.

5.1   Experimental Setup

We use three evaluation metrics to assess the performance of the SUPER-Movelets approach: (i) computational cost, (ii) classification Accuracy (ACC) and (iii) F-Score. The datasets were split using stratified holdout with 70% of the data for training and 30% for test. After movelet extraction, we used the Multilayer-Perceptron (MLP), as it is commonly used and achieved the best results in [5,9]. The model was implemented using Python language, and the `keras`[4] package. The MLP has a fully-connected hidden layer with 100 units, a Dropout Layer rate of 0.5, learning rate of 10^{-3} and Output Layer with softmax activation. The network was trained using Adam Optimization to improve the learning time and to avoid categorical cross entropy loss, with 200 of batch size, and a total of 200 epochs for each training.

In the experiments, we calculated the distance for each dimension by using: (i) euclidean distance for the space, (ii) difference for the numerical, and (iii) simple equality (if is equal or not) for the semantics. The experiments were performed in an *Intel(R) Core(TM) i7-6700 CPU* @ 3.40 GHz, with 4 cores and main memory of **32 GB**.

5.2   Results and Discussion

After running MASTERMovelets for movelet extraction with and without using SUPERMovelets, we compare the time spent for completing trajectory classification, the number of *movelet candidates* generated, the number of movelets extracted and the accuracy and F-Score obtained in the trajectory classification. We tested the movelet extraction with and without using the natural to log limit the size of subtrajectories (denoted by the *-Log* suffix) which was originally designed to improve the movelets extraction speed. Table 2 summarizes the results. As we can see, the main conclusion is that the SUPERMovelets

---

[2] https://developer.foursquare.com/.
[3] https://www.wunderground.com/weather/api/.
[4] https://keras.io/.

method makes the movelet extraction at least 70% faster than the cases in which MASTERMovelets is used alone and at least 50% faster than the MASTER-Movelets-Log, while maintaining the same accuracy level.

Table 2. Comparison between the number of candidates, computational time, accuracy and F-Score of each technique.

Dataset		MASTERMovelets	MASTERMovelets-Log	SUPERMovelets	SUPERMovelets-Log
Brightkite	Candidates	13,997,595	4,264,230	757,488	213,002
	Movelets	54,739	56,739	5,514	3,540
	Time	17 h 45 m	9 h 45 m	01 h 29 m	00 h 35 m
	ACC	95.09	94.58	95.63	95.38
	F-Score	95.01	94.37	95.81	95.52
Gowalla	Candidates	11,409,135	3,239,865	628,864	211,090
	Movelets	52,460	54,739	5,576	3,702
	Time	07 h 20 m	3 h 20 m	00 h 55 m	00 h 21 m
	ACC	91.72	91.72	91.32	90.91
	F-Score	92.22	91.84	92.09	91.61
Foursquare NY	Candidates	187,853,655	40,680,660	33,922,970	18,819,626
	Movelets	35,824	36,874	9,444	8,320
	Time	54 h 33 m	18 h 41 m	14 h 40 m	9 h 11 m
	ACC	97.27	96.88	97.18	97.76
	F-Score	96.53	96.48	95.86	96.72
Foursquare generic	Candidates	45,723,636	9,981,405	6,257,312	3,191,411
	Movelets	20,068	23,975	6,861	5,399
	Time	07 h 11 m	02 h 50 m	02 h 04 m	01 h 21 m
	ACC	74.20	74.29	74.59	74.39
	F-Score	74.56	74.47	74.60	73.22

The most expressive difference in processing time is in the Brightkite dataset, where the SUPERMovelets-Log completed the task in 35 m, 94% faster than the MASTERMovelets-Log that made it in 9 h 45. The less expressive difference is found in the Foursquare NY dataset, where the SUPERMovelets-Log took 9 h 11 to finish the task, while the MASTERMovelets-Log took 18 h 41, a reduction of 50.84% in time. An important observation is that besides of the time reduction obtained, and the reduced number of movelet candidates, for any of the classification results the accuracy loss was less than 1.00% for every dataset. The fastest method, the SUPERMovelets-Log even increased the accuracy by 0.9% in the Foursquare NY dataset.

Movelet Candidate Generation Analysis. We evaluated the number of *movelet candidates* generated by each technique, and the subsequently number of movelets found. The most expressive difference is in the Brightkite dataset that SUPERMovelets reduced in 94%, reducing at least 53% (Foursquare NY) in number of generated *movelet candidates* (Table 2).

The proposed method reduce the number of *movelets*, which are the attributes that are used for training the classifier, in at least 65% in comparison to the MASTERMovelets. For example, in the Brightkite dataset, MASTER-Movelets produced a total of 54, 739 movelets, while SUPERMovelets outputted 5, 514, a reduction of 89.92%, while the SUPERMovelets-Log produced 3, 540, a reduction of 93.76% compared to MASTERMovelets-Log. This means that besides reducing the movelet extraction time, it reduced also the dimensionality of the classification task, which is inherently high when using movelets. Indeed, the smallest dimensionality reduction achieved by the SUPERMovelets was 65, 81% in the Foursquare Generic dataset, in which the MASTERMovelets-Log method produced 20, 068 movelets, against 6, 861 generated by SUPER-Movelets, and the SUPERMovelets-Log produced 5, 399, a reduction of 77, 48%.

Scalability Performance Comparison. We evaluate how SUPERMovelets improves the scalability of movelet extraction over different dataset configurations. We used three synthetic datasets with different characteristics, as designed in [5]: (i) with fixed number of 200 trajectories and variation in the trajectory size from 10 to 400 points with 1 dimension, (ii) with fixed trajectory size of 50 points and variation in the number of trajectories in the dataset from 100 to 4000 trajectories with also 1 dimension, and (iii) with fixed number of 200 trajectories and fixed trajectory size of 50 points, but with variation in the number of dimensions, from 1 to 5.

Figure 5 shows the scalability results for each dataset. In any of the datasets, we can see that the time spent by MASTERMovelets increases faster than the others, requiring computational power in a greater scale as the dataset increases. On the other hand, the time spent by SUPERMovelets-Log increases in a smaller scale in every dataset, and is the faster compared technique, as it demands less computation time, with a time reduction of 98.0% in comparison to the MASTERMovelets, and 80.48% in comparison to the MASTERMovelets-Log.

The MASTERMovelets-Log and SUPERMovelets are in middle-term, where the MASTER Movelets-Log outperforms the SUPERMovelets in the dataset of variational trajectory size presented in Fig. 5 (a), as this is the most effective scenario for limiting the *movelet candidate* size to the ln of the trajectory size. It is noteworthy that SUPERMovelets outperforms the MASTERMovelets-Log in the dataset of fixed trajectory size presented in Fig. 5 (b) and has a similar behavior in the dataset of varied trajectory dimensions of Fig. 5 (c).

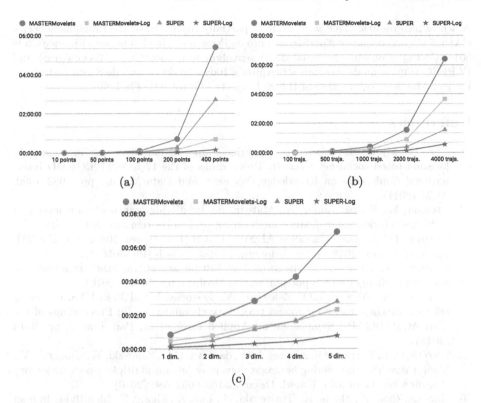

Fig. 5. Scalability of running time varying (a) the size of the trajectories, (b) the number of trajectories and (c) the number of dimensions.

6   Conclusion

In this paper we proposed a new method for extracting movelets, as the use of the best subtrajectories is currently a promising approach for multiple aspect trajectory classification. Both Movelets and MASTERMovelets generate subtrajectories of any size and from every position of the trajectories for finding the *movelets*. Our method does not generate the distance matrix of a trajectory point to all other trajectory points in the dataset, but to the points of the trajectories of a single class. It extracts *movelets* only from subtrajectories that occur more frequently in the trajectories of the same class, that are the SUPER-Pivots. Indeed, it limits the number of trajectory dimensions in each movelet candidate by counting the most frequent number of dimensions that appear in the class. Experimental results show that the proposed method is much faster for both extracting the *movelets* and building the classification models. Furthermore, scalability experiments show how well the proposed method scales compared to the state-of-the-art method for multiple aspect trajectory classification.

Acknowledgments. This work has been partially supported by the Brazilian agencies CAPES (Coordenação de Aperfeiçoamento de Pessoal de Nível Superior - Finance Code 001), CNPQ (Conselho Nacional de Desenvolvimento Científico e Tecnológico) and FAPESC (Fundação de Amparo a Pesquisa e Inovação do Estado de Santa Catarina - Project Match - Co-financing of H2020 Projects - Grant 2018TR 1266).

References

1. Cho, E., Myers, S.A., Leskovec, J.: Friendship and mobility: user movement in location-based social networks. In: Proceedings of the 17th ACM SIGKDD International Conference on Knowledge Discovery and Data Mining, pp. 1082–1090. ACM (2011)
2. Etemad, M., Soares Júnior, A., Matwin, S.: Predicting transportation modes of GPS trajectories using feature engineering and noise removal. In: Bagheri, E., Cheung, J.C.K. (eds.) Canadian AI 2018. LNCS (LNAI), vol. 10832, pp. 259–264. Springer, Cham (2018). https://doi.org/10.1007/978-3-319-89656-4_24
3. Ferrero, C.A., Alvares, L.O., Bogorny, V.: Multiple aspect trajectory data analysis: research challenges and opportunities. In: GeoInfo, pp. 56–67 (2016)
4. Ferrero, C.A., Alvares, L.O., Zalewski, W., Bogorny, V.: MOVELETS: exploring relevant subtrajectories for robust trajectory classification. In: Proceedings of the 33rd ACM/SIGAPP Symposium on Applied Computing, Pau, France, pp. 9–13 (2018)
5. Ferrero, C.A., Petry, L.M., Alvares, L.O., da Silva, C.L., Zalewski, W., Bogorny, V.: MasterMovelets: discovering heterogeneous movelets for multiple aspect trajectory classification. Data Min. Knowl. Discov. **34**(3), 652–680 (2020)
6. Gao, Q., Zhou, F., Zhang, K., Trajcevski, G., Luo, X., Zhang, F.: Identifying human mobility via trajectory embeddings. In: Proceedings of the 26th International Joint Conference on Artificial Intelligence, pp. 1689–1695. AAAI Press (2017)
7. Lee, J.G., Han, J., Li, X., Cheng, H.: Mining discriminative patterns for classifying trajectories on road networks. IEEE Trans. Knowl. Data Eng. **23**(5), 712–726 (2011)
8. Lee, J., Han, J., Li, X., Gonzalez, H.: TraClass: trajectory classification using hierarchical region-based and trajectory-based clustering. Proc. VLDB Endow. **1**(1), 1081–1094 (2008)
9. Leite da Silva, C., May Petry, L., Bogorny, V.: A survey and comparison of trajectory classification methods. In: 2019 8th Brazilian Conference on Intelligent Systems (BRACIS), pp. 788–793, October 2019
10. Mello, R.d.S., et al.: MASTER: a multiple aspect view on trajectories. Trans. GIS **23**, 805–822(2019)
11. Niemeyer, G.: Geohash (2008). https://en.wikipedia.org/wiki/Geohash
12. Petry, L.M., Silva, C.L.D., Esuli, A., Renso, C., Bogorny, V.: MARC: a robust method for multiple-aspect trajectory classification via space, time, and semantic embeddings. Int. J. Geogr. Inf. Sci. **0**(0), 1–23 (2020)
13. Sugiyama, M.: Local fisher discriminant analysis for supervised dimensionality reduction. In: Proceedings of the 23rd International Conference on Machine Learning, pp. 905–912 (2006)
14. Tragopoulou, S., Varlamis, I., Eirinaki, M.: Classification of movement data concerning user's activity recognition via mobile phones. In: Proceedings of the 4th International Conference on Web Intelligence, Mining and Semantics (WIMS14), p. 42. ACM (2014)

15. Varlamis, I.: Evolutionary data sampling for user movement classification. In: 2015 IEEE Congress on Evolutionary Computation (CEC). IEEE (2015)

16. Vicenzi, F., Petry, L.M., da Silva, C.L., Alvares, L.O., Bogorny, V.: Exploring frequency-based approaches for efficient trajectory classification. In: The 35th ACM/SIGAPP Symposium on Applied Computing, online event, SAC 2020, Brno, Czech Republic, 30 March–3 April 2020, pp. 624–631. ACM (2020)

17. Yang, D., Zhang, D., Zheng, V.W., Yu, Z.: Modeling user activity preference by leveraging user spatial temporal characteristics in LBSNs. IEEE Trans. Syst. Man Cybern. Syst. 45(1), 129–142 (2015)

18. Zhang, Z., Zhang, H., Wen, Y., Zhang, Y., Yuan, X.: Discriminative extraction of features from time series. Neurocomputing 275, 2317–2328 (2018)

19. Zhou, F., Gao, Q., Trajcevski, G., Zhang, K., Zhong, T., Zhang, F.: Trajectory-user linking via variational autoencoder. In: IJCAI, pp. 3212–3218 (2018)

Interpreting Classification Models Using Feature Importance Based on Marginal Local Effects

Rogério Luiz Cardoso Silva Filho[1,2](✉) [ID], Paulo Jorge Leitão Adeodato[2],
and Kellyton dos Santos Brito[3] [ID]

[1] Instituto Federal do Norte de Minas Gerais – IFNMG, Montes Claros, Brazil
rogerio.luiz@ifnmg.edu.br
[2] Centro de Informática, Universidade Federal de Pernambuco – UFPE, Recife, Brazil
[3] Universidade Federal Rural de Pernambuco – UFRPE, Recife, Brazil

Abstract. Machine learning models are widespread in many different fields due to their remarkable performances in many tasks. Some require greater interpretability, which often signifies that it is necessary to understand the mechanism underlying the algorithms. Feature importance is the most common explanation and is essential in data mining, especially in applied research. There is a frequent need to compare the effect of features over time, across models, or even across studies. For this, a single metric for each feature shared by all may be more suitable. Thus, analysts may gain better first-order insights regarding feature behavior across these different scenarios. The β-coefficients of additive models, such as logistic regressions, have been widely used for this purpose. They describe the relationships among predictors and outcomes in a single number, indicating both their direction and size. However, for black-box models, there is no metric with these same characteristics. Furthermore, even the β-coefficients in logistic regression models have limitations. Hence, this paper discusses these limitations together with the existing alternatives for overcoming them, and proposes new metrics of feature importance. As with the coefficients, these metrics indicate the feature effect's size and direction, but in the probability scale within a model-agnostic framework. An experiment conducted on openly available breast cancer data from the UCI Archive verified the suitability of these metrics, and another on real-world data demonstrated how they may be helpful in practice.

Keywords: Feature importance · Explainable artificial intelligence · ALE plots

1 Introduction

Explainable artificial intelligence (XAI) is an emerging research area that enables black-box models to become trustworthy for humans. With a growing interest in explaining machine learning (ML) models to fill the gap between interpretability and prediction performance, over the past few years, many techniques have been proposed, and explainability has become an essential subfield of ML [1]. This combination has helped the spread of ML in applied research areas even more, such as in education, healthcare, finance, and social media.

© Springer Nature Switzerland AG 2021
A. Britto and K. Valdivia Delgado (Eds.): BRACIS 2021, LNAI 13073, pp. 484–497, 2021.
https://doi.org/10.1007/978-3-030-91702-9_32

For instance, simply classifying a patient in a hospital into a particular health status is not particularly helpful. It is more desirable to investigate the conditions that have contributed to this [2] and would even become compulsory should any legal matters arise. Additionally, in the education domain, discovering why a student will drop out is more valuable than just predicting it [3, 4] because, as in medicine, the treatment depends on the probable cause. Similarly, auditing the behavior of machine learning bot detectors in social media is valuable in order to improve the models for new kinds of bots [5, 6].

Explanations may be expressed in many forms. For classification problems, specifically, feature importance is widely used [7, 8], and demonstrates the global impact of single features in the model predictions. A wide variety of different methods with different feature importance representations have been proposed for this purpose [9–11]. Despite these advances, there is still a lack of understanding as to how these methods are related and whether one method is preferable over another [12].

In applied research it is often necessary to track feature importance over time, across models or even across studies. Therefore, a method that enables the global feature contribution to be represented by a single metric is more suitable than multiple metrics or graphical representation, otherwise interpretability may be challenging to understand when handling several features in several models.

A standard single metric of feature importance is the coefficients of additive models such as linear and logistic regressions. This coefficient represents the weight of each feature in the additive function, which describes the relationship among features and outcome. However, for generalized linear models (GLM), which involve transformations of this linear predictor into other discrete outcomes, such as logistic regressions, the coefficient interpretability is not straightforward. Moreover, the coefficients are highly sensitive to unobserved heterogeneity [13] and data scale [14].

To circumvent these limitations, marginal effects (MEs) have long been proposed [14] in the traditional statistical literature. Marginal effects use the prediction function to calculate the differences in probabilities of the outcome when the features partially change from one specified value to another. However, MEs fail to isolate the feature effect when data are correlated [15]. This problem arises when the computation of the feature effect uses conditional distribution. Thus, since correlated features move in tandem, it is unable to distinguish which feature value changes influence the model probabilities.

For black-box models however, the permutation feature importance (PFI) derived from tree-based algorithms is a standard single metric used to report feature contribution in classification problems. However, this kind of metric is linked to model error, which cannot be a metric of interest for analysts [15]. Furthermore, it does not report the direction of the feature effect, which may be critical for actionable research. Recently, SHapley Additive exPlanations (SHAP) [12] have been pointed out as the most common explainability technique in organizations [1]. This technique uses game theory (Shapley Values) to measure the contribution of each data point to each feature value. Thus, it can deliver explanations in fine grain, and by means of the average, report the global feature contributions, including their direction. Unfortunately, calculating the exact Shapley value is computationally expensive [15], and several approximations have been proposed [16].

Moreover, both PFI and SHAP are permutation-based techniques, and so, they are able to randomly sample from the marginal distribution considering unrealistic data points that are not present on training data. Therefore, they extrapolate in areas where the model was trained with either little or no training data, which may cause misleading explanations [17].

Recent advances in the interpretable machine learning field, such as accumulated local effects (ALE) plots, have put forward relevant contributions in this direction. They have shed light on detecting a more reliable feature effect with low computational cost when features are correlated. However, ALEs have only been used to visualize the feature effects across different values by plots, which are not visually friendly when the analyst compares the feature importance across multiple models.

Thus, this paper proposes new metrics of feature importance as a valuable option when compared to those already in existence. They allow direct interpretation in the probability scale, are more realistic when dealing with correlated data and are modeled in a model-agnostic framework. Although these metrics may be extrapolated for any class of supervised models, in this paper they are focused only on binary classification. Experiments use open-access data from the UCI Archive to introduce differences among provided metrics for logistic regression coefficients and random forest permutation feature importance. Lastly, real-world data are used to demonstrate how they may be helpful in a practical problem.

The remainder of this paper is organized as follows: Sect. 2 provides a background of the theory related to this work. Section 3 introduces and explains the proposed metrics. Section 4 presents the experiments, results, and interpretation, and lastly, Sect. 5 summarizes the main findings, future work, limitations, and conclusions.

2   Background

There are several goals for explaining prediction models. In this paper, the main goal of this paper is to support applied research providing single metrics that, in a more realistic scenario, are able to report the overall contribution of model features. Hence, this section defines feature importance, and reports the main existing metrics and methods for this goal, which are directly related to this work.

2.1   Feature Importance

The most common explanation for the classification model is feature importance. Also known as the feature-level interpretation or saliency method, the method is the most well-studied explainability technique. It explains the decision of an algorithm by assigning values that reflect the importance of input components in their contribution to the decisions. Regardless of the mechanism to calculate it, its common meaning is related to the individual contribution of the corresponding feature for a particular classifier [8].

This individual contribution may be derived from the global perspective, where the feature importance is related to the whole model, and from the local perspective, where the importance is derived for a specific data point [18]. Moreover, it may be internal to the model (intrinsically) as the coefficients of linear models, or by applying methods that analyze the model after training (post hoc). Another criterion to classify these methods is related to their generalizability, whether they are model-specific or model-agnostic. While every intrinsic method is specific, all model-agnostic work is in a post hoc framework [15].

2.2 Marginal Effects

Marginal effects are a general concept and have different meanings depending on the discipline. This Subsection defines it according to econometrics, as the "additional" effect. On the other hand, in Subsect. 2.3, the word "marginal" is related to the probability distribution of a feature as well.

Given a features set $X = (x_1, x_2)$, and the predicted function $\hat{y} = \hat{f}(X)$, the marginal effects (ME) with respect to x_1 at a specific value corresponds to the changes in the outcome $\hat{y}$, given that x_1 changes in one unit. In other words, it is the first derivative of $\hat{f}(X)$ with respect to a x_1 at a specified value of the input space. However, if x_1 is discrete or binary, the computation is more straightforward, and the finite difference is applied, rather than the derivative [4].

For linear and additive models, MEs are constant across the feature values and are exactly the same as the regression estimated slopes (coefficients) [14]. However, for GLM models MEs take different values across feature distribution. For a logistic regression model, MEs reflects the logit shape, and are small when the probability is close to 0 or 1 and relatively large when close to 0.5 [18]. Thus, to summarize the MEs of x_1 the average of all MEs (AMEs) is commonly used [14]. Moreover, there are other alternatives that may be more suitable depending on the researcher's questions.

Summary Metrics of ME

Average Marginal Effects (AMEs) are how much the outcome y changes on average when x_1 varies in small changes. Thus, the derivative is computed for every small change on x_1 for every data point and averaged. In practice, the numerical derivative is implemented across the observed values of x_1. A step (h) is defined for continuous features, and the MEs become close to their theoretical value on the limit, as h tends to 0. The Equations below demonstrate this beyond the simplified computation of AMEs.

$$ME = \hat{f}(x) = \lim_{h \to 0} \left(\frac{f(x+h) - f(x)}{h} \right) \tag{1}$$

$$AMEs = \frac{1}{n} \sum_{i=1}^{n} ME_i$$

Marginal Effect at the Mean (MEM) is simply the computation of the MEs around the mean of the feature distribution. In practice, MEM is close to the AME if $\hat{f}(x)$ is not too noisy and more feasible to compute, since evaluating the derivatives at the means is easier than taking the mean of each derivative [19].

Marginal Effect at the Representative Value (MER) is a simplification of MEM calculation for a value that could be an interesting operation point for the research domain. The marginal effect is calculated for each variable at a particular combination of

X values. Thus, MER provides a means to understand and communicate model estimates at theoretically important combinations of feature values [20].

These metrics were essential to shed light on those proposed in this paper since they are based on solid statistical theory [14]. In addition to XAI advances, it is possible to report a less-biased feature effect which could play an important role for ML applied researchers.

2.3 Marginal Local Effects

The concept of local effects was brought in [21] and is a fundamental part of their accumulated local effects (ALEs) plots. ALE plots were presented as an alternative for visualizing the effects of features in black-box supervised learning models instead of partial dependence plots (PDPs).

The PDPs, introduced by Friedman [22], are widely used to visualize the influence of features in supervised ML models, and have even been considered a causal interpreter for black-box models [23]. For a prediction function $\hat{y} = \hat{f}(X)$, where $\hat{y}$ is a scalar response variable and $X = (x_1, x_2)$, PDPs illustrate the relationship between x_1 and the outcome, marginalizing $\hat{f}(X)$ over the distribution of x_2. Hence, the PDPs function at a particular value of x_1 represent the average prediction from $\hat{f}(X)$ if all data points take that value for x_1. This process takes into account unlikely combinations of X, building unrealistic plots when data are dependent.

As with MEs, ALEs use the conditional instead of marginal distribution. Thus, to overcome the intrinsic problems of MEs, as mentioned in Subsect. 2.2, ALEs use the averaged differences in $\hat{f}(X)$ across intervals of the training data (local effect). This hack allows the extraction of isolated effects of features within the intervals. Lastly, it accumulates this averaged local effect and center subtracting the mean using the equations below.

$$\widetilde{f_{j,ALE}}(x) = \sum_{k=1}^{k(x_1)} \frac{1}{n(k)} \sum_{i:x_j^{(i)} \in N(k)} \left[f\left(z_{k,j}, x_1^{(i)}\right) - f\left(z_{k-1,j}, x_1^{(i)}\right) \right] \qquad (2)$$

$$\widetilde{f_{j,ALE\,Centered}}(x) = \widetilde{f_{j,ALE}}(x) - \frac{1}{n}\sum_{i=1}^{n} \widetilde{f_{j,ALE}}\left(x_j^{(i)}\right) \qquad (3)$$

where k is the interval of data and $n_j(k)$ is the neighborhood. Hence, the ALE method calculates the differences in predictions, whereby the features of interest are replaced by grid values of z. The difference in prediction is the effect that features have for an individual instance in a specific interval. The sum on the right in (2) adds up the effects of all instances, which is divided by the number of instances in the interval k to produce the average. Finally, the ALE is vertically centered (3) in the sense that the mean ALEs of x_j and x_l on $\widetilde{f_{j,lALE}}(x_j, x_l)$ are both zero.

Figure 1 presents a better insight into the computation of the local effect for the function $\hat{f}(x_1, x_2)$. The feature range of x_1 was subdivided into k bins with roughly the same number of points indexed by $N(k)$. Focusing on bin $N(4)$, for each point falling into this bin, $\hat{f}(x_1, x_2)$ have their x_1 held by the left and right endpoints of the interval z_3 and z_4. Next, the differences of the predictions of these points were averaged by dividing their sum by the number of points in $N(4)$. The same was carried out for all intervals and summed up. Finally, the expectation over $p(x_1)$ was subtracted.

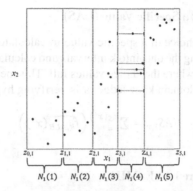

Fig. 1. Illustration of the ALE estimation. Excerpt from [21]

2.4 Summary of Metrics for Explainability

Marginal effects have been discussed in traditional statistical literature as an alternative for the coefficients of the GLM models due to their direct interpretability on the probability scale. In addition, they are less sensitive to the differences of data scale and unobserved heterogeneity than coefficients [10]. Summary metrics of MEs use the conditional distribution and are not robust against correlated data.

In the field of explainable machine learning field, PDPs have been widely used to report the feature effect. However, they extrapolated their results when using the density function and took into account an unreliable combination of data in their computation. ALE plots recently brought the new concept of local MEs and overcame both the aforementioned problems. Thus, this paper claims that the ALEs theory [21] is a good starting point together with the existing traditional metrics of MEs [14] to propose new global features importance which is able to fill the gap of robustness regarding the size and direction of feature effects on a binary classification model.

3 Proposed Metrics

This section proposes four new metrics that are suitable for applied research when comparing the feature effects across multiple models. These metrics are based on the ME and the shape of ALE plots that may be used to report the feature contribution on binary classification problems. Three of them possess explainability in terms of size and direction of the feature effect, while one accounts only for the contribution amount.

3.1 Average Uncentered ALE (AUA)

Average is a natural metric to summarize a distribution (first-order momentum) and leverages great insights regarding the size and directions of the feature mean effects.

$$AUA_x = \sum_{k=1}^{k(x^l)} \left(\widehat{f_{k,ALE}}(x^l) \right) \tag{4}$$

This is close to the MEs for linear models and AMEs for GLMs since it uses the average conditional distribution of the observed training data. However, it accounts for each local effect and may be somewhat different for noisy models.

3.2   Uncentered ALE at a Specific Value (UAS)

The UAS is an arbitrary choice of a specific value to calculate the uncentered ALE. In practice, it requires splitting the data into k intervals and calculating the uncentered ALE with (3) up to the interval where the choice values fall. This metric may be helpful when the analyst has sufficient domain knowledge or is verifying hypotheses.

$$UAS_{S,x} = \sum_{k=1}^{K_S(x^i)} \left(\widehat{f_{k,ALE}}(x^i) \right) \tag{5}$$

3.3   Maximum Uncentered ALE (MUA)

This metric is more actionable and consists of extracting the maximum change in predicting the outcome for the feature x_i. As the maximum may be positive or negative related to the outcome, it requires a previous absolute comparison in order to achieve the highest value.

$$MUA_x = MAX \left(\left| \widehat{f_{J,ALE}}(x) \right| \right) \tag{6}$$

3.4   ALE Absolute Average (AABSA)

This metric is a non-directional metric and highlights the overall feature importance. It measures, on average, how far the prediction changes away from the ALE average. Unlike the others, the centered ALE is considered, which has a mean zero.

$$AABSA_x = \frac{1}{k} \sum_{k=1}^{k(x^i)} \left(\left| \widehat{f_{k,ALE}}(x^i) \right| - E(x) \right) \tag{7}$$

3.5   Summary of the Novelty of the Proposal

In order to clarify the novelty and issues addressed by the proposal, Table 1 summarizes the main differences of each proposed metric and of those that already exist. In particular, the β- coefficients of logistic regression (LR) and permutation feature importance (PFI) from tree-based algorithms were considered, both widely used in the machine learning field as a global metrics and reported by a single parameter.

4   Experiments

To introduce the new metrics, two experiments were conducted[1]. The goal of the experiments was two-fold. First, to compare the proposed metrics in this paper with intrinsic model metrics. More specifically, it considered the widespread β-coefficients and the permutation feature importance derived from LR and random forest (RF), respectively.

---

[1] The implementation code can be found in this repository: https://github.com/rogerioluizsi/summary_ale.git.

Table 1. Differences among proposed methods and similar existing methods in the ML field

Metric	Report		Concern about		
	Size	Direction	Dependence data	Report in probability scale	Model-agnostic
AUA	*v*	*v*	*v*	*v*	*v*
UAS	*v*	*v*	*v*	*v*	*v*
MUA	*v*	*v*	*v*	*v*	*v*
AABSA	*v*		*v*	*v*	*v*
β – LR	*v*	*v*			
PFI	*v*				

Thus, it was possible to evaluate whether or not the metrics highlight the features in a similar manner. An open-access breast cancer dataset was used. This dataset is available ready for modeling binary classification and is a linearly separable problem with highly correlated data [24].

Second, a dataset was used from the National Brazilian Test for Secondary Education (ENEM), and the School Census from the 2009–2019 period. The goal here was to demonstrate how the proposed metrics may be helpful in a real-world problem. Thus, a data mining solution was developed to explore which and how the most important variables associated with school performance behave over the years. The report aimed at yielding valuable and actionable results for decision-makers through a single feature importance metric. These summary measures may enhance the analytical ability of the researcher when comparing the feature effects across supervised models, whichever the algorithm chooses to fit the data.

The ALIBI package [25], which has implemented the ALE plots, supports the computation of the metrics proposed in this paper. Therefore, all the mechanisms intrinsic to the ALEs theory, such as the interval definition, numerical derivation, and the computation of the local effects, follow the software implementation. In this paper, the performance of the models was not reported since the goals were limited to analyzing the model explicability.

4.1 Breast Cancer Data

The breast cancer data included benign and malignant cell samples from 369 patients, 212 with cancer, and 157 with fibrocystic breast masses. Each sample contained thirty features, and LR and RF predicted the patient class in a 5-fold cross-validation setting with random sampling stratified by the target class. Therefore, both algorithms were applied for the same folds, and the mean was adopted as feature importance for each metric.

Figure 2 shows the LR coefficients in red and the four proposed metrics. For the metrics that illustrate the direction of the relationship, there is a high correlation and close magnitude. However, there were some discrepancies. This could have been due to highly paired correlated data (*e.g.* perimeter *vs.* radius), and so logistic regression arbitrarily chose one of these (*e.g.* radius) to highlight the coefficient [26]. Also, the *p-values* were not checked in this experiment, and maybe some highlighted coefficients were statistically insignificant. However, neither of these are of concern for our metrics.

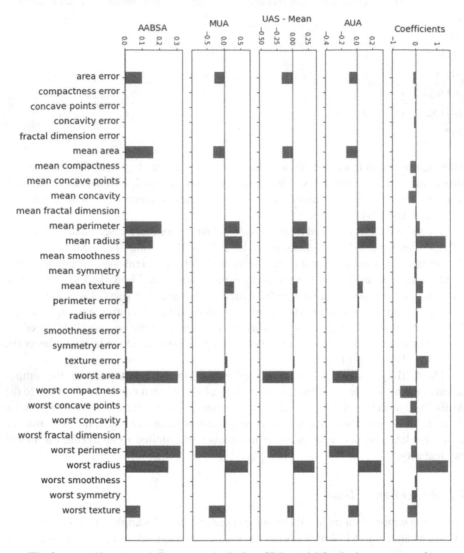

Fig. 2. β-coefficients and the proposed metrics of LR model for the breast cancer data

Figure 3 illustrates the permutation feature importance from the RF and the proposed metrics. The features set highlighted by all metrics is similar with a high correlation. In addition, surprisingly, the AABSA is fairly close to the permutation feature importance. Both are only positive and only account for the amount of feature contributions. However, they are built differently. While AABSA metrics the average change from the expected

ALEs mean (ALE 0), permutation is related to increasing the prediction error after permuting a feature. Hence, it may be due to the simplicity of the classification task on the cancer data [24].

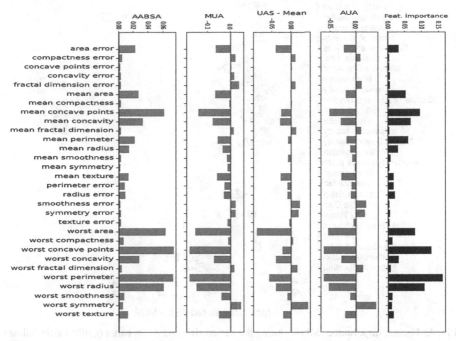

Fig. 3. Feature importance and the proposed metrics of RF model for the breast cancer data

4.2 Brazilian Secondary Educational Data

The second part of the experiment demonstrates how single feature importance may be helpful in practice. The data was taken from the 2009–2019 period of the largest test for secondary education in Brazil. The dataset contained demographic and socioeconomic information on students, and school characteristics over 32 features. The data included more than ten million students and was preprocessed to the school grain. The school was classified as good or bad in relation to the average scores achieved by their students in the test. To highlight the model-agnostic framework, two tree-based classifiers (RF and AdaBoost (AB)) were applied combined in a 10-fold cross-validation setting, and the overall mean was adopted for each metric.

Thus, systematized temporal data mining evaluated how the main features related to the performance of the school had behaved over the years. Due to space limitations, only the Max Uncentered ALE - MUA was reported by three plots presenting the outputs, as discussed below.

Figure 4 presents a box plot with the MUA for each feature. The clarity of colored dots illustrates the evolutionary directions of the variable over the years. The income per

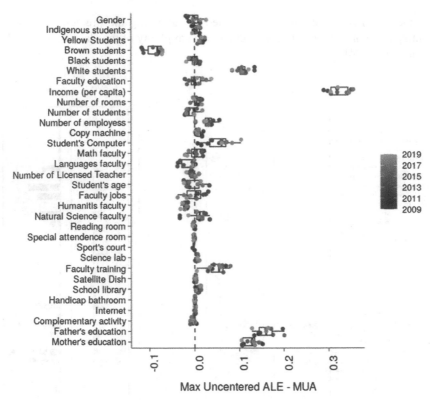

Fig. 4. Feature importance (MUA) during the period (RF and AB means) (Color figure online)

capita of students is the feature with the highest importance during the whole period. This is followed by parent's education and the students' race (brown students' negative effects). The importance of student computers seems to be a growing tendency over the years.

Figure 5 separates a selected set of features to obtain a better understanding of their behavior during the period. The computer lab has a higher positive slope, while faculty jobs (the number of schools where teachers work) have a higher negative.

Lastly, in Fig. 6, temporality was disregarded, and the features were organized for an overview of their importance in the following groups: non-actionable features (race and gender), school features (infrastructure), student features (parent's education and income) and teacher features. In general, the group of features related to students demonstrated more potential to improve the quality of schools than others. Additionally, non-actionable features had a strong influence, both negative and positive.

5 Discussion and Conclusion

This paper has proposed new model-agnostic metrics of feature importance in an attempt to circumvent the drawbacks and constraints of the existing methods, such as the β- coefficients of additive models and feature importance from tree-based algorithms, widely

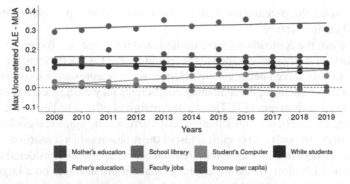

Fig. 5. Evolution of the feature importance (MUA) of a selected features set (RF and AB means)

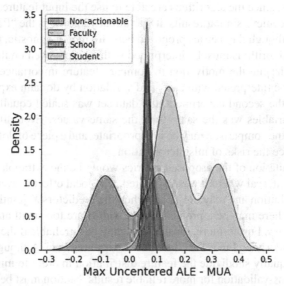

Fig. 6. Feature importance (MUA) by groups during the period (RF and AB means)

used for this purpose. This paper has proposed other options with a number of advantages, such as being agnostic to models, interpretable in the probability scale, and more reliable under correlated data.

The accumulated local effects are the key trick for isolating the main effect of the variable even in correlated data. The four proposed metrics were validated in two experiments that illustrated their suitability to actual data mining applications.

In the first experiment, breast cancer data were used to compare the proposed metrics with the coefficients of logistic regression and the permutation feature importance of random forest. The results illustrated that the proposed metrics are robust when facing correlated data and did not suffer the effects experienced by logistic regression (LR). All the proposed metrics captured the desired aspects of the attributes and were highly correlated. The AABSA proposed metric, which is directly comparable to the random forest (RF), since the permutation feature importance is only positive, captured very similar attribute importance.

Nevertheless, the comparisons were limited, and more tests are required to evaluate the metric behavior in other situations. For example, the LR coefficients are known to be sensitive to unobserved heterogeneity. Hence, when features are added to the model and improve the predictions, the remaining coefficients may change, even if this feature is not correlated with others [13]. Despite the marginal effects (a key to the proposed metrics) being more robust in this situation [26], there was no empirical evidence of this in our context. Additionally, an empirical test of the robustness in a scenario of correlated data compared to existing metrics is required since this paper only considered the theory inherited from ALE plots. Thus, we intend to make more analyses on a large scale in future work, including other XAI approaches.

It should be mentioned that this paper has not yet compared the proposed metrics across algorithms, since the algorithms are able to use the input features in a totally different manner to achieve similar results, it was already known as the "Rashomon" effect [27]. Thus, even though the metrics proposed here are model-agnostic, the comparisons across different algorithms must be interpreted with caution, even on the same data.

In addition, despite the motivation to compare feature importance across models, the results must be interpreted with care, and validation by domain experts is required. For example, in the second experiment, the data set was scaled equally for each year, and the set of variables was the same with the same values. Even after these careful transformations, the comparison may be inappropriate, and a piece of domain knowledge may help to reduce the risks of misinterpretation.

The main limitation of the proposed metrics would be the extrapolation of the local effect out of the interval where it was computed. The local effect is averaged across the conditional distribution and may only hold when the predictors X jointly fall within the same bin. Thus, there may be a problem if bin widths are too small and the predictive function is too noisy. Furthermore, local effects may be unreliable if the quantity of data points into the underlying bin is very small on the training data. Although this paper has used deciles to equally subdivide the feature distribution in order to minimize this risk, together with cross-validation for more reliable results, caution must be always taken in the interpretation with the validation of the domain expert.

References

1. Bhatt, U., Xiang, A., Sharma, S., et al.: Explainable machine learning in deployment. In: Proceedings of the 2020 Conference on Fairness, Accountability, and Transparency, pp. 648–657. ACM, New York (2020)
2. Razavian, N., Blecker, S., Schmidt, A.M., et al.: Population-level prediction of type 2 diabetes from claims data and analysis of risk factors. Big Data **3**, 277–287 (2015). https://doi.org/10.1089/big.2015.0020
3. Pellagatti, M., Masci, C., Ieva, F., Paganoni, A.M.: Generalized mixed-effects random forest: a flexible approach to predict university student dropout. Stat. Anal. Data Min., 1–17 (2021). https://doi.org/10.1002/sam.11505
4. Berens, J., Schneider, K., Görtz, S., et al.: Early Detection of Students at Risk-Predicting Student Dropouts Using Administrative Student Data from German Universities and Machine Learning Methods (2019)
5. Yang, K.C., Varol, O., Davis, C.A., et al.: Arming the public with artificial intelligence to counter social bots. Hum. Behav. Emerg. Technol. **1**, 48–61 (2019). https://doi.org/10.1002/hbe2.115

6. Leite, M.A.G.L., Guelpeli, M.V.C., Santos, C.Q.: Um Modelo Baseado em Regras para a Detecção de bots no Twitter, pp. 37–48 (2020). https://doi.org/10.5753/brasnam.2020.11161
7. Barredo Arrieta, A., Díaz-Rodríguez, N., del Ser, J., et al.: Explainable Artificial Intelligence (XAI): concepts, taxonomies, opportunities and challenges toward responsible AI. Inf. Fusion **58**, 82–115 (2020). https://doi.org/10.1016/j.inffus.2019.12.012
8. Saarela, M., Jauhiainen, S.: Comparison of feature importance measures as explanations for classification models. SN Appl. Sci. **3**(2), 1–12 (2021). https://doi.org/10.1007/s42452-021-04148-9
9. Shrikumar, A., Greenside, P., Kundaje, A.: Learning important features through propagating activation differences. In: 34th International Conference on Machine Learning, ICML 2017, vol. 7, pp. 4844–4866 (2017)
10. Štrumbelj, E., Kononenko, I.: Explaining prediction models and individual predictions with feature contributions. Knowl. Inf. Syst. **41**(3), 647–665 (2013). https://doi.org/10.1007/s10115-013-0679-x
11. Ribeiro, M.T., Singh, S., Guestrin, C.: Why should i trust you?" explaining the predictions of any classifier. In: Proceedings of the ACM SIGKDD International Conference on Knowledge Discovery and Data Mining 13-17-August, pp. 1135–1144 (2016). https://doi.org/10.1145/2939672.2939778
12. Lundberg, S.M., Lee, S.-I.: A unified approach to interpreting model predictions. In: Proceedings of the 31st International Conference on Neural Information Processing Systems, pp. 4768–4777. Curran Associates Inc., Red Hook (2017)
13. Mood, C.: Logistic regression: uncovering unobserved heterogeneity, pp. 1–25 (2017)
14. Long, J.S., Long, J.S.: Regression Models for Categorical and Limited Dependent Variables. Sage, New York (1997)
15. Molnar, C.: Interpretable Machine Learning (2019)
16. Bhatt, U., Ravikumar, P., Moura, J.M.F.: Towards aggregating weighted feature attributions (2019)
17. Hooker, G., Mentch, L.: Please stop permuting features: an explanation and alternatives, pp. 1–15 (2019)
18. Guidotti, R., Monreale, A., Ruggieri, S., et al.: A survey of methods for explaining black box models. ACM Comput. Surv. **51** (2018). https://doi.org/10.1145/3236009
19. Bartus, T.: Estimation of marginal effects using margeff. Stata J. **5**, 309–329 (2005). https://doi.org/10.1177/1536867x0500500303
20. Leeper, T.J.: Interpreting Regression Results using Average Marginal Effects with R's margins (2021). https://cran.r-project.org/web/packages/margins/vignettes/TechnicalDetails.pdf32
21. Apley, D.W., Zhu, J.: Visualizing the effects of predictor variables in black box supervised learning models. J. R. Stat. Soc. Ser. B Stat. Methodol. **82**, 1059–1086 (2020). https://doi.org/10.1111/rssb.12377
22. Friedman, J.H.: Greedy function approximation: a gradient boosting machine. Ann. Stat. **29**, 1189–1232 (2001). https://doi.org/10.1214/aos/1013203451
23. Zhao, Q., Hastie, T.: Causal interpretations of black-box models. J. Bus. Econ. Stat. **39**, 272–281 (2021). https://doi.org/10.1080/07350015.2019.1624293
24. Dua, D., Graff, C.: UCI Machine Learning Repository. University of California, Irvine, School of Information and Computer Sciences (2017). http://archive.ics.uci.edu/ml
25. Klaise, J., Van Looveren, A., Vacanti, G., Coca, A.: Alibi explain: Algorithms for explaining machine learning models. J. Mach. Learn. Res. **22**(181), 1–7 (2021). http://jmlr.org/papers/v22/21-0017.html
26. Mood, C.: Logistic regression: why we cannot do what we think we can do, and what we can do about it. Eur. Sociol. Rev. **26**, 67–82 (2010). https://doi.org/10.1093/esr/jcp006
27. Fisher, A., Rudin, C., Dominici, F.: All models are wrong, but many are useful: learning a variable's importance by studying an entire class of prediction models simultaneously. J. Mach. Learn. Res. **20**, 1–81 (2019)

On the Generalizations of the Choquet Integral for Application in FRBCs

Giancarlo Lucca[1]($\boxtimes$), Eduardo N. Borges[1]🆔, Rafael A. Berri[1]🆔,
Leonardo Emmendorfer[1]🆔, Graçaliz P. Dimuro[1,2]🆔, and Tiago C. Asmus[1,2]🆔

[1] Universidade Federal do Rio Grande, Rio Grande, RS, Brazil
{giancarlo.lucca,eduardoborges,rafaelberri,leonardoemmendorfer,
gracalizdimuro,tiagoasmus}@furg.br
[2] Universidad Publica de Navarra, Pamplona, Navarra, Spain

Abstract. An effective way to cope with classification problems, among others, is by using Fuzzy Rule-Based Classification Systems (FRBCSs). These systems are composed by two main components, the Knowledge Base (KB) and the Fuzzy Reasoning Method (FRM). The FRM is responsible for performing the classification of new examples based on the information stored in the KB. A key point in the FRM is how the information given by the fired fuzzy rules is aggregated. Precisely, the aggregation function is the component that differs from the two most widely used FRMs in the specialized literature. In this paper we provide a revision of the literature discussing the generalizations of the Choquet integral that has been applied in the FRM of a FRBCS. To do so, we consider an analysis of different generalizations, by t-norms, copulas, and by F functions. Also, the main contributions of each generalization are discussed.

Keywords: Choquet integral · Generalizations choquet integral · Pre-aggregation function

1 Introduction

A classification problem [1] is a research field in the area of data mining [2], which can be tackled in two different ways. An approach to deal with this problem is known as supervised learning, where a function (classifier) is generated from the available and labeled data (classes). Then, when a new example needs to be classified, the learned classifier is responsible to perform the prediction.

In the literature it is possible to find several methods that aim to cope with these problems using supervised learning, such as Support Vector Machines (SVM) [3], decisions trees [4] and neural networks [5]. Here, the focus is on

---

Supported by PNPD/CAPES (process. 464880/2019-00), FAPERGS (19/2551-0001279-9, 19/2551-0001660), CNPq (301618/2019-4), the Spanish Ministry of Science and Technology (TIN2016-77356-P, PID2019-108392GB I00 (AEI/10.13039/501100 011033)).

© Springer Nature Switzerland AG 2021
A. Britto and K. Valdivia Delgado (Eds.): BRACIS 2021, LNAI 13073, pp. 498–513, 2021.
https://doi.org/10.1007/978-3-030-91702-9_33

Fuzzy Rule-Based Classification Systems (FRBCCs) [6], because they provide the user with interpretable models by using linguistic labels [7] in their rules. Another reason is related with their accurate results and versatility, as shown in the many different fields where they have been applied like health [8], security [9], economy [10], food industry [11].

An important role in any FRBCS is played by the Fuzzy Reasoning Method (FRM) [12]. This method is responsible to perform the classification of new examples. For that, it makes usage of the information available in the rule base and the database. Moreover, in order to perform the classification, this mechanism uses an aggregation operator in order to aggregate, by classes, the information provided by the fired fuzzy rules when classifying new examples.

A widely used FRM considers the function Maximum as aggregation operator. By using this aggregation function, for each class, the FRM performs the selection of the best fired rule since it has the highest compatibility with the example [13]. The issue of this inference method is that the information provided by the remainder fired fuzzy rules is ignored. The Maximum is an averaging aggregation operator, since the obtained result is within the range between the minimum and the maximum of the aggregated values (in this case, obviously, the result is always the maximum).

To avoid the problem of ignoring information, it was proposed a FRM that applies the normalized sum [12] to perform the aggregation of the available information given by the fired rules. In this way, for each class, all information is taken into account in the aggregation step. This aggregation operator is considered as non-averaging since the result of this function can leave the range minimum–maximum.

In [14] the authors introduced a FRM considering the usage of the Choquet integral (CI) [15], which is an averaging operator. In this way, this approach mixes the characteristics of the previous FRMs considering an averaging operator that uses the information provided by all the fired rules of the system. Moreover, the CI is defined in terms of a fuzzy measure, which provides it with the nice properties to take into account the interaction among the data to be aggregated [15].

The objective of this paper is to discuss different methodologies that change the aggregation step performed in the FRM, when considering different generalizations of the CI, which are supported by solid theoretical studies [16], varying from the generalization by t-norms (C_T-integrals) [17], by copulas (CC-integrals [18–20]) and functions F (C_F-integrals [21] and $C_{F_1F_2}$-integrals [22,23]). Moreover, for each generalization it is provided provide a discussion of the main obtained results of each study (we highlight that our focus here are related with the main conclusions and not the specific obtained results of each approach).

This paper is organized as follows. Section 2 present the main components of a FRBC, showing an example of how the aggregation function is used in this context. Sections 3–6 discuss the theoretical and applied contributions of different generalizations of the CI. Section 8 is the conclusion.

2   The Role of Aggregation Functions in the FRM

Fuzzy Rule-Based Classification Systems (FRBCSs) [6] are extensions of the
rule-based system by using fuzzy sets in the antecedents of the rules. The best-
known FRBCSs are the ones defined by Takagi-Sugeno-Kang (TSK) [24] and
Mamdani [25], which is the one that it is adopted. The standard architecture of
the Mamdani method is presented in Fig. 1.

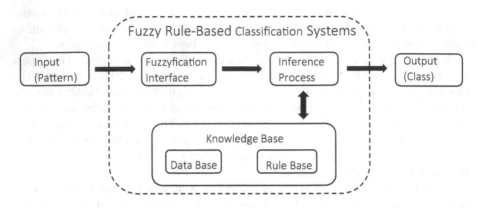

Fig. 1. A structure of FRBCS of the Mamdani type.

Where the Knowledge Base (KB) is composed by:
Data Base (DB) – Stores the membership functions associated with the linguistic labels considered in the fuzzy rules.
Rule Base (RB) – Is composed by a collection of linguistic fuzzy rules that
are joined by a connective (operator and). Here we consider that a classification
problem ins composed by t training patterns $x_p = (x_{p1}, \ldots, x_{pm}), p = 1, 2, \ldots, t$.
where x_{pi} is the i-th attribute and with the rules having the following structure:

$$\text{Rule } R_j : \text{ If } x_1 \text{ is } A_{j1} \text{ and } \ldots \text{ and } x_n \text{ is } A_{jn} \tag{1}$$
$$\text{then Class is } C_j \text{ with } RW_j,$$

where R_j is the label of the j-th rule, A_{ji} is a fuzzy set modeling a linguistic term,
modeled by a triangular shaped function. C_j is the class label and $RW_j \in [0, 1]$
is the rule weight [26].

The fuzzyfication interface converts the inputs (real values) into fuzzy values. In case of categorical variables, each value is modeled by a singleton and,
consequently, its membership value is either 1 or 0. Once the input is fuzzified,
the inference process is the mechanism responsible for the use of the information
stored in the KB to determine the class in which the example will be classified.
The generalizations discussed in this paper are applied at this point.

Once the knowledge has been learnt and a new example $x_p = x_{p1}, \ldots, x_{pn}$ has to be classified, the FRM [27] is applied to perform this task, where M is the number of classes of the problem and L is the number of rules that compose the RB. The stages of the FRM are:

Matching Degree: It represents the importance of the activation of the if-part of the rules for the example to be classified x_p, using a t-norm as conjunction operator:

$$\mu_{A_j}(x) = T(\mu_{A_{j1}}(x_1), \ldots, \mu_{A_{jn}}(x_n)). \tag{2}$$

with $j = 1, \ldots, L$. and $\mu_{A_{j1}}$ as the membership function with relation to a membership function.

Association Degree: For each rule, the matching degree is weighted by its rule weight:

$$b_j^k(x) = \mu_{A_j}(x) \cdot RW_j^k, \tag{3}$$

with $k = Class(R_j)$ and $j = 1, \ldots, L$.

Example Classification Soundness Degree for All Classes: For each class k, the positive information $b_j^k(x) > 0$, given by the fired fuzzy rules of the previous step, is aggregated by an aggregation function $\mathbb{A}$:

$$S_k(x) = \mathbb{A}_k\left(b_1^k(x), \ldots, b_L^k(x)\right), \tag{4}$$

with $k = 1, \ldots, M$.

In what follows, three different well-known FRMs are presented. Observe that their main difference is in the use of a different aggregation function to perform the aggregation of the information provided by the rules:

Winning Rule (WR) – For each class, it only considers the rule having the maximum compatibility with the example.

$$S_k(x) = \max_{R_{j_k} \in RB;} b_j(x). \tag{5}$$

Additive Combination (AC) – It aggregates all the fired rules, for each class k, by using the normalized sum.

$$S_k(x) = \frac{\sum_{j=1}^{R_{j_k} \in RB} b_j(x)}{f_{1_{max}}}, \tag{6}$$

where $f_{1_{max}} = \max_{k=1,\ldots,M} \sum_{j=1}^{R_{j_k} \in RB} b_j(x)$.

The Choquet integral (CI) – It is the function $\mathfrak{C}_m : [0,1]^n \rightarrow [0,1]$, defined, for all of $x \in [0,1]^n$, by:

$$\mathfrak{C}_m(x) = \sum_{i=1}^{n} \left(x_{(i)} - x_{(i-1)}\right) \cdot m\left(A_{(i)}\right), \tag{7}$$

Table 1. Association degrees for each class.

	C_1	C_2	C_3
R_a	0.94	0.15	0.89
R_b	0.10	0.40	0.88
R_c	0.25	0.10	0.85

where $N = \{0, \ldots, n\}$, $\mathfrak{m} : 2^N \to [0, 1]$ is a fuzzy measure[1], $(x_{(1)}, \ldots, x_{(n)})$ is an increasing permutation on the input $\boldsymbol{x}$, that is, $0 \leq x_{(1)} \leq \ldots \leq x_{(n)}$, with $x_{(0)} = 0$, and $A_{(i)} = \{(i), \ldots, (n)\}$ is the subset of indices corresponding to the $n - i + 1$ largest components of $\boldsymbol{x}$. Then:

$$S_k(x) = \sum_{j=1}^{R_{j_k} \in RB} \mathfrak{C}_{\mathfrak{m}}(b_j(x)). \tag{8}$$

where $\mathfrak{C}$ is the standard CI and $\mathfrak{m}$ the fuzzy measure.

Classification: For the final decision, the class that maximizes all the example classification soundness degrees is considered, using the function $F : [0, 1]^M \to \{1, \ldots, M\}$:

$$F(S_1, \ldots, S_M) = arg \max_{k=1,\ldots,M} (S_k). \tag{9}$$

To exemplify the role of different aggregation operator in the FRM, consider a classification problem composed by 3 classes (C_1, C_2 and C_3). For each one, 3 generic fuzzy rules, R_a, R_b and R_c are fired when classifying a new example (they can be different for each class). We present the information about this problem in Table 1. Notice that the numbers in this table represent the positive association degree (Step 2 of the FRM) obtained for each fired rule. Having into account that three fuzzy rules are fired for each class, by columns, three aggregations have to be computed (one for each class).

Since the CI is defined with respect to a fuzzy measure, in this example the standard cardinality (see [28]) is considered as fuzzy measure. The values computed for each class using these three FRMs are the following ones:

– C_1
  - WR = 0.94
  - AC = $\frac{0.94 + 0.1 + 0.25}{2.62}$ = 0.49
  - Choquet = $((0.1 - 0) \frac{3}{3}) + ((0.25 - 0.1) \frac{2}{3}) + ((0.94 - 0.25) \frac{1}{3})$ = 0.43
– C_2
  - WR = 0.4
  - AC = $\frac{0.15 + 0.4 + 0.1}{2.62}$ = 0.24
  - Choquet =0 $((0.1 - 0) \frac{3}{3}) + ((0.15 - 0.1) \frac{2}{3}) + ((0.4 - 0.15) \frac{1}{3})$ = 0.21

---

[1] A fuzzy measure $\mathfrak{m}$ is an increasing function on 2^N such that $\mathfrak{m}(\emptyset) = 0$ and $\mathfrak{m}(N) = 1$.

– C_3
- WR = 0.89
- AC = $\frac{0.89 + 0.88 + 0.85}{2.62}$ = 1.0
- Choquet = $\left((0.85 - 0)\frac{3}{3}\right) + \left((0.88 - 0.85)\frac{2}{3}\right) + \left((0.89 - 0.88)\frac{1}{3}\right)$ = 0.87

Once the example classification soundness degree for each class has been computed, the predicted class is the one associated with the largest value (step 4 of the FRM):

– WR = arg max[**0.94**, 0.4, 0.89] = C_1
– AC = arg max[0.49, 0.24, **1.0**] = C_3
– Choquet = arg max[0.43, 0.21, **0.87**] = C_3

It is observable that the usage of the maximum as an aggregation operator predicts class 1, since it only considers the information provided by one fuzzy rule (having the maximum compatibility). However, if we look in detail at the association degrees presented in Table 1, this prediction may not be ideal, since that class 1 has one rule having high compatibility whereas class 3 has three rules having high compatibilities (slightly less than that of class 1). Then, class 3 seems to be the most appropriated option. This fact is taken into account by the CI and the AC, since the information given by all the fuzzy rules and not only by the best one is considered and, consequently, the prediction assigns class 3.

In this example, it is noticeable the non-averaging behavior of AC. Observe that the result of this function for class C_3 is greater than the maximum value. This fact does not occur for averaging functions. In the case of WR, the result is always the maximum, meanwhile for the CI the result is a value between the minimum and the maximum. Another interesting point that raises with this example, is that the usage of different aggregation in the FRM is directly related with the performance of the classifier.

3   The C_T-integral and Pre-aggregations

This study was originally based on [14], where the authors modified the FRM of the Chi et al. algorithm [29] by applying the CI to aggregate all available information for each class. Furthermore, they introduced a learning method using a genetic algorithm in which the most suitable fuzzy measure for each class was computed. We highlight that this fuzzy measure is considered in all the applications of the generalizations of the Choquet integral.

For the first proposed generalization, the product operator of the standard CI was replaced by different other t-norms [30]. In this way, the manner how the information was aggregated would be different, consequently leading into different FRMs that could present performances even more accurately. The Choquet integral generalized by t-norms T, known as C_T-integral [17], is defined as:

Definition 1. *[17] Let* $m : 2^N \to [0,1]$ *be a fuzzy measure and* $T : [0,1]^2 \to [0,1]$ *be an t-norm. A* C_T*-integral is the function* $\mathfrak{C}_m^T : [0,1]^n \to [0,1]$, *defined, for all* $\boldsymbol{x} \in [0,1]^n$, *by*

$$\mathfrak{C}_m^T(\boldsymbol{x}) = \sum_{i=1}^{n} T\left(x_{(i)} - x_{(i-1)}, m\left(A_{(i)}\right)\right),\tag{10}$$

where $x_{(i)}$ *and* $A_{(i)}$ *are defined as in Eq. (7).*

Observe that some C_T-integrals are not aggregation function. E.g., take the minimum t-norm $T_M(x,y) = \min(x,y)$ and the cardinality measure (see [17,28]), and consider $\mathbf{x_1} = (0.05, 0.2, 0.7, 0.9)$ and $\mathbf{x_2} = (0.05, 0.1, 0.7, 0.9)$, where $\mathbf{x_1} > \mathbf{x_2}$. However, $\mathfrak{C}_m^{T_M}(\mathbf{x_1}) = 0.7$ and $\mathfrak{C}_m^{T_M}(\mathbf{x_2}) = 0.8$. Thus, the primordial condition of increasingness of any aggregation function is not fulfilled by $\mathfrak{C}_m^{T_M}$.

Yet, it is noticeable that the monotonicity property is not crucial for aggregation functions. Take for example a well-known statistical tool, the mode. It is not considered as an aggregation since the monotonicity of this function is not fulfilled, although it is useful. In [31], Bustince et al. introduced the notion of directional monotonicity, which allows monotonicity to be fulfilled along (some) fixed ray. So, with this in mind, the concept of pre-aggregation functions was introduced in [17]. These functions respect the boundary condition as any aggregation function, however, they are directional increasing:

Definition 2. *[31] Let* $\vec{r} = (r_1, \dots, r_n)$ *be a real n-dimensional vector,* $\vec{r} \neq \vec{0}$. *A function* $F : [0,1]^n \to [0,1]$ *is directionally increasing with respect to* $\vec{r}$ *($\vec{r}$-increasing, for short) if for all* $(x_1, \dots, x_n) \in [0,1]^n$ *and* $c > 0$ *such that* $(x_1 + cr_1, \dots, x_n + cr_n) \in [0,1]^n$ *it holds that*

$$F(x_1 + cr_1, \dots, x_n + cr_n) \geq F(x_1, \dots, x_n).\tag{11}$$

Similarly, one defines an $\vec{r}$*-decreasing function.*

Now, as the Chi algorithm it is not a state-of-the-art fuzzy classifier, the C_T-integrals were applied in the FRM of a powerful fuzzy classifier like FARC-HD [32]. The quality of the proposal was analyzed by applying these generalizations to cope with 27 classification problems. The considered datasets are available in KEEL [33] dataset repository. When comparing the different generalizations among themselves, it can be noticed that the one based on Hamacher t-norm was superior to the remaining ones. This fact occurred with four out the five considered fuzzy measures. The best accuracy was obtained when combining the Hamacher product with the power measure. To evaluate the quality of this best generalization, the study has compared it against the classical FRM of WR, since both FRMs apply averaging aggregation functions. In this comparison, it was empirically demonstrated that this generalization is statistically superior to WR and the standard CI.

4   Copulas and CC-integrals

The usage of the generalizations of the CI in a powerful fuzzy classifier has produced satisfactory results to cope with classification problems. However, these generalizations were pre-aggregation functions, that is, the monotonicity is not satisfied. Then, with this in mind, generalizations that are idempotent and averaging aggregation functions were developed. For that, in Eq. (7), firstly the distributivity property of the product operation is considered with the subtraction and then replaced the two instances of the product by copulas [30], obtaining the CC-integrals [18]:

Definition 3. *Let* $\mathfrak{m} : 2^N \to [0,1]$ *be a fuzzy measure and* $C : [0,1]^2 \to [0,1]$ *be a bivariate copula. The CC-integral is defined as a function* $\mathfrak{C}_\mathfrak{m}^C : [0,1]^n \to [0,1]$, *given, for all* $x \in [0,1]^n$, *by*

$$\mathfrak{C}_\mathfrak{m}^C(x) = \sum_{i=1}^{n} C\left(x_{(i)}, \mathfrak{m}\left(A_{(i)}\right)\right) - C\left(x_{(i-1)}, \mathfrak{m}\left(A_{(i)}\right)\right), \qquad (12)$$

where $x_{(i)}$ *and* $A_{(i)}$ *are defined as in Eq. (7).*

To demonstrate the efficiency of the CC-integrals to tackle classification problems, an experimental study considering 30 numerical datasets is considered. This study was conducted in two different ways. The first one was focused on comparisons per family of copulas (t-norms, overlap functions [34,35] and specific copulas), in order to find the function that presented the best generalization. Then, this best generalization is compared with 1) the classical FRM of WR (considering that both functions are averaging); 2) to the standard CI and 3) the best pre-aggregation function achieved in the previous study (C_T-integral), the one based on the Hamacher t-norm. The best CC-integral is the CMin-integral, constructed with the Minimum copula[2]. The obtained results showed that the CMin-Integral is statistically equivalent to the CI and the C_T-integral and superior than the WR.

5 C_F-integrals

The acquired knowledge from the previous studies shows that the function responsible to generalize the CI is very important. At this point only generalizations with averaging characteristics were presented. Having this in mind, the CI was generalized by special functions, in order to produce more competitive generalizations, allowing to produce non-averaging integrals. To achieve it, its used a family of left 0-absorbing aggregation functions F, which satisfy: (LAE) $\forall y \in [0,1] : F(0,y) = 0$. Moreover, the following two basic properties are also important:
(RNE) Right Neutral Element: $\forall x \in [0,1] : F(x,1) = x$;
(LC) Left Conjunctive Property: $\forall x,y \in [0,1] : F(x,y) \leq x$;
    Any bivariate function $F : [0,1]^2 \to [0,1]$ satisfying both **(LAE)** and **(RNE)** is called left 0-absorbent **(RNE)**-function.

---

[2] Examples of special CC-integrals were studied in [19,20].

Then, the so-called C_F-integral [21] is defined as:

Definition 4. *[21] Let $F : [0,1]^2 \to [0,1]$ be a bivariate function and $\mathfrak{m} : 2^N \to [0,1]$ be a fuzzy measure. The C_F-integral is the function $\mathfrak{C}^F_\mathfrak{m} : [0,1]^n \to [0,1]$, defined, for all $\boldsymbol{x} \in [0,1]^n$, by*

$$\mathfrak{C}^F_\mathfrak{m}(\boldsymbol{x}) = \min\left\{1, \sum_{i=1}^n F\left(x_{(i)} - x_{(i-1)}, \mathfrak{m}\left(A_{(i)}\right)\right)\right\}, \tag{13}$$

where $x_{(i)}$ and $A_{(i)}$ are defined as in Eq. (7).

In [21, Theorems 1 and 2], it was proved that the set of conditions that the function F should fulfill for the C_F-integral to be a pre-aggregation function is one of the following ones: Theorem 1 ((LAE) and (RNE)) or Theorem 2 ((LAE), $F(1,1) = 1$ and $(1,0)$-increasingness). Moreover, for the C_F-integral to be averaging F must satisfy (RNE) and (LC). This means that there exist a lot of non-averaging C_F-integrals.

The quality of the C_F-integrals to cope with classification problems was tested considering 33 different datasets. The experimental study was conducted considering C_F-integrals with and without averaging characteristics. Considering the non-averaging functions, six C_F-integrals were studied. In order to support the quality of this approach, a comparison with the best non-averaging C_F-integral with the FRM of AC and a FRM considering the probabilistic sum - PS (since it is an operator with non-averaging characteristics) is provided. The results showed that the non-averaging C_F-integrals-integrals, as expected, offer a performance superior than the averaging ones, and the best C_F-integral, based on the function FNA2[3] provides results that are statistically superior than all classical FRMs, and also, very competitive with the classical non-averaging FRMs like AC or PS.

6 $C_{F_1 F_2}$-integrals

The previous study demonstrated that the generalization of the standard Choquet integral by functions F resulted in satisfactory results. Then, this study combine the ideas of previous approaches, precisely, it take the same idea of CC-integrals, generalizing the each of the two instances of copulas by a pair of functions F, called F_1 and F_2, as consequence obtaining the $C_{F_1 F_2}$-integrals [22]:

Definition 5. *Let $\mathfrak{m} : 2^N \to [0,1]$ be a symmetric fuzzy measure and $F_1, F_2 : [0,1]^2 \to [0,1]$ be two fusion functions fulfilling:*

(i) *F_1-dominance (or, equivalently, F_2-Subordination): $F_1 \geq F_2$;*
(ii) *F_1 is $(1,0)$-increasing,*

---

[3] The function $FNA2 : [0,1]^2$ is defined, for all $x, y \in [0,1]$ by $F(0,y) = 0$, $F(x,y) = \frac{x+y}{2}$ if $0 < x \leq y$ and $F(x,y) = \min(\frac{x}{2}, y)$, otherwise, which satisfies the conditions of [21, Theorems 2].

A $C_{F_1F_2}$-integral is defined as a function $\mathfrak{C}_m^{(F_1,F_2)} : [0,1]^n \to [0,1]$, *given, for all* $\boldsymbol{x} \in [0,1]^n$, *by*

$$\mathfrak{C}_m^{(F_1,F_2)}(\boldsymbol{x}) = \tag{14}$$

$$\min\left\{1, x_{(1)} + \sum_{i=2}^{n} F_1\left(x_{(i)}, \mathfrak{m}\left(A_{(i)}\right)\right) - F_2\left(x_{(i-1)}, \mathfrak{m}\left(A_{(i)}\right)\right)\right\},$$

where $x_{(i)}$ and $A_{(i)}$ are defined as in Eq. (7).

In this paper, twenty-three different functions, F, were considered. As consequence, 201 different pairs of functions that could be used as F_1 and F_2 could be combined, respecting the dominance property. An important question that could appear is related to the choice of the function to be selected as F_1 and the one to act as F_2. Therefore, a methodology to reduce the scope of the study have been proposed by using the concept of Dominance and Subordination Strength degree, DSt and SSt respectively.

Definition 6. *Let $\mathcal{F} = \{F_1, \ldots, F_m\}$ be a set of m fusion functions. The dominance and subordination strength degrees, DSt and SSt, of a fusion function $F_i \in \mathcal{F}$ are defined, respectively, for $j \in \{1, \ldots, m\}$, by as follows:*

$$DSt(F_i) = \frac{1}{m} \sum_{j=1}^{m} \begin{cases} 1 \ if \ F_i \geq F_j, \\ 0 \ otherwise \end{cases} \cdot 100\%$$

$$SSt(F_i) = \frac{1}{m} \sum_{j=1}^{m} \begin{cases} 1 \ if \ F_i < F_j, \\ 0 \ otherwise. \end{cases} \cdot 100\%$$

The generalizations provided in this study are non averaging. Moreover, they satisfy the boundary conditions of any (pre) aggregation function. However, considering the monotonicity, we observed that these functions are neither increasing nor directional increasing. In fact, they are Ordered Directionally (OD) monotone functions [36]. These functions are monotonic along different directions according to the ordinal size of the coordinates of each input.

The $C_{F_1F_2}$-integrals were used to cope with classification problems in 33 different datasets. When analyzing the results that were obtained by the usage of these generalizations, it is noticeable that the combination of a function having a high dominance as F_1 combined with a function with high subordination as F_2 presented the best results of this study (from the top ten of the best global accuracies from the 81 pairs, eight have this characteristic). We also observed that the opposite, for each function F_2, is also true and that its best results are achieved when using a F_1 with a high dominance.

The performance of this proposal is analyzed by comparing them against distinct state-of-the-art FRBCSs, namely: FARC-HD [32], FURIA [37], IVTURS [38], a classical non-averaging aggregation operator like the probabilistic sum, P^*, and, the best C_F-integral that was selected from the previous study,

F_{NA2}. In this comparison, FURIA was the fuzzy classifier that achieved the highest accuracy mean, however, our new approach achieved a close classification rate. Furthermore, the number of specific datasets where the performance of our generalization is the worst among all the methods in the comparison is less than that of FURIA. The function representing the C_F-integrals also achieved good results, meanwhile the remainder cases (IVTURS, P^* and FARC-HD) where inferior and similar among themselves.

The 81 pairs of combinations considered to construct $C_{F_1 F_2}$-integrals were compared against IVTURS, P^*, FARC-HD and F_{NA2}. The results highlighted the quality of our new method because an equal or greater average result was obtained by 39, 36, 34 and 12 different combinations in these comparisons.

Finally, from the considered pairs, it was observed that five different $C_{F_1 F_2}$-integrals were considered as control variable in the statistical test in which all methods are compared, including FURIA. The last generalization only presented statistical differences with respect to FARC-HD. However, for any remaining pair, it is statistically equivalent when compared to FURIA and to F_{NA2} and superior to IVTURS, P^* and FARC-HD.

7   Detailed Results

In this section the results obtained by the usage of different aggregation operator are shown. We highlight that these results consider the same 33 datasets as in [21, 22] and [28]. Also, the results are related with the power measure, as mentioned previously, take into consideration the 5-fold cross validation technique [2] and are applied in the FRM of the FARC-HD [32] fuzzy classifier[4].

The results are provided in Table 2 where each cell correspond to the mean accuracy among all folds, the rows are related with the different considered datasets and the columns are the results obtained by classical FRMs such as: of the Additive Combination (AC), Probabilistic Sum (PS), Winning Rule (WR), Choquet integral (CI), C_T-integrals (due to lack of space were summarized to int, in all integrals) with is defined by the Hamacher product t-norm, CC-integral that use the copula of the minimum, C_F-integral considering the FNA_2 function and $C_{F_1 F_2}$-integrals using the pair $GM-FBPC$.

From the detailed results, we can noticed that classic FRM of the WR is the one that achieved the lowest global mean, indicating that the usage of all information related with the problem is an interesting alternative. Moreover, it is also observable that all non-averaging generalizations (AC, PS, C_F-integrals and $C_{F_1 F_2}$-integrals) presents superior results when compared against the averaging ones (WR, IC, C_T-integral and CC-integral).

The results also showed that the generalizations of the CI (C_T, CC, C_F, $C_{F_1 F_2}$-integrals) provided a superior performance in comparison to the standard CI. Finally, as mentioned before, the largest performance is obtained when the $C_{F_1 F_2}$-integral is used to cope with classification problems.

---

[4] The considered datasets and the fuzzy classifier are available in KEEL repository. Available at https://www.keel.es.

Table 2. Detailed results achieved in test by different generalizations of the CI.

Dataset	AC	PS	WR	CI	C_T-int	CC-int	C_F-int	$C_{F_1 F_2}$-int
App	83.03	85.84	83.03	80.13	82.99	85.84	85.84	86.80
Bal	85.92	87.20	81.92	82.40	82.72	81.60	88.64	89.12
Ban	85.30	84.85	83.94	86.32	85.96	84.30	84.60	84.79
Bnd	68.28	68.82	69.40	68.56	72.13	71.06	70.48	71.30
Bup	67.25	61.74	62.03	66.96	65.80	61.45	64.64	66.96
Cle	56.21	59.25	56.91	55.58	55.58	54.88	56.55	56.22
Con	53.16	52.21	52.07	51.26	53.09	52.61	53.16	54.72
Eco	82.15	80.95	75.62	76.51	80.07	77.09	80.08	81.86
Gla	65.44	64.04	64.99	64.02	63.10	69.17	66.83	68.25
Hab	73.18	69.26	70.89	72.52	72.21	74.17	71.87	72.53
Hay	77.95	77.95	78.69	79.49	79.49	81.74	79.43	78.66
Ion	88.90	88.32	90.03	90.04	89.18	88.89	89.75	88.33
Iri	94.00	95.33	94.00	91.33	93.33	92.67	94.00	94.00
Led	69.60	69.20	69.40	68.20	68.60	68.40	69.80	70.00
Mag	80.76	80.39	78.60	78.86	79.76	79.81	79.70	80.86
New	94.88	94.42	94.88	94.88	95.35	93.95	96.28	96.74
Pag	95.07	94.52	94.16	94.16	94.34	93.97	94.15	95.25
Pen	92.55	93.27	91.45	90.55	90.82	91.27	92.91	92.91
Pho	81.70	82.51	82.29	82.98	83.83	82.94	81.44	81.42
Pim	74.74	75.91	74.60	74.60	73.44	75.78	74.61	75.38
Rin	90.95	90.00	90.00	90.95	88.78	87.97	89.86	91.89
Sah	68.39	69.69	68.61	69.69	70.77	70.78	70.12	71.43
Sat	79.47	80.40	79.63	79.47	80.40	79.01	80.41	79.47
Seg	93.12	92.94	93.03	93.46	93.33	92.25	92.42	93.29
Shu	95.59	94.85	96.00	97.61	97.20	98.16	97.15	96.83
Son	78.36	82.24	77.42	77.43	79.34	76.95	83.21	85.15
Spe	77.88	77.90	77.90	77.88	76.02	78.99	79.77	79.39
Tit	78.87	78.87	78.87	78.87	78.87	78.87	78.87	78.87
Two	90.95	90.00	86.49	84.46	85.27	85.14	92.57	92.30
Veh	68.56	68.09	66.67	68.44	68.20	69.86	68.08	68.20
Win	96.03	94.92	96.60	93.79	96.63	93.83	96.08	95.48
Wis	96.63	97.22	96.34	97.22	96.78	95.90	96.78	96.78
Yea	58.96	59.03	55.32	55.73	56.53	57.01	57.08	58.56
Mean	80.12	80.07	79.15	79.22	79.69	79.58	80.52	81.02

8 Conclusions

The application of the Choquet integral (CI) in the Fuzzy Reasoning Method (FRM) of Fuzzy Rule-Based Classification Systems (FRBCSs) modified the way in which the information was used and enhanced the system quality. After that, many generalizations of the CI were proposed and also applied in FRM, obtaining success as well. In this paper the main contributions, theoretical and applied, of the generalizations are summarized and discussed.

The first generalization was built by the replacement of the product operator of the standard CI by different t-norms. These generalizations were supported by an important theoretical concept known as pre-aggregation functions. Differently from a simple aggregation function, a pre-aggregation function is monotonic only in a determined direction. This first generalization produced averaging functions and its applications to cope with classification problems showed that the generalization by the Hamacher product t-norm was superior than the FRM of the Winning Rule (WR) and the CI.

The second step aimed in generalizations of the CI that produce aggregation functions. To do so, the IC was used in its expanded form and generalized by copula functions, introducing the concept of Choquet-like Copula-Based aggregation functions, the so called CC-integrals. These functions also present averaging characteristics. The results of their applications demonstrated that the classical WR was statistically overcame.

It is observable that up to this point only generalizations with averaging characteristics were presented. On the otter hand, fuzzy classifiers known as state-of-the-art take into account the usage of non-averaging functions. Thus, to produce more competitive generalizations, a family of fusion functions F were introduced. The generalization of the Choquet integral by F functions introduced the concept of C_F-integrals. This generalization has averaging and non-averaging characteristics, it depends on the considered function. It was observed that the application of any non-averaging function statistically overcome any averaging one. Also, the developed operators outperforms the classical WR and Additive Combination (AC).

The generalization of the expanded CI by two functions F, F_1 and F_2, introduced the concept of $C_{F_1 F_2}$-integrals. These functions present an Ordered Directional increasing functions (OD increasing) and, therefore, represent a different level of aggregation operators. The summit of the performance in the classification problems was reached in this generalization. To do so, a methodology to select different functions as F_1 and F_2 were presented, based on the concept of degrees of dominance and subordination. For the considered $C_{F_1 F_2}$-integrals, in five different cases the generalizations are equivalent, or even superior, in comparison with fuzzy classifiers found in the literature.

Taking as basis the analysis provided by this paper, some interesting research points emerge. For example, the application of these generalizations in the FRM of different fuzzy classifiers. Also, considering that the generalizations are based on the Choquet integral, the usage of a different operator, such as the Sugeno integral can produce even more powerful operators. Finally, the combinations with different fuzzy measures are an alternative with great potential.

References

1. Alpaydin, E.: Introduction to Machine Learning, 2nd edn. MIT Press, Cambridge (2010)
2. Tan, P.-N., Steinbach, M., Kumar, V.: Introduction to Data Mining, 1st edn. Addison-Wesley Longman Publishing Co., Inc., Boston (2005)
3. Cortes, C., Vapnik, V.: Support vector networks. Mach. Learn. **20**, 273–297 (1995). https://doi.org/10.1007/BF00994018
4. Quinlan, J.: C4.5: Programs for Machine Learning. Morgan Kauffman, San Francisco (1993)
5. García-Pedrajas, N., García-Osorio, C., Fyfe, C.: Nonlinear boosting projections for ensemble construction. J. Mach. Learn. Res. **8**, 1–33 (2007)
6. Ishibuchi, H., Nakashima, T., Nii, M.: Classification and Modeling with Linguistic Information Granules, Advanced Approaches to Linguistic Data Mining. Advanced Information Processing, Springer, Heidelberg (2005). https://doi.org/10.1007/b138232
7. Zadeh, L.A.: The concept of a linguistic variable and its application to approximate reasoning - I. Inf. Sci. **8**(3), 199–249 (1975)
8. Unold, O.: Diagnosis of cardiac arrhythmia using fuzzy immune approach. In: Dobnikar, A., Lotrič, U., Šter, B. (eds.) ICANNGA 2011, Part II. LNCS, vol. 6594, pp. 265–274. Springer, Heidelberg (2011). https://doi.org/10.1007/978-3-642-20267-4_28
9. Vitoriano, B., Rodrígues, J.T., Tirado, G., Mart ín-Campo, F.J., Ortuão, M.T., Montero, J.: Intelligent decision-making models for disaster management. Hum. Ecol. Risk Assess. Int. J. **21**(5), 1341–1360 (2015)
10. Sanz, J.A., Bernardo, D., Herrera, F., Bustince, H., Hagras, H.: A compact evolutionary interval-valued fuzzy rule-based classification system for the modeling and prediction of real-world financial applications with imbalanced data. IEEE Trans. Fuzzy Syst. **23**(4), 973–990 (2015)
11. Goel, N., Sehgal, P.: Fuzzy classification of pre-harvest tomatoes for ripeness estimation - an approach based on automatic rule learning using decision tree. Appl. Soft Comput. **36**, 45–56 (2015)
12. Cordon, O., del Jesus, M.J., Herrera, F.: Analyzing the reasoning mechanisms in fuzzy rule based classification systems. Mathw. Soft Comput. **5**(2–3), 321–332 (1998)
13. Chi, Z., Yan, H., Pham, T.: Fuzzy Algorithms: With Applications to Image Processing and Pattern Recognition. World Scientific Publishing Co., Inc., River Edge (1996)
14. Barrenechea, E., Bustince, H., Fernandez, J., Paternain, D., Sanz, J.A.: Using the Choquet integral in the fuzzy reasoning method of fuzzy rule-based classification systems. Axioms **2**(2), 208–223 (2013)
15. Choquet, G.: "Theory of capacities," Annales de l'Institut Fourier, vol. 5, pp. 131–295, (1953–1954)
16. Dimuro, G.P., et al.: The state-of-art of the generalizations of the Choquet integral: From aggregation and pre-aggregation to ordered directionally monotone functions. Inf. Fusion **57**, 27–43 (2020)
17. Dimuro, G., et al.: Pre-aggregation functions: construction and an application. IEEE Trans. Fuzzy Syst. **24**(2), 260–272 (2016)
18. Lucca, G., et al.: CC-integrals: Choquet-like copula-based aggregation functions and its application in fuzzy rule-based classification systems. Knowl-Based Syst. **119**, 32–43 (2017)

19. Dimuro, G.P., Lucca, G., Sanz, J.A., Bustince, H., Bedregal, B.: CMin-integral: a Choquet-like aggregation function based on the minimum t-norm for applications to fuzzy rule-based classification systems. In: Torra, V., Mesiar, R., De Baets, B. (eds.) AGOP 2017. AISC, vol. 581, pp. 83–95. Springer, Cham (2018). https://doi.org/10.1007/978-3-319-59306-7_9

20. Lucca, G., Sanz, J.A., Dimuro, G.P., Bedregal, B., Bustince, H.: A proposal for tuning the the α parameter in $C_\alpha C$-integrals for application in fuzzy rule-based classification systems. Nat. Comput. **19**, 533–546 (2020)

21. Lucca, G., Sanz, J.A., Dimuro, G.P., Bedregal, B., Bustince, H., Mesiar, R.: CF-integrals: a new family of pre-aggregation functions with application to fuzzy rule-based classification systems. Inf. Sci. **435**, 94–110 (2018)

22. Lucca, G., Dimuro, G.P., Fernandez, J., Bustince, H., Bedregal, B., Sanz, J.A.: Improving the performance of fuzzy rule-based classification systems based on a nonaveraging generalization of CC-integrals named $C_{F_1 F_2}$-integrals. IEEE Trans. Fuzzy Syst. **27**(1), 124–134 (2019)

23. Dimuro, G.P., et al.: Generalized cf1f2-integrals: from Choquet-like aggregation to ordered directionally monotone functions. Fuzzy Sets Syst. **378**, 44–67 (2020)

24. Takagi, T., Sugeno, M.: Fuzzy identification of systems and its applications to modeling and control. IEEE Trans. Syst. Man Cybern. **SMC-15**(1), 116–132 (1985)

25. Mamdani, E.H.: Application of fuzzy algorithms for control of simple dynamic plant. In: Proceedings of the Institution of Electrical Engineers, vol. 121, issue number 12, pp. 1585–1588 (1974)

26. Ishibuchi, H., Nakashima, T.: Effect of rule weights in fuzzy rule-based classification systems. IEEE Trans. Fuzzy Syst. **9**(4), 506–515 (2001)

27. Cordón, O., del Jesus, M.J., Herrera, F.: A proposal on reasoning methods in fuzzy rule-based classification systems. Int. J. Approximate Reasoning **20**(1), 21–45 (1999)

28. Lucca, G., Sanz, J.A., Dimuro, J.A., Borges, E.N., Santos, H., Bustince, H.: Analyzing the performance of different fuzzy measures with generalizations of the Choquet integral in classification problems. In: 2019 IEEE International Conference on Fuzzy Systems (FUZZ-IEEE), pp. 1–6, June 2019

29. Chi, Z., Yan, H., Pham, T.: Fuzzy Algorithms: With Applications to Image Processing and Pattern Recognition. World Scientific, Singapore (1996)

30. Alsina, C., Frank, M.J., Schweizer, B.: Associative Functions: Triangular Norms and Copulas. World Scientific Publishing Company, Singapore (2006)

31. Bustince, H., Fernandez, J., Kolesárová, A., Mesiar, R.: Directional monotonicity of fusion functions. Eur. J. Oper. Res. **244**(1), 300–308 (2015)

32. Alcala-Fdez, J., Alcala, R., Herrera, F.: A fuzzy association rule-based classification model for high-dimensional problems with genetic rule selection and lateral tuning. IEEE Trans. Fuzzy Syst. **19**(5), 857–872 (2011)

33. Alcalá-Fdez, J., et al.: KEEL: a software tool to assess evolutionary algorithms for data mining problems. Soft Comput. **13**(3), 307–318 (2009). https://doi.org/10.1007/s00500-008-0323-y

34. Bustince, H., Fernandez, J., Mesiar, R., Montero, J., Orduna, R.: Overlap functions. Nonlinear Anal. Theory Methods Appl. **72**(3–4), 1488–1499 (2010)

35. Dimuro, G.P., Bedregal, B.: On the laws of contraposition for residual implications derived from overlap functions. In: 2015 IEEE International Conference on Los Alamitos Fuzzy Systems (FUZZ-IEEE), pp. 1–7. IEEE, August 2015

36. Bustince, H., et al.: Ordered directionally monotone functions. justification and application. IEEE Trans. Fuzzy Syst. 1 (2017, In press). Corrected proof

37. Hühn, J., Hüllermeier, E.: FURIA: an algorithm for unordered fuzzy rule induction. Data Min. Knowl. Discov. **19**(3), 293–319 (2009)
38. Sanz, J., Fernández, A., Bustince, H., Herrera, F.: IVTURS: a linguistic fuzzy rule-based classification system based on a new interval-valued fuzzy reasoning method with tuning and rule selection. IEEE Trans. Fuzzy Syst. **21**(3), 399–411 (2013)

Optimizing Diffusion Rate and Label Reliability in a Graph-Based Semi-supervised Classifier

Bruno Klaus de Aquino Afonso$^{(\boxtimes)}$ and Lilian Berton$^{(\boxtimes)}$

Institute of Science and Technology, University of São Paulo, São Paulo, Brazil
{bruno.klaus,lberton}@unifesp.br

Abstract. Semi-supervised learning has received attention from researchers, as it allows one to exploit the structure of unlabeled data to achieve competitive classification results with much fewer labels than supervised approaches. The Local and Global Consistency (LGC) algorithm is one of the most well-known graph-based semi-supervised (GSSL) classifiers. Notably, its solution can be written as a linear combination of the known labels. The coefficients of this linear combination depend on a parameter α, determining the decay of the reward over time when reaching labeled vertices in a random walk. In this work, we discuss how removing the self-influence of a labeled instance may be beneficial, and how it relates to leave-one-out error. Moreover, we propose to minimize this leave-one-out loss with automatic differentiation. Within this framework, we propose methods to estimate label reliability and diffusion rate. Optimizing the diffusion rate is more efficiently accomplished with a spectral representation. Results show that the label reliability approach competes with robust ℓ_1-norm methods and that removing diagonal entries reduces the risk of overfitting and leads to suitable criteria for parameter selection.

Keywords: Machine learning · Leave-one-out · Semi-supervised learning · Graph-based approaches · Label propagation · Eigendecomposition

1 Introduction

Machine learning (ML) is the subfield of Computer Science that aims to make a computer learn from data [11]. The task we'll be considering is the problem of *classification*, which requires the prediction of a discrete label corresponding to a *class*.

Before the data can be presented to an ML model, it must be represented in some way. Our input is nothing more than a collection of n examples. Each object of this collection is called an *instance*. For most practical applications, we may consider an instance to be a vector of d dimensions. Graph-based semi-supervised learning (GSSL) relies a lot on matrix representations, so we will be representing the observed instances as an *input matrix*

$$\mathbf{X} \in \mathbb{R}^{n \times d}. \tag{1}$$

This study was financed in part by the Coordenação de Aperfeiçoamento de Nível Superior - Brasil (CAPES) - Finance Code 001, and São Paulo Research Foundation (FAPESP) grant #18/01722-3.

© Springer Nature Switzerland AG 2021
A. Britto and K. Valdivia Delgado (Eds.): BRACIS 2021, LNAI 13073, pp. 514–527, 2021.
https://doi.org/10.1007/978-3-030-91702-9_34

A ML classifier should learn to map an input instance to the desired output. To do this, it needs to know the **labels** (i.e. the output) associated with instances contained in the training set. In **semi-supervised learning**, only some of the labels are known in advance [3]. Once again, it is quite convenient to use a matrix representation, referred to as the *(true) label matrix*:

$$\mathbf{Y}_{ij} = \begin{cases} 1 & \text{if the } i\text{-th instance is associated with the } j\text{-th class} \\ 0 & \text{otherwise} \end{cases} \tag{2}$$

Most approaches are *inductive* in nature so that we can predict the labels of instances not seen before deployment. However, most GSSL methods are *transductive*, which simply means that we are only interested in the fixed but unknown set of labels corresponding to unlabeled instances [17]. Accordingly, we may represent this with a *classification matrix*:

$$\mathbf{F} \in \mathbb{R}^{n \times c} \tag{3}$$

In order to separate the labeled data $\mathcal{L}$ from the unlabeled data $\mathcal{U}$, we divide our matrices as following:

$$\mathbf{X} = \left[\mathbf{X}_{\mathcal{L}}^{\top}, \mathbf{X}_{\mathcal{U}}^{\top} \right]^{\top} \tag{4}$$

$$\mathbf{Y} = \left[\mathbf{Y}_{\mathcal{L}}^{\top}, \mathbf{Y}_{\mathcal{U}}^{\top} \right]^{\top} \tag{5}$$

$$\mathbf{F} = \left[\mathbf{F}_{\mathcal{L}}^{\top}, \mathbf{F}_{\mathcal{U}}^{\top} \right]^{\top} \tag{6}$$

The idea of SSL is appealing for many reasons. One of them is the possibility to integrate the toolset developed for unsupervised learning. Namely, we may use unlabeled data to measure the density $P(\mathbf{x})$ within our d-dimensional input space. Once that is achieved, the only thing left is to take advantage of this information. To do this, we have to make use of *assumptions* about the relationship between the input density $P(\mathbf{x})$ and the conditional class distribution $P(y \mid \mathbf{x})$. If we are not assuming that our datasets satisfy any kind of assumption, SSL can potentially cause a significant decrease compared to baseline performance [14]. This is currently an active area of research: *safe semi-supervised learning* is said to be attained when SSL never performs worse than the baseline, for any choice of labels for the unlabeled data. This is indeed possible in some limited circumstances, but also provably impossible for others, such as for a specific class of margin-based classifiers [9].

In Fig. 1, we illustrate which kind of dataset is suitable for GSSL. There are two clear spirals, one corresponding to each class. In a bad dataset for SSL, we can imagine the spiral structure to be a red herring, i.e. something misguiding. A very common assumption for SSL is the *smoothness assumption*, which is one of the cornerstones for GSSL classifiers. It states that "If two instances $\mathbf{x}_1$, $\mathbf{x}_2$ in a high-density region are close, then so should be the corresponding outputs y_1, y_2" [3]. Another important assumption is that the (high-dimensional) data lie (roughly) on a low-dimensional manifold, also known as the *manifold assumption* [3]. If the data lie on a low-dimensional manifold, the local similarities will approximate the manifold well. As a result, we can, for example, increase the resolution of our image data without impacting performance.

Graph-based semi-supervised learning has been used extensively for many different applications across different domains. In computer vision, these include car plate character recognition [19] and hyperspectral image classification [13]. GSSL is particularly appealing if the underlying data has a natural representation as a graph. As such, it has been a promising approach for drug property prediction from the structure of molecules [8]. Moreover, it has had much application in knowledge graphs, such as the development of web-scale recommendation systems [16].

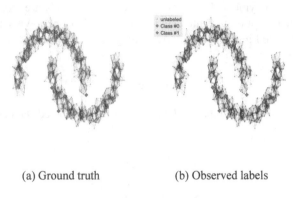

(a) Ground truth   (b) Observed labels

Fig. 1. An ideal scenario for semi-supervised learning

GSSL methods put a greater emphasis on using *geodesics* by expressing *connectivity* between instances through the creation of a graph. Many successful deep semi-supervised approaches use a similar yet slightly weaker assumption, namely that **small perturbations** in input space should cause little corresponding perturbation on the output space [12].

It turns out that we can best express our concepts by defining a measure of *similarity*, instead of distance. In particular, we search for an **affinity matrix** $\mathbf{W} \in \mathbb{R}^{n \times n}$, such that

$$\mathbf{W}_{ij} = \begin{cases} w(\mathbf{x}_i, \mathbf{x}_j) \in \mathbb{R} & \text{if } \mathbf{x}_i \text{ and } \mathbf{x}_j \text{ are considered neighbors} \\ 0 & \text{otherwise} \end{cases} \quad (7)$$

where w is some function determining the similarity between any two instances $\mathbf{x}_i, \mathbf{x}_j$. When constructing an affinity matrix in practice, instances are not considered neighbors of themselves, i.e. we have $\forall i \in \{1..n\} : \mathbf{W}_{ii} = 0$.

The specification of an affinity matrix is a necessary step for any GSSL classifier, and its sparsity is often crucial for reducing computational costs. There are many ways to choose a neighborhood. Most frequently, it is constructed by looking at the **K-Nearest neighbors (KNN)** of a given instance.

One last important concept to GSSL is that of the graph Laplacian operator. This operator is analogue to the Laplace-Beltrami operator on manifolds. There are a few graph Laplacian variants, such as the **combinatorial Laplacian**

$$\mathbf{L}_{\mathbb{C}} = \mathbf{D} - \mathbf{W} \qquad (8)$$

where $\mathbf{D}$, called the *degree matrix*, is a diagonal matrix whose entries are the sum of each row of $\mathbf{W}$. There is also the **normalized Laplacian**, whose diagonal is the *identity matrix* $\mathbf{I}$:

$$\mathbf{L}_{\mathbb{N}} = \mathbf{I} - \mathbf{D}^{-\frac{1}{2}} \mathbf{W} \mathbf{D}^{-\frac{1}{2}} = \mathbf{D}^{-\frac{1}{2}} \mathbf{L}_{\mathbb{C}} \mathbf{D}^{-\frac{1}{2}} \qquad (9)$$

Each graph Laplacian $\mathbf{L}$ induces a measure of **smoothness with respect to the graph** on a given classification matrix $\mathbf{F}$, namely

$$\widetilde{S}_{\mathbf{L}}(\mathbf{F}) = \frac{1}{2} \sum_{k=1}^{c} (\mathbf{F}_{[:,k]})^{\top} \mathbf{L}(\mathbf{F}_{[:,k]}) = \frac{1}{2} tr(\mathbf{F}^{\top} \mathbf{L} \mathbf{F}) \qquad (10)$$

where tr is the trace of the matrix and c the number of classes. If we consider each column $\mathbf{f}$ of $\mathbf{F}$ individually, then we can express graph smoothness of each graph Laplacian as

$$\mathbf{f}^{\top} \mathbf{L}_{\mathbb{C}} \mathbf{f} = \sum_{1 \leq i,j \leq n} \mathbf{W}_{ij} (\mathbf{f}_i - \mathbf{f}_j)^2 \qquad (11)$$

$$\mathbf{f}^{\top} \mathbf{L}_{\mathbb{N}} \mathbf{f} = \sum_{1 \leq i,j \leq n} \left(\frac{\mathbf{W}_{ij}}{\sqrt{\mathbf{D}_{ii}}} \mathbf{f}_i - \frac{\mathbf{W}_{ij}}{\sqrt{\mathbf{D}_{jj}}} \mathbf{f}_j \right)^2 \qquad (12)$$

Each graph Laplacian also has an eigendecomposition:

$$\mathbf{L} = \mathbf{U} \mathbf{\Lambda} \mathbf{U}^{\top} \qquad (13)$$

where the set of columns of $\mathbf{U}$ is an orthonormal basis of eigenfunctions, with $\mathbf{\Lambda}$ a diagonal matrix with the eigenvalues. As $\mathbf{U}$ is a unitary matrix, any real-valued function on the graph may be expressed as a linear combination of eigenfunctions. We map a function to the spectral domain by pre-multiplying it by $\mathbf{U}^{\top}$ (also called **graph Fourier transform**, also known as GFT). Additionally, pre-multiplying by $\mathbf{U}$ gives us the inverse transform. This spectral representation is very useful, as eigenfunctions that are smooth with respect to the graph have smaller eigenvalues. Outright restricting the amount of eigenfunctions is known as **smooth eigenbasis pursuit** [6], a valid strategy for semi-supervised regularization.

In this work, we explore the problem of parameter selection for the Local and Global Consistency model. This model yields a propagation matrix, which itself depends on a fixed parameter α determining the diffusion rate. We show that, by removing the diagonal in the propagation matrix used by our baseline, a leave-one-out criterion can be easily computed. Our first proposed algorithm attempts to calculate label reliability by optimizing the label matrix, subject to constraints. This approach is shown to be competitive with robust ℓ_1-norm classifiers. Then, we consider the problem of optimizing the diffusion rate. Doing this in the usual formulation of LGC is impractically expensive. However, we show that the spectral representation of the problem can be exploited to easily solve an approximate version of the problem. Experimental results show that minimizing leave-one-out error leads to good generalization.

The remaining of this paper is organized as follows. Section 2 summarizes the key ideas and concepts that are related to our work. Section 3 presents our two proposed algorithms: LGC_LVO_AutoL for determining label reliability, and LGC_LVO_AutoD for determining the optimal diffusion rate parameter. Our methodology is detailed in Sect. 4, going over the basic framework and baselines. The results are presented in Sect. 5. Lastly, concluding remarks are found in Sect. 6.

2   Related Work

In this section, we present some of the algorithms and concepts that are central to our approach. We describe the inner workings of our baseline algorithm, and how eliminating diagonal entries of its propagation matrix may lead to better generalization.

2.1   Local and Global Consistency

The *Local and Global Consistency* (LGC) [17] algorithm is one of the most widely known graph-based semi-supervised algorithms. It minimizes the following cost:

$$Q(\mathbf{F}) = \frac{1}{2}\left(tr(\mathbf{F}^{\top}\mathbf{L_N}\mathbf{F}) + \mu\|\mathbf{F} - \mathbf{Y}\|^2\right) \tag{14}$$

LGC addresses the issue of label reliability by introducing the parameter $\mu \in (0, \infty)$. This parameter controls the trade-off between fitting labels, and achieving high graph smoothness.

LGC has an analytic solution. To see this, we take the partial derivative of the cost with respect to $\mathbf{F}$:

$$\frac{\partial Q}{\partial \mathbf{F}} = \frac{1}{2}\frac{\partial tr(\mathbf{F}^{\top}\mathbf{L_N}\mathbf{F})}{\partial \mathbf{F}} + \frac{1}{2}\mu\frac{\partial \|\mathbf{F} - \mathbf{Y}\|^2}{\partial \mathbf{F}} \tag{15}$$

$$= ((1 + \mu)\mathbf{I} - \mathbf{S})\mathbf{F} - \mu\mathbf{Y} \tag{16}$$

where $\mathbf{S} = \mathbf{D}^{-\frac{1}{2}}\mathbf{W}\mathbf{D}^{-\frac{1}{2}} = \mathbf{I} - \mathbf{L_N}$. By dividing the above by $(1 + \mu)$, we observe that this derivative is zero exactly when

$$(I - \alpha\mathbf{S})\mathbf{F} = \beta\mathbf{Y} \tag{17}$$

with

$$\alpha = \frac{1}{1 + \mu} \in (0, 1) \tag{18}$$

and

$$\beta = 1 - \alpha \tag{19}$$

The matrix $(\mathbf{I} - \alpha\mathbf{S})$ can be shown to be positive-definite and therefore invertible, so the optimal $\mathbf{F}$ can be obtained as

$$\mathbf{F} = \beta(I - \alpha\mathbf{S})^{-1}\mathbf{Y} \tag{20}$$

We hereafter refer to $(I - \alpha S)^{-1}$ as the **propagation matrix P**. Each entry $\mathbf{P}_{ij}$ represents the amount of label information from X_j that X_i inherits. It can be shown that the inverse is a result of a diffusion process, which is calculated via iteration:

$$F(0) = \mathbf{Y} \tag{21}$$

$$F(t+1) = \alpha \mathbf{S}F(t) + (1-\alpha)\mathbf{Y} \tag{22}$$

Moreover, it can be shown that the closed expression for F at any iteration is

$$F(t) = (\alpha \mathbf{S})^{t-1}\mathbf{Y} + (1 - \alpha) \sum_{i=0}^{t-1}(\alpha \mathbf{S})^i \mathbf{Y} \tag{23}$$

S is similar to $D^{-1}W$, whose eigenvalues are always in the range $[-1, 1]$ [17]. This ensures the first term vanishes as t grows larger, whereas the second term converges to PY. Consequently, P can be characterized as

$$\mathbf{P} = (1 - \alpha) \lim_{t \to \infty} \sum_{i=0}^{t}(\alpha \mathbf{S})^i \tag{24}$$

$$= (1 - \alpha) \lim_{t \to \infty} \sum_{i=0}^{t} \alpha^i \mathbf{D}^{\frac{1}{2}}(\mathbf{D}^{-1}\mathbf{W})^i \mathbf{D}^{-\frac{1}{2}} \tag{25}$$

The **transition probability matrix** $\widetilde{W} = \mathbf{D}^{-1}\mathbf{W}$ makes it so we can interpret the process as a random walk. Let us imagine a particle walking through the graph according to the transition matrix. Assume it began at a labeled vertex v_a, and at step i it reaches a labeled vertex v_b, initially labeled with class c_b. When this happens, v_a receives a **confidence boost** to class c_b. Alternatively, one can say that this boost goes to the entry which corresponds to the contribution from v_b to v_a, i.e. $\mathbf{P}_{ab}$. This boost is proportional to α^i. This gives us a good intuition as to the role of α. More precisely, the contribution of vertices found later in the random walk decays exponentially according to α^i.

2.2 Self-influence and Leave-One-Out-Error

There is one major problem with LGC's solution: the diagonal of $\mathbf{P}$. At first glance, we would think that "fitting the labels" means looking for a model that explains our data very well. In reality, this translates to memorizing the labeled set. The main problem resides within the **diagonal** of the propagation matrix. Any entry $\mathbf{P}_{ii}$ stores the **self-influence** of a vertex, which is calculated according to the expected reward obtained by looping around and visiting itself. The optimal solution w.r.t. label fitting occurs when α tends to zero. For labeled instances, an initial reward is given for the starting vertex itself, and the remaining are essentially ignored.

We argue that the diagonal is directly related to **overfitting**. It essentially tells the model to rely on the label information it knows. There are a few analogies to be made: say that we are optimizing the number of neighbors k for a KNN classifier. The analog of "LGC-style optimal label fitting" would be to include each labeled instance as a neighbor to itself, and set $k = 1$. This is obviously not a good criterion. The answer that maximizes a *proper* "label fitting" criteria, in this case, is selecting k that minimizes classification error with the extremely important caveat: directly using each instance's own label is prohibited.

In spite of the problems we have presented, the family of LGC solutions remains very interesting to consider. We will have to eliminate diagonals, however. Let

$$\mathbf{H}(\alpha) := \left((\mathbf{I} - \alpha\mathbf{S})^{-1} - diag((\mathbf{I} - \alpha\mathbf{S})^{-1})\right)_{\mathcal{L}} \mathbf{Y}_{\mathcal{L}} \qquad (26)$$

By eliminating the diagonal, we obtain, for each label, its classification if it were not included in the label propagation process. As such, it can be argued that minimizing

$$\left\|\mathbf{H}(\alpha) - \mathbf{Y}_{\mathcal{L}}\right\|^2 \qquad (27)$$

also minimizes the *leave-one-out* (LOO) error. There is an asterisk: each instance is still used as unlabeled data, but this effect should be insignificant. It is also interesting to **row-normalize** (a small constant ϵ may be added for stabilization) the rows, so that we end up with **classification probabilities**.

Previously, we developed a semi-supervised leave-one-out filter [2,4]. We also managed to reduce the amount of storage used by only calculating a **propagation sub-matrix**. The LOO-inspired criterion encourages label information to be redundant, so labels that are incoherent with the implicit model are removed.

The major drawback of our proposal is that it needs an extra parameter r, which is the number of labels to remove. The optimal r is usually around the number of noisy labels, which is unknown to us. This was somewhat addressed in [4]: we can instead use a threshold, which tells us how much labels can deviate from the original model. Nonetheless, it is desirable to solve this problem in a way that removes such a parameter. We will do this by introducing a new optimization problem.

3 Proposal

In this work, we consider the optimal value of α for our LOO-inspired loss [4]. The objective is to develop a method to minimize **surrogate losses** based on LOO error for the LGC GSSL classifier and evaluate the generalization of its solutions.

We use the term "surrogate loss" [9] to denote losses such as squared error, cross-entropy and so on. We use this term to purposefully remind that the solution $\mathbf{H}(\alpha)$ that minimizes the loss on the test set does not also necessarily be the one that maximizes accuracy. We use **automatic differentiation** as our optimization procedure. As such, we call our approach LGC_LVO_AutoL and LGC_LVO_AutoD when learning label reliability and diffusion rate, respectively. We exploit the fact that we can compute the gradients of our loss.

3.1 LGC_LVO_AutoL: Automatic Correction of Noisy Labels

Let $\mathbf{P}$ be the propagation matrix, whose submatrix $\mathbf{P}_{\mathcal{LL}}$ corresponds to kernel values between labeled data only. Then, the modified LGC solution is given by $\widetilde{\mathbf{P}}\mathbf{Y}_{\mathcal{L}}$, where

$$\widetilde{\mathbf{P}} := (\mathbf{P}_{\mathcal{LL}} - diag(\mathbf{P}_{\mathcal{LL}})) \qquad (28)$$

Our optimization problem is to optimize a diagonal matrix Ω indicating the updated reliability of each label:

$$\min_{\Omega} \text{LOSS}(deg(\widetilde{\mathbf{P}}\Omega\mathbf{Y}_{\mathcal{L}})^{-1}\widetilde{\mathbf{P}}\Omega\mathbf{Y}_{\mathcal{L}}, \ \mathbf{Y}_{\mathcal{L}}) \quad \text{such that} \ \ \forall i : \Omega_{ii} \geq 0 \quad (29)$$

where LOSS denotes some loss function such as squared error or cross-entropy.

3.2 LGC_LVO_AutoD: Automatic Choice of Diffusion Rate

In [4], we use a modified version of the power method to calculate the submatrix $\mathbf{P}_{\mathcal{L}}$. This is enough to give us the answer in a few seconds for a fixed α, but would quickly turn into a huge bottleneck if we were to constantly update α. Our new approach uses the graph Fourier transform. In other words, we adapt the idea of smooth eigenbasis pursuit to this particular problem. Let us write the propagation matrix using eigenfunctions:

$$\mathbf{P} = \sum_{t=0}^{\infty} \alpha^t \mathbf{S}^t \quad (30)$$

$$= \sum_{t=0}^{\infty} \alpha^t (\mathbf{I} - \mathbf{L}_{\mathbb{N}})^t \quad (31)$$

Using the graph Fourier transform, we have that

$$\mathbf{L}_{\mathbb{N}} = \mathbf{U}\Lambda\mathbf{U}^{\top} \quad (32)$$

$$\mathbf{I} = \mathbf{U}\mathbf{I}\mathbf{U}^{\top} \quad (33)$$

So it follows that

$$\mathbf{P} = \mathbf{U}\widetilde{\Lambda}\mathbf{U}^{\top} \quad (34)$$

where $\forall i \in \{1..N\}$:

$$\widetilde{\Lambda}_{ii} = \sum_{t=0}^{\infty} (\alpha(1 - \Lambda_{ii}))^t \quad (35)$$

$$\{\|\alpha(1 - \Lambda_{ii})\| < 1\} = \frac{1}{1 - (\alpha(1 - \Lambda_{ii}))} \quad (36)$$

$$= \frac{1}{(1 - \alpha) \times 1 + (\alpha) \times \Lambda_{ii}} \quad (37)$$

In practice, we can assume that $\mathbf{U}$ is an $l \times p$ matrix, with l the number of labeled instances and p the chosen amount of eigenfunctions. The diagonal entries are given by

$$\mathbf{P}_{ii} = (\mathbf{U}\widetilde{\Lambda}\mathbf{U}^{\top})_{ii} \quad (38)$$

$$= (\mathbf{U}^{\top})_{[:,i]}\widetilde{\Lambda}(\mathbf{U}^{\top})_{[:,i]} \quad (39)$$

$$= \Sigma_{i=1}^{p} \mathbf{U}_{ik}^2 \widetilde{\Lambda}_{kk} \quad (40)$$

$$= (\mathbf{U}_{[i,:]})(\mathbf{U}_{[i,:]}\widetilde{\Lambda})^{\top} \quad (41)$$

Next, we'll analyze the complexity of calculating the leave-one-out error for a given diffusion rate α, given that we have stored the first p eigenfunctions in the matrix $\mathbf{U}$. Let $\mathbf{U}_{\mathcal{L}} \in \mathbb{R}^{p \times l}$ be the matrix of eigenfunctions with domain restricted to labeled instances. Exploiting the fact that $\mathbf{Y}_{\mathcal{U}} = 0$, it can be shown that

$$\mathbf{P}_{\mathcal{L}\mathcal{L}} = \mathbf{U}_{\mathcal{L}} \widetilde{\mathbf{\Lambda}} (\mathbf{U}_{\mathcal{L}}^{\top} \mathbf{Y}_{\mathcal{L}}) \tag{42}$$

We can precompute $(\mathbf{U}_{\mathcal{L}}^{\top} \mathbf{Y}_{\mathcal{L}}) \in \mathbb{R}^{p \times c}$ with $\mathcal{O}(plc)$ multiplications. For each diffusion rate candidate α, we obtain $\mathbf{U}_{\mathcal{L}} \widetilde{\mathbf{\Lambda}} \in \mathbb{R}^{l \times p}$ by multiplying each restricted eigenfunction with its new eigenvalue. The only thing left is to post-multiply it by the pre-computed matrix. As a result, we can compute the leave-one-out error for arbitrary α with $\mathcal{O}(plc)$ operations.

We have shown that, by using the propagation submatrix, we can re-compute the propagation submatrix $\mathbf{P}_{\mathcal{L}\mathcal{L}}$ in $\mathcal{O}(plc)$ time. In comparison, the previous approach [4] requires, for each diffusion rate, a total of $O(tknlc)$ operations, assuming t iterations of the power method and a sparse affinity matrix with average node degree equal to k. Even if we use the full eigendecomposition ($p = n$), this new approach is significantly more viable for different learning rates. Moreover, we can lower the choice of p to use a faster, less accurate approximation.

4   Methodology

Basic Framework. We have a **configuration dispatcher** which enables us to vary a set of parameters (for example, the chosen dataset and the parameters for affinity matrix generation).

We start out by reading our dataset, including features and labels. We use the random seed to select the sampled labels, and create the affinity matrix $\mathbf{W}$ necessary for LGC. For automatic label correction, we assume that α is given as a parameter. For choosing the automatic diffusion rate, we need to extract the p smoothest eigenfunctions as a pre-processing step. Next, we repeatedly calculate the gradient and update either α or $\mathbf{\Omega}$ to minimize LOO error. After a set amount of iterations, we return the final classification and perform an evaluation on unlabeled examples.

The programming language of choice is Python 3, for its versatility and support. We also make use of the *tensorflow-gpu* [1] package, which massively speeds up our calculations and also enables automatic differentiation of loss functions. We use a Geforce GTX 1070 GPU to speed up inference, and also for calculating the k-nearest neighbors of each instance with the *faiss-gpu* package [7].

Evaluation and Baselines. In this work, our datasets have a roughly equal number of labels for each class. As such, we will report the mean accuracy, as well as its standard deviation. However, one distinction is that we calculate the accuracy independently on labeled and unlabeled data. This is done to better assess whether our algorithms are improving classification on instances outside the labeled set, or if it outperforms its LGC baseline only when performing diagnosis of labels.

Our approach, LGC_LVO_Auto, is compatible with any differentiable loss function, such as *mean squared error* (MSE) or *cross-entropy* (xent). The chosen optimizer was Adam, with a learning rate of 0.7 and 5000 iterations. For the approximation of the propagation matrix in LGC_LVO_AutoL, we used $t = 1000$ iterations throughout.

Perhaps the most interesting classifier to compare our approach to is the LGC algorithm, as that is the starting point and the backbone of our own approach. We will also be comparing our results with the ones reported by [6]. These include: Gaussian Fields and Harmonic Functions (GFHF) [18], Graph Trend Filtering [15], Large-Scale Sparse Coding (LSSC) [10] and Eigenfunction [5].

5 Results

This section presents the results of employing our two approaches LGC_LVO_AutoL and LGC_LVO_AutoD on ISOLET and MNIST datasets, compared to other graph-based SSL algorithms from the literature.

5.1 Experiment 1: LGC_LVO_AutoL on ISOLET

Experiment Setting. In this experiment, we compared LGC_LVO_AutoL to the baselines reported in [6], specifically for the ISOLET dataset. Unlike the authors, our 20 different seeds also control both the **label selection and noise processes**. The graph construction was performed exactly as in [6], a symmetric 10-nearest neighbors graph with the width σ of the RBF kernel set to 100. We emphasize that the reported results by the authors correspond to the best-performing parameters, divided for each individual noise level. In [6], λ_1 is set to 10^5, 10^2, 10^2 and 10^2 for the respective noise rates of 0%, 20%, 40% and 60%; λ_2 is kept to 10, and the number of eigenfunctions is $m = 30$. We could not find any implementation code for SIIS, so we had to manually reproduce it ourselves. As for parameter selection for LGC_LVO_AutoL, we simply set $\alpha = 0.9$ (equivalently, $\mu = 0.1111$). We reiterate that **having a single parameter** is a **strength of our approach**. In future work, we will try to combine LGC_LVO_AutoL with LGC_LVO_AutoD to fully eliminate the need for parameter selection.

Experiment Results. The results are contained in Tables 1a and b. With respect to the accuracy on **unlabeled examples**, we observed that:

- SIIS appeared to have a slight edge in the noiseless scenario.
- **LGC's own inherent robustness was evident**. When 60% noise was injected, it went from 84.72 to 70.69, a decrease of 16.55%. In comparison, SIIS had a decrease of 14.39%; GFHF a decrease of 22.08%; GTF a decrease of 21.82%.
- With 60% noise, LGC_LVO_AutoL(xent) decreased its accuracy by 11.52%, so **LGC_LVO_AutoL(xent) had the lowest percentual decrease**.
- **LGC_LVO_AutoL(MSE) disappointed** for both labeled and unlabeled instances.
- **LGC_LVO_AutoL was not noticeably superior to LGC when there was less than 60% noise**.

With respect to the accuracy on **unlabeled examples**, we observed that:

- **LGC was unable to correct noisy labels**.
- LGC_LVO_AutoL($xent$) discarded only 5% of the labels for the noiseless scenario, which is **better than SIIS and LSSC**.
- Moreover, **LGC_LVO_AutoL**($xent$) had the **highest average accuracy on labeled instances** for 20%, 40%, 60% label noise.

Table 1. Accuracy on ISOLET dataset

(a) Accuracy on **unlabeled examples** only

Dataset	Noise level	LSSC	GTF	GFHF	SIIS
	0%	84.8 ± 0.0	70.1 ± 0.0	86.5 ± 0.0	**85.4 ± 0.0**
ISOLET	20%	82.8 ± 0.3	69.9 ± 0.2	81.6 ± 0.4	**84.9 ± 0.6**
(1040/7797 labels)	40%	78.5 ± 0.6	59.8 ± 0.3	79.7 ± 1.0	**80.2 ± 1.3**
reported results	60 %	67.5 ± 1.8	54.8 ± 0.5	67.4 ± 1.5	**74.9 ± 1.4**
Dataset	Noise Level	LGC	LGC_LVO_AutoL(MSE)	LGC_LVO_AutoL(XENT)	SIIS
	0%	84.71 ± 0.56	84.21±0.4	84.22±0.45	**85.24 ± 0.32**
ISOLET	20%	82.89 ± 0.59	81.6±0.63	82.56±0.62	**83.69 ± 0.33**
(1040/7797 labels)	40%	79.33 ± 0.92	77.73±0.96	80.23±0.74	**80.88 ± 0.77**
our results	60%	70.69 ± 1.01	68.98±1.81	**74.51±1.75**	72.97 ± 1.16

(b) Accuracy on **labeled examples** after label correction

Dataset	Noise level	LSSC	GTF	GFHF	SIIS
	0%	89.9 ± 0.0	95.8 ± 0.0	**100.00 ± 0.0**	91.1 ± 0.0
ISOLET	20%	87.7 ± 0.3	79.8 ± 0.7	80.00 ± 0.00	**90.5 ± 0.8**
(1040/7797 labels)	40%	82.9 ± 0.9	63.3 ± 0.4	60.00 ± 0.00	**83.6 ± 1.0**
reported results	60 %	71.8 ± 1.7	55.3 ± 0.6	40.00 ± 0.00	**77.4 ± 1.0**
Dataset	Noise level	LGC	LGC_LVO_AutoL(MSE)	LGC_LVO_AutoL(XENT)	SIIS
	0%	**99.9 ± 0.02**	97.36±0.52	95.01±0.58	90.24 ± 0.69
ISOLET	20%	80.84 ± 0.27	90.93±1.28	**91.52±0.79**	88.5 ± 1.07
(1040/7797 labels)	40%	60.24 ± 0.20	82.54 ± 0.92	**87.19±0.89**	85.25 ± 0.96
our results	60%	40.00 ± 0.04	71.16 ± 1.79	**79.14±1.72**	76.34 ± 1.53

5.2 Experiment 2: **LGC_LVO_AutoL** on MNIST

Experiment Setting. This experiment was based on [10], where a few classifiers were tested on the MNIST dataset subject to label noise. In that paper, the parameters for the graph were tuned to minimize cross-validation errors. Moreover, an anchor graph was used, which is a large-scale solution. We did not use such a graph, as **our TensorFlow iterative implementation of LGC_LVO_AutoL was efficient enough to perform classification on MNIST in just a few seconds**. As we also included the results for LGC (without anchor graph), it is interesting to observe that its accuracy decreases similarly to the previously reported results: the main difference is better performance for the noiseless scenario, which is to be expected (anchor graph is an approximation).

Once again, we simply set $\alpha = 0.9$ for LGC_LVO_AutoL. We used a symKNN matrix with $k = 15$ neighbors, and a heuristic sigma $\sigma = 423.57$ obtained by taking one-third of the mean distance to the 10th neighbor (as in [3]).

Table 2. Accuracy on MNIST dataset

(a) Accuracy on **unlabeled examples** only

Dataset	Noise level	LSSC*	Eigenfunction*	LGC* (anchor graph)
MNIST	0%	**93.1 ± 0.7**	73.8 ± 1.6	90.4 ± 0.7
(100/70000 labels)	15%	**91.1 ± 2.0**	68.6 ± 2.8	83.5 ± 1.6
reported results	30%	**89.0 ± 3.6**	61.9 ± 4.0	74.4 ± 2.8
Dataset	Noise level	LGC_LVO_AutoL(MSE)	LGC_LVO_AutoL(XENT)	LGC
MNIST	0%	91.7 ± 0.7	92.69 ± 1.19	**93.09 ± 0.92**
(100/70000 labels)	15%	86.48 ± 2.59	**90.45 ± 2.13**	85.40 ± 1.66
our results	30%	81.33 ± 4.43	**84.46 ± 3.89**	74.58 ± 2.6

(b) Accuracy on **labeled examples** after label correction

Dataset	Noise level	LGC_LVO_AutoL(MSE)	LGC_LVO_AutoL(XENT)	LGC
MNIST	0%	99.5 ± 0.59	98.05 ± 0.59	**100.00 ± 0.0**
(100/70000 labels)	15%	95.05 ± 2.01	**96.10 ± 1.3**	85.00 ± 0.0
our results	30%	85.55±4.88	**89.75 ± 4.06**	70.00 ± 0.0

Experiment Results. The results are found in Tables 2a and b. With respect to the accuracy on **unlabeled examples**, we observed that:

- **LGC_LVO_AutoL with cross-entropy improved the LGC baseline significantly on unlabeled instances.** For 30% label noise, mean accuracy increases **from 74.58% to 84.46%**.
- The **mean squared error loss is once again consistently inferior to cross-entropy when there is noise.**
- LGC_LVO_AutoL was not able to obtain better results than LSSC.

With respect to the accuracy on **labeled examples**, we observed that:

- **LGC was not able to correct the labeled instances.**
- **LGC_LVO_AutoL with cross-entropy improved the LGC baseline significantly on labeled instances.** With 30% noise, accuracy goes **from 70.00% to 89.75%**.

5.3   Experiment 3: LGC_LVO_AutoD on MNIST

Experiment settings were kept the same as Experiment 2 (without noise), and $p = 300$ eigenfunctions were extracted. In Fig. 2, we show accuracy on the labeled set as red, and on the unlabeled set as purple. The x values on the horizontal axis relate to α as following: $\alpha = 2^{-1/x}$. Looking at Fig. 2b, we can see that, if we do not remove the diagonal, there is much overfitting and the losses reach their minimum value much earlier than does the accuracy on unlabeled examples. When the diagonal is removed, the loss-minimizing estimates for α get much closer to the optimal one (Fig. 2a). Moreover, by removing the diagonal we obtain a much better estimation of the accuracy on unlabeled data. Therefore, we can simply select the α corresponding to the best accuracy on known labels.

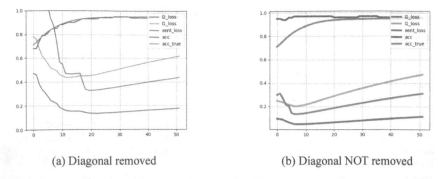

(a) Diagonal removed   (b) Diagonal NOT removed

Fig. 2. `LGC_LVO_AutoD` on MNIST. The x coordinate on the horizontal axis determines diffusion rate $\alpha = 2^{-1/x}$ for $x \geq 1$. The orange, blue and green curves represent the surrogate losses: mean absolute error, mean squared error, cross-entropy. Moreover, the accuracy on the known labeled data is shown in red, and the accuracy on unlabeled data in purple. (Color figure online)

6   Concluding Remarks

We have proposed the `LGC_LVO_Auto` framework, based on leave-one-out validation of the LGC algorithm. This encompasses two methods, `LGC_LVO_AutoL` and `LGC_LVO_AutoD`, for estimating label reliability and diffusion rates. We use automatic differentiation for parameter estimation, and the eigenfunction approximation is used to derive a faster solution for `LGC_LVO_AutoD`, in particular when having to recalculate the propagation matrix for different diffusion rates.

Overall, `LGC_LVO_AutoL` produced interesting results. For the ISOLET dataset, it was **very successful at the task of label diagnosis**, being able to detect and remove labels with overall better performance than every ℓ_1-norm method. On the other hand, **did not translate too well for unlabeled instance classification**. For the MNIST dataset, **performance is massively boosted for unlabeled instances as well**. In spite of outperforming its LGC baseline by a wide margin, it could not match the reported results of LSSC on MNIST. Preliminary results showed that **`LGC_LVO_AutoD`** is a viable way to get a good estimate of the optimal diffusion rate, and removing the diagonal entries proved to be the crucial step for avoiding overfitting.

For future work, we will be further evaluating `LGC_LVO_AutoD`. We will also try to integrate `LGC_LVO_AutoD` and `LGC_LVO_AutoL` together into one single algorithm. Lastly, we will aim to extend `LGC_LVO_AutoD` to a broader class of graph-based kernels, in addition to the one resulting from the LGC baseline.

References

1. Abadi, M., et al.: TensorFlow: Large-scale machine learning on heterogeneous systems (2015). http://tensorflow.org/, software available from tensorflow.org
2. de Aquino Afonso, B.K.: Analysis of Label Noise in Graph-Based Semi-supervised Learning. Master's thesis (2020)

3. Chapelle, O., Schölkopf, B., Zien, A. (eds.): Semi-supervised Learning. MIT Press, Cambridge (2006). http://www.kyb.tuebingen.mpg.de/ssl-book
4. de Aquino Afonso, B.K., Berton, L.: Identifying noisy labels with a transductive semi-supervised leave-one-out filter. Pattern Recognit. Lett. **140**, 127–134 (2020). https://doi.org/10.1016/j.patrec.2020.09.024. http://www.sciencedirect.com/science/article/pii/S0167865520303603
5. Fergus, R., Weiss, Y., Torralba, A.: Semi-supervised learning in gigantic image collections. In: Advances in Neural Information Processing Systems, pp. 522–530 (2009)
6. Gong, C., Zhang, H., Yang, J., Tao, D.: Learning with inadequate and incorrect supervision. In: 2017 IEEE International Conference on Data Mining (ICDM), pp. 889–894. IEEE (2017)
7. Johnson, J., Douze, M., Jégou, H.: Billion-scale similarity search with GPUs. IEEE Trans. Big Data (2019)
8. Kearnes, S., McCloskey, K., Berndl, M., Pande, V., Riley, P.: Molecular graph convolutions: moving beyond fingerprints. J. Comput.-Aided Mol. Des. **30**(8), 595–608 (2016). https://doi.org/10.1007/s10822-016-9938-8
9. Krijthe, J.H.: Robust semi-supervised learning: projections, limits and constraints. Ph.D. thesis, Leiden University (2018)
10. Lu, Z., Gao, X., Wang, L., Wen, J.R., Huang, S.: Noise-robust semi-supervised learning by large-scale sparse coding. In: AAAI, pp. 2828–2834 (2015)
11. Mitchell, T.: Machine Learning. McGraw-Hill, New York (1997)
12. Miyato, T., Maeda, S.I., Ishii, S., Koyama, M.: Virtual adversarial training: a regularization method for supervised and semi-supervised learning. IEEE Trans. Pattern Anal. Mach. Intell. **41**, 1979–1993 (2018)
13. Shao, Y., Sang, N., Gao, C., Ma, L.: Probabilistic class structure regularized sparse representation graph for semi-supervised hyperspectral image classification. Pattern Recognit. **63**, 102–114 (2017)
14. Van Engelen, J.E., Hoos, H.H.: A survey on semi-supervised learning. Mach. Learn. **109**(2), 373–440 (2020). https://doi.org/10.1007/s10994-019-05855-6
15. Wang, Y.X., Sharpnack, J., Smola, A.J., Tibshirani, R.J.: Trend filtering on graphs. J. Mach. Learn. Res. **17**(1), 3651–3691 (2016)
16. Ying, R., He, R., Chen, K., Eksombatchai, P., Hamilton, W.L., Leskovec, J.: Graph convolutional neural networks for web-scale recommender systems. In: Proceedings of the 24th ACM SIGKDD International Conference on Knowledge Discovery and Data Mining, pp. 974–983 (2018)
17. Zhou, D., Bousquet, O., Lal, T.N., Weston, J., Schölkopf, B.: Learning with local and global consistency. In: Advances in Neural Information Processing Systems, pp. 321–328 (2004)
18. Zhu, X., Ghahramani, Z., Lafferty, J.: Semi-supervised learning using gaussian fields and harmonic functions. In: Proceedings of the Twentieth International Conference on International Conference on Machine Learning, pp. 912–919. AAAI Press (2003)
19. Catunda, J.P.K., da Silva, A.T., Berton, L.: Car plate character recognition via semi-supervised learning. In: 2019 8th Brazilian Conference on Intelligent Systems (BRACIS), pp. 735–740. IEEE (2019)

Tactical Asset Allocation Through Random Walk on Stock Network

Washington Burkart Freitas[✉] and João Roberto Bertini Junior

School of Technology, University of Campinas, Limeira, SP, Brazil
w230317@dac.unicamp.br, bertini@ft.unicamp.br

Abstract. Tactical asset allocation is an essential method for defining a profitable portfolio for a given period. An analyst usually creates a tactical asset portfolio through technical analysis, a process subjective to the analyst's knowledge and interpretations. Another aspect that can directly influence the quality of a portfolio is the number of assets considered for analysis. Human analysts tend to focus on a pre-defined group of assets, limiting choices, and, consequently, the possibility of better results. This work proposes the Stock Network Portfolio Allocation (SNPA) algorithm for the automatic recommendation of a stock portfolio, aiming to maximize profit and minimize risk. The proposed method considers a possibly large set of assets represented as a complex network. In which the nodes represent assets, and the edges stand for the correlation between their returns. Portfolio allocation is done through a random walk on the stock network, selecting, in the end, the most visited nodes (stocks). We conducted investment simulations on Brazilian stocks from the IBrX100 index, for 24 month periods, from Jan. 2018 to Dec. 2019. We compare the results with portfolio strategies: Ibovespa index (IBOV), classic Markowitz's mean-variance portfolio (MV), Mean Absolute Deviation (MAD) portfolio, Conditional Value at Risk (CVaR), and Hierarchical Risk Parity (HRP). The Shape Ratio (SR), Maximum Drawdown (MDD), and Cumulated Wealth (CW) were used as performance metrics. The SNPA algorithm demonstrated its effectiveness, presenting a CW of 236.3%, being 203.5% above MV portfolio; 181.7% above CVaR portfolio; 175.6% above MAD portfolio, 184.6% above IBOV index, and 165.1% above HRP. SNPA also surpassed the benchmarks considering the performance metrics SR and MDD, with values 0.67 and −1.37 respectively, the best results among the benchmarks were observed by the HRP strategy, with 0.48 in SR index, and MV with −1.39 in MDD index.

Keywords: Tactical asset allocation · Stock networks · Random walks · Portfolio management · Markowitz portfolio

1 Introduction

Portfolio management is one of the major problems in today's competitive financial environment [4,16]. Due to stock market volatility as well as diverse factors related to human, political, and economic behavior, creating and managing portfolios successfully becomes an increasingly challenging practice. Since the precursor

© Springer Nature Switzerland AG 2021
A. Britto and K. Valdivia Delgado (Eds.): BRACIS 2021, LNAI 13073, pp. 528–542, 2021.
https://doi.org/10.1007/978-3-030-91702-9_35

works of Markowitz [13,14] referring to portfolio selection and modern portfolio theory, many researchers have been striving to find an ideal solution to create an efficient asset portfolio that maximizes return and minimizes risks.

To cope with market instability, researchers use different methods such as modeling the correlation between asset pricing, listening to human interactions in specialized forums, clustering assets according to their performance and, analyzing machine learning predictions [2,11]. Among the portfolio allocation methods, Tactical Asset Allocation (TAA) is an asset management investment method that consists of readjusting the proportions of each category of assets, based on a signal that indicates which asset class will perform best in the upcoming period [1]. Particularly, systematic TAA strategies involve a quantitative, data-driven investment model to identify trends and perform forecasts to aid portfolio management.

One aspect that can directly influence the quality of a portfolio is the number of assets considered for analysis. Human analysts tend to focus on a pre-defined group of assets, limiting choices, and, consequently, the possibility of better results. While most automated portfolio management systems may exhibit prohibitive computational costs associated with optimization procedures when dealing with a large set of assets [18].

In this paper, we propose the Stock Network Portfolio Allocation (SNPA) algorithm for the automatic recommendation of a stock portfolio, aiming to maximize profit and minimize risk. The motivation for this study is to efficiently cover a large number of assets by modeling them as a complex network. In this network, each node represents an asset, and edges are established according to the correlation between their returns. Portfolio selection is then made through a random walk on the network, selecting, in the end, the most visited assets. Investment simulations have shown a return higher than the benchmarks: the Ibovespa, classic Markowitz's mean-variance portfolio (MV), Mean Absolute Deviation (MAD) portfolio, Conditional Value at Risk (CVaR) portfolio, and Hierarchical Risk Parity (HRP).

The remainder of this paper is organized as follows. In Sect. 2, we present a literature review on some topics related to this work as methods for tactical asset allocation, stock correlation analyzes, and stock networks. In Sect. 3, we detail the SNPA algorithm. In Sect. 4, we conduct some analysis with the hyperparameters of the model. Section 5 we describe the portfolio strategies found in the literature and used in the present work to compare with the proposed algorithm. The performance metrics used in the comparisons are also presented. We present and discuss the result of experiments in Sect. 6. Finally, Sect. 7 concludes the paper.

2   Related Works

Network analysis is commonly used to describe the characteristics or behavior of complex networks. Some research has been conducted to model the stock market using networks [15], where the stock market is presented through a minimal spanning tree (MST) obtained from the matrix of correlation coefficients calculated between all pairs of stocks in a portfolio, considering stock returns. MST-type networks have also been implemented to demonstrate the impact of the new coronavirus pandemic on the financial market [26].

George and Changat [7] emphasize that stock network analysis can play an important role in the stock market study. They created a network of stocks with a set of data from the period of one year, their objective was to identify the structural properties of the market. His work was able to reveal important inputs for decision making, such as, for example, creating a portfolio of more relevant sectors.

Ben-Jacob et al. [20] analyzed stock index correlations, nested stock correlations, and correlations after subtraction of the index return from the stocks' returns, they showed that the behavior of the stock market could not be understood and predicted only based on individual stocks; rather, system-level analysis methods need to be devised to analyze the stock market as a whole. Kim and Jeong [8] proposed improved methods to identify stock groups using the correlation matrix of stock price changes, they identified the multiple groups of stocks from the empirical correlation matrix of stock price changes in the New York Stock Exchange.

Chen et al. [5] discuss the stock network construction problem under simultaneous consideration of linear and nonlinear relations between stocks. The results showed that the proposed multi-layer network better balances the relation between prediction accuracy and the number of predictable nodes. Hong et al. [10] consider cross-correlations among stock prices in the Korean stock market. They used the daily Korean stock market prices of KOSPI200 for four years from Jan. 2000 to Dec. 2004. They observed the behavior of the stock network specifying a threshold for the correlation between the nodes of that network, so a scale-free network was obtained by the cross-correlation coefficient.

3   The Stock Network Portfolio Allocation Algorithm

Prior to detail the Stock Network Portfolio Allocation (SNPA) algorithm, we shall present some concepts about asset return, risk, portfolio return, and weighted stock network.

Measuring the asset return is the way to determine its performance over time. Let $R_{t,i}$ be the observed return of the asset i at time $t \in \{1, 2, \ldots, \tau\}$ considering historical data over the time horizon τ. Here, t can represent a day, a week, a month, a quarter or a year, according to the study interval.

From time $t - 1$ to time t, the returns from a set of N stocks accessible to investors is denoted as $\mathbf{R}_t = [R_{t,1}, \ldots, R_{t,i}, \ldots, R_{t,N}]$. The return $R_{t,i}$ for the i-th asset is calculated by Eq. (1).

$$R_{t,i} = \frac{pr_{t,i} - pr_{t-1,i}}{pr_{t-1,i}} \tag{1}$$

where $pr_{t,i}$ and $pr_{t-1,i}$ represent the closing prices of the ith asset at times t and $t - 1$, respectively.

The cumulative wealth of a given asset i during the analyzed period t is denoted as $cw_{t,i}$, and can be calculated according Eq. (2).

$$cw_{t,i} = \prod_{t=1}^{\tau} (1 + R_{t,i}) - 1 \tag{2}$$

The standard deviation, σ_i, and the variance, σ_i^2, of the return for asset i, can be used to measure the risk of a given asset. Let $\mu_i = \frac{1}{\tau} \sum_{i=1}^{\tau} R_{t,i}$ be the average of returns for asset i over time horizon τ, the variance of the returns for the asset i is given by Eq. (3)

$$\sigma_i^2 = \frac{1}{\tau} \sum_{i=1}^{\tau} (R_{t,i} - \mu_i)^2 \tag{3}$$

We denote $\boldsymbol{\omega}_t = [\omega_{t,1}, \ldots, \omega_{t,i}, \ldots, \omega_{t,N}]$ as the weights from set of N stocks of the portfolio at time t, such that $\sum \boldsymbol{\omega}_t = 1$; $\omega_{t,i}$ represents the allocation percentage of the i-th asset in the portfolio. Having obtained the set of returns and the set of weights of the assets in the portfolio, we can calculate the return of the portfolio in time t, noted as r_t through Eq. (4).

$$r_t = \sum \mathbf{R}_t \boldsymbol{\omega}_t \tag{4}$$

A weighted stock network is denoted as $G = (V, E, Z, W)$, where V is the set with n nodes, E is the set of m edges, Z is the set of z nodes weights and W is the set of w edges weights. The network G is obtained through a correlation matrix $\mathbf{M}_t$, as shown in Eq. (5), and obtained by Eq. (6).

$$\mathbf{M}_t = \begin{bmatrix} c_{11} & c_{12} & c_{13} & \cdots & c_{1N} \\ c_{21} & c_{22} & c_{23} & \cdots & c_{2N} \\ c_{31} & c_{32} & c_{33} & \cdots & c_{3N} \\ \vdots & \vdots & \vdots & \ddots & \vdots \\ c_{N1} & c_{N2} & c_{N3} & \cdots & c_{NN} \end{bmatrix} \tag{5}$$

$$c_{ij} = \frac{\sum_{t=1}^{\tau} (R_{t,i} - \mu_i)(R_{t,j} - \mu_j)}{\sqrt{\sum_{t=1}^{\tau} (R_{t,i} - \mu_i)^2 \sum_{t=1}^{\tau} (R_{t,j} - \mu_j)^2}} \tag{6}$$

To suggest a portfolio for the following month, the algorithm represents the set of assets as an undirected complex network. The most visited assets compose the portfolio after a random walk takes place. Therefore, the proposed approach can be addressed into two parts, building the network from a set of assets and traversing the network to find out the portfolio.

Given a set of assets $A = \{1, \ldots, N\}$, the network is built as follows:

1. each asset $i \in A$ is abstracted to a node v_i.
2. given the correlation between assets i and j, $c_{i,j}$ (Eq. (6)), each pair of nodes v_i and v_j is connected if $c_{ij} < \lambda_n$ or $c_{ij} > \lambda_p$.

In item 2, λ_n and λ_p are thresholds on the value of the correlation of the returns. The rationale of those parameters is to avoid values of correlations close to zero, keeping only significant values of correlation between returns. Notice that their values directly interfere with the structure of the network, consequently, they need to be chosen considering the algorithm performance and restraining to values that generate connected networks.

SNPA uses the asset cumulative wealth, Eq. (5), of the analyzed period as a node weight. So node v_i has as an associated weight, z_i, its cumulative wealth $cw_{t,i}$. In this work, the historical series of the last five years prior to the date on which the portfolio is to be generated is used. Let Γ_i be the set of neighbors of vertex v_i. The way the network is built, for each node v_i, Γ_i may have nodes whose correlation to v_i is positive or negative. Supposing $v_j \in \Gamma_i$, the probability of the particle to visit v_j, noted by P_{ij}, can be calculated by the softmax function as in Eq. (7). The reason to consider the softmax function is that $cw_{t,j}$ may be negative.

$$P_{ij} = \frac{e^{cw_{t,i}}}{\sum_{v_j \in \Gamma_i} e^{cw_{t,j}}} \tag{7}$$

Figure 1(a) shows the paths that can be taken by the particle starting from node v_1 and the cumulative wealth $cw_{i,j}$ for each asset represented in node v_j. Figure 1(b) shows the transition probabilities from node v_1 to each of its neighbors v_2, v_3 and v_4 as given by Eq. (7).

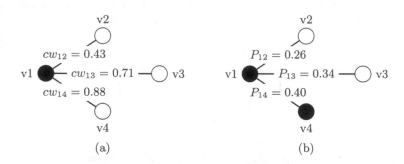

(a)   (b)

Fig. 1. Example of selecting a neighbor using cumulative wealth probability

Algorithm 1 presents the details of the proposed method. The algorithm has as inputs: the set of stocks A, the number of iterations k, the thresholds on the correlation values λ_p, and λ_n. The number of stocks in the portfolio, q, can be either given as an input and in this way, the algorithm will search for a portfolio with q stocks. Or, if it is not given, the algorithm outputs the portfolio with the highest return considering any possible value of q. The most costly part of the algorithm is to calculate the transition probabilities, which requires to access every edge of the graph.

Algorithm 1: THE SNPA ALGORITHM

Input:
A: set of stocks
k: number of iterations
λ_p: positive correlation threshold
λ_n: negative correlation threshold
q number of stocks in the portfolio (optional)
Output: A portfolio S and the assets' weights ω, (S, ω)

```
 1 begin
 2 |    Calculate the correlation between all asset in A obtaining M (Eq. (5)).
 3 |    Build the network, G, from M, with respect to λn and λp
 4 |    for all i ∈ A do
 5 |    |    for all j ∈ Γi do
 6 |    |    |    Calculate Pij as in Eq. (7)
 7 |    |    end
 8 |    end
 9 |    Select a random node vi in G to start
10 |    while k > 0 do
11 |    |    Proceed to the next node, say vj, according to Pij
12 |    |    Increment the visits number ηj to vj, i.e., ηj = ηj + 1
13 |    |    j = i
14 |    |    k = k − 1
15 |    end
16 |    Sort the nodes in descending order according to their number of visits.
17 |    if q has been given then
18 |    |    Define S = Sq as the set of stocks represented by the q top visited nodes
19 |    |    Define ωj = ηj / Σi∈Sq ηi as the weights for each asset j in portfolio S
20 |    end
21 |    else
22 |    |    for q = 1 → N do
23 |    |    |    Calculate the return of portfolio Sq, with the stocks represented by
       |    |    |    the top q visited nodes
24 |    |    end
25 |    |    Define S = max(Sq) as the set of stocks represented by the portfolio
       |    |    with the highest return
26 |    |    Define ωj = ηj / Σi∈Sq ηi as the weights for each asset j in portfolio S
27 |    end
28 |    return (S, ω).
29 end
```

Once all the probabilities have been defined, a random walk takes place. The walk proceeds in the network by following the nodes according to the transition probabilities, while the algorithm keeps track of the number of visits each node had. In the end, the stocks represented by the most visited nodes are chosen to compose the portfolio, and the weight of each asset corresponds to the proportion of the total number of visits. The selected stocks tend to have a high return and be highly correlated to each other (positively and negatively). At the same time, the randomness of the algorithm allows us to consider different but also profitable portfolios.

4   Hyperparameter Analysis

To define the hyperparameters values of the algorithm, experiments were carried out with available stocks from IBrX100 index (approximately 100 stocks). Ten experiments were performed to mitigate the effect of randomization. The experiments consider build a portfolio for Jan. 2020 using stock data of the period from Jan. 2019 to Dec. 2019. The datasets used in this work were built with information from the *Yahoo Finance* [https://finance.yahoo.com/lookup].

First, consider analysing how each parameter relates to each other and to portfolio return. For this analysis, experiments to find a portfolio have been conducted considering several parameters settings. The number of iterations k varied in the range $\{50, 100, \ldots, 1000\}$, threshold values λ_p and λ_n were evaluated in $\{0.1, 0.2, \ldots, 1\}$. The number of assets in the portfolio, q, was given by the algorithm along with the best portfolio return. To each parameter setting, risk has been calculated as the standard deviation of returns among ten runs. Node average degree in the stock network have also been considered for analysis.

Figure 2 shows some pairwise correlations, considering hyperparameters, return, risk and node average degree. It is possible to observe a strong positive correlation between risk and return (0.63), i.e., the higher the return, the higher the risk. Also, a strong negative correlation between λ_p and node average degree in the network, because as the value of λ_p as increases, less nodes are connected. It is also interesting to note the positive correlation between λ_n and risk and return, the greater the value of λ_n the greater the value of the return, although the risk also increases, the increase in risk is less than the increase in return.

Table 1 displays a summary of the main statistical measures obtained through the experiments. The most relevant are the average return and standard deviation, 0.164 ± 0.106. Another relevant information is about the hyperparameter q, although the range was between 2 and 29, most portfolios were composed of 2 assets, as can be seen in the measurement of the median.

With the data from the experiments it is possible to identify the best values for hyperparameters. Figure 3 presents the mean return against the standard deviation of the returns for each value of k and varying λ_p and λ_n in the set $\{0.1, 0.2, \ldots, 1\}$. Each result is the average of 10 realizations.

For $k = 950$ we have a value with less risk and for $k = 350$ portfolios with higher mean return. It is a trade-off matter between risk and return. Whereas we aim to maximize the return and the difference in risk is small between the two values, the value of 350 was chosen for the hyperparameter k.

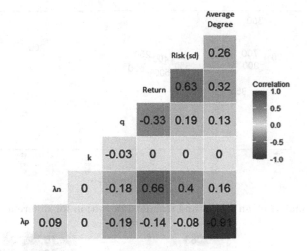

Fig. 2. Hyperparameters correlations

Table 1. Statistical summary

	Range	Mean	Median	Sd
λ_p	0.1/1	0.55	0.5	±0.29
λ_n	−1/−0.1	−0.51	−0.5	±0.26
k	50/1000	525	525	±288
q	2/29	2.5	2	±1.43
Return	−0.089/0.363	0.164	0.141	±0.106
Average degree	2.1/141.2	48.9	37.1	±41.0
Visited nodes	2/52	9.5	7	±6.9

Having defined the value of k, we can obtain the values of λ_p and λ_n. Figure 4 shows the portfolio average returns for $k = 350$, considering different combination of λ_p and λ_n. In the figure, only those return values that are within the range 0.164 ± 0.106 are shown, in order to stay close to average returns and avoid outliers.

As can be seen in Fig. 4, the best result obtained with the experiment is the mean return for $\lambda_p = 0.9$ and $\lambda_n = -0.3$. These are the values of the hyperparameters that will be used in Sect. 6 about experimental result. Hyperparameters λ_p and λ_n are very relevant in the decision process of the algorithm, by defining these hyperparameters, the number of paths in the network and the interconnection between nodes are defined. Setting hyperparameters at random can result in an ineffective portfolio at the end process.

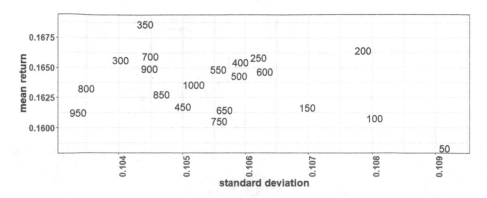

Fig. 3. Estimated mean return and standard deviation for different values of k.

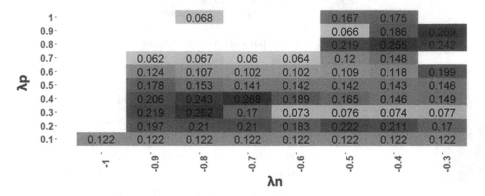

Fig. 4. Estimated portfolio return for when varying λ_p and λ_n and $k = 350$.

5   Methodology

In this section, we describe the main characteristics of the benchmarks used in the present work, which will be compared with the proposed algorithm: Markowitz's classic mean-variance (MV) [14] portfolio; MAD portfolio (Mean Absolute Deviation) [9]; CVaR portfolio (Conditional Value-at-Risk) [19] and Hierarchical Risk Parity (HRP) portfolio [17]. The performance metrics methodologies that are used to compare these portfolio strategies with the proposed algorithm are also presented.

5.1   Portfolio Allocation Methods

Markowitz's Classic Mean-Variance (MV): Markowitz has developed a mathematical model showing a reduction in the volatility of a portfolio through the combination of investments [13,14]. He assumed that portfolios could be entirely characterized by their mean return and variance (or risk).

Mean Absolute Deviation (MAD): In 1991, Konno & Yamazaki [9] suggested a linear programming model in which portfolio risk is measured with the Mean Absolute Deviation (MAD) instead of Markowitz's variance-based portfolio. The purpose of the proposal was to remove most of the difficulties associated with the classical Markowitz's model.

Conditional Value-at-Risk (CVaR): In 2000, Rockafellar & Uryasev [19] introduced a new approach to optimizing a portfolio to reduce the risk of high losses. Value-at-Risk (VaR) played a relevant role in their approach, but the emphasis was on conditional value-at-risk (CVaR) [19,25].

Hierarchical Risk Parity (HRP): portfolio optimization method suggested by Lopez de Prado [17] in 2016. HRP applies modern concepts on graph theory and machine learning techniques to build a diversified portfolio based on the information contained in the asset return matrix.

5.2 Portfolio Performance Metrics

To measure the performance of stock portfolios, some methods can be used, such as: the Sharpe ratio, the maximum drawdown (difference between the highest and lowest value in a given period) and the cumulative wealth. These indicators are widely used in works involving asset portfolio performance analysis and can be found in [3,6,22,24].

Sharpe Ratio (SR): was created by the Nobel William Forsyth Sharpe [21], it is the most widely adopted index for measuring the risk and return ratio of a portfolio. The SR provides a measure of risk-adjusted return for a portfolio strategy [23]. The higher the SR index the better.

The SR can be obtained using Eq. (8), where $\bar{r}_\tau$ is the average of r_t returns from portfolios for the time horizon τ (Eq. (9)), and σ_τ is the standard deviation of portfolio returns (Eq. (10)).

$$SR = \frac{\bar{r}_\tau}{\sigma_\tau} \tag{8}$$

$$\bar{r}_\tau = \frac{1}{\tau} \sum_{t=1}^{\tau} r_t \tag{9}$$

$$\sigma_\tau = \sqrt{\sigma_\tau^2}, \sigma_\tau^2 = \frac{1}{\tau} \sum_{t=1}^{\tau} (r_t - \bar{r}_\tau)^2 \tag{10}$$

Maximum drawdown (MDD) the maximum drawdown is a risk measure widely used by fund managers, as a very high drawdown usually triggers fund redemptions by investors [12]. To obtain the value of the MDD, the Eq. (11) can be used, where $min(r_t)$ is the lowest value observed in the portfolio returns and $max(r_t)$ the highest value found. The lower the MDD index the better.

$$MDD = \frac{max(r_t) - min(r_t)}{max(r_t)} \tag{11}$$

Cumulative Wealth (CW): is the cumulative wealth of stock portfolio over the period τ. Measures the aggregate amount that the investment has gained or lost over time. CW can be calculated by Eq. (12). The higher the CW index the better.

$$CW = \prod_{t=1}^{\tau} (1 + r_t) - 1 \qquad (12)$$

6   Experimental Result

The experiments were performed using the hyperparameters values presented in Table 2, and found as described in Sect. 4.

Table 2. Hyperparameters summary for experiments

Hyperparameter	Description	Value
k	Iterations	350
λ_p	Positive correlation threshold	0.9
λ_n	Negative correlation threshold	−0.3
A	Set of stocks	≈96

The hyperparameter k is the number of iterations that will be performed on the network. The set of stocks A is formed by the stocks in the IBrX100 index. This index is composed by the 100 most traded stocks on the Brazilian stock market. For the purpose of this work, however, the number of stocks used in the experiment was 96 stocks, due to data availability.

The experiments consist in finding a portfolio for the next month, $t + 1$, giving the stock information (time series) from the previous five years. Experiments were analyzed with the base period January 2018 to December 2019. When analyzing a period with these reference months, for example, October 2019, the time series analyzed was from September 2014 to September 2019, and the month that wanted the best result is October 2019.

The results is the month return averaged for 10 times. To evaluate the proposed approach, the SNPA algorithm is compared with the following benchmarks: the Ibovespa index (IBOV), classic Markowitz's mean-variance portfolio (MV), Mean Absolute Deviation (MAD) portfolio, Conditional Value at Risk (CVaR) portfolio and Hierarchical Risk Parity (HRP). The results are separated into two sections; Sect. 6.1 focuses on returns, while Sect. 6.2 presents performance analysis.

6.1   Return-Based Results

Table 3 shows the summary results for each portfolio analyzed. As can be seen, the SNPA algorithm obtained the highest mean return when compared to the benchmarks, it also presents a balanced risk value (sd - standard deviation). The value $\bar{q}$ is the average number of assets each portfolio has suggested.

Table 3. Summary results

	Mean return	Risk(sd)	$\bar{q}$
IBOV	1.88%	±5.16%	75
MV	1.28%	±4.40%	7.3
CVaR	2.17%	±8.70%	9.5
MAD	2.23%	±7.04%	12.3
HRP	2.41%	±5.23%	59.5
SNPA	5.48%	±8.17%	8.8

Table 4 shows the monthly returns obtained by the algorithm under comparison. As can be seen, SNPA presented higher results in 11 out of the 24 analysed months. Conversely, SNPA have performed worse than all the other in 2 out of the 24 analysed months. It is important to note that in August 2018 the algorithm obtained a positive return, while all the others presented negative values. At last, the SNPA algorithm obtained negative results in 5 of the 24 months evaluated, that is, in 21% of the cases. The benchmarks were, on average, negative 9 months, that is, 37.5%.

Figure 5 shows the cumulative portfolio's return to the analyzed period. In the cumulative wealth, the algorithm SNPA demonstrated its effectiveness, presenting a total return of 236.3%, being 203.5% above MV portfolio; 181.7% above CVaR portfolio; 175.6% above MAD portfolio, 184.6% above IBOV index and 165.1% above HRP.

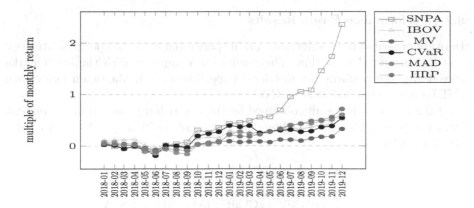

Fig. 5. Portfolio's cumulative return

Table 4. Monthly returns for the algorithms under comparison. Best results are bold faced and worse results are highlighted in red

$t + 1$	IBOV	MV	CVaR	MAD	HRP	SNPA
2018-01	**11.14%**	6.04%	2.67%	7.71%	4.33%	2.38%
2018-02	**0.67%**	−3.72%	−1.68%	−0.50%	−1.06%	−3.59%
2018-03	−0.13%	−1.10%	−6.44%	−5.17%	**0.79%**	−3.73%
2018-04	0.88%	−0.23%	**5.18%**	−0.22%	0.06%	3.53%
2018-05	−10.87%	**−4.90%**	−6.70%	−12.83%	−9.79%	−8.41%
2018-06	−5.20%	−4.03%	−12.71%	−3.51%	−4.97%	**−1.83%**
2018-07	8.87%	8.28%	**24.66%**	9.25%	6.42%	15.03%
2018-08	−3.21%	−0.57%	−0.55%	−6.83%	−2.01%	**2.59%**
2018-09	**3.47%**	−4.27%	−3.24%	−3.15%	−2.46%	3.30%
2018-10	10.19%	9.30%	23.15%	**21.56%**	12.71%	21.43%
2018-11	2.38%	2.56%	**3.43%**	0.31%	3.09%	−3.97%
2018-12	−1.81%	2.19%	2.88%	3.91%	1.39%	**7.78%**
2019-01	10.82%	0.13%	9.46%	**14.64%**	13.32%	5.37%
2019-02	−1.86%	−1.55%	−1.79%	**4.27%**	−2.49%	1.40%
2019-03	−0.18%	0.77%	1.67%	−3.13%	−0.27%	**2.42%**
2019-04	0.98%	0.23%	−10.25%	−1.21%	3.46%	**4.82%**
2019-05	0.70%	−1.08%	2.29%	4.08%	**4.13%**	0.46%
2019-06	4.06%	4.68%	1.92%	4.98%	4.89%	**8.62%**
2019-07	0.84%	−0.01%	1.06%	4.71%	5.45%	**15.00%**
2019-08	−0.66%	−1.74%	−3.20%	2.77%	1.79%	**5.49%**
2019-09	3.57%	**3.94%**	1.02%	0.24%	3.81%	1.34%
2019-10	2.36%	2.61%	5.97%	3.18%	1.25%	**17.76%**
2019-11	0.77%	0.65%	1.70%	1.77%	3.84%	**11.51%**
2019-12	7.33%	12.51%	11.61%	6.58%	10.06%	**22.80%**

6.2   Performance-Based Results

Here are presented the results based on the performance of each portfolio strategy and the suggested algorithm. The results are compared with the benchmarks considering the performance metrics: Shape Ratio (SR), Maximum Drawdown (MDD) and cumulated wealth (CW).

Table 5 shows the results obtained by the algorithm proposed in the present work comparing with the benchmarks. The numbers highlighted were the best observed values.

Table 5. Summary of portfolio performance

	IBOV	MV	CVaR	MAD	HRP	SNPA
SR	0.36	0.29	0.25	0.32	0.48	**0.67**
MDD	−1.98	−1.39	−1.52	−1.60	−1.73	**−1.37**
CW	51.8%	32.8%	54.6%	60.8%	71.2%	**236.3%**

It is possible to observe in Table 5 that the performance of the algorithm proposed is predominant, obtaining the best values considering all the performance metrics evaluated: SR, MDD and CW with the values 0.67, −1.37 and 236.3% respectively.

The second best performance was obtained by the HRP strategy, considering the performance metrics SR and CW, with the values 0.48 and 71.2% respectively. In the case of MDD, the second best performance was obtained by the MV strategy, with a value of −1.39.

7 Conclusion

We propose the Stock Network Portfolio Allocation (SNPA) algorithm to recommend a profitable stock portfolio. SNPA can efficiently handle a large set of assets by representing them as a complex network, where nodes represent assets and edges how their returns correlate. A random walk process carried in this network defines a portfolio. Simulated investment results show the proposed algorithm presented superior performance when compared to the benchmarks: Ibovespa index (IBOV), classic Markowitz's mean-variance portfolio (MV), Mean Absolute Deviation (MAD) portfolio, Conditional Value at Risk (CVaR) and Hierarchical Risk Parity (HRP). The results were compared with the benchmarks considering the performance metrics: Shape Ratio (SR), Maximum Drawdown (MDD), and Cumulated Wealth (CW). The performance of the algorithm proposed was predominant, obtaining the best values considering all the performance metrics. SNPA demonstrated its effectiveness, presenting a CW of 236.3%, being 203.5% above MV portfolio; 181.7% above CVaR portfolio; 175.6% above MAD portfolio, 184.6% above IBOV index and 165.1% above HRP. The proposed algorithm also surpassed the benchmarks considering the performance metrics SR and MDD, with values 0.67 and −1.37 respectively, the best results among the benchmarks observed by the HRP strategy, with 0.48 and −1.73 respectively. The downside of the proposed algorithm is its computational complexity due to the calculation of the return matrix and the length of the random walk. Future work includes analyzing larger periods of time considering all stocks from a particular market, for more markets. And, also consider machine learning predictions of the return in place of historical data.

References

1. Amenc, N., Le Sourd, V.: Portfolio Theory and Performance Analysis. Wiley, Chichester (2003)
2. Baser, P., Saini, J.R.: Agent based stock clustering for efficient portfolio management. Int. J. Comput. Appl. **116**, 35–41 (2015)
3. Brandt, M.: Portfolio choice problems. In: Handbook of Financial Econometrics vol. 1 (2010)
4. Brugière, P.: Quantitative Portfolio Management. STBE, Springer, Cham (2020). https://doi.org/10.1007/978-3-030-37740-3

542 W. B. Freitas and J. R. Bertini Junior

5. Chen, W., Jiang, M., Jiang, C.: Constructing a multilayer network for stock market. Soft Comput. 1–17 (2019). https://doi.org/10.1007/s00500-019-04026-y
6. DeMiguel, V., Garlappi, L., Uppal, R.: Optimal versus naive diversification: how inefficient is the 1/n portfolio strategy? Rev. Financ. Stud. **22**, 1915–1953 (2009)
7. George, S., Changat, M.: Network approach for stock market data mining and portfolio analysis. In: 2017 International Conference on Networks and Advances in Computational Technologies (NetACT), pp. 251–256 (2017)
8. Kim, D.H., Jeong, H.: Systematic analysis of group identification in stock markets. Phys. Rev. **E72**, 046133 (2005)
9. Konno, H., Yamazaki, H.: Mean-absolute deviation portfolio optimization and its applications to Tokyo stock market. Manag. Sci. **37**(5), 519–531 (1991)
10. Lee, K., Lee, J., Hong, H.: Complex networks in a stock market. Comput. Phys. Commun. **177**, 186 (2007)
11. Lee, Y.J., Cho, H.G., Woo, G.: Analysis on stock market volatility with collective human behaviors in online message board. In: Proceedings of the IEEE International Conference on Computer and Information Technology, pp. 482–489 (2014)
12. Magdon-Ismail, M., Atiya, A.F., Pratap, A., Abu-Mostafa, Y.S.: On the maximum drawdown of a Brownian motion. J. Appl. Probab. **41**, 147–161 (2004)
13. Makowitz, H.M.: Portfolio Selection: Efficient Diversification of Investments. Wiley, New York (1951)
14. Makowitz, H.M.: Porfolio selection. J. Finance **7**(1), 77–91 (1952)
15. Mantegna, R.: Hierarchical structure in financial markets. Eur. Phys. J. B Condens. Matter Complex Syst. **11**, 193–197 (1999). https://doi.org/10.1007/s100510050929
16. Pareek, M.K., Thakkar, P.: Surveying stock market portfolio optimization techniques. In: 2015 5th Nirma University International Conference on Engineering (NUiCONE), pp. 1–5 (2015)
17. Building diversified portfolios that outperform out of sample: Lopez de Prado, M. J. Portfolio Manag. **42**, 59–69 (2016)
18. Raudys, S.: Portfolio of automated trading systems: complexity and learning set size issues. IEEE Trans. Neural Netw. Learn. Syst. **24**, 448–459 (2013)
19. Rockafellar, R.T., Uryasev, S.: Optimization of conditional value-at-risk. J. Risk **2**, 21–41 (2000)
20. Shapira, Y., Kenett, D., Ben-Jacob, E.: The index cohesive effect on stock market correlations. Eur. Phys. J. **B72**, 657 (2009). https://doi.org/10.1140/epjb/e2009-00384-y
21. Sharpe, W.: Mutual fund performance. J. Bus. **39**, 657–669 (1965)
22. Shen, W., Wang, B., Pu, J., Wang, J.: The Kelly growth optimal portfolio with ensemble learning. In: Proceedings of the AAAI Conference on Artificial Intelligence, vol. 33, pp. 1134–1141 (2019)
23. Shen, W., Wang, J.: Portfolio blending via Thompson sampling. In: Proceedings of the Twenty-Fifth International Joint Conference on Artificial Intelligence (IJCAI-16) (2016)
24. Shen, W., Wang, J., Jiang, Y.G., Zha, H.: Portfolio choices with orthogonal bandit learning. In: Proceedings of the Twenty-Fourth International Joint Conference on Artificial Intelligence (IJCAI-15) (2015)
25. Uryasev, S., Krokhmal, P., Palmquist, J.: Portfolio optimization with conditional value-at-risk objective and constraints. J. Risk **7**, 43–68 (2002)
26. Zhang, D., Hu, M., Ji, Q.: Financial markets under the global pandemic of COVID-19. Finance Res. Lett. **36**, 101528 (2020)

Author Index

Printed in the United States
by Baker & Taylor Publisher Services

Printed in the United States
by Baker & Taylor Publisher Services